SOFTWARE DESIGN FOR SIX SIGMA

SOFTWARE DESIGN FOR SIX SIGMA

A Roadmap for Excellence

BASEM EL-HAIK
ADNAN SHAOUT

WILEY

A JOHN WILEY & SONS, INC., PUBLICATION

Published by John Wiley & Sons, Inc., Hoboken, New Jersey.
Published simultaneously in Canada.

Library of Congress Cataloging-in-Publication Data

El-Haik, Basem.
Software design for six sigma : a roadmap for excellence / Basem S. El-Haik, Adnan Shaout.
p. cm.
ISBN 978-0-470-40546-8 (hardback)
1. Computer software–Quality control. 2. Six sigma (Quality control standard) I. Shaout, Adnan, 1960– II. Title.
QA76.76.Q35E45 2010
005.1–dc22 2010025493

To our parents, families, and friends for their continuous support.

CONTENTS

PREFACE

Information technology (IT) quality engineering and quality improvement methods are constantly getting more attention from world corporate leaders, all levels of management, design engineers, and academia. This trend can be seen easily by the widespread of "Six Sigma" initiatives in many Fortune IT 500 companies. For a Six Sigma initiative in IT, software design activity is the most important to achieve significant quality and reliability results. Because design activities carry a big portion of software development impact, quality improvements done in design stages often will bring the most impressive results. Patching up quality problems in post-design phases usually is inefficient and very costly.

During the last 20 years, there have been significant enhancements in software development methodologies for quality improvement in software design; those methods include the Waterfall Model, Personal Software Process (PSP), Team Software Process (TSP), Capability Maturity Model (CMM), Software Process Improvement Capability Determination (SPICE), Linear Sequential Model, Prototyping Model, RAD Model, and Incremental Model, among others.[1] The historical evolution of these methods and processes, although indicating improvement trends, indicates gaps that each method tried to pick up where its predecessors left off while filling the gaps missed in their application.

Six Sigma is a methodology to manage process variations that use data and statistical analysis to measure and improve a company's operational performance. It works by identifying and eliminating defects in manufacturing and service-related processes. The maximum permissible defects are 3.4 per one million opportunities.[2]

[1] See Chapters 2 and 4.
[2] See Chapter 6.

Although Six Sigma is manufacturing-oriented, its application to software problem solving is undisputable because as you may imagine, there are problems that need to be solved in software and IT domains. However, the real value is in prevention rather than in problem solving, hence, software Design For Six Sigma (DFSS).

DFSS is very vital to software design activities that decide quality, cost, and cycle time of the software and can be improved greatly if the right strategy and methodologies are used. Major IT corporations are training many software design engineers and project leaders to become Six Sigma Black Belts, or Master Black Belts, enabling them to play the leader role in corporate excellence.

Our book, *Software Design For Six Sigma: A Roadmap for Excellence*, constitutes an algorithm of software design[3] using the design for Six Sigma thinking, tools, and philosophy to software design. The algorithm also will include conceptual design frameworks, mathematical derivation for Six Sigma capability upfront to enable design teams to disregard concepts that are not capable upfront . . . learning the software development cycle and saving developmental costs.

DFSS offers engineers powerful opportunities to develop more successful systems, software, hardware, and processes. In applying Design for Six Sigma to software systems, two leading experts offer a realistic, step-by-step process for succeeding with DFSS. Their clear, start-to-finish road map is designed for successfully developing complex high-technology products and systems.

Drawing on their unsurpassed experience leading DFSS and Six Sigma in deployment in Fortune 100 companies, the authors cover the entire software DFSS project life cycle, from business case through scheduling, customer-driven requirements gathering through execution. They provide real-world experience for applying their techniques to software alone, hardware alone, and systems composed of both. Product developers will find proven job aids and specific guidance about what teams and team members need to do at every stage. Using this book's integrated, systems approach, marketers and software professionals can converge all their efforts on what really matters: addressing the customer's true needs.

The uniqueness of this book is bringing all those methodologies under the umbrella of design and giving a detailed description about how those methods, QFD,[4] robust design methods,[5] software failure mode and effect analysis (SFMEA),[6] Design for X,[7] and axiomatic design[8] can be used to help quality improvements in software development, what kinds of different roles those methods play in various stages of design, and how to combine those methods to form a comprehensive strategy, a design algorithm, to tackle any quality issues during the design stage.

This book is not only helpful for software quality assurance professionals, but also it will help design engineers, project engineers, and mid-level management to

[3]See Chapter 11.

[4]Chapter 12.

[5]Chapter 18.

[6]Chapter 16.

[7]Design for X-ability includes reliability, testability, reusability, availability, etc. See Chapter 14 for more details.

[8]Chapter 13.

gain fundamental knowledge about software Design for Six Sigma. After reading this book, the reader could gain the entire body knowledge for software DFSS. So this book also can be used as a reference book for all software Design for Six Sigma-related people, as well as training material for a DFSS Green Belt, Black Belt, or Master Black Belt.

We believe that this book is coming at the right time because more and more IT companies are starting DFSS initiatives to improve their design quality.

Your comments and suggestions to this book are greatly appreciated. We will give serious consideration to your suggestions for future editions. Also, we are conducting public and in-house Six Sigma and DFSS workshops and provide consulting services.

Dr. Basem El-Haik can be reached via e-mail:

basem.haik@sixsigmapi.com

Dr. Adnan Shaout can be reached via e-mail:

shaout@umich.edu

ACKNOWLEDGMENTS

In preparing this book we received advice and encouragement from several people. For this we are thankful to Dr. Sung-Hee Do of ADSI for his case study contribution in Chapter 13 and to the editing staff of John Wiley & Sons, Inc.

CHAPTER 1

SOFTWARE QUALITY CONCEPTS

1.1 WHAT IS QUALITY

The American Heritage Dictionary defines quality as "a characteristic or attribute of something." Quality is defined in the International Organization for Standardization (ISO) publications as the totality of characteristics of an entity that bear on its ability to satisfy stated and implied needs.

Quality is a more intriguing concept than it seems to be. The meaning of the term "Quality" has evolved over time as many concepts were developed to improve product or service quality, including total quality management (TQM), Malcolm Baldrige National Quality Award, Six Sigma, quality circles, theory of constraints (TOC),Quality Management Systems (ISO 9000 and ISO 13485), axiomatic quality (El-Haik, 2005), and continuous improvement. The following list represents the various interpretations of the meaning of quality:

- "Quality: an inherent or distinguishing characteristic, a degree or grade of excellence" (American Heritage Dictionary, 1996).
- "Conformance to requirements" (Crosby, 1979).
- "Fitness for use" (Juran & Gryna, 1988).
- "Degree to which a set of inherent characteristic fulfills requirements" ISO 9000.

Software Design for Six Sigma: A Roadmap for Excellence, By Basem El-Haik and Adnan Shaout

- "Value to some person" (Weinberg).
- "The loss a product imposes on society after it is shipped" (Taguchi).
- "The degree to which the design vulnerabilities do not adversely affect product performance" (El-Haik, 2005).

Quality is a characteristic that a product or service must have. It refers to the perception of the degree to which the product or service meets the customer's expectations. Quality has no specific meaning unless related to a specific function or measurable characteristic. The dimensions of quality refer to the measurable characteristics that the quality achieves. For example, in design and development of a medical device:

- Quality supports safety and performance.
- Safety and performance supports durability.
- Durability supports flexibility.
- Flexibility supports speed.
- Speed supports cost.

You can easily build the interrelationship between quality and all aspects of product characteristics, as these characteristics act as the qualities of the product. However, not all qualities are equal. Some are more important than others. The most important qualities are the ones that customers want most. These are the qualities that products and services must have. So providing quality products and services is all about meeting customer requirements. It is all about meeting the needs and expectations of customers.

When the word "quality" is used, we usually think in terms of an excellent design or service that fulfil's or exceeds our expectations. When a product design surpasses our expectations, we consider that its quality is good. Thus, quality is related to perception. Conceptually, quality can be quantified as follows (El-Haik & Roy, 2005):

$$Q - \frac{\sum P}{\sum E} \tag{1.1}$$

where Q is quality, P is performance, and E is an expectation.

In a traditional manufacturing environment, conformance to specification and delivery are the common quality items that are measured and tracked. Often, lots are rejected because they do not have the correct documentation supporting them. Quality in manufacturing then is conforming product, delivered on time, and having all of the supporting documentation. In design, quality is measured as consistent conformance to customer expectations.

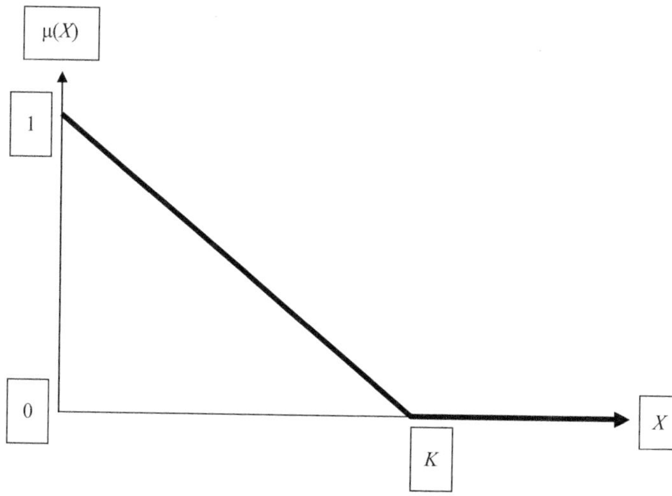

FIGURE 1.1 A membership function for an affordable software.[1]

In general, *quality*[2] is a fuzzy linguistic variable because quality can be very subjective. What is of a high quality to someone might not be a high quality to another. It can be defined with respect to attributes such as cost or reliability. It is a degree of membership of an attribute or a characteristic that a product or software can or should have. For example, a product should be reliable, or a product should be both reliable and usable, or a product should be reliable or repairable. Similarly, software should be affordable, efficient, and effective. These are some characteristics that a good quality product or software must have. In brief, quality is a desirable characteristic that is subjective. The desired qualities are the ones that satisfy the functional and nonfunctional requirements of a project. Figure 1.1 shows a possible membership function, $\mu(X)$, for the affordable software with respect to the cost (X).

When the word "quality" is used in describing a software application or any product, it implies a product or software program that you might have to pay more for or spend more time searching to find.

1.2 QUALITY, CUSTOMER NEEDS, AND FUNCTIONS

The quality of a software product for a customer is a product that meets or exceeds requirements or expectations. Quality can be achieved through many levels (Braude,

[1] where K is the max cost value of the software after which the software will be not be affordable ($\mu(K) = 0$).
[2] J. M. Juran (1988) defined quality as "fitness for use." However, other definitions are widely discussed. Quality as "conformance to specifications" is a position that people in the manufacturing industry often promote. Others promote wider views that include the expectations that the product or service being delivered 1) meets customer standards, 2) meets and fulfills customer needs, 3) meets customer expectations, and 4) will meet unanticipated future needs and aspirations.

2001). One level for attaining quality is through inspection, which can be done through a team-oriented process or applied to all stages of the software process development. A second level for attaining quality is through formal methods, which can be done through mathematical techniques to prove that the software does what it is meant to do or by applying those mathematical techniques selectively. A third level for attaining quality is through testing, which can be done at the component level or at the application level. A fourth level is through project control techniques, which can be done through predicting the cost and schedule of the project or by controlling the artifacts of the project (scope, versions, etc.). Finally, the fifth level we are proposing here is designing for quality at the Six Sigma level, a preventive and proactive methodology, hence, this book.

A quality function should have the following properties (Braude, 2001):

- Satisfies clearly stated functional requirements
- Checks its inputs; reacts in predictable ways to illegal inputs
- Has been inspected exhaustively in several independent ways
- Is thoroughly documented
- Has a confidently known defect rate, if any

The American Society for Quality (ASQ) defines quality as follows: "A subjective term for which each person has his or her own definition." Several concepts are associated with quality and are defined as follows[3]:

- *Quality Assurance*: Quality assurance (QA) is defined as a set of activities whose purpose is to demonstrate that an entity meets all quality requirements usually after the fact (i.e., mass production). We will use QA in the Verify & Validate phase of the Design For Six Sigma (DFSS) process in the subsequent chapters. QA activities are carried out to inspire the confidence of both customers and managers that all quality requirements are being met.
- *Quality Audits*: Quality audits examine the elements of a quality management system to evaluate how well these elements comply with quality system requirements.
- *Quality Control*: Quality control is defined as a set of activities or techniques whose purpose is to ensure that all quality requirements are being met. To achieve this purpose, processes are monitored and performance problems are solved.
- *Quality Improvement*: Quality improvement refers to anything that enhances an organization's ability to meet quality requirements.
- *Quality Management*: Quality management includes all the activities that managers carry out in an effort to implement their quality policy. These activities

[3]See ISO 13485, 2003.

include quality planning, quality control, quality assurance, and quality improvement.

- *Quality Management System (QMS)*: A QMS is a web of interconnected processes. Each process uses resources to turn inputs into outputs. And all of these processes are interconnected by means of many input–output relationships. Every process generates at least one output, and this output becomes an input for another process. These input–output relationships glue all of these processes together—that's what makes it a system. A quality manual documents an organization's QMS. It can be a paper manual or an electronic manual.
- *Quality Planning*: Quality planning is defined as a set of activities whose purpose is to define quality system policies, objectives, and requirements, and to explain how these policies will be applied, how these objectives will be achieved, and how these requirements will be met. It is always future oriented. A quality plan explains how you intend to apply your quality policies, achieve your quality objectives, and meet your quality system requirements.
- *Quality Policy*: A quality policy statement defines or describes an organization's commitment to quality.
- *Quality Record*: A quality record contains objective evidence, which shows how well a quality requirement is being met or how well a quality process is performing. It always documents what has happened in the past.
- *Quality Requirement*: A quality requirement is a characteristic that an entity must have. For example, a customer may require that a particular product (entity) achieve a specific dependability score (characteristic).
- *Quality Surveillance*: Quality surveillance is a set of activities whose purpose is to monitor an entity and review its records to prove that quality requirements are being met.
- *Quality System Requirement*: A quality is a characteristic. A system is a set of interrelated processes, and a requirement is an obligation. Therefore, a quality system requirement is a characteristic that a process must have.

1.3 QUALITY, TIME TO MARKET, AND PRODUCTIVITY

The time to market of a software product is how fast a software company can introduce a new or improved software products and services to the market. It is very important for a software company to introduce their products in a timely manner without reducing the quality of their products. The software company that can offer their product faster without compromising quality achieve a tremendous competitive edge with respect to their competitors.

There are many techniques to reduce time to market, such as (El-Haik, 2005):

- Use the proper software process control technique(s), which will reduce the complexity of the software product

- Concurrency: Encouraging multitasking and parallelism
- Use the Carnegie Mellon Personal Software Process (PSP) and Team Software Process (TSP) with DFSS (El-Haik & Roy, 2005)
- Project management: Tuned for design development and life-cycle management

Using these techniques and methods would increase the quality of the software product and would speed up the production cycle, which intern reduces time to market the product.

1.4 QUALITY STANDARDS

Software system quality standards according to the IEEE Computer Society on Software Engineering Standards Committee can be an object or measure of comparison that defines or represents the magnitude of a unit. It also can be a characterization that establishes allowable tolerances or constraints for categories of items. Also it can be a degree or level of required excellence or attainment.

Software quality standards define a set of development criteria that guide the way software is engineered. If the criteria are not followed, quality can be affected negatively. Standards sometimes can negatively impact quality because it is very difficult to enforce it on actual program behavior. Also standards used to inappropriate software processes may reduce productivity and, ultimately, quality.

Software system standards can improve quality through many development criteria such as preventing idiosyncrasy (e.g., standards for primitives in programming languages) and repeatability (e.g., repeating complex inspection processes). Other ways to improve software quality includes preventive mechanisms such as Design for Six Sigma (design it right the first time), consensus wisdom (e.g., software metrics), cross-specialization (e.g., software safety standards), customer protection (e.g., quality assurance standards), and badging (e.g., capability maturity model [CMM] levels).

There are many standards organizations. Table 1.1 shows some of these standard organizations.

Software engineering process technology (SEPT) has posted the most popular software Quality standards.[4] Table 1.2 shows the most popular software Quality standards.

1.5 SOFTWARE QUALITY ASSURANCE AND STRATEGIES

Professionals in any field must learn and practice the skills of their professions and must demonstrate basic competence before they are permitted to practice their professions. This is not the case with the software engineering profession (Watts,

[4] http://www.12207.com/quality.htm.

TABLE 1.1 Shows Some Standard Organizations

Organization	Notes
ANSI	American National Standards Institute (does not itself make standards but approves them)
AIAA	American Institute of Aeronautics and Astronautics
EIA	Electronic Industries Association
IEC	International Electro technical Commission
IEEE	Institute of Electrical and Electronics Engineers Computer Society Software Engineering Standards Committee
ISO	International Organization for Standardization

1997). Most software engineers learn the skills they need on the job, and *this* is not only expensive and time consuming, but also it is risky and produces low-quality products.

The work of software engineers has not changed a lot during the past 30 years (Watts, 1997) even though the computer field has gone through many technological advances. Software engineers uses the concept of modular design. They spend a large portion of their time trying to get these modules to run some tests. Then they test and integrate them with other modules into a large system. The process of integrating and testing is almost totally devoted to finding and fixing more defects. Once the software product is deployed, then the software engineers spend more time fixing the defects reported by the customers. These practices are time consuming, costly, and retroactive in contrast to DFSS. A principle of DFSS quality is to build the product right the first time.

The most important factor in software quality is the personal commitment of the software engineer to developing a quality product (Watts, 1997). The DFSS process can produce quality software systems through the use of effective quality and design methods such as axiomatic design, design for X, and robust design, to name few.

The quality of a software system is governed by the quality of its components. Continuing with our fuzzy formulation (Figure 1.1), the overall quality of a software system (μ_{Quality}) can be defined as

$$\mu_{\text{Quality}} = \min(\mu_{Q1}, \mu_{Q2}, \mu_{Q3}, \ldots \mu_{Qn})$$

where $\mu_{\text{Q1}}, \mu_{\text{Q2}}, \mu_{\text{Q3}}, \ldots, \mu_{\text{Qn}}$ are the quality of the *n* parts (modules) that makes up the software system, which can be assured by the QA function.

QA includes the reviewing, auditing, and reporting processes of the software product. The goal of quality assurance is to provide management (Pressman, 1997) with the data needed to inform them about the product quality so that the management can control and monitor a product's quality. Quality assurance does apply throughout a software design process. For example, if the water fall software design process is followed, then QA would be included in all the design phases (requirements and analysis, design, implementation, testing, and documentation). QA will be included in the requirement and analysis phase through reviewing the functional and

TABLE 1.2 Shows the Most Popular Software Quality Standards

Quality Standard	Name and Use
AIAA R-013	Recommended Practice for Software Reliability
ANSI/IEEE Std 730-1984 and 983-1986	Software Quality Assurance Plans
ANSI/AAMI/ISO 13485:2003	Medical Devices—Quality Management Systems—Requirements for Regulatory Purposes
ASME NQA-1	Quality Assurance Requirements for Nuclear Facility Applications
EIA/IS 632	Systems Engineering
IEC 60601-1-4	Medical Electrical Equipment—Part 1: General Requirements for Safety—4. Collateral Standard: Programmable Electrical Medical Systems
IEC 60880	Software for Computers in the Safety Systems of Nuclear Power Stations
IEC 61508	Functional Safety Systems
IEC 62304	Medical Device Software—Software Life Cycle Processes
IEEE 1058.1–1987	Software Project Management Plans
IEEE Std 730	Software Quality Assurance Plans
IEEE Std 730.1	Guide for Software Assurance Planning
IEEE Std 982.1	Standard Dictionary of Measures to Produce Reliable Software
IEEE Std 1059–1993	Software Verification and Validation Plans
IEEE Std 1061	Standard for a Software Quality Metrics Methodology
IEEE Std 1228-1994	Standard for Software Safety Plans
IEEE Std 1233–1996	Guide for Developing System Requirements Specifications
IEEE Std 16085	Software Life Cycle Processes—Risk Management
IEEE Std 610.12:1990	Standard Glossary of Software Engineering Terminology
ISO/IEC 2382-7:1989	Vocabulary—part 7: Computer Programming
ISO 9001:2008	Quality Management Systems—Requirements
ISO/IEC 8631:1989	Program Constructs and Conventions for their Representation
ISO/IEC 9126-1	Software Engineering—Product Quality—Part 1: Quality Model
ISO/IEC 12119	Information Technology—Software Packages—Quality Requirements and Testing
ISO/IEC 12207:2008	Systems and Software Engineering—Software Life Cycle Processes
ISO/IEC 14102	Guideline For the Evaluation and Selection of CASE Tools
ISO/IEC 14598-1	Information Technology—Evaluation of Software Products—General Guide
ISO/IEC WD 15288	System Life Cycle Processes
ISO/IEC 20000-1	Information Technology—Service Management—Part 1: Specification
ISO/IEC 25030	Software Engineering—Software Product Quality Requirements and Evaluation (SQuaRE)—Quality Requirements
ISO/IEC 90003	Software Engineering. Guidelines for the Application of ISO 9001:2000 to Computer Software

nonfunctional requirements, reviewing for conformance to organizational policy, reviews for configuration management plans, standards, and so on. QA in the design phase may include reviews, inspections, and tests. QA would be able to answer questions like, "Does the software design adequately meet the quality required by the management?" QA in the implementation phase may include a review provision for QA activities, inspections, and testing. QA would be able to answer questions like, "Have technical disciplines properly performed their roles as part of the QA activity?" QA in the testing phase would include reviews, and several testing activities. QA in the maintenance phase could include reviews, inspections, and tests as well. The QA engineer serves as the customer's in-house representative (Pressman, 1997). The QA engineer usually is involved with the inspection process. Ideally, QA should (Braude, 2001) be performed by a separate organization (independent) or engineers can perform QA functions on each other's work.

The ANSI/IEEE Std 730-1984 and 983-1986 software quality assurance plans[5] provide a road map for instituting software quality assurance. Table 1.3 shows the ANSI/IEEE Std 730-1984 and 983-1986 software quality assurance plans. The plans serve as a template for the QA activates that are instituted for each software project.

The QA activities performed by software engineering team and the QA group are controlled by the plans. The plans identify the following (Pressman, 1997):

- Evaluations to be performed
- Audits and reviews to be performed
- Standards that are applicable to the project
- Procedures for error reporting and tracking
- Documents to be produced by the QA group
- Amount of feedback provided to the software project team

To be more precise in measuring the quality of a software product, statistical quality assurance methods have been used. The statistical quality assurance for software products implies the following steps (Pressman, 1997):

1. Information about software defects is collected and categorized.
2. An attempt is made to trace each defect to its cause.
3. Using the Pareto principle, the 20% of the vital causes of errors that produce 80% of the defects should be isolated.
4. Once the vital causes have been identified, the problems that have caused the defects should be corrected.

1.6 SOFTWARE QUALITY COST

Quality is always deemed to have a direct relationship to cost—the higher the quality standards, the higher the cost. Or so it seems. Quality may in fact have an inverse

[5]Software Engineering Standards (1994 edition), IEEE Computer Society.

TABLE 1.3 ANSI/IEEE Std 730-1984 and 983-1986 Software Quality Assurance Plans

I. Purpose of the plan
II. References
III. Management
  a. Organization
  b. Tasks
  c. Responsibilities
IV. Documentation
  a. Purpose
  b. Required software engineering documents
  c. Other documents
V. Standards, practices, and conventions
  a. Purpose
  b. Conventions
VI. Reviews and audits
  a. Purpose
  b. Review requirements
    i. Software requirements review
    ii. Design reviews
    iii. Software verification and validation reviews
    iv. Functional audit
    v. Physical audit
    vi. In-process audit
    vii. Management reviews
VII. Test
VIII. Problem reporting and corrective action
IX. Tools, techniques, and methodologies
X. Code control
XI. Media control
XII. Supplier control
XIII. Records collection, maintenance, and retention
XIV. Training
XV. Risk management

relationship with cost in that deciding to meet high-quality standards at the beginning of the project/operation ultimately may reduce maintenance and troubleshooting costs in the long term. This a Design for Six Sigma theme: Avoid design–code–test cycles.

Joseph Juran, one of the world's leading quality theorists, has been advocating the analysis of quality-related costs since 1951, when he published the first edition of his *Quality Control Handbook* (Juran & Gryna, 1988). Feigenbaum (1991) made it one of the core ideas underlying the TQM movement. It is a tremendously powerful tool for product quality, including software quality.

Quality cost is the cost associated with preventing, finding, and correcting defective work. The biggest chunk of quality cost is the cost of poor quality (COPQ), a Six Sigma terminology. COPQ consists of those costs that are generated as a result of producing defective software. This cost includes the cost involved in fulfilling the gap between the desired and the actual software quality. It also includes the cost of lost opportunity resulting from the loss of resources used in rectifying the defect. This cost includes all the labor cost, recoding cost, testing costs, and so on. that have been added to the unit up to the point of rejection. COPQ does not include detection and prevention cost.

Quality costs are huge, running at 20% to 40% of sales (Juran & Gryna, 1988). Many of these costs can be reduced significantly or avoided completely. One key function of a Quality Engineer is the reduction of the total cost of quality associated with a product. Software quality cost equals the sum of the prevention costs and the COPQ as defined below (Pressman, 1997):

1. Prevention costs: The costs of activities that specifically are designed to prevent poor quality. Examples of "poor quality" include coding errors, design errors, mistakes in the user manuals, as well as badly documented or unmaintainable complex code. Note that most of the prevention costs does not fit within the testing budget, and the programming, design, and marketing staffs spend this money. Prevention costs include the following:
   a. DFSS team cost
   b. Quality planning
   c. Formal technical reviews
   d. Test equipment
   e. Training
2. Appraisal costs (COPQ element): The are costs of activities that are designed to find quality problems, such as code inspections and any type of testing. Design reviews are part prevention and part appraisal to the degree that one is looking for errors in the proposed software design itself while doing the review and an appraisal. The prevention is possible to the degree that one is looking for ways to strengthen the design. Appraisal cost are activities to gain insight into product condition. Examples include:
   a. In-process and interprocess inspection
   b. Equipment calibration and maintenance
   c. Testing
3. Failure costs (COPQ elements): These costs result from poor quality, such as the cost of fixing bugs and the cost of dealing with customer complaints. Failure costs disappear if no defects appeared before shipping the software product to customers. It includes two types:
   a. Internal failure costs—the cost of detecting errors before shipping the product, which includes the following:
      i. Rework

ii. Repair
iii. Failure mode analysis

b. External failure costs—the cost of detecting errors after shipping the product. Examples of external failure costs are:
   i. Complaint resolution
   ii. Product return and replacement
   iii. Help-line support
   iv. Warranty work

The costs of finding and repairing a defect in the prevention stage is much less that in the failure stage (Boehm, 1981; Kaplan et al. 1995).

Internal failure costs are failure costs that originate before the company supplies its product to the customer. Along with costs of finding and fixing bugs are many internal failure costs borne outside of software product development. If a bug blocks someone in the company from doing one's job, the costs of the wasted time, the missed milestones, and the overtime to get back onto schedule are all internal failure costs. For example, if the company sells thousands of copies of the same program, it will probably require printing several thousand copies of a multicolor box that contains and describes the program. It (the company) will often be able to get a *much* better deal by booking press time with the printer in advance. However, if the artwork does not get to the printer on time, it might have to pay for some or all of that wasted press time anyway, and then it also may have to pay additional printing fees and rush charges to get the printing done on the new schedule. This can be an added expense of many thousands of dollars. Some programming groups treat user interface errors as low priority, leaving them until the end to fix. This can be a mistake. Marketing staff needs pictures of the product's screen long before the program is finished to get the artwork for the box into the printer on time. User interface bugs—the ones that will be fixed later—can make it hard for these staff members to take (or mock up) accurate screen shots. Delays caused by these minor design flaws, or by bugs that block a packaging staff member from creating or printing special reports, can cause the company to miss its printer deadline. Including costs like lost opportunity and cost of delays in numerical estimates of the total cost of quality can be controversial. Campanella (1990) did not include these in a detailed listing of examples. Juran and Gryna (1988) recommended against including costs like these in the published totals because fallout from the controversy over them can kill the entire quality cost accounting effort. These are found very useful, even if it might not make sense to include them in a balance sheet.

External failure costs are the failure costs that develop after the company supplies the product to the customer, such as customer service costs, or the cost of patching a released product and distributing the patch. External failure costs are huge. It is much cheaper to fix problems before shipping the defective product to customers. The cost rules of thumb are depicted in Figure 1.2. Some of these costs must be treated with care. For example, the cost of public relations (PR) efforts to soften the publicity effects of bugs is probably not a huge percentage of the company's PR budget. And thus the entire PR budget cannot be charged as a quality-related cost. But any money

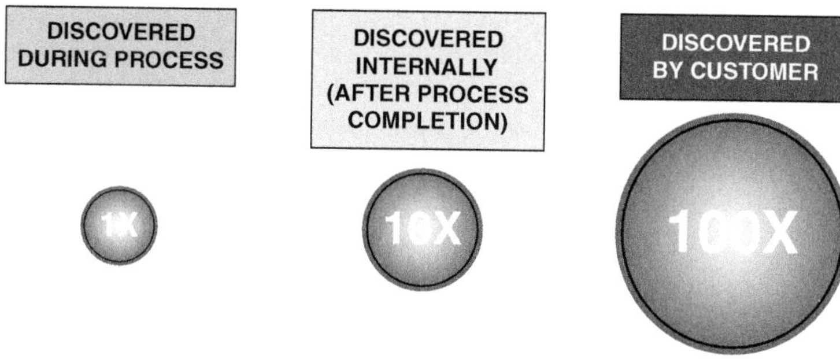

FIGURE 1.2 Internal versus external quality cost rules of thumb.

that the PR group has to spend to cope specifically with potentially bad publicity because of bugs is a failure cost. COPQ ***is the*** sum of appraisal, internal and external quality costs (Kaner, 1996).

Other intangible quality cost elements usually are overlooked in literature (see Figure 1.3). For example, lost customer satisfaction and, therefore, loyalty, lost sales, longer cycle time, and so on. These type of costs can alleviate the total COPQ, which handsomely can be avoided via a thorough top-down DFSS deployment approach. See DFSS deployment chapters for further details (Chapter 8).

1.7 SOFTWARE QUALITY MEASUREMENT

The software market is growing continuously, and users often are dissatisfied with software quality. Satisfaction by users is one of the outcomes of software quality and quality of management.

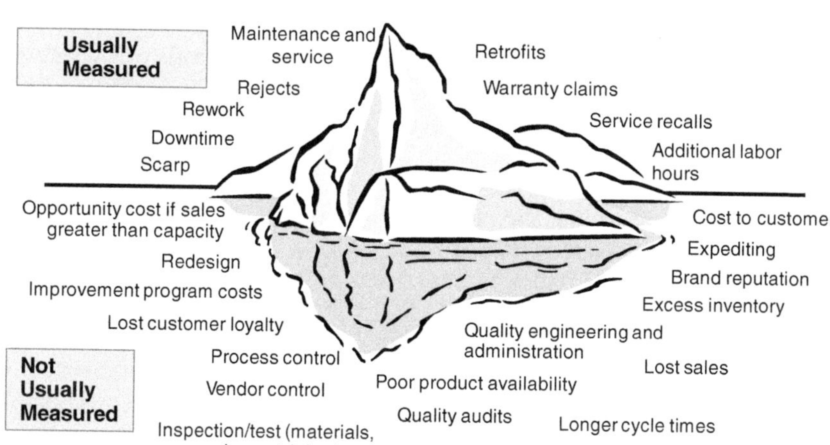

FIGURE 1.3 Measured and not measured quality cost elements.

Quality can be defined and measured by its attributes. A proposed way that could be used for measuring software quality factors is given in the following discussion.[6] For every attribute, there is a set of relevant questions. A membership function can be formulated based on the answers to these questions. This membership function can be used to measure the software quality with respect to that particular attribute. It is clear that these measures are fuzzy (subjective) in nature.

The following are the various attributes that can be used to measure software quality:

1.7.1 Understandability

Understandability can be accomplished by requiring all of the design and user documentation to be written clearly. A sample of questions that can be used to measure the software understandability:

Do the variable names describe the functional property represented? (V1)
Do functions contain adequate comments? (C1)
Are deviations from forward logical flow adequately commented? (F1)
Are all elements of an array functionally related? (A1)
Are the control flow of the program used adequately? (P1)

The membership function for measuring the software quality with respect to understandability can be defined as follows:

$$\mu_{\text{Understandability}} = f1\,(\text{V1, C1, F1, A1, P1})$$

1.7.2 Completeness

Completeness can be defined as the presence of all necessary parts of the software system, with each part fully developed. This means that[7] if the code calls a module from an external library, the software system must provide a reference to that library and all required parameters must be passed. A sample of questions that can be used to measure the software completeness:

The membership function for measuring the software quality with respect to completeness can be defined as follows:

$$\mu_{\text{Completeness}} = f2\,(\text{C2, P2, S2, E2})$$

[6] http://en.wikipedia.org/wiki/Software_quality.
[7] http://en.wikipedia.org/wiki/Software_quality.

Are all essential software system components available? (C2)
Does any process fail for lack of resources? (P2)
Does any process fail because of syntactic errors? (S2)
Are all potential pathways through the code accounted for, including proper error handling? (E2)

1.7.3 Conciseness

Conciseness means to minimize the use of redundant information or processing. A sample of questions that can be used to measure the software conciseness:

Is all code reachable? (C3)
Is any code redundant? (R3)
How many statements within loops could be placed outside the loop, thus reducing computation time? (S3)
Are branch decisions too complex? (B3)

The membership function for measuring the software quality with respect to conciseness can be defined as follows:

$$\mu_{\text{Conciseness}} = \text{f3}(\text{C3}, \text{R3}, \text{S3}, \text{B3})$$

1.7.4 Portability

Portability can be the ability to run the software system on multiple computer configurations or platforms. A sample of questions that can be used to measure the software portability:

Does the program depend upon system or library routines unique to a particular installation? (L4)
Have machine-dependent statements been flagged and commented? (M4)
Has dependency on internal bit representation of alphanumeric or special characters been avoided? (R4)
How much effort would be required to transfer the program from one hardware/software system or environment to another? (E4)

The membership function for measuring the software quality with respect to portability can be defined as follows:

$$\mu_{\text{Portability}} = f4(\text{L4}, \text{M4}, \text{R4}, \text{E4})$$

1.7.5 Consistency

Consistency means the uniformity in notation, symbols, appearance, and terminology within the software system or application. A sample of questions that can be used to measure the software consistency:

Is one variable name used to represent different logical or physical entities in the program? (V5)
Does the program contain only one representation for any given physical or mathematical constant? (P5)
Are functionally similar arithmetic expressions similarly constructed? (F5)
Is a consistent scheme used for indentation, nomenclature, the color palette, fonts and other visual elements? (S5)

The membership function for measuring the software quality with respect to consistency can be defined as follows:

$$\mu_{\text{Consistency}} = f5(\text{V5}, \text{P5}, \text{F5}, \text{S5})$$

1.7.6 Maintainability

Maintainability is to provide updates to satisfy new requirements. A maintainable software product should be well documented, and it should not be complex. A maintainable software product should have spare capacity of memory storage and processor utilization and other resources. A sample of questions that can be used to measure the software maintainability:

Has some memory capacity been reserved for future expansion? (M6)
Is the design cohesive (i.e., does each module have distinct, recognizable functionality)? (C6)
Does the software allow for a change in data structures? (S6)
Is the design modular? (D6)
Was a software process method used in designing the software system? (P6)

The membership function for measuring the software quality with respect to maintainability can be defined as follows:

$$\mu_{\text{Maintainability}} = f6(\text{M6}, \text{C6}, \text{S6}, \text{D6}, \text{P6})$$

1.7.7 Testability

A software product is testable if it supports acceptable criteria and evaluation of performance. For a software product to have this software quality, the design must not be complex. A sample of questions that can be used to measure the software testability:

Are complex structures used in the code? (C7)
Does the detailed design contain clear pseudo-code? (D7)
Is the pseudo-code at a higher level of abstraction than the code? (P7)
If tasking is used in concurrent designs, are schemes available for providing adequate test cases? (T7)

The membership function for measuring the software quality with respect to testability can be defined as follows:

$$\mu_{\text{Testability}} = f7(C7, D7, P7, T7)$$

1.7.8 Usability

Usability of a software product is the convenience and practicality of using the product. The easier it is to use the software product, the more usable the product is. The component of the software that influence this attribute the most is the graphical user interface (GUI).[8] A sample of questions that can be used to measure the software usability:

Is a GUI used? (G8)
Is there adequate on-line help? (H8)
Is a user manual provided? (M8)
Are meaningful error messages provided? (E8)

The membership function for measuring the software quality with respect to usability can be defined as follows:

$$\mu_{\text{Usability}} = f8(G8, H8, M8, E8)$$

1.7.9 Reliability

Reliability of a software product is the ability to perform its intended functions within a particular environment over a period of time satisfactorily. A sample of questions that can be used to measure the software reliability:

Are loop indexes range-tested? (L9)
Is input data checked for range errors? (I9)
Is divide-by-zero avoided? (D9)
Is exception handling provided? (E9)

[8] http://en.wikipedia.org/wiki/Software_quality.

The membership function for measuring the software quality with respect to reliability can be defined as follows:

$$\mu_{\text{Reliability}} = f9(\text{L9}, \text{I9}, \text{D9}, \text{E9})$$

1.7.10 Structuredness

Structuredness of a software system is the organization of its constituent parts in a definite pattern. A sample of questions that can be used to measure the software structuredness:

Is a block-structured programming language used? (S10)
Are modules limited in size? (M10)
Have the rules for transfer of control between modules been established and followed? (R10)

The membership function for measuring the software quality with respect to structuredness can be defined as follows:

$$\mu_{\text{Structuredness}} = f10(\text{S10}, \text{M10}, \text{R10})$$

1.7.11 Efficiency

Efficiency of a software product is the satisfaction of goals of the product without waste of resources. Resources like memory space, processor speed, network bandwidth, time, and so on. A sample of questions that can be used to measure the software efficiency:

Have functions been optimized for speed? (F11)
Have repeatedly used blocks of code been formed into subroutines? (R11)
Has the program been checked for memory leaks or overflow errors? (P11)

The membership function for measuring the software quality with respect to efficiency can be defined as follows:

$$\mu_{\text{Efficiency}} = f11(\text{F11}, \text{R11}, \text{P11})$$

1.7.12 Security

Security quality in a software product means the ability of the product to protect data against unauthorized access and the resilience of the product in the face of malicious or inadvertent interference with its operations. A sample of questions that can be used to measure the software security:

Does the software protect itself and its data against unauthorized access and use? (A12)
Does it allow its operator to enforce security policies? (S12)
Are security mechanisms appropriate, adequate, and correctly implemented? (M12)
Can the software withstand attacks that can be anticipated in its intended environment? (W12)
Is the software free of errors that would make it possible to circumvent its security mechanisms? (E12)
Does the architecture limit the potential impact of yet unknown errors? (U12)

The membership function for measuring the software quality with respect to security can be defined as follows:

$$\mu_{\text{Security}} = \text{f12}(\text{A12}, \text{S12}, \text{M12}, \text{W12}, \text{E12}, \text{U12})$$

There are many perspectives within the field on software quality measurement. Some believe that quantitative measures of software quality are important. Others believe that contexts where quantitative measures are useful are they rare, and so prefer qualitative measures.[9] Many researchers have written in the field of software testing about the difficulty of measuring what we truly want to measure (Pressman, 2005, Crosby, 1979).

In this section, the functions f1 through f12 can be linear or nonlinear functions. They can be fuzzy measures. The function f_i can be a value within the unit interval (f_i € [0, 1]), where $f_{i=1}$ means that the software quality with respect to the attribute i is the highest, and $f_{i=} 0$ means that the software quality with respect to the attribute i is the lowest; otherwise the software quality will be relative to the value of $f_{i.}$

1.8 SUMMARY

Quality is essential in all products and systems, and it is more so for software systems because modern computer systems do execute millions of instructions per second, and a simple defect that would occur once in a billion times can occur several times a day.

High-quality software would not only decrease cost but also reduce the production time and increase the company's competence within the software production world.

Achieving a high quality in software systems demands changing and improving the process. An improved process would include defining the quality goal, measuring the software product quality, understanding the process, adjusting the process, using the adjusted process, measuring the results, comparing the results with the goal, and recycling and continue improving the process until the goal is achieved. It also can be achieved by using DFSS as will be discussed in the following chapters.

[9] http://en.wikipedia.org/wiki/Software_quality.

Many quality standards can be used to achieve high-quality software products. Standards can improve quality by enforcing a process and ensuring that no steps are skipped. The standards can establish allowable tolerance or constraints for levels of items. It can achieve a degree of excellence.

REFERENCES

American Heritage Dictionary (1996), 6th Ed., Houghton Mifflin, Orlando, Florida.

Boehm, Barry (1981), *Software Engineering Economics*, Prentice Hall, Upper Saddle River, NJ.

Braude, J. Eric (2001), *Software Engineering—An Object-Oriented Perspective*, John Wiley & Sons, New York.

Jack Campanella, (1990), *Principles of Quality Costs*, 2nd Ed., ASQC Quality Press, Milweaskeej WI.

Crosby, Philip (1979), *Quality is Free*, McGraw-Hill, New York.

El-Haik, Basem S. (2005), *Axiomatic Quality: Integrating Axiomatic Design with Six[Sigma, Reliability, and Quality*, Wiley-Interscience, New York.

El-Haik, B. and Roy, D. (2005), *Service Design for Six Sigma: A Roadmap for Excellence*, John Wiley, New York.

Feigenbaum, Armaund V. (1991, *"Chapter 7," Total Quality Control* 3rd Ed. Revised, McGraw-Hill, New York.

Juran, Joseph M. and Gryna, Frank M. (1988), *Juran's Quality Control Handbook*, 4th Ed., McGraw-Hill, New York. pp. 4.9–4.12.

Kaner, Cem (1996), "Quality cost analysis: Benefits and risks." *Software QA*, Volume *3*, #1, p. 23.

Kaplan, Craig, Raph Clark, and Tang, Victor (1995), *Secrets of Software Quality: 40 Inventions from IBM*, McGraw Hill, New York.

Pressman, S. Roger (1997), *Software Engineering—A Practitioner's Approach*, 4th Ed., McGraw-Hill, New York.

Pressman, S. Roger (2005), *Software Engineering: A Practitioner's Approach*, 6th Ed. McGraw-Hill, New York, p. 388.

Taguchi, G. Elsayed, E.A. and Thomas, C. Hsiang (1988), *Quality Engineering in Production Systems* (Mcgraw Hill Series in Industrial Engineering and Management Science), Mcgraw-Hill College, New York.

Watts, S. Humphrey (1997), *Introduction to Personal Software Process*, Addison Wesley, Boston, MA.

Weinberg, G.M. (1991), *Quality Software Management: Systems Thinking*, 1st Ed., Dorset House Publishing Company, New York.

CHAPTER 2

TRADITIONAL SOFTWARE DEVELOPMENT PROCESSES[1]

2.1 INTRODUCTION

More and more companies are emphasizing formal software processes and requesting diligent application. For the major organizations, businesses, government agencies, and the military, the biggest constraints are cost, schedule, reliability, and quality for a given software product. The Carnegie Mellon Software Engineering Institute (SEI) has carried out the refined work for Personal Software Process (PSP), Team Software Process (TSP), Capability Mature Model (CMM), and Capability Maturity Model Integration (CMMI). We will discuss software design techniques focusing on real-time operating systems (RTOS) in the next chapter to complement, and in some cases zoom in, on certain concepts that are introduced here.

A goal of this chapter is to present the various existing software processes and their pros and cons, and then to classify them depending on the complexity and size of the project. For example, *Simplicity (or complexity) and size (Small size, Medium size, or Large Size)* attributes will be used to classify the existing software developmental processes, which could be useful to a group, business, or organization. This classification can be used to understand the pros and cons of the various software processes at a glance and its suitability to a given software development project. A few automotive software application examples will be presented to justify the needs for including Six Sigma in the software process modeling techniques in Chapter 10.

[1] In the literature, software development processes also are known as models (e.g., the Waterfall Model).

In a big organization for a given product, usually there are lots of different people who are working within a group/team for which an organized effort is required to avoid repetition and to get a quality end product. A software process is required to be followed, in addition to coordination within a team(s), that will be elaborated further in PSP or TSP (Chapter 10).

Typically, for big and complex projects, there are many teams working for one goal, which is to deliver a final quality product. Design and requirements are required to be specified among the teams. Team leaders[2] along with key technical personnel are responsible for directing each team to prepare their team product to interface with each other's requirements. Efforts are required to coordinate hardware, software, and system level among these teams as well as for resolving issues among these team efforts at various levels. To succeed with such a high degree of complex projects, a structured design process is required.

2.2 WHY SOFTWARE DEVELOPMENTAL PROCESSES?

Software processes enable effective communication among users, developers, managers, customers, and researchers. They enhance management's understanding, provide a precise basis for process automation, and facilitate personnel mobility and process reuse.

A "process" is the building element of any value-added domain. In any field, process development is time consuming and expensive. Software development processes evolution provides an effective means for learning a solid foundation for improvement. Software developmental processes aid management and decision making where both requires clear plans and a precise, quantified data to measure project status and make effective decisions. Defined developmental processes provide a framework to reduce cost, increase reliability, and achieve higher standards of quality.

Quite often while dealing with larger, more complex, and safety-oriented software systems, predictable time schedules are needed. Without adopting a software process, the following may not happen[3]:

– Improved communication among the persons involved in the project
– Uniform procedure in public authorities and commissioned industry
– Insurance of better product quality
– Productivity increase because of the reduction of familiarization and training times
– More precise calculation of new projects cycle time using standardized procedures
– Less dependencies on persons and companies

[2] Usually Six Sigma Belts in our context.
[3] The V-Model as the Software Development Standard—the effective way to develop high-quality software—IABG Industrieanlagen—Betriebsgesellschaft GmbH Einsteinstr. 20, D-85521 Ottobrunn, Release 1995.

2.2.1 Categories of Software Developmental Process

The process could possess one or more of the following characteristics and could be categorized accordingly:

Ad hoc: The software process is characterized as ad hoc and occasionally even chaotic. Few processes are defined, and success depends on individual effort, skills, and experience.

Repeatable: Basic project management processes are established to track, cost, schedule, and functionality. The necessary process discipline is in place to repeat earlier successes on software projects with similar applications.

Defined: The software process for both management and engineering activities is documented, standardized, and integrated into a standard software process for the organization. All projects use an approved, tailored version of the organization's standard software process for developing and maintaining software.

Managed: Detailed measures of software process and product quality are collected. Both the software development process and products are understood quantitatively and controlled.

Optimized: Continuous process improvement is enabled by quantitative feedback from the process and from piloting innovative ideas and technologies.

2.3 SOFTWARE DEVELOPMENT PROCESSES

What is to be determined here is which activities have to be carried out in the process of the development of software, which results have to be produced in this process and what are the contents that these results must have. In addition, the functional attributes of the project and the process need to be determined. Functional attributes include an efficient software development cycle, quality assurance, reliability assurance, configuration management, project management and cost-effectiveness. They are called Critical-To-Satisfaction (CTSs) in Six Sigma domain (Chapters: 7, 8, 9, and 11).

2.3.1 Different Software Process Methods in Practice

Below is a list of software development process methods that are either in use or were used in past, for various types of projects in different industries. Also, while going through these processes and their pros and cons, we will discuss their advantages, disadvantages and suitability to different complexities and sizes of software for industrial applications.

1. PSP and TSP[4]
2. Waterfall
3. Sashimi Model

[4]Will be discussed in Chapter 9.

4. V-Model
5. V-Model XT
6. Spiral
7. Chaos Model
8. Top Down and Bottom Up
9. Joint Application Development
10. Rapid Application Development
11. Model Driven Engineering
12. Iterative Development Process
13. Agile Software Process
14. Unified Process
15. eXtreme Process (XP)
16. LEAN method (Agile)
17. Wheel and Spoke Model
18. Constructionist Design Methodology

In this book, we are developing the Design for Six Sigma (DFSS)[5] as a replacement for the traditional software the development processes discussed here by formulating for methodologies integration, importing good practices, filling gaps, and avoiding failure modes and pitfalls that accumulated over the years of experiences.

2.3.1.1 *PSP and TSP.* The PSP is a defined and measured software development process designed to be used by an individual software engineer. The PSP was developed by Watts Humphrey (Watts, 1997). Its intended use is to guide the planning and development of software modules or small programs; it also is adaptable to other personal tasks. Like the SEI CMM, the PSP is based on process improvement principles. Although the CMM is focused on improving organizational capability, the focus of the PSP is the individual software engineer. To foster improvement at the personal level, PSP extends process management and control to the practitioners. With PSP, engineers develop software using a disciplined, structured approach. They follow a defined process to plan, measure, track their work, manage product quality, and apply quantitative feedback to improve their personal work processes, leading to better estimating and to better planning and tracking. More on PSP and TSP is presented in Chapter 11.

2.3.1.2 *Waterfall Process* The Waterfall Model (2008) is a popular version of the ***systems development life-cycle model*** for software engineering. Often considered the classic approach to the systems development life cycle, the Waterfall Model describes a development method that is linear and sequential. Waterfall development has distinct goals for each phase of development. Imagine a waterfall on the cliff of

[5]See Chapters 10 and 11.

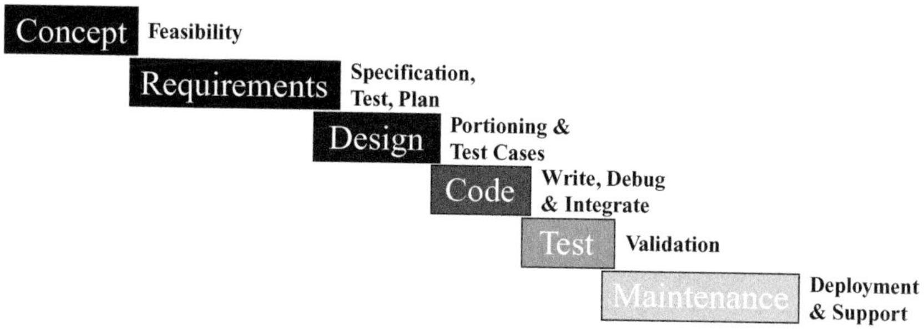

FIGURE 2.1 The steps in the Waterfall Model (2008).

a steep mountain. Once the water has flowed over the edge of the cliff and has begun its journey down the side of the mountain, it cannot turn back. It is the same with waterfall development. Once a phase of development is completed, the development proceeds to the next phase and there is no turning back. This is a classic methodology were the life cycle of a software project has been partitioned into several different phases as specified below:

1. Concepts
2. Requirements
3. Design
4. Program, Code, and Unit testing
5. Subsystem testing and System testing
6. Maintenance

The term "waterfall" is used to describe the idealized notion that each stage or phase in the life of a software product occurs in time sequence, with the boundaries between phases clearly defined as shown in Figure 2.1.

This methodology works well when complete knowledge of the problem is available and do not experiences change during the development period. Unfortunately, this is seldom the case. It is difficult and perhaps impossible to capture everything in the initial requirements documents. In addition, often the situation demands working toward a moving target. What was required to build a year ago is not what is needed now. Often, it is seen in projects that the requirements continually change. The Waterfall Process is most suitable for small projects with static requirements.

Development moves from concept, through design, implementation, testing, installation, and troubleshooting, and ends up at operation and maintenance. Each phase of development proceeds in strict order, without any overlapping or iterative steps. A schedule can be set with deadlines for each stage of development, and a product can proceed through the development process like a car in a carwash and, theoretically, be delivered on time.

2.3.1.2.1 Advantage. An advantage of waterfall development is that it allows for departmentalization and managerial control. However, for simple, static/frozen requirements and a small project this method might prove effective and cheaper.

2.3.1.2.2 Disadvantage. A disadvantage of waterfall development is that it does not allow for much reflection or revision. Once an application is in the testing stage, it is very difficult to go back and change something that was not well thought out in the concept stage. For these reasons, the classic waterfall methodology usually breaks down and results in a failure to deliver the needed product for complex and continuously changing requirements.

2.3.1.2.3 Suitability. Alternatives to the Waterfall Model include Joint Application Development (JAD), Rapid Application Development (RAD), Synch and Stabilize, Build and Fix, and the Spiral Model. For more complex, continuously changing, safety-critical, and large projects, use of the spiral method is proven to be more fruitful.

2.3.1.3 Sashimi Model. The Sashimi Model (so called because it features overlapping phases, like the overlapping fish of Japanese sashimi) was originated by Peter DeGrace (Waterfall Modelt, 2008). It is sometimes referred to as the "waterfall model with overlapping phases" or "the waterfall model with feedback." Because phases in the Sashimi Model overlap, information on problem spots can be acted on during phases that would typically, in the pure Waterfall Model, precede others. For example, because the design and implementation phases will overlap in the Sashimi Model, implementation problems may be discovered during the design and implementation phase of the development process.

2.3.1.3.1 Advantage. Information on problem spots can be acted on during phases that would typically, in the pure Waterfall Model, precede others.

2.3.1.3.2 Disadvantage. May not by very efficient for complex applications and where requirements are constantly changing.

2.3.1.3.3 Suitability. For small-to-moderate-size applications and for applications where requirements are not changing continually.

2.3.1.4 V-Model. The life-cycle process model (V-Model) is described as the standard for the first level. It regulates the software development process in a uniform and binding way by means of activities and products (results), which have to be taken into consideration during software development and the accompanying activities for quality assurance, configuration management, and project management.[6] The

[6]The V-Model as Software Development Standard—the effective way to develop high-quality software—IABG Industrieanlagen—Betriebsgesellschaft GmbH Einsteinstr. 20, D-85521 Ottobrunn, Release 1995.

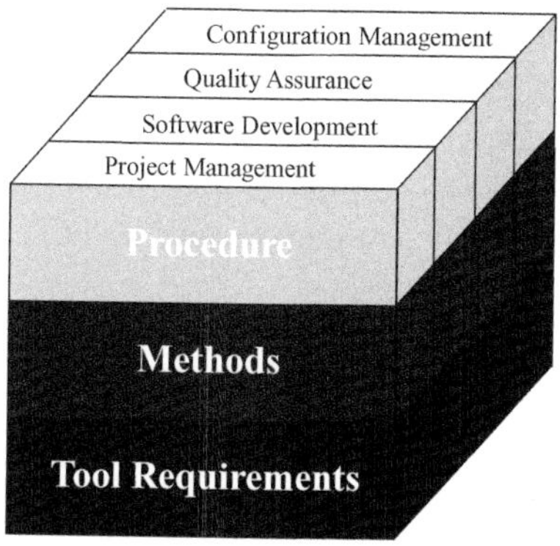

FIGURE 2.2 V-Model.

V-Model[7] is a software development process, which can be presumed to be the extension of the Waterfall Model. Instead of moving down in a linear way, the process steps are bent upward after the coding phase, to form the typical V shape. The V-Model demonstrates the relationships between each phase of the development life cycle and its associated phase of testing.

The V-Model is structured into functional parts, so-called **submodels**, as shown in Figure 2.2. They comprise software development (SWD), quality assurance (QA), configuration management (CM), and project management (PM). These four submodels are interconnected closely and mutually influence one another by exchange of products/results.

- **PM** plans, monitors, and informs the submodels SWD, QA, and CM.
- **SWD** develops the system or software.
- **QA** submits quality requirements to the submodels SWD, CM, and PM, test cases, and criteria and unsures the products and the compliance of standards.
- **CM** administers the generated products.

The V-Model describes in detail the interfaces between the submodels SWD and QA, as software quality can only be ensured by the consequent application of quality assurance measures and by checking if they are complied with standards. Of particular relevance for software is the *criticality*, that is, the classification of software with respect to reliability and security. In the V-Model, this is considered a quality requirement and is precisely regulated. Mechanisms are proposed to how the expenditure for development and assessment can be adapted to the different levels of criticality of the software.

[7] V-Model (software development). (2008, July 7). In Wikipedia. the Free Encyclopedia. Retrieved 13:01, July 14, 2008. http://en.wikipedia.org/w/index.php?title=V-Model_%28software_development%29&oldid=224145058.

2.3.1.4.1 Advantages

- Decrease in maintenance cases resulting from improved product quality.
- Decrease in the maintenance effort resulting in the existence of adequate software documentation and an easier understanding because of the uniform structure.

2.3.1.4.2 Disadvantages

- It is resource heavy and very costly to implement.
- The V-Model is not complete. It says that the submodels cover all activity while it is done at too abstract level.
- It is hard to find out whether peer reviews and inspections are done in the V-Model.
- It is difficult to find out whether the self-assessment activity is conducted before product is passed on to the QA for acceptance.

2.3.1.4.3 Suitability. The V-Model was originally intended to be used as a standard development model for information technology (IT) projects in Germany, but it has not been adapted to innovations in IT since 1997.

2.3.1.5 V-Model XT. The V-Model represents the development standard for public-sector IT systems in Germany. For many companies and authorities, it is the way forward for the organization and implementation of IT planning, such as the development of the Bundestag's new address management, the police's new IT system "Inpol-neu," and the Eurofighter's on-board radar (V-Model XT, 2008). More and more IT projects are being abandoned before being completed, or suffer from deadlines and budgets being significantly overrun, as well as reduced functionality. This is where the V-Model comes into its own and improves the product and process quality by providing concrete and easily implementable instructions for carrying out activities and preformulated document descriptions for development and project documentation (V-Model XT, 2008).

The current standard, the V-Model 97, has not been adapted to innovations in information technology since 1997. It was for this reason that the Ministry of Defense/Federal Office for Information Management and Information Technology and Interior Ministry Coordination and Consultancy Office for Information Technology in Federal Government commissioned the project *Further Development of the Development Standard for IT Systems of the Public sector Based on the V-Model 97* from the Technical University of Munich (TUM) and its partners IABG, EADS, Siemens AG, 4Soft GmbH, and TU Kaiserslautern (V-Model XT, 2008). The new V-Model XT (eXtreme Tailoring) includes extensive empirical knowledge and suggests improvements that were accumulated throughout the use of the V-Model 97 (V-Model

XT, 2008). In addition to the updated content, the following specific improvements and innovations have been included:

- Simplified project-specific adaptation—tailoring
- Checkable project progress steps for minimum risk project management
- Tender process, award of contract, and project implementation by the customer
- Improvement in the customer–contractor interface
- System development taking into account the entire system life cycle
- Cover for hardware development, logistics, system security, and migration
- Installation and maintenance of an organization-specific procedural model
- Integration of current (quasi) standards, specifications, and regulations
- View-based representation and user-specific access to the V-Model
- Expanded scope of application compared with the V-Model 97

2.3.1.5.1 Advantages

- Decisive points of the project implementation strategies predefine the overall project management framework by the logical sequencing of project completion.
- Detailed project planning and management are implemented based on the processing and completion of products.
- Each team member is allocated explicitly a role for which it is responsible.
- The product quality is checkable by making requests of the product and providing an explicit description of its dependence on other products.

2.3.1.5.2 Disadvantages. None that we can spot. It is a fairly new model mostly used in Germany and hence yet to find out its disadvantages.

2.3.1.5.3 Suitability. With the V-Model XT (2008), the underlying philosophy also has developed further. The new V-Model makes a fundamental distinction in customer–contractor projects. The focus is on the products and not, as before, on the activities. The V-Model XT thus describes a target and results-oriented approach (V-Model XT, 2008).

2.3.1.6 Spiral Model. Figure 2.3 shows the Spiral Model, which is also known as the spiral life-cycle Model. It is a systems development life-cycle model. This model of development combines the features of the Prototyping Model and the Waterfall Model.

The steps in the Spiral Model can be generalized as follows (Watts, 1997):

1. The new system requirements are defined in as much detail as possible. This usually involves interviewing several users representing all the external or internal users and other aspects of the existing system.

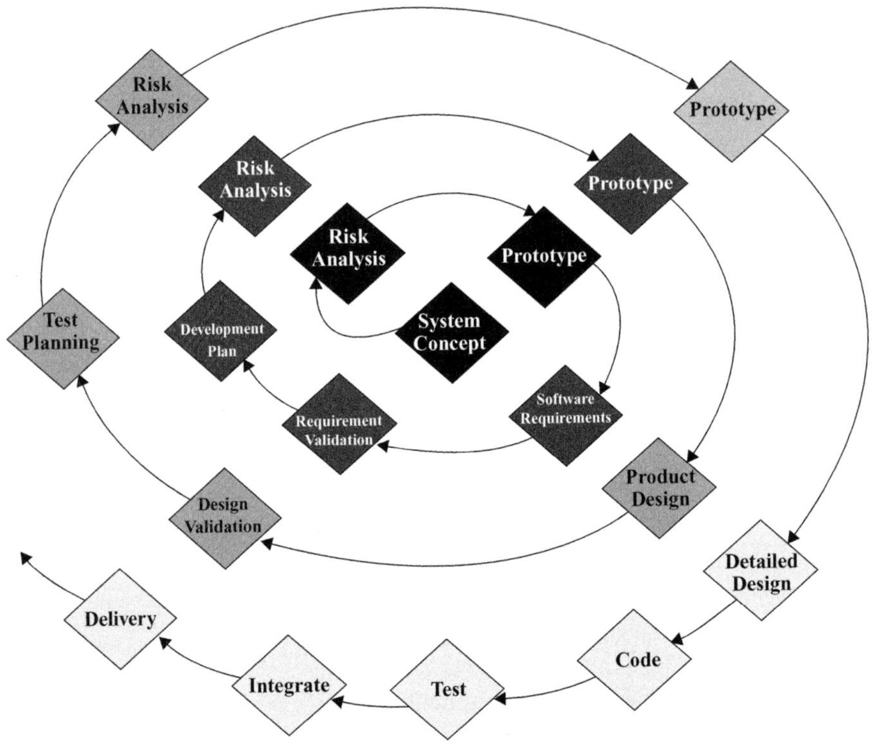

FIGURE 2.3 Spiral model.

2. A preliminary design is created for the new system.
3. A first prototype of the new system is constructed from the preliminary design. This is usually a scaled-down system, and it represents an approximation of the characteristics of the final product.
4. A second prototype is evolved by a fourfold procedure: 1) evaluating the first prototype in terms of its strengths, weaknesses, and risks; 2) defining the requirements of the second prototype; 3) planning and designing the second prototype; and 4) constructing and testing the second prototype.
5. At the customer's option, the entire project can be aborted if the risk is deemed too great. Risk factors might involve development cost overruns, operating-cost miscalculation, or any other factor that could, in the customer's judgment, result in a less-than-satisfactory final product.
6. The existing prototype is evaluated in the same manner as was the previous prototype, and if necessary, another prototype is developed from it according to the fourfold procedure outlined.
7. The preceding steps are iterated until the customer is satisfied that the refined prototype represents the final product desired.

8. The final system is constructed, based on the refined prototype.
9. The final system is evaluated thoroughly and tested. Routine maintenance is carried out on a continuing basis to prevent large-scale failures and to minimize downtime.

2.3.1.6.1 Advantages

- Decisive points of the project implementation strategies predefine the overall project management framework by the logical sequencing of project completion.
- This model of development combines the features of the Prototyping Model and the simplicity of the Waterfall Model.

2.3.1.6.2 Disadvantages

- It could become very costly and time consuming.

2.3.1.6.3 Suitability. This model for development is good for the prototyping or importantly iterative process of prototyping projects. Although, the Spiral Model is favored for large, expensive, and complicated projects (Watts, 1997), if practiced correctly, it could be used for small- or medium-size projects and/or organization.

2.3.1.7 Chaos Model. In computing, the Chaos Model (2008) is a structure of software development that extends the Spiral Model and the Waterfall Model. The Chaos Model notes that the phases of the life cycle apply to all levels of projects, from the whole project to individual lines of code.

- The whole project must be defined, implemented, and integrated.
- Systems must be defined, implemented, and integrated.
- Modules must be defined, implemented, and integrated.
- Functions must be defined, implemented, and integrated.
- Lines of code are defined, implemented, and integrated.

One important change in perspective is whether projects can be thought of as whole units or must be thought of in pieces. Nobody writes tens of thousands of lines of code in one sitting. They write small pieces, one line at a time, verifying that the small pieces work. Then they build up from there. The behavior of a complex system emerges from the combined behavior of the smaller building block. There are several tie-ins with chaos theory.

- The Chaos Model may help explain why software tends to be so unpredictable.
- It explains why high-level concepts like architecture cannot be treated independently of low-level lines of code.
- It provides a hook for explaining what to do next, in terms of the chaos strategy.

2.3.1.7.1 Advantages

- Building complex system through building of small blocks.

2.3.1.7.2 Disadvantages

- Lines of code, functions, modules, system, and project must be defined a priori.

2.3.1.7.3 Suitability

- Mostly suitable in computing application.

2.3.1.8 Top-Down and Bottom-Up. **Top-down** and **bottom-up** are strategies of information processing and knowledge ordering, mostly involving software but also involving other humanistic and scientific theories. In practice, they can be seen as a style of thinking and teaching. In many cases, *top-down* is used as a synonym for analysis or decomposition, and *bottom-up* is used as a synonym for *synthesis*.

A **top-down** approach is essentially breaking down a system to gain insight into its compositional subsystems. In a top-down approach, an overview of the system is first formulated, specifying but not detailing any first-level subsystems. Each subsystem is then refined in yet greater detail, sometimes in many additional subsystem levels, until the entire specification is reduced to base elements. A top-down model is often specified with the assistance of "black boxes" that make it easier to manipulate. However, black boxes may fail to elucidate elementary mechanisms or be detailed enough to validate realistically the model (Top down bottom up, 2008).

A **bottom-up** approach is essentially piecing together systems to give rise to grander systems, thus making the original systems subsystems of the emergent system. In a bottom-up approach, the individual base elements of the system are first specified in great detail. These elements then are linked together to form larger subsystems, which then in turn are linked, sometimes in many levels, until a complete top-level system is formed. This strategy often resembles a "seed" model, whereby the beginnings are small but eventually grow in complexity and completeness. However, "organic strategies" may result in a tangle of elements and subsystems, developed in isolation and subject to local optimization as opposed to meeting a global purpose (Top down bottom up, 2008). In the software development process, the **top-down** and **bottom-up** approaches play a key role.

Top-down approaches emphasize planning and a complete understanding of the system. It is inherent that no coding can begin until a sufficient level of detail has been reached in the design of at least some part of the system (Top down bottom up, 2008). The top-down approach is done by attaching the stubs in place of the module. This, however, delays testing of the ultimate functional units of a system until significant design is complete. Bottom-up emphasizes coding and early testing, which can begin as soon as the first module has been specified. This approach, however, runs the risk that modules may be coded without having a clear idea of how they link to other parts of the system, and that such linking may not be as easy as first thought. Reusability of code is one of the main benefits of the bottom-up approach.

Top-down design was promoted in the 1970s by IBM researcher Harlan Mills and Niklaus Wirth (Top down bottom up, 2008). Harlan Mills developed structured programming concepts for practical use and tested them in a 1969 project to automate the New York Times Morgue Index (Top down bottom up, 2008). The engineering and management success of this project led to the spread of the top-down approach through IBM and the rest of the computer industry. Niklaus Wirth, among other achievements the developer of the Pascal programming language, wrote the influential paper, "Program Development by Stepwise Refinement." (Top down bottom up, 2008) As Niklaus Wirth went on to develop languages such as Modula and Oberon (where one could define a module before knowing about the entire program specification), one can infer that top-down programming was not strictly what he promoted. Top-down methods were favored in software engineering until the late 1980s, and object-oriented programming assisted in demonstrating the idea that both aspects of top-down and bottom-up programming could be used (Top down bottom up, 2008).

Modern software design approaches usually combine both top-down and bottom-up approaches. Although an understanding of the complete system is usually considered necessary for good design, leading theoretically to a top-down approach, most software projects attempt to make use of existing code to some degree. Preexisting modules give designs a bottom-up flavor. Some design approaches also use an approach in which a partially functional system is designed and coded to completion, and this system is then expanded to fulfill all the requirements for the project.

Top-down starts with the overall design. It requires finding modules and interfaces between them, and then going on to design class hierarchies and interfaces inside individual classes. Top-down requires going into smaller and smaller detail until the code level is reached. At that point, the design is ready and one can start the actual implementation. This is the classic sequential approach to the software process.

Top-down programming is a programming style, the mainstay of traditional procedural languages, in which design begins by specifying complex pieces and then dividing them into successively smaller pieces. Eventually, the components are specific enough to be coded and the program is written. This is the exact opposite of the bottom-up programming approach, which is common in object-oriented languages such as C++ or Java. The technique for writing a program using top-down methods is to write a main procedure that names all the major functions it will need. Later, the programming team looks at the requirements of each of those functions and the process is repeated. These compartmentalized subroutines eventually will perform actions so simple they can be coded easily and concisely. When all the various subroutines have been coded, the program is done.

By defining how the application comes together at a high level, lower level work can be self-contained. By defining how the lower level objects are expected to integrate into a higher level object, interfaces become defined clearly (Top down bottom up, 2008).

Bottom-up means to start with the "smallest things." For example, if there is a need for a custom communication protocol for a given distributed application, then start by writing the code for that. Then, for example, let's say the software programmer may write database code and then UI code and finally something to glue them all together. The overall design becomes apparent only when all the modules are ready.

In a bottom-up approach, the individual base elements of the system first are specified in great detail. These elements then are linked together to form larger subsystems, which then in turn are linked, sometimes in many levels, until a complete top-level system is formed. This strategy often resembles a "seed" model, whereby the beginnings are small, but eventually they grow in complexity and completeness (Top down bottom up, 2008).

2.3.1.8.1 Advantages

- A **bottom-up** approach is essentially piecing together systems to give rise to grander systems, thus making the original systems subsystems of the emergent system, which is a nice way to deal with complexity.
- Reusability of code is one of the main benefits of the bottom-up approach.

2.3.1.8.2 Disadvantages

- In top-down, black boxes may fail to elucidate elementary mechanisms or to be detailed enough to validate the model realistically.
- In bottom-up, "organic strategies" may result in a tangle of elements and subsystems, developed in isolation and subject to local optimization as opposed to meeting a global purpose (Top down bottom up, 2008).
- In top-down, stubs are attached in place of the module. This, however, delays testing of the ultimate functional units of a system until significant design is complete. It requires bigger picture to understand first.
- Bottom-up emphasizes coding and early testing, which can begin as soon as the first module has been specified. This approach, however, runs the risk that modules may be coded without having a clear idea of how they link to other parts of the system, and that such linking may not be as easy as first thought.
- Bottom-up projects are hard to manage. With no overall vision, it is hard to measure progress. There are no milestones. The total budget is guesswork. Schedules mean nothing. Teamwork is difficult, as everyone tends to work at their own pace and in isolation.

2.3.1.8.3 Suitability. Although suitable to any kind of project, in the case of software controls projects, it could be done completely top-down or bottom-up. It is important for control engineers, therefore, to understand the two approaches and to apply them appropriately in the hybrid approach. Even when an engineer is working alone, the hybrid approach helps keep the project organized and the resulting system useable, maintainable, and extensible (Masi, 2008).

2.3.1.9 Joint Application Development (JAD). JAD is a methodology that involves the client or end user in the design and development of an application through a succession of collaborative workshops called JAD sessions. Chuck Morris and Tony Crawford, both of IBM, developed JAD in the late 1970s and began teaching the approach through workshops in 1980 (JAD, 2008). The results were encouraging, and JAD became a well-accepted approach in many companies.

The JAD approach, in comparison with the more traditional practice, is thought to lead to faster development times and to greater client satisfaction because the client is involved throughout the development process. In comparison, in the traditional approach to systems development, the developer investigates the system requirements and develops an application, with client input consisting of only a series of interviews. A variation on JAD, Rapid Application Development (RAD) creates an application more quickly through such strategies as using fewer formal methodologies and reusing software components.

2.3.1.9.1 Advantages

- Faster development times and greater client satisfaction because the client is involved throughout the development process.
- Many companies find that JAD allows key users to participate effectively in the requirements modeling process. When users (customers) participate in the systems development process, they are more likely to feel a sense of ownership in the results and support for the new system. This is a DFSS best practice as well.
- When properly used, JAD can result in a more accurate statement of system requirements, a better understanding of common goals, and a stronger commitment to the success of the new system.

2.3.1.9.2 Disadvantages

- Compared with traditional methods, JAD may seem more expensive and can be cumbersome if the group is too large relative to the size of the project.
- A drawback of JAD is that it opens up a lot of scope for interpersonal conflict.

2.3.1.9.3 Suitability. JAD is popular in information technology (IT) applications. It is a process used in the systems development life cycle (SDLC) to collect business requirements while developing new information systems for a company.

2.3.1.10 Rapid Application Development (RAD). RAD (2008) is a process that helps develop products faster and of higher quality through the use of one or more of the following methods:

- Gathering requirements using workshops or focus groups
- Prototyping and early reiterative user testing of designs

- Reusing of software components
- Setting a rigidly paced schedule that defers design improvements to next product version

In RAD, the quality of a system is defined as the degree to which the system meets business requirements (or user requirements) at the time it begins operation. This is fundamentally different from the more usual definition of quality as the degree to which a system conforms to written specifications (Rapid Application Development, 1997). Rapid development, high quality, and lower costs go hand in hand if an appropriate development methodology is used. Some companies offer products that provide some or all of the tools for RAD software development. These products include requirements gathering tools, prototyping tools, computer-aided software engineering tools, language development environments such as those for the Java (Sun Microsystems, Santa Clara, CA) platform, groupware for communication among development members, and testing tools (Top down bottom up, 2008). RAD usually embraces object-oriented programming methodology, which inherently fosters software reuse. The most popular object-oriented programming languages, C++ and Java, are offered in visual programming packages often described as providing Rapid Application Development (Top down bottom up, 2008).

2.3.1.10.1 Advantages

- Inherently fosters software reuse.
- Creates an application more quickly through such strategies as using fewer formal methodologies and reusing software components.
- Can be applied to hardware development as well.
- Rapid development, high quality, and lower costs go hand in hand if an appropriate development methodology is used.
- Less formality in reviews and other team communication. Quality is a primary concept in the RAD environment.
- Systems developed using the RAD development path meet the needs of their users effectively and have low maintenance costs.

2.3.1.10.2 Disadvantages

- There is a danger inherent in rapid development. Enterprises often are tempted to use RAD techniques to build stand-alone systems to solve a particular business problem in isolation. Such systems, if they meet user needs, then become institutionalized. If an enterprise builds many such isolated systems to solve particular problems, the result is a large, undisciplined mass of applications that do not work together.

2.3.1.10.3 Suitability. RAD is used widely in the IT domain, where a carefully planned set of architectures is used to lesson IT productivity problems. RAD is one such path that could be used for rapid development of a stand-alone system. And thus the design of the architectures is a matter of primary strategic importance to the enterprise as a whole because it directly affects the enterprise's ability to seize new business opportunities (Rapid Application Development, 1997).

2.3.1.10.4 Model-Driven Engineering (MDE). **Model-driven engineering (MDE)** focuses on creating models that capture the essential features of a design. A modeling paradigm for MDE is considered effective if its models make sense from the point of view of the user and can serve as a basis for implementing systems. The models are developed through extensive communication among product managers, designers, and members of the software development team. As the models approach completion, they enable the development of software and systems.

The best-known MDE initiative is the Object Management Group (OMG) initiative Model-Driven Architecture (MDA), which is a registered trademark of OMG (Needham, MA) (Leveson, 2004). Another related acronym is Model-Driven Development (MDD), which also is an OMG trademark (Leveson, 2004), (Schmidt, 2006).

2.3.1.10.5 Advantages

- MDE is a very promising technique that can be used to improve the current processes of system engineering.
- Using MDD, software can become more verifiable, scalable, maintainable, and cheaper.

2.3.1.10.6 Disadvantages

- Challenges in modeling languages, separation of concerns, model management, and model manipulation.
- Too many questions left on the table about actual implementation of model management and model manipulation in day-to-day operations.
- The user must have a good working knowledge about the models that are input. This might not always be true and may result in errors from the merging process because the user chose the incorrect merge.

2.3.1.10.7 Suitability

- More recent research is being pored into the methodology for further development.

2.3.1.11 Iterative Development Processes. **Iterative development** (Pressman, 2000) prescribes the construction of initially small but even larger portions of a software project to help all those involved to uncover important issues early before problems or faulty assumptions can lead to disaster. Commercial developers prefer iterative processes because they allow customers who do not know how to define what they want to reach their design goals.

The Waterfall Model has some well-known limitations. The biggest drawback with the Waterfall Model is that it assumes that requirements are stable and known at the start of the project. Unchanging requirements, unfortunately, do not exist in reality, and requirements do change and evolve. To accommodate requirement changes while executing the project in the Waterfall Model, organizations typically define a change management process, which handles the change requests. Another key limitation is that it follows the "big bang" approach—the entire software is delivered in one shot at the end. No working system is delivered until the end of the process. This entails heavy risks, as the users do not know until the very end what they are getting (Jalote et al., 2004).

To alleviate these two key limitations, an iterative development model can be employed. In iterative development, software is built and delivered to the customer in iterations. Each iteration delivers a working software system that is generally an increment to the previous delivery. Iterative enhancement and spiral are two well-known process models that support iterative development. More recently, agile and XP methods also promote iterative development.

2.3.1.11.1 Advantages

- With iterative development, the release cycle becomes shorter, which reduces some of the risks associated with the "big bang" approach.
- Requirements need not be completely understood and specified at the start of the project; they can evolve over time and can be incorporated into the system in any iteration.
- Incorporating change requests also is easy as any new requirements or change requests simply can be passed on to a future iteration.

2.3.1.11.2 Disadvantages

- It is hard to preserve the simplicity and integrity of the architecture and the design.

2.3.1.11.3 Suitability

- Overall, iterative development can handle some of the key shortcomings of the Waterfall Model, and it is well suited for the rapidly changing business world, despite having some of its own drawbacks.

2.3.2 Agile Software Development

With the advent of the World Wide Web in the early 1990s, the *agile* software design methodologies [also referred to as *light weight*, *lean*, *Internet-speed*, *flexible*, and *iterative* (Kaner, 1996), (Juran & Gryna, 1988] were introduced in an attempt to provide the lighter, faster, nimbler software development processes necessary for survival in the rapidly growing and volatile Internet software industry. Attempting to offer a "useful compromise between no process and too much process" (Juran & Gryna, 1988), the *agile* methodologies provide a novel, yet sometimes controversial, alternative for software being built in an environment with vague and/or rapidly changing requirements (Agile Journal, 2006).

Agile software development is a methodology for software development that promotes development iterations, open collaboration, and adaptability throughout the life cycle of the project. There are many agile development methods; most minimize risk by developing software in short amounts of time. Software developed during one unit of time is referred to as an iteration, which typically lasts from two to four weeks. Each iteration passes through a full software development cycle, including planning, requirements analysis, design, writing unit tests, and then coding until the unit tests pass and a working product is finally demonstrated to stakeholders. Documentation is no different than software design and coding. It, too, is produced as required by stakeholders. The iteration may not add enough functionality to warrant releasing the product to market, but the goal is to have an available release (without bugs) at the end of the iteration. At the end of the iteration, stakeholders re-evaluate project priorities with a view to optimizing their return on investment.

Agile software development processes are built on the foundation of iterative development to that foundation. They add a lighter, more people-centric viewpoint than traditional approaches. Agile processes use feedback, rather than planning, as their primary control mechanism. The feedback is driven by regular tests and releases of the evolving software (Agile Journal, 2006). Figure 2.4 shows the conceptual comparison of the Waterfall Model, iterative method, and an iterative time boxing method.

2.3.2.0.4 Advantages (Stevens et al., 2007)

- The agile process offers the advantage of maximizing a product's innovative features.
- The agile process can produce a product that has the optional to be highly successful in the market.
- The agile development process minimizes upfront investment and provides options for incorporating customer learning before, during, and after product launch.

2.3.2.0.5 Disadvantages (Stevens et al., 2007)

- The process is an open-ended program plan.
- It may create cost and schedule overruns that could impact a company's entire operational stability.

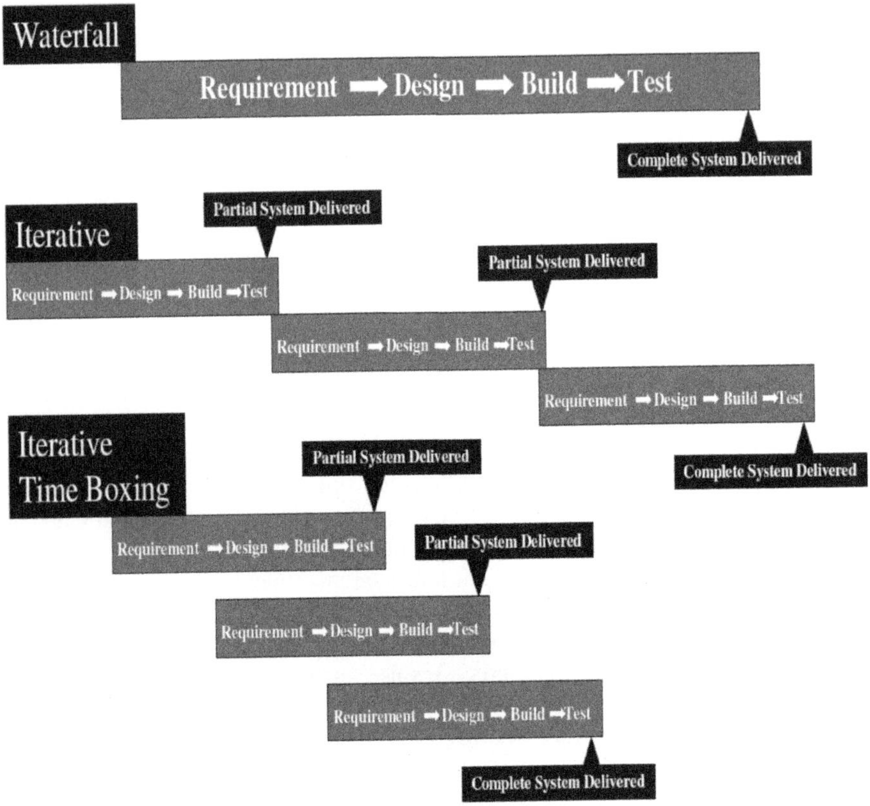

FIGURE 2.4 Agile software development process (Agile Journal, 2006).

2.3.2.0.6 Suitability

- Suitable to emerging products that are examples of extreme nonlinear systems where slight variations in assumptions can lead to drastic changes in outcomes, which can be caused by unknown variation from tolerances, wear, and environment (Stevens et al., 2007).

2.3.2.1 Unified Process. The **Unified Software Development Process** or **Unified Process** (UP) is a popular iterative and incremental software development process framework. The best-known and extensively documented refinement of the Unified Process is the Rational Unified Process (RUP). The Unified Process is not simply a process but an extensible framework, which should be customized for specific organizations or projects. The RUP is, similarly, a customizable framework (Unified Process, 2008). As a result, it often is impossible to say whether a refinement of the process was derived from UP or from RUP, and so the names tend to be used interchangeably (Unified Process, 2008). The name *Unified Process* (as opposed to

Rational Unified Process) generally is used to describe the generic process, including those elements that are common to most refinements (Unified Process, 2008). The *Unified Process* name also is used to avoid potential issues of copyright infringement because *Rational Unified Process* and *RUP* are trademarks of IBM (Unified Process, 2008). Since 2008, various authors unaffiliated with Rational Software have published books and articles using the name *Unified Process*, whereas authors affiliated with Rational Software have favored the name Rational Unified Process (Unified Process, 2008).

The Unified Process is an iterative and incremental development process. The Elaboration, Construction and Transition phases are divided into a series of time-boxed iterations. (The Inception phase also may be divided into iterations for a large project.) Each iteration results in an *increment*, which is a release of the system that contains added or improved functionality compared with the previous release. Although most iterations will include work in most process disciplines (e.g., Requirements, Design, Implementation, and Testing) the relative effort and emphasis will change over the course of the project. The number of Unified Process refinements and variations is countless. Organizations using the Unified Process invariably incorporate their own modifications and extensions. The following is a list of some of the better known refinements and variations (Unified Process, 2008):

- Agile Unified Process (AUP), a lightweight variation developed by *Scott W. Ambler.*
- Basic Unified Process (BUP), a lightweight variation developed by IBM and a precursor to OpenUP.
- Enterprise Unified Process (EUP), an extension of the Rational Unified Process.
- Essential Unified Process (EssUP), a lightweight variation developed by Ivar Jacobson.
- Open Unified Process (OpenUP), the Eclipse Process Framework software development process.
- Rational Unified Process (RUP), the IBM/Rational Software development process.
- Oracle Unified Method (OUM), the Oracle development and implementation process.
- Rational Unified Process-System Engineering (RUP-SE), a version of RUP tailored by Rational Software for System Engineering.

2.3.2.1.1 Advantages

- It provides a disciplined approach to assigning tasks and responsibilities within a development organization.
- Unified Process is architecture-centric, and the Unified Process prescribes the successive refinement of an executable architecture.
- Risks are mitigated earlier.

- Change is more manageable.
- Higher level of reuse.
- The project team can learn along the way.
- Better overall quality.

2.3.2.1.2 Disadvantages

- Extensive knowledge is required—Someone needs initially to learn and understand the Unified Process so that he or she can develop, tailor, or enhance the Unified Process for new type of project, situation, and requirements.
- Contradictory advice—The new version may be in contradiction with the Unified Process or RUP, or other process materials, at certain points. Having the source material available "as is" may cause confusion unless people understand that you have overridden portions of it. An effective approach is to set a specific design scheme for your pages and then make sure that everyone is aware that your pages are official and that all other pages are simply reference.
- Complexity—Providing a process in which people must understand the base description and then understand the changes to it at another location may be confusing for some people.

2.3.2.1.3 Suitability

- The Unified Process with several different flavors (enhancements) from IBM, Oracle, and Agile are used more commonly in IT; however, the could be curtailed to the specific need. For example, the Rational Unified Process provides a common language and process for business engineering and software engineering communities, as well as shows how to create and maintain direct traceability between business and software models. Yet the Basic Unified Process was an enhancement to the Unified Process that is more suited for small and simple projects.

2.3.2.2 eXtreme Programming. Although many *agile* methodologies have been proposed during the past decade (e.g., ASD: Adaptive Software Development, the Crystal Family; DSDM: Dynamic Systems Development Method; FDD: Feature-Driven Development; ISD: Internet-Speed Development; PP: Pragmatic Programming; and SCRUM, RUP: Rational Unified Programming) (Abrahamsson et al., 2003), (Highsmith, 2001), here the focus is on the best known and most widely used of the agile software development methodologies: Extreme Programming (Baird, 2003), (Van Cauwenberghe, 2003).

In the early 1990s, the concept of a simple, yet efficient, approach to software development was already under consideration by Kent Beck and Ward Cunningham (Wells, 2001). In early 1996, in a desperate attempt to revive the Chrysler Comprehensive Compensation (C3) project, the Chrysler Corporation hired Beck as a consultant; his recommendation was to throw away all of their existing code and abandon their current Waterfall methodology. During the next 14 months, Beck, along with the help of Ron Jeffries and Martin Fowler, restarted the C3 payroll project from scratch (keeping only the existing GUIs), employing his new software development concepts along the way. By mid-1997, his informal set of software engineering practices had been transformed into an agile *methodology* known as Extreme Programming[8] (Anderson, 1998) (Beck, 1999). With respect to his newly introduced Extreme Programming methodology, Kent Beck stated, "Extreme Programming turns the conventional software process sideways. Rather than planning, analyzing, and designing for the far-flung future, XP programmers do all of these activities—a little at a time—throughout development" (Beck, 1999, p. 70).

In surveys conducted by Ganssle (2001) very few companies have actually adopted the Extreme Programming methodology for their embedded applications; however, there was a fair amount of interest in doing so (Grenning, 2002). Having made its debut as a software development methodology only seven years ago, Extreme Programming is a relatively immature software development methodology. In general, academic research for agile methodologies is lacking, and most of what has been published involves case studies written by consultants or practitioners (Abrahamsson et al., 2002, p. 1). According to Paulk, agile methods are the "programming methodology of choice for the high-speed, volatile world of Internet software development" and are best suited for "software being built in the face of vague and/or rapidly changing requirements" (Paulk, 2002, p. 2).

2.3.2.2.1 Advantages

- XP is also very productive and produces high-quality software.
- Project Restrictions—There is a small set of project environments that the XP methodology to which successfully can be applied—software only, small teams, and a clearly definable cooperative customer. It is a nonscalable process as a whole and claims it needs to be whole to reap the most benefit.
- Local Optimization—Ian Alexander (2001, p. 1) states that "Maxims like, do the simplest thing that could possibly work, do not necessarily lead to optimal solutions."
- Process versus Process Improvements—For example, the Capability Maturity Model Integration (CMMI) models emphasize complete coverage of the "what" of the model, but the "how" is left to the organization or project and needs to make business sense. XP emphasizes complete coverage of the process specifying the "how" and it does not fit nondetrimentally within as many business environments.

[8]Wiki (The Portland Pattern Repository). Hosted by Ward Cunningham. Embedded Extreme Programming. http://c2.com/cgi/wiki?Embedded Extreme Programming.

2.3.2.2.2 Disadvantages

- XP is framed as trying to solve the problem of software development risk with a solution of people in the environment of a small project. XP's approach is fragile and can fail if the project environment changes or the people change.

2.3.2.2.3 Suitability

- Extreme programming is targeted toward small-to-medium-sized teams building software in the face of vague and/or rapidly changing requirements.

2.3.2.3 Wheel and Spoke Model. The **Wheel and Spoke Model** is a sequential parallel software development model. It is essentially a modification of the Spiral Model that is designed to work with smaller initial teams, which then scale upward and build value faster. It is best used during the design and prototyping stages of development. It is a bottom-up methodology. The Wheel and Spoke Model retains most of the elements of the Spiral Model, on which it is based.

As in the Spiral Model, it consists of multiple iterations of repeating activities:

1. New system requirements are defined in as much detail as possible from several different programs.
2. A preliminary common application programming interface (API) is generated that is the greatest common denominator across all the projects.
3. The implementation stage of a first prototype.
4. The prototype is given to the first program where it is integrated into their needs. This forms the first spoke of the Wheel and Spoke Model.
5. Feedback is gathered from the first program and changes are propagated back to the prototype.
6. The next program can now use the common prototype, with the additional changes and added value from the first integration effort. Another spoke is formed.
7. The final system is the amalgamation of common features used by the different programs—forming the wheel, and testing/bug-fixes that were fed back into the code-base—forming the spokes.

Every program that uses the common code eventually sees routine changes and additions, and the experience gained by developing the prototype for the first program is shared by each successive program using the prototype (Wheel and Spoke Model, 2008). The wheel and spoke is best used in an environment where several projects have a common architecture or feature set that can be abstracted by an API. The core team developing the prototype gains experience from each successful program that adapts the prototype and sees an increasing number of bug-fixes and a general rise in code quality. This knowledge is directly transferable to the next program because the core code remains mostly similar.

2.3.2.3.1 Advantages

- Decisive points of the project implementation strategies predefine the overall project management framework by the logical sequencing of project completion.
- Presents low initial risk.
- Since one is developing a small-scale prototype instead of a full-blown development effort, much fewer programmers are needed initially.
- If the effort is deemed successful, the model scales very well by adding new people as the scope of the prototype is expanded.
- Gained expertise could be applicable across different programs.

2.3.2.3.2 Disadvantages

- No data from any business or industry are available at this point.

2.3.2.3.3 Suitability

- It is suitable in an environment where several projects have a common architecture or feature set that can be abstracted by an API, and it is best used during the design and prototyping stages of development.

2.3.2.4 Constructionist Design Methodology. This is a methodology for designing and implementing interactive intelligences. The Constructionist Design Methodology (CDM)—so called because it advocates modular building blocks and incorporation of prior work—addresses factors that can be perceived as key to future advances in artificial intelligence (AI) including interdisciplinary collaboration support, coordination of teams, and large-scale systems integration. Inspired to a degree by the classic LEGO bricks, this methodology, which is known as the Constructionist Approach to AI, puts modularity at its center. The functionalities of the system are broken into individual software modules, which are typically larger than software classes (i.e., objects and methods) in object-oriented programming but smaller than the typical enterprise application. The role of each module is determined in part by specifying the message types and information content that needs to flow between the various functional parts of the system. Using this functional outline, one can then define and develop, or select, components for perception, knowledge representation, planning, animation, and other desired functionalities. There is essentially nothing in the Constructionist Approach to AI that lends it more naturally to behavior-based

AI or "classical" AI—its principles sit beside both (Thórisson et al., 2004). In fact, because CDM is intended to address the integration problem of very broad cognitive systems, it must be able to encompass all variants and approaches to date. It is unlikely that a seasoned software engineer will find any of the principles presented objectionable, or even completely novel for that matter. But these principles are custom-tailored to guide the construction of large cognitive systems that could be used, extended, and improved by many others over time.

2.3.2.4.1 Advantages

- Modularity at its center, where functionalities of the system are broken into individual software modules.
- CDM's principle strength is in simplifying the modeling of complex, multi-functional systems requiring architectural experimentation and exploration of subsystem boundaries, undefined variables, and tangled data flow and control hierarchies.

2.3.2.4.2 Disadvantages

- Not proliferated into other industry or areas other than AI.

2.3.2.4.3 Suitability

- CDM is a methodology for designing and implementing interactive intelligences, and it is mostly suitable for building large cognitive robotics systems, communicative humanoids, facial animation, interdisciplinary collaboration support, coordination of teams, and large-scale systems integration in AI. It is most applicable for systems with ill-defined boundaries between subsystems, and where the number of variables to be considered is relatively large. In the current state of science, primary examples include ecosystems, biological systems, social systems, and intelligence.

2.4 SOFTWARE DEVELOPMENT PROCESSES CLASSIFICATION

The classification of traditional software development processes can be done in many different ways; however, here the models discussed in Section 2.2 are viewed from complexity and size of a project. Table 2.1 shows the classification based on the suitability of size and complexity of project. The gray areas shown in Table 2.1 depict the nonsuitability of the given software process depending on the size and complexity of the project. This does not mean the process cannot be used, but knowing the nature of the process or model's best results may not be obtained.

TABLE 2.1 Classification Based on the Suitability of Size and Complexity of Project

Software Process	Simple and Small	Moderate and Medium	Complex and Large
Waterfall Model **Sashimi Model** **Chaos Model**	1. It allows for departmentalization and managerial control. 2. A schedule can be set with deadlines for each stage of development, and a product can proceed through the development process and, theoretically, be delivered on time. 3. Development moves from concept, through design, implementation, testing, installation, troubleshooting, and ends up at operation and maintenance. Each phase of development proceeds in strict order, without any overlapping or iterative steps. 4. For Simple, Static/Frozen requirements and Small Project. These methods might prove effective and cheaper. 5. The disadvantage of Waterfall development is that it does not allow for much reflection or revision. 6. Once an application is in the testing stage, it is very difficult to go back and change something that was not well thought out in the concept stage. 7. Classic Waterfall methodology usually breaks down and results in a failure to deliver the needed product for complex and continuously changing requirements.		

(*Continued*)

TABLE 2.2 ***(Continued)***

Software Process	Simple and Small	Moderate and Medium	Complex and Large
V-Model			1. It is resource heavy and very costly to implement, suited for large organization and government projects. 2. The V-Model is not complete because the activities are done at too abstract a level. It is hard to find out whether peer reviews and inspections are done in the V-model. It is difficult to find out whether the self-assessment activity is conducted before product is passed on to the QA for acceptance.
V-Model XT		Defense and Safety Critical IT Early Phase of introduction	1. Defense and Safety Critical IT Early Phase of introduction. 2. It was introduced in 2006 and until now mostly used in Germany in government and military applications with very limited information available.
Spiral		It is a good approach for safety-critical systems, but may endure very high cost.	1. Suited for Safety Critical Systems, but high chance of becoming extremely costly and time consuming. 2. This model of development combines the features of the Prototyping Model and the Waterfall Model. 3. The Spiral Model is favored for large, expensive, and complicated projects. 4. The entire project can be aborted if the risk is deemed too great. Risk factors might involve development cost overruns, operating-cost miscalculation, or any other factor that could, in the customer's judgment, result in a less-than-satisfactory final product.

Top-Down Bottom-Up	1. A top-down approach is essentially breaking down a system to gain insight into its compositional subsystems. 2. Top-down approaches emphasize planning and a complete understanding of the system. It is inherent that no coding can begin until a sufficient level of detail has been reached in the design of at least some part of the system. 3. A top-down model is often specified with the assistance of "black boxes" that make it easier to manipulate. 4. A bottom-up approach is essentially piecing together systems to give rise to grander systems, thus making the original systems subsystems of the emergent system. 5. The reusability of code is one of the main benefits of the bottom-up approach. 6. Black boxes may fail to elucidate elementary mechanisms or be detailed enough to validate the model realistically. 7. The top-down approach is done by attaching the stubs in place of the module. This, however, delays testing of the ultimate functional units of a system until significant design is complete. 8. In a bottom-up approach, the individual base elements of the system are first specified in great detail. 9. Bottom-up emphasizes coding and early testing, which can begin as soon as the first module has been specified. 10. This approach, however, runs the risk that modules may be coded without having a clear idea of how they link to other parts of the system, and that such linking may not be as easy as first thought. Although suitable to any kind of project, in the case of controls projects, it could be done completely top-down or bottom-up. It is important for control engineers, therefore, to understand the two approaches and apply them appropriately in the hybrid approach. Even when an engineer is working alone, the hybrid approach helps keep the project organized and the resulting system usable, maintainable, and extensible.

(*Continued*)

TABLE 2.3 ***(Continued)***

Software Process	Simple and Small	Moderate and Medium	Complex and Large
Joint Application Development (JAD)		In comparison with the more traditional practice, it is thought to lead to faster development times and greater client satisfaction, because the client is involved throughout the development process.	
Rapid Application Development (RAD)		A variation on JAD, Rapid Application Development (RAD) creates an application more quickly through such strategies as using fewer formal methodologies and reusing software components.	
Six Sigma[9]		1. Six Sigma DMAIC was mostly concerned with problem solving to enhance processes by reducing defects and variation that would cause customer dissatisfaction for existing products. 2. Six Sigma DFSS was created to address low yields in high-volume electronics manufacturing, which required near perfect levels of quality. The process starts with and is guided by conformance to customer needs and product specifications. Six Sigma provides infrastructure, including Green Belts, Black Belts and Master Black Belts, to enable team-based problem solving to work outside the normal work processes and minimize disruptions to normal operations (except when warranted).	
Model-Driven Engineering (MDE)	1. It focuses on creating models that capture the essential features of a design. 2. A modeling paradigm for MDE is considered effective if its models make sense from the point of view of the user and can serve as a basis for implementing systems. 3. The models are developed through extensive communication among product managers, designers, and members of the development team.		

[9]See Chapter 7.

	4. As the models approach completion, they enable the development of software and systems.	
Iterative Development Process		1. In an iterative development, software is built and delivered to the customer in iterations—each iteration delivering a working software system that is generally an increment to the previous delivery. 2. With iterative development, the release cycle becomes shorter, which reduces some risks associated with the "big bang" approach. 3. Requirements need not be completely understood and specified at the start of the project—they can evolve over time and can be incorporated in the system in any iteration. 4. It is hard to preserve the simplicity and integrity of the architecture and the design.
Agile Software Process	1. Agile software development processes are built on the foundation of iterative development. To that foundation they add a lighter, more people-centric viewpoint than traditional approaches. 2. Agile processes use feedback, rather than planning, as their primary control mechanism.	
Unified Process		1. The Unified Process is not simply a process but an extensible framework, which should be customized for specific organizations or projects. 2. The Unified Process is an iterative and incremental development process. Each iteration results in an increment, which is a release of the system that contains added or improved functionality compared with the previous release. Although most iteration will include work in most process disciplines (e.g., Requirements, Design, Implementation, and Testing), the relative effort and emphasis will change over the course of the project. 3. The Elaboration, Construction, and Transition phases are divided into a series of time-boxed iterations. (The Inception phase may also be divided into iterations for a large project.)

(*Continued*)

TABLE 2.4 ***(Continued)***

Software Process	Simple and Small	Moderate and Medium	Complex and Large
eXtreme Programming (Agile)	1. Extreme programming is targeted toward small-to-medium-sized teams building software in the face of vague and/or rapidly changing requirements. 2. Although it is true that embedded systems development may not be the most common application for agile software methodologies, several detailed and well-written exist published by those who have successfully done so. 3. Heavily dependent on customer interface, focuses on features and key processes while making last minute changes.		
Wheel and Spoke Model	1. The Wheel and Spoke Model is a sequentially parallel software development model. 2. It is essentially a modification of the Spiral Model that is designed to work with smaller initial teams, which then scale upward and build value faster. 3. It is best used during the design and prototyping stages of development. It is a bottom-up methodology. 4. Low initial risk. As one is developing a small-scale prototype instead of a full-blown development effort, much fewer programmers are needed initially. 5. Also, gained expertise could be applicable across different programs.		
Constructionist Design Methodology			1. Advocates modular building blocks and incorporation of prior work. 2. Principles are custom-tailored to guide the construction of communicative humanoids, facial animation, and large robotic cognitive systems in AI that could be used, extended, and improved by many others over time.

2.5 SUMMARY

This chapter presented the various existing software processes and their pros and cons, and then classified them depending on the complexity and size of the project. For example, ***Simplicity (or complexity) and size (Small size, Medium size, or Large Size)*** attributes were used to classify the existing software processes that could be useful to a group, business, and/or organization. This classification can be used to understand the pros and cons of the various software processes at a glance and its suitability to a given software development project.

REFERENCES

Abrahamsson, Pekka, Outi, Salo, Jussi, Ronkainen, Juhani, Warsta (2002), *Agile Software Development Methods: Review and Analysis*, VTT Publications 478. espoo, Finland, pp. 1–108.

Abrahamsson, Pekka, Juhani, Warsta, , Mikko T. Siponen,, and Jussi, Ronkainen (2003), *New Directions on Agile Methods: A Comparative Analysis*, IEEE, Piscataway, NJ.

Journal Agile (2006), Agile Survey Results: Solid Experience And Real Results. www.**agilejournal**.com/home/site-map.

Alexander, Ian (2001), *"The Limits of eXtreme Programming," eXtreme Programming Pros and Cons: What Questions Remain*? IEEE Computer Society Dynabook. http://www.computer.org/SEweb/Dynabook/AlexanderCom.htm.

Anderson, Ann (1998), *Case Study*: Chrysler Goes to "Extremes," pp. 24–28. Distributed Computing. http://www.DistributedComputing.com.

Baird, Stewart (2003), *Teach Yourself Extreme Programming in 24 Hours*, Sams, Indianapolis, IN.

Beck, Kent (1999), "Embracing change with extreme programming." *Computer*, Volume 32, #10, pp. 70–77.

(Chaos Model 2008), In *Wikipedia. the Free Encyclopedia*. http://en.wikipedia.org/wiki/Chaos_model.

Ganssle, Jack (2001), *Extreme Embedded*. The Ganssle Group. http://www.ganssle.com.

Grenning, James (2002), *Extreme Programming and Embedded Software Development*. XP and Embedded Systems Development, Parlorsburg, WV.

Highsmith, Jim (2001), *Agile Methodologies: Problems, Principles, and Practices*. Cutter Consortium, PowerPoint presentation, slides 1-49. Information Architects, Inc, Toronto, Canada.

JAD (2008), In *Wikipedia. The Free Encyclopedia*. http:// searchsoftwarequality.techtarget.com/sDefinition/0,,sid92_gci820966,00.html.

Jalote, Pankaj, Patil, Aveejeet, Kurien, Priya, and Peethamber, V. T. (2004), Timeboxing: A process model for iterative software development. Journal of Systems and Software Volume *70*, #1–2, pp. 117–127.

Juran, Joseph M., and Gryna, Frank M. (1988), "Quality Costs," *Juran's Quality Control Handbook*, 4th. McGraw-Hill, New York, pp. 4.9–4.12.

Kaner, Cem (1996), "Quality cost analysis: Benefits and risks." *Software QA*, Volume *3*, # 1, p. 23.

Leveson, Nancy (2004), "A new accident model for engineering safer systems." *Safety Science*, Volume *42*, #4, pp. 237–270.

Masi, C. (2008), What are top-down and bottom-up design methods?. Controls Engineering, http://www.controleng.com/blog/820000282/post/960021096.html. (February 4, 2008).

Paulk, Mark C (2002), Agile Methodologies and Process Discipline. STSC Crosstalk. http://www.stsc.hill.af.mil/crosstalk/2002/10/paulk.html.

Pressman, Roger S. (2000), *Software Engineering (A Practitioner's Approach)* 5th ed., McGraw-Hill Education, New York.

RAD (2008), In *Wikipedia, The Free Encyclopedia*. http://searchsoftwarequality.techtarget.com/search/1,293876,sid92,00.html?query=RAD.

Rapid Application Development (1997). Application Development Methodology by Davis, University of California, built on May 29, 1997. http://sysdev.ucdavis.edu/WEBADM/document/rad-archapproach.htm.

Schmidt, Douglas C. (2006), "Model-driven engineering." IEEE Computer, Volume *39* #2.

Siviy Jeamine M., Penn M. Lynn, and Stoddard, Robert W. (2007), *CMMI and Six Sigma: Partners in Process Improvement*, Addison-Wesley, Boston, MA.

Stevens, Robert A., and Lenz Deere, Jim et al. (2007), "CMMI, Six Sigma, and Agile: What to Use and When for Embedded Software Development," Presented at SAE International—Commercial Vehicle Engineering Congress and Exhibition Rosemont, Chicago, IL Oct. 30-Nov. 1, 2007.

Tayntor, Christine (2002), *Six Sigma Software Development*, CRC Press, Boca Raton, FL.

Chowdhury, Subir (2002), *Design For Six Sigma: The Revolutionary Process for Achieving Extraordinary Profits*, Dearborn Trade Publishing, Chicago, IL

Thórisson, Kristinn R., Hrvoje, Benko, Abramov, Denis, Andrew, Arnold, Maskey, Sameer, and Vaseekaran, Aruchunan (2004), Constructionist Design Methodology for Interactive Intelligences, A.I. Magazine, Volume 25, #4.

Top down bottom up (2008), In *Wikipedia. the Free Encyclopedia*. http://en.wikipedia.org/wiki/Top-down.

Unified Process Software Development (2008), *Wikipedia, The Free Encyclopedia*. http://en.wikipedia.org/w/index.php?title=V-Model_%28software_development%29&oldid=224145058.

Van Cauwenberghe, Pascal (2003), *Agile Fixed Price Projects, part 2*: "*Do You Want Agility With That?*" Volume 3.2, pp. 1–7.

V-Model XT (2008), http://www.iabg.de/presse/aktuelles/mitteilungen/200409_V-Model_XT_en.php (retrieved 11:54, July 15, 2008).

Waterfall Model 2008, In *Wikipedia. the Free Encyclopedia*. http://en.wikipedia.org/wiki/Waterfall_model.

Watts, S. Humphrey (1997), *Introduction to the Personal Software Process*, Addison Wesley, Boston, MA.

Wells, Don (2001), *Extreme Programming*: A Gentle Introduction. http://www.ExtremeProgramming.org.

Wheel and Spoke Model (2008), In Wikipedia, http://en.wikipedia.org/wiki/Wheel_and_spoke_model.

White, Robert V. (1992), "An Introduction to Six Sigma with a Design Example," APEC '92 Seventh Annual Applied Power Electronics Conference and Exposition, Feb. pp. 28–35.

CHAPTER 3

DESIGN PROCESS OF REAL-TIME OPERATING SYSTEMS (RTOS)

3.1 INTRODUCTION

This chapter discusses different processes and features that are included in real-time operating system (RTOS) designs. It complements Chapter 2, which discusses the traditional development processes. We also cover in this chapter the common design techniques of the past, present, and future. Real-time operating systems differ from general-purpose operating systems in that resources are usually limited in real-time systems so the operating system usually only has features that are needed by the application.

A real-time software is a major part of existing software applications in the industry. Applications of real-time software are in automotive systems, consumer electronics, control systems, communication systems, and so on. Real-time software systems demand special attention between they use special design techniques that are time sensitive.

Because of the industry movement toward multiprocessor and multicore systems, new challenges are being introduced. The operating system must now address the needs of two processors, scheduling tasks on multiple cores and protecting the data of a system whose memory is being accessed from multiple sources. New issues are being uncovered, and the need for reliable solutions is needed. This chapter will cover many of the design issues for real-time software.

In addition to hardware evolution impacting real-time operating system designs, another factor is the need for efficient and cheap systems. Many companies are

Software Design for Six Sigma: A Roadmap for Excellence, By Basem El-Haik and Adnan Shaout

finding that commercial real-time operating systems are expensive to purchase and support. Future RTOS designs will be developed in-house and leverage the vast amount of open-source code available for real-time systems, which will demand the use of Design for Six Sigma (DFSS) to optimize their design. In addition to the features found in standard operating systems such as memory management, task scheduling, and peripheral communication, the operating system must provide a method for ensuring time deadlines are met. This is not to say that all real-time systems will always meet their deadlines because other factors need to be considered, factors that are out of the control of the operating system. The real-time operating system has additional features such as timers and preemption.

A real-time operating system must have a deterministic kernel, which means that system calls that are handled by the operating system must complete within a predetermined and known time (Kalinsky, 2003). If a task makes a system call, the time to perform the system call should be consistent, but the worst-case time to perform the system call must be known. This is essential for programmers to ensure that the task will always meet their deadlines. If a system uses an operating system that is nondeterministic, there is no time guarantee that a call will finish in time to allow the task to complete by its deadline.

3.2 RTOS HARD VERSUS SOFT REAL-TIME SYSTEMS

There are three types of real-time systems: soft, hard, and firm. Hard systems are defined as ones that experience catastrophic failure if deadlines are not meant. Failure is deemed catastrophic if the system cannot recover from such an event. A hard real-time system would not be able to recover if deadlines were missed and the effects could be disastrous. Examples of this are vehicle and flight controllers; if a deadline were missed in these systems, the vehicle or plan may crash causing devastating damage and people may lose their lives.

Soft systems are those that can sustain some missed deadlines and the system will not cause devastating results. For example, a machine that records television programs is a real-time system because it must start and stop at a certain time in order to record the appropriate program. But, if the system does not start/stop the recording at the correct time, it may be annoying but will not cause catastrophic damage. An operating system must be designed so that it can meet the requirements of the type of system in which it is used.

A firm system falls somewhere in between soft and hard, where occasional failures may be tolerated. But if the issue persists, the system may experience failures because deadlines that are repeatedly missed may not be recoverable. This may indicate a system that is overused. If system utilization is occurring, meaning that the central processing unit (CPU) is overused and unable to support the task deadlines, before new hardware is purchased there may be optimization techniques that can be performed on the system and improve efficiencies (Furr, 2002).

3.2.1 Real Time versus General Purpose

There are two main categories of operating systems, real time and general purpose. The difference between the two is given in the word "time", time is what separates an (RTOS), from a general purpose operating system (GPOS). An RTOS must meet time deadlines that are specified in the requirements of the system.

This design of an RTOS is such that tasks may have priorities and scheduling is based on time, and it may be partial, meaning the system will give preference to a task that has a higher priority. A GPOS makes no such guarantees and may treat all tasks as equals, meaning they get equal CPU time. The time at which a task runs is of little significance to a GPOS; each task is allowed its time slice, and then it moves on to the next task.

In addition, the kernel of a GPOS is generally not *preemptible*. Once the thread begins execution, another process cannot interrupt it because it has higher priority. Some kernels, such as Linux 2.6, have been modified to allow some preemption, but not to the degree that would support a hard real-time system. Real-time systems require a *preemptible* kernel, one that has been designed to allow system calls to be interrupted so a higher priority task can execute (Leroux, 2005).

3.3 RTOS DESIGN FEATURES

3.3.1 Memory Management

An RTOS must include a method for handling memory for both the program and the data. Program memory is more straightforward because it usually is located in some static form such as flash or Electrically Erasable Programmable Read-Only Memory (EEPROM).

Memory allocated for data can be in cache or RAM and is accessible by the whole application. Many desktop processors have a memory management unit (MMU) that can switch to supervisor mode for system calls, thus preventing data corruption by a task-modifying system memory. Because an MMU requires additional hardware, most embedded systems do not have one and this responsibility lies with the operating system. The operating system may prevent tasks from modifying data belonging to other tasks so that their data are protected from rogue processes (Kumar, et al., 2007). Memory protection is perhaps even more important to real-time systems because many times those systems are safety critical and data corruption can lead to catastrophic failure.

Dynamic memory allocation is a service provided by the operating system that allows tasks to borrow memory from the heap (Taksande, 2007). Because dynamic memory allocation is nondeterministic, it has not been good practice to use with real-time systems and it was not a standard feature in RTOS. However, because of its benefits, there has been significant research on this topic so that it can be used with real-time systems. The research is focused on developing algorithms that provide an upper bound limit for allocation and deallocation times. Dynamic memory allocation also requires a defragmentation or garbage collection algorithm that maintains the operating system memory heap. These algorithms are a necessary part of dynamic memory allocation because as memory is requested and released, it becomes fragmented. Because the defragmentation algorithm is not deterministic, it is not suitable for real-time systems and usually pointless to offer such a service in the operating system.

However, some real-time kernels do provide dynamic memory allocation services, and there are a couple of allocation algorithms that maintain that their allocation and deallocation times are constant. These algorithms are called half-fit and two-level segregated fit (TLSF). But equally important to consistent allocation and deallocation times is keeping fragmentation to a minimum. An independent analysis was performed on these two allocations algorithms, and it was found that although both half-fit and TLSF have consistent upper bound response times, only TSLF had minimal fragmentation. Although dynamic memory allocation is not recommended for use with real-time systems, if it is necessary, TLSF may offer a possible solution (Masmano et al., 2006).

The physical memory of system refers to the actual memory that exists in a system. Each physical memory address represents a real location in memory. This memory can include RAM, ROM, EEPROM, flash, and cache. The operating system is responsible for managing the memory for use by the application. The application needs access to memory to read program instructions and variables.

An operating system may have virtual memory. Virtual memory, like its name suggests, is not physical memory, but it instead is a technique an operating systems uses to give the illusion to a process or task that there is more memory than actually exists in the system and that the memory is contiguous. The purpose of this was to take off the burden of addressing memory from the programmer and have the operating system provide a way so that the memory locations are adjacent and easier for programmers (D'Souza, 2007). Virtual memory usually is not supported or recommended for use in real-time operating systems because a real-time system needs predictable data return times, and with virtual memory, the time can vary depending on the actual location of the data. However, some new embedded operating systems, such as Windows CE, support virtual memory (Wang et al., 2001). But it is still not recommended for use with hard real-time systems because if a page fault occurs, the memory access time is nondeterministic.

However, significant research has been done on this topic in recent years, and some real-time applications would like to realize the benefit of using virtual memory. In desktop systems that use virtual memory, they typically use a translation look-aside buffer (TLB). The TLB maps the virtual address used by the program to a physical address in memory. Most real-time systems do not have the option of including a TLB in their architecture. One new method of using virtual memory in real-time systems proposes a way to calculate the physical address by simple arithmetic computation, thus replacing the need for a TLB (Zhou & Petrov, 2005).

Another area in memory that is often considered separate from both program memory and RAM is called the run-time stack. The run-time stack maintained by the operating system is responsible for keeping track of routines and subroutines that have been interrupted and still need to complete execution. When a program is executing, if it is interrupted by another routine, the original's program return address is pushed onto the stack and the other subroutine executes. When the subroutine is finished, the run-time stack pops the address of the previous routine and it continues with its execution. The operating system is responsible for allocating memory for use by the run-time stack. A stack is a data structure that follows a last-in, first-out of data return. In other words, the information that is stored on the stack most recently is returned first. Table 3.1 shows a comparison for several memory management design options.

3.3.2 Peripheral Communication (Input / Output)

There are several different ways for a system to communicate with its peripherals. Peripherals are considered external to the system, but either input or output provides vital information to the system or takes data from the system and performs a task with it. With an embedded system, there is a microprocessor performing the tasks for the system, but many times, it requires data from outside the system. These data can be provided by analog sensors such as voltage or current sensors. Some sensors may measure brightness or wind speed. Depending on the purpose of the embedded system, a variety of sensors and/or actuators may be required. Although sensors are input devices, meaning the data are inputted into the microprocessor, other devices such as switches and actuators are output devices. Output devices are controlled by the microprocessor and the microprocessor controls these outputs by sending different signals to it.

Real-time operating systems provide different methods to communicate with peripherals; these methods include interrupts, polling, and direct memory access (DMA). Depending on the operating system design, an operating system may offer one or all of these methods.

Arguably, one of the most popular methods of notifying the system that hardware requires service is interrupts. The operating system must be prepared to handle interrupts as they occur, and most hardware interrupts occur asynchronously or at any time. The operating system must store the data in memory so it can be processed by the application at a later time. There are two main types of interrupts, hardware and software. With hardware interrupts, the operating system is not responsible for executing code to handle the interrupt. Instead the CPU usually handles the interrupt without the assistance of any software. However, the operating system does handle two things for the interrupt; it loads the program counter with the memory address of the Interrupt Service Routine (ISR), and when the ISR completes, it loads the program counter with the next instruction of the task it interrupted. An interrupt vector is needed when there are more than one hardware interrupt lines in the system. The addresses of the interrupt service routines are stored in the interrupt vector, and when a particular interrupt occurs, the vector points to its corresponding service routine. In a system with only one hardware interrupt, an interrupt vector is not needed and control is passed to the one service routine.

Hardware interrupts can be either edge-triggered or level-triggered. An edge-triggered interrupt is when the interrupt is recognized during a transition from high to low or vice versa. The device that needs to cause an interrupt sends a pulse on the

TABLE 3.1 Memory Management Design Options: A Comparison

	Purpose	Advantages	Disadvantages	Efficiency	Implementation
Run-Time Stack	Points to the memory locations of programs waiting to run	Supports reentrancy; each task has their own stack	Only supports first-in, last-out	Fast	Easy
Dynamic Memory Allocation	Service Provided by the operating system allowing tasks to borrow memory from the heap	Allows the program to request memory	Does not allow for deterministic operating system	Very slow, takes too much time to allocate and deallocate for real-time systems	Difficult
Memory Protection	Protect system memory	Is necessary for memory validity	For system calls, tasks must give up control to the operating system	Relatively fast	Mildly difficult
Virtual Memory	Gives the illusion of contiguous memory	Makes programming easier and allows programs that require more memory than physically available to run	Nondeterministic memory access times	Can be slow if memory is on disk instead of RAM	Difficult and not recommended for real-time operating systems

line. The pulse needs to be long enough for the system to recognize it; otherwise, the interrupt may be overlooked by the system and it will not get serviced. Level-triggered interrupts are requested by the device setting the line to either high or low, whichever one will indicate an interrupt on the system. The level-triggered interrupt method is often preferred over the edge-triggered method because it holds the line active until serviced by the CPU.[1] Even though line sharing is allowed with level-triggered interrupts, it is not recommended for real-time operating system design because this leads to nondeterministic behavior. A concern regarding the hardware-triggered interrupt is interrupt overload. Hardware interrupts that are triggered by external events, such as user intervention, can cause unexpected load on the system and put task deadlines at risk. The design of the operating system can include special scheduling algorithms that can address an unexpected increase in hardware interrupts. One such method suggested ignoring some interrupts when experiencing a higher than normal arrival rate. It was argued that it is better to risk slight degradation in performance than risking overloading the whole system, especially in the case where the interrupt frequency is drastically higher than what was estimated (Regehr & Duongsaa, 2005).

A software interrupt is one that has an instruction associated with it, and it is executed by the CPU. The instruction may be for a system call or caused by a trap. A process or task may cause a software interrupt so that the CPU will go into supervisor mode so that I will execute and access protected memory. A trap occurs when an unexpected or unintended event happens that causes an error with the system. Some examples are divide-by-zero errors or register overflow.

When an interrupt occurs, the control is transferred to the Interrupt Service Routine or ISR. A context switch occurs when information specific to the current process, such as registers and the program counter, are saved off to the stack and the new process information is loaded. The latency of an ISR must be both minimized and determined statistically for use with real-time operating systems. Interrupts are usually disabled while the code inside of the ISR is being executed; this is another reason why the ISR latency must be minimized so the system does not miss any interrupts while servicing another interrupt.

Polling is another method an operating system may use to determine whether a device needs servicing. Polling differs from interrupts in that instead of the device notifying the system that it needs service, the service will keep checking on the device to see whether it needs service. These "checks" are usually set up on regular time intervals, and a clock interrupt may trigger the operating system to poll the device. Polling is generally viewed as wasted effort because the device may not need to be serviced as often as it is checked or it may be sitting for some time waiting to be serviced before its time quantum is up and serviced. However, devices that are not time critical may be polled in the idle loop, and this can make the system more efficient because it cuts down on the time to perform the context switch. Hence, there may be some benefits to having an RTOS that supports' polling in addition to interrupts.[2]

[1] http://en.wikipedia.org/wiki/Interrupt.
[2] FreeBSD Manual Reference Pages - POLLING, February 2002.

A third method for peripherals to communicate with the system is through direct memory access or DMA. DMA usually is supported through the hardware, not the operating system. But it can alleviate some overhead in an operating system by providing a means to transfer data from device memory to system memory or RAM. Typically DMA requires a separate hardware controller that handles the memory transfer. The CPU does not perform the transfer; instead it hands control over to the DMA controller. A common use for DMA is transferring data to and from peripheral memory, such as analog-to-digital converters or digital-to-analog converters. A benefit of DMA is that the CPU does not need to handle the data transfer, allowing it to execute code. However, because DMA is using the data lines, if the CPU needs to transfer data to memory, it must wait for the DMA transfer to complete. Because DMA frees up the CPU, it can add efficiency to the system, but it also adds cost because additional hardware is required. Most cheap real-time systems cannot afford this luxury so it is up to the operating system to manage the peripheral data transfer. Table 3.2 shows peripheral communication design options and comparison for some input/output (I/O) synchronizing methods.

3.3.3 Task Management

A real-time system has tasks that are time sensitive, meaning they must be completed by a certain predetermined time in order for the system to be correct. Some real-time systems support tasks that are both real-time and non-real-time and the systems resources must be shared between both task types. Most importantly to hard real-time systems is that the task deadlines are satisfied and that they meet the requirements of the system.

In real-time systems, tasks may have different priorities assigned to them and a task that has a higher priority may preempt a running task with a lower priority. A task may be preempted when its time quantum has expired and the next task is scheduled to run. Because tasks in real-time systems are usually time sensitive, the operating system must be designed to allow for preemption of tasks. It must have a method to arbitrate between tasks that want to run at the same time. This is usually handled by assigning priorities to each of the tasks and the priorities may be static, meaning they never change. Or they may be dynamic, meaning they may change based on the state of the system.

In addition to priorities, tasks are usually in one of the following states: running (executing), ready, and suspended (blocked). An operating system puts tasks in certain states to organize them and let the scheduler know which tasks are ready to run on the processor. A task that is "running" means that its code is currently being executed on the CPU. In a single processor system, only one task at a time can be in the "running" state. A task in the "ready" state is a task that is ready to run on the CPU but is not currently running. Tasks in the "suspended" state are waiting for something external to occur, many times related to peripheral communication, such as disk read/write or memory access (Rizzo, et al., 2006). Also, when a task completes, it also moves to the suspended state until it is time for it to run again. A task is considered "dormant" if it exists in a system that has a fixed number of task control blocks (TCBs) and

TABLE 3.2 Peripheral Communication Design and Comparison

	Purpose	Advantages	Disadvantages	Efficiency	Implementation
Interrupts	Lets the operating system know that the hardware is ready to be serviced	The operating system does not need to waste time checking the hardware	Can be complicated to implement	Efficient since the hardware notifies the operating system as soon as it's ready	Requires special hardware that supports interrupts
Polling	The operating system checks to see whether the hardware is ready	Does not require special hardware	Wastes CPU time checking hardware that may not be ready. Hardware must wait for poll even if it's ready	Time is wasted when poll is performed and hardware is not ready	Easy
DMA	The hardware writes data directly to memory	Does not need CPU; it is freed up for task execution	The operating system is not notified when the hardware is ready; the application must check the memory	Efficient because it does not require CPU, but operating system is not notified	Requires special hardware that handles the DMA transfer of data

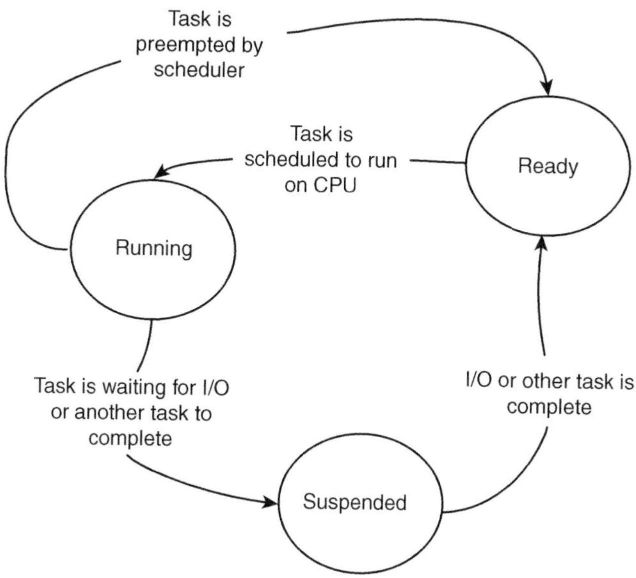

FIGURE 3.1 State diagram showing possible task states along with their transitions.

it "is best described as a task that exists but is unavailable to the operating system" (Laplante, 2005). Figure 3.1 shows a state diagram with possible task states along with their transitions.

A context switch occurs when a task that has not completed is preempted by another task. This can happen because the task running has a lower priority or its scheduled execution time has expired. It also can refer to when the flow of control is passed from the application to the kernel. The "context" of the task must be switched from the current task's information to the new task's information. Task-specific information commonly includes register information and the current program counter. The task information that is saved is determined by the operating system. It takes time to save off the data from the current task and to load the data associated with the new task. This latency is considerable, and it is the responsibility of the operating system to minimize this time as much as possible to maintain the efficiency of the system. A context switch occurs whenever the flow of control moves from one task to another or from task to kernel. Assuming we are dealing with a single processor system, there can only be one task that has control of the processor at a time.

With a multitasking environment, each task has a scheduled time slice where it is allowed to run on the processor. If the task has not completed when its time has expired, the timer causes an interrupt to occur and prompts the scheduler to switch in the next task. Tasks may be scheduled in a round-robin fashion, where each of the tasks has equal priority and a determined amount of time to run. Another method is where tasks are assigned various priorities and the tasks with the highest priorities are given preference to run over lower priority tasks.

With an interrupt handling system, a peripheral piece of hardware may cause an interrupt to occur on the system. The operating system will then save the data from the interrupt and schedule the task that processes the data. When going from user to kernel mode, the data specific to a task usually is saved to a task control block or TCB. When a task is scheduled to run, the information contained in the TCB is loaded into the registers and program counter. This puts the system in the same state as when the task finished running. The TCB is an alternative to the stack approach. A drawback of the stack approach is its rigid, first-in last-out structure. If the scheduling of tasks requires more flexibility, it may be beneficial to design the operating system to manage task scheduling by TCB rather than by a stack. Each TCB points to the next TCB that is scheduled to execute. If during execution of the current task, the execution order needs to change, it easily can be accomplished by changing the address of the next task in the TCB. Table 3.3 shows task management design options and a comparison.

3.4 TASK SCHEDULING: SCHEDULING ALGORITHMS

In real-time embedded systems, usually only one application is running on a microprocessor. However, there may be many tasks that make up an application and the operating system must have a method for scheduling tasks so that the overall needs of the system are met. The real-time system is responsible for performing a certain function. For example, with motor controls, the purpose of the embedded system is to control an electric motor. Many subroutines or tasks contribute to the motor control application. But the responsibilities of the application usually are broken down functionally into smaller pieces; these pieces are referred to as tasks. Going back to the example of a motor control application, one task may be responsible for controlling the current going to the motor where another task may be responsible for controlling the state of the system. And yet another task may be responsible for diagnostics. Each of these tasks has varied priorities and may need to run at different task rates. Some tasks may need to run more often than others, and tasks may need different priorities assigned to them. If the system has periodic tasks that run at certain intervals such as every 1 ms, 10 ms, or 100 ms, two or more tasks may need to run at the same time. The operating system uses priorities to determine which task should be allowed to execute on the CPU. This provides a method for the operating system to arbitrate between multiple tasks that are requesting the CPU. Task scheduling is very important to the success of a system, and an operating system must provide at least one method of scheduling tasks.

3.4.1 Interrupt-Driven Systems

Interrupt-driven systems for real-time applications are one of the most prevalent designs used in operating systems. Because time is critical to the success of the system, interrupts allow the system to perform tasks on regular intervals, commonly called periodic tasks. They address immediate needs that occur randomly, called

TABLE 3.3 Task Management Design Options and Comparison

	Purpose	Advantages	Disadvantages	Efficiency	Implementation
Task States	Organizes tasks for the operating system	The operating system is aware of the state of each task so they can be scheduled appropriately	Requires additional complexity on the operating system	Improves efficiency because the operating system only schedules tasks that are ready to run	Adds complexity in the operating system because it must assign, read, and keep track of task states
Reentrancy	Allows tasks to be reexecuted concurrently	Allows reuse of existing code	Each instance of the task or process requires its own data structure and run-time stack	The code is more efficient, because it can be used multiple times	Adds complexity to the operating system and application
Context Switching	Provides a method of saving of data from the current task so a new task can be executed	Allows for preemption in a multitasking environment	Takes time to switch between tasks	Can improve overall efficiency by allowing higher priority tasks to run first, but takes time to switch in and out of task-specific data	Is complex to implement in operating system; the operating system must support multitasking, preemption, and provide a method to save and retrieve data
TCB (Task Control Block)	Saves data specific to task, such as registers and program counter	Keeps all data specific to a tasks together in a structure	A predetermined size of memory must be set aside for each task	Can improve efficiency because all task data are kept together	The operating system must include data structures for tasks

aperiodic tasks. Because interrupts allow for this flexibility, they are very popular among real-time operating system designs. An interrupt is a signal to the system that something needs to be addressed. If a task is in the middle of execution and an interrupt occurs, depending on the type of scheduling implemented, the task may be preempted so that the new task can run.

There are a couple types of interrupt-driven systems; they usually are referred to as foreground, background, or foreground/background systems. With a foreground system, all tasks are scheduled into periodic tasks that execute at regular intervals: 1 ms, 2 ms, 10 ms, and so on. A background system is one where there are no periodic tasks and everything runs from the main program. A foreground/background system is a hybrid between the two. There is a background task, often referred to as the idle loop. Also, there are periodic tasks that are executed based on their rate. The background task usually is reserved for gathering statistically information regarding system utilization, whereas the foreground tasks run the application.

3.4.2 Periodic versus Aperiodic Tasks

Tasks may be scheduled periodically, or they may occur aperiodically. A periodic task is one that occurs during regular time intervals; for example, a task may execute every 1 ms or every 2 ms. An aperiodic task is one that happens randomly as a result of an outside request or an exception. An example of an outside request is a user typing on a keyboard. The task may be initiated when a user presses down on a key and the purpose of the task may be to determine which key has been pressed. An example of an exception is a divide-by-zero error. The system must satisfy the deadlines of the periodic tasks and service the aperiodic tasks as soon as possible (Lin & Tarng, 1991). This can be difficult because the frequency of aperiodic tasks many times are not known during the design of the system. They must be estimated as closely as possible so that the system utilization is at a safe level, allowing periodic tasks to complete safely before their deadline. At the same time, there should not be a noticeable delay for the servicing of aperiodic tasks.

A significant amount of research has been performed on this topic, and new algorithms have been developed to address the concern of mixing aperiodic tasks with periodic ones. The *Slack Stealing Algorithm*, designed by Lehoczky and Thuel is one such algorithm. The methods in their algorithm "provide a unified framework for dealing with several related problems, including reclaiming unused periodic and aperiodic execution time, load shedding, balancing hard and soft aperiodic execution time and coping with transient overloads"(Lehoczky & Thuel, 1995).

3.4.3 Preemption

Preemption occurs when a task that currently is being executed is evicted by the scheduler so that another task may run on the CPU. Tasks may be preempted because another task, one that has a higher priority, is ready to execute its code. In a multitasking environment, most operating systems allow each task o run for a predetermined time quantum. This provides the appearance that multiple tasks are running simultaneously. When the time quantum has expired, the schedule preempts the current task allowing the next task to run.

The operating system kernel also must allow preemption in a real-time environment. For example, a task with a lower priority currently may be executing and it performs a system call. Then a higher priority task tries to interrupt the current task so that it can execute. The operating system must be able to allow for the new task to run within a certain amount of time; otherwise there is no guarantee that the new task will meet its deadline.

Because time is of the essence, the worsts case execution time (WCET) must be calculated for all tasks. This is especially difficult when tasks are preempted, but the operating system kernel must provide WCET required for system calls before it allows preemption to occur (Tan & Mooney, 2007). Table 3.4 shows task scheduling design options and comparison.

3.4.4 Static Scheduling

A multitasking operating system must include a method to schedule tasks. One of the basic methods of scheduling tasks is static scheduling. With static scheduling, the priorities assigned to tasks does not change; it stays constant throughout the execution of the program.

One of the most common and oldest static scheduling algorithms is called round-robin. With round-robin, all tasks are treated as equals and each is allowed a predetermined time quantum in which they can use the CPU to execute their instructions. When their time quantum expires, an interrupt occurs and the old task is switched out and the new task is switched in. Although simple to implement, the round-robin task scheduler does not give preference to tasks that are more important than other tasks. These tasks may be more critical to the system, but round-robin does not give preferential treatment.

Another type of scheduling algorithm is called rate monotonic (RM). With RM, tasks are assigned a fixed priority based on the frequency of which they run. For example, if there are three tasks, that run at 1 ms, 2 ms, and 10 ms. The task running at 1 ms would have higher priority, and the one running at 10 ms would have the lowest priority. This type of scheduling is the most efficient for fixed priority, meaning that if a system cannot meet its deadlines with this algorithm, there is no other fixed priority algorithm that would. A disadvantage to the RM scheduling method is that the processor cannot be used fully and even on relatively low utilization, such as 70%, tasks may miss their deadline (Steward & Barr, 2002). However, research over the past few years has been performed, and the algorithm has been modified to allow for maximum processor utilization. The name of this modified algorithm is called the delayed rate monotonic (DRM), and it has been proven that, in some cases, systems that run safely on DRM are unsafe on RM (Naghibzadeh, 2002). In summary, RM scheduling is the most optimal static scheduling algorithm. It is easy to implement, and the concept is easy to understand. Many users are familiar with the algorithm, and it is implemented on many multitasking, interrupt-driven systems. Table 3.5 shows static scheduling design options and comparison.

TABLE 3.4 Task Scheduling Design Options & Comparison

	Purpose	Advantages	Disadvantages	Efficiency	Implementation
Periodic Tasks	Usually uses a timer to perform regular maintenance tasks	Ideal for tasks that must be performed at regular time intervals	Code may be executed more often than required	Mostly efficient, although context switching can cause latency	Easy with statically scheduling systems; complicated if dynamic
Aperiodic Tasks	Can occur at any time, usually triggered by something external to the system	Good for use when the system only needs to respond when an event occurs	May increase WCET of the system	Mostly efficient, although there is latency for context switching	Relatively easy
Interrupt Driven	A timer causes an interrupt signaling the operating system to execute a task	Provides an efficient method of notifying operating system the that it is time for the task to execute	Must have an operating system and hardware in place to support interrupts	Usually more effective than other alternatives, but there can be significant latency if not implemented properly	Implementing code to handle context switches efficiently can be moderately difficult
Preemptive	Allows tasks to interrupt a task that is executing	Without this, all tasks must execute until completion, difficult to support in multitasking real-time system	It takes time to switch out tasks	Depending on the implementation, the time to switch tasks can be minimized	Relatively difficult to implement, and the time to perform the switch must be known

TABLE 3.5 Static Scheduling Design Options and Comparison

	Purpose	Advantages	Disadvantages	Efficiency	Implementation
Round Robin	Allows multiple tasks to execute on a unipro-cessor system	Ease of implemen-tation and adequate for simple systems where all tasks are equal	Does not give preference to more critical tasks	Can be efficient if the correct time quantum is selected	Easy to implement
Rate Mono-tonic (RM)	Assigns fixed priorities to tasks	Easy to implement and simple concept; faster tasks have higher priority	Even with low utilization, ~70%, tasks can miss deadlines	Is the most efficient static scheduling algorithm	More complicated than round-robin but less than dynamic scheduling

3.4.5 Dynamic Scheduling

An alternative to static scheduling is dynamic scheduling. Dynamic scheduling is when the priorities of tasks can change during run time. The reasons for dynamic schedule varies; it could be that a task may miss its deadline or that a task may need a resource that another, lower priority task currently has. When using static scheduling and the CPU is highly used (greater than 70%), there is a high likelihood that a task may miss its deadline. However, dynamic scheduling allows the CPU to reach much higher utilization, but it comes at a price—dynamic scheduling is complex.

A common dynamic scheduling technique is called priority inversion. This type of scheduling is used in an interrupt-driven system that has priorities assigned to each of the periodic tasks. However, if a lower priority task has a resource that is needed by a higher priority task, the lower priority task is allowed to complete its execution until it releases the resource, even if the higher priority task is scheduled to run. The reasoning behind this type of scheduling technique is that it makes the resource available for the high-priority task as soon as possible. If the control switched out from the lower priority task while it was still holding the resource, the higher priority task would be blocked anyway, thus increasing its overall time to execute. In summary, priority inversion has its benefits because it frees up resources quickly for the high-priority task so that it can have access to the resource and execute its code. But it can be difficult to determine when priority inversion may occur, and therefore, worst-case execution time can be difficult to calculate. The overall efficiency of the system is better than static scheduling algorithms, but it can be difficult to implement and not all systems would benefit from this type of algorithm.

TABLE 3.6 Dynamic Scheduling Design Options and Comparison

	Purpose	Advantages	Disadvantages	Efficiency	Implementation
Priority Inversion	Frees up a resource that is held by a low-priority task so that a high-priority task can run	Frees up resources quickly	The WCET can be difficult to calculate	Can be more efficient than static scheduling algorithms	Difficult
Earliest Deadline First (EDF)	Gives highest priority to the task that must finish first	Allows for higher CPU utilization (up to 100%)	If overutilized, it is difficult to predict which tasks will meet their deadline	Can be very efficient	Difficult

Another type of dynamic scheduling algorithm is called the earliest deadline first (EDF) algorithm. This algorithm allows for very high utilization of the CPU, up to 100%. To ensure tasks finish by their deadline, the scheduler places all tasks in a queue and keeps track of their deadlines. The task with the closest deadline is given highest priority for execution. This means that the tasks priorities can change based on their deadline time. However, this type of scheduling is not practical for systems that require tasks to execute at regular time intervals. If a current sensor must be read every 100 μs, or as close as possible to it, this type of algorithm does not guarantee that the task will execute at a certain designated time. It instead guarantees that the task will finish before its deadline; consistency is not important. This type of scheduling, is not used very often because of the complexity involved in its implementation. Most commercial RTOS do not support this type of scheduling and the cost associated with developing this in-house does not make it a popular choice. However, if the system becomes overused and purchasing new hardware is not an option, the EDF algorithm may be a good choice. Table 3.6 shows dynamic scheduling design options and comparison.

3.5 INTERTASK COMMUNICATION AND RESOURCE SHARING

In a multitasking system running on a single processor, tasks usually need to communicate with each other. Data produced in one task may be consumed by another task or one task may be responsible for calculating a value that is then required for another tasks calculations. Protecting this data and preserving its integrity is extremely important because without valid data, the system will behave unpredictable and fail. One of the most basic principles for data integrity is task reentrancy. Tasks containing global data must be reentrant, which means that a task may be interrupted and the data will not be compromised. Critical sections in the code must be protected, and there are different methods for protecting data, such as semaphores and disabling interrupts. Depending on the requirements of the system, one method may be more suitable than others. These methods will be discussed in greater detail in the following sections.

Shared variables commonly are referred to as global data because it can be viewed by all tasks. Variables that are specific to a task instance or local are referred to as local or static variables. An example of when data integrity becomes an issue is when global data are being modified by a task and another task preempts the first task and reads that data before the modification is complete.

In addition to data integrity, resources often are limited and must be shared among the tasks in the system. Control of these resources usually is the job of the operating system. The design of a real-time operating system may include several methods for protecting data and sharing of resources. Some methods include semaphores, read/write locks, mailboxes, and event flags/signals.

3.5.1 Semaphores

Operating systems commonly use semaphores as a method to signal when a resource is being used by a task. Use of semaphores in computer science is not a new concept, and papers have been published on the topic since the early 1970s. Today, it maintains a popular way for operating systems to allow tasks to request resources and signal to other tasks that the resource is being used. Two main functions make up a semaphore, *wait* and *signal*. The usual implementation of a semaphore is to protect a critical section, of code; before the task enters the critical section, it checks to see whether the resource is available by calling the *wait* function. If the resource is not available, the task will stay inside the *wait* function until it is available. Once it becomes available, the task requests the resource and therefore makes it unavailable to other tasks. Once the task is finished with the resource, it must release it by using the *signal* function so other tasks may use it. There are two main types of semaphores, binary and counting. Binary usually is sufficient, but counting semaphores are nice when there are more than one resource.

Although semaphores are a relatively easy concept, issues can develop if they are not implemented and used properly. With binary and counting semaphores, a race condition can occur if code that is responsible for reserving the resource is not protected until the request is complete. There are, however, a couple different approaches on how to eliminate race conditions from the *wait* function. One method, presented by Hemendinger in comments on "A correct implementation of general semaphores" discusses a common race condition and provides a simple solution. This solution was further improved on by Kearns in "A correct and unrestrictive implementation of general semaphores" (1988) as Kearns had found another possible race condition within the solution.

Another issue that can occur with semaphores, or any method where a task must wait until a resource is freed, is called deadlock. Deadlock usually is avoidable in real-time applications. Four conditions must be present for deadlock to occur.

Once deadlock occurs on a system, it will stay in that condition unless there is outside intervention and the easiest way to deal with deadlock is to avoid it. The four conditions are as follows: mutual exclusion, circular wait, no preemption, and hold and wait. If the rules for requesting a resource are modified so that one of these conditions can never occur, then deadlock will not occur. Some conditions are easier to remove than others; for example, if there is only one resource, the mutual exclusion condition cannot be removed. However, the hold and wait condition can be avoided by implementing a rule that requires a task to request all resources if they are available. If one of the resources is not available, then the task does not request any. The section of code where the task is requesting resources is a critical section of code because it must not be interrupted until it has all resources.

3.6 TIMERS

These include the watchdog timer and the system timer.

3.6.1 Watchdog Timer

Timers are an essential part of a real-time system. One of the most critical timers is called the watchdog timer. This timer is responsible for making sure that tasks are being serviced by their deadlines. The watchdog timer can be implemented in hardware, where counters are increasing until an upper limit is reached. This upper limit value depends on the system requirements. For example, if all tasks must complete within a 100 ms time limit, the upper limit can be set at 100 ms. If the limit is reached, it causes a system reset. To avoid system reset, the timer must be cleared. The clearing of the watchdog timer can occur at the end of the longest task because this would indicate that all tasks have completed execution.

3.6.2 System Timer

Other timers in real-time systems cause a task to begin execution. If a task is scheduled to run every 1 ms, there must be a timer associated with this task that initiates the task to run after the time has expired.

With round-robin scheduling, each task has a certain time quantum in which it has to execute its instructions. The timer begins when the task is scheduled, and after its time quantum has expired, an interrupt occurs causing a context switch and the task is replaced by the next scheduled task.

3.7 CONCLUSION

This chapter has addressed the past and present design techniques for real-time systems, but future designs tend to be moving toward Network-on-Chip or moving some tasks that usually reside solely on the microprocessor to a field-programmable gate array (FPGA) device.

As a result of the industry moving toward multiprocessor and multicore systems, new challenges are being introduced. The operating system must now address the needs of two processors, scheduling tasks on multiple cores and protecting the data of a system whose memory is being accessed from multiple sources. New issues are being uncovered, and the need for solutions is great. One such problem is found on the Ada95 microprocessor. It was designed to support priority inversion; however, because of limitations in the software, it does not support unbounded priority inversion (Naeser, 2005).

The future of RTOS design will depend greatly on the hardware designs. New hardware many times requires new software, including operating systems. As the industry is moving toward more processor cores on one chip, this will present challenges for real-time operating systems that have been developed for only one core.

In addition to hardware evolution impacting real-time operating system designs, another factor is the need for efficient and cheap systems. Many companies are finding that commercial real-time operating systems are expensive to purchase and support. They often include features that are not required by the system and use valuable resources. Future RTOS designs will be developed in-house and leverage the vast amount of open-source code available for real-time systems.

REFERENCES

D'Souza, L. (2007), "Virtual memory—designing virtual memory systems." *Embedded Technology.* http://www.embeddedtechmag.com/component/content/article/6114?start=5

Furr, Steve (2002), "What is real time and why do I need it?" *QNX Software Systems.* http://community.qnx.com/sf/docman/do/downloadDocument/projects.core_os/docman.root.articles/doc1161

Kalinsky, David (2003), "Basic concepts of real-time operating systems." *LinuxDevices.com.* http://www.jmargolin.com/uavs/jm_rpv2_npl_16.pdf

Kearns, Phil (1988), "A correct and unrestrictive implementation of general semaphores." *SIGOPS Operating Systems Review*, Volume 22, #4.

Kumar, Ram, Singhania, Akhilesh, Castner, Andrew, Kohler, Eddie, and Srivastava, Mani (2007), "A System for Coarse Grained Memory Protection in Tiny Embedded Processors," ACM DAC '07: Proceedings of the 44th annual conference on Design Automation June.

Laplante, Phillip A. (2005), "*Real-Time Systems Design and Analysis*," 3rd Ed., IEEE Press, New York.

Lehoczky, John P., and Thuel, Sandra R. (1995), "Scheduling periodic and aperiodic tasks using the slack stealing algorithm," *Advances in Real-Time Systems*, Prentice-Hall, Sang H. Son, Ed., Englewood Cliffs, NJ.

Leroux, Paul (2005), RTOS versus GPOS: What is best for embedded development? "*Embedded Computing Design*,"

Lin, Tein, and Tarng, Wernhuar (1991), "Scheduling periodic and aperiodic tasks in hard real-time computing systems," *ACM Sigmetrics Performance Evaluation Review*, Department of Electrical and Computer Engineering, State University of New York at Buffalo, New York.

Masmano, Miguel, Ripoll, Ismael, and Crespo, Alfons (2006), "A Comparison of Memory Allocators for Real-Time Applications," ACM JTRES '06: Proceedings of the 4th international workshop on Java technologies for real-time and embedded systems, July.

Naeser, Gustaf (2005), "*Priority Inversion in Multi Processor Systems due to Protected Actions,*" Department of Computer Science and Engineering, Malardalen University, Sweden.

Naghibzadeh, Mahmoud (2002), "A modified version of the rate-monotonic scheduling algorithm and its efficiency assessment," Object-Oriented Real-Time Dependable Systems, IEEE - Proceedings of the Seventh International Workshop, pp. 289–294.

Regehr, John, and Duongsaa, Usit (2005), "Preventing interrupt overload," ACM SIGPLAN Notices, Volume 40, #7.

Rizzo, L., Barr, Michael, and Massa, Anthony (2006), "Programming embedded systems," *O'Reilly*.

Steward, David, and Barr, Michael (2002), "Rate monotonic scheduling (computer programming technique)," *Embedded Systems Programming*, p. 79.

Taksande Bipin (2007), "Dynamic memory allocation." *WordPress.com*. http://belhob.wordpress.com/2007/10/21/dynamic-memory-allocation/

Tan, Yudong, and Vincent Mooney (2007), "Timing analysis for preemptive multitasking real-time systems with caches," ACM Transactions on Embedded Computing Systems (TECS), Georgia Institute of Technology, Feb.

Wang, Catherine L., Yao, B., Yang, Y., and Zhu, Zhengyong (2001), "*A Survey of Embedded Operating System*." Department of Computer Science, UCSD.

Zhou, Xiangrong, and Petrov Peter (2005), "Arithmetic-Based Address Translation for Energy Efficient Virtual Memory Support in Low-Power, Real-Time Embedded Systems," SBCCI '05: Proceedings of the 18th annual symposium on Integrated circuits and system design, University of Maryland, College Park, Sept.

CHAPTER 4

SOFTWARE DESIGN METHODS AND REPRESENTATIONS

4.1 INTRODUCTION

A software design method typically is defined as a systematic approach for carrying out a design and describes a sequence of steps for producing a software design (Gomaa, 1989). There are certainly several ways to design software, but a designer must use certain types of established practices when preparing software. Different types of approaches to software designs may be used depending on the type of problem being encountered. Moreover, different types of software design methods each have unique advantages and disadvantages one another. Many people think that software engineering is a creative activity that does not need a structured approach; however, it is important to note that an informal approach toward software development does not build a good software system.

Dividing software design methodologies into classifications aids in the understanding of software design methodologies (Khoo, 2009). The main design approaches that will be discussed are as follows: level-oriented, data-flow-oriented, data-structure-oriented, and object-oriented.

4.2 HISTORY OF SOFTWARE DESIGN METHODS

This section will discuss the past, present, and future of software design methods and will consider how each software design method compares with each other. Also this

---

section discusses the history of software design methods. In particular, an overview of how software designs methods came to be, and how they have evolved since the late 1960s will be presented. The main design approaches by defining each design method in detail and discussing the advantages and disadvantages of using each one also will be presented. Comparing the different types of software design methodologies and as well as discussing which methodologies may be best will be discussed. Finally, this section will discuss the future of software design methods. The software development field is a rapidly changing area of technology, as it seems that every decade or so there is a shift in software design strategies. When compared with other engineering disciplines, such as, for example, metallurgy, software engineering is a relatively new field that was almost nonexistent until approximately 50 years ago.

Primitive types of software development started around the late 1940s and early 1950s, with the first stored-program computer, the Cambridge EDSAC. By the late 1960s, software had become part of many products. However, there was no real metric to determine the quality of software, which led to many safety issues. This particular situation became known as the *software crisis*. In response, software manufacturing has to be based on the same types of foundations traditionally used in other types of engineering.[1] During the early 1970s, structured design and software development models evolved. Researchers started focusing on software design to develop more complex software systems. In the 1980s and 1990s, software engineering shifted toward software development processes.

Although object-oriented programming initially was developed around the late 1960s, this type of programming did not become especially popular until the late 1980s and 1990s (Barkan, 1992), (Urlocker, 1989). Object-orientation programming can be traced back to the late 1960s with the development of Simula and Smalltalk, which are types of object-oriented programming languages. However, object-oriented programming did not become extremely popular until the mid-1990s, as the Internet became more popular.

During the 1990s, object orientation also was modified with class responsibilities collaborators (CRC) cards. Moreover, methods and modeling notations that came out of the structured design movement were making their way into object-oriented modeling. During this time, an integrated approach to design was becoming needed in an effort to manage large-scale software systems and developed into the Unified Modeling Language (UML). UML integrates modeling concepts and notations from many methodologists.[2]

UML is a widely used, generalized type of modeling language, and falls under an object-oriented approach. The UML approach was started around the early to mid-1990s and was developed by James Rumbaugh and Grady Booch of Rational Software Corporation.[3] At that time, Rational was the source for the two most

[1] "An Introduction to Software Architecture." http://media.wiley.com/product_data/excerpt/69/04712288/0471228869.pdf.

[2] "An Introduction to Software Architecture." http://media.wiley.com/product_data/excerpt/69/04712288/0471228869.pdf.

[3] http://en.wikipedia.org/wiki/Unified_Modeling_Language.

popular object-oriented modeling approaches of the day: Rumbaugh's OMT, which was known for object-oriented analysis (OOA), and Grady Booch's Booch method, which was known for object-oriented design (OOD). Rumbaugh and Booch attempted to combine their two approaches and started work on a Unified Method.

Another popular approach that started to develop around the same time was the use of design patterns.[4] A design pattern is a reusable solution used to solve commonly occurring problems in software design. In other words, a design pattern is not a finished design that can be transformed directly into code but a template for how to solve a problem. Originally design patterns emerged as an architectural concept in the late 1970s. It was not until the late 1980s that design patterns were considered in programming. However, design patterns did not start to become extremely popular until around 1994, after the book *Design Patterns: Elements of Reusable Object-Oriented Software* was published. That same year the first Pattern Languages of Programming Conference was held.[5] In 1995, the Portland Pattern Repository was set up for documentation of design patterns.

4.3 SOFTWARE DESIGN METHODS

When a software problem occurs, a software engineer usually will try and group problems with similar characteristics together. This particular approach is called a problem domain. For each type of software design methodology there is a corresponding problem domain. Some criteria that can be used to classify software design methods include the characteristics of the systems to be designed as well as the type of software representation (Khoo, 2009). As best explained by the Software Engineering Institute, there can be three distinct views of a system:

> The basic view of the system taken by a design method, and hence captured by a design based on that method, can be functional, structural, or behavior. With the functional view, the system is considered to be a collection of components, each performing a specific function, and each function directly answering a part of the requirement. The design describes each functional component and the manner of its interaction with the other components. With the structural view, the system is considered to be a collection of components, each of a specific type, each independently buildable and testable, and able to be integrated into a working whole. Ideally, each structural component is also a functional component. With the behavioral view, the system is considered to be an active object exhibiting specific behaviors, containing internal state, changing state in response to inputs, and generating effects as a result of state changes (Khoo, 2009, p. 4).

Indeed, grouping software design methodologies into different approaches helps not only in the explanation of software design but also will aid a designer in selecting the best available methodology to use. This section discusses the main design

[4] http://en.wikipedia.org/wiki/Design_pattern_(computer_science).
[5] http://en.wikipedia.org/wiki/Software_design_pattern.

approaches that are available, including object-oriented design, level-oriented, data-flow-oriented, and data-structure-oriented. Below is a detailed explanation of what each software design method is, what they entail, as well as any benefits and drawbacks of using that particular design method.

4.3.1 Object-Oriented Design

Object-oriented design uses objects that are black boxes used to send and receive messages. These objects contain code as well as data. This approach is noteworthy because traditionally code is kept separated from the data that it acts upon. For example, when programming in C language, units of code are called "functions" and units of data are called "structures". Functions and structures are not connected formally in C (Software Design Consultants, 2009).

Proponents of object-oriented design argue that this type of programming is the easiest to learn and use, especially for those who are relatively inexperienced in computer programming because the objects are self-contained, easily identified, and simple. However, some drawbacks to object-oriented design are that it takes more memory and can be slow. Several object-oriented programming languages are on the market; however, the most popular object-oriented languages are C++, Java, and Smalltalk.

In object-oriented software, objects are defined by classes. Classes are a way of grouping the objects based on the characteristics and operations of an object. Defining classes can be complicated, as a poorly chosen class can complicate an application's reusability and hinder maintenance.[6]

The main components of object-oriented programming are encapsulation, inheritance, polymorphism, and message passing. The first component, encapsulation, can be defined as hiding implementation. That is, encapsulation is the process of hiding all the details of the object that do not contribute to the essential characteristics of the object, and only shows the interface.[7] Inheritance is a way to form new classes by using classes that already have been defined. These new classes are sometime called "derived classes." Inheritance can be useful because one can recycle and reuse code this way, which is high desirable. Polymorphism is the ability to assign different meanings to something in different contexts. That is, polymorphism allows an entity such as a variable, a function, or an object to have more than one form.[8] Finally, message passing allows for objects to communicate with one another, and to support the methods that they are supposed to be running

The main benefit of using object-oriented software is that it can be reused with relative ease. Indeed, software systems are subject to almost nearly continuous change. As a result, it must be built to be able to withstand constant revisions. Four basic

[6] http://www.codeproject.com/KB/architecture/idclass.aspx.

[7] http://www.fincher.org/tips/General/SoftwareEngineering/ObjectOrientedDesign.shtml.

[8] http://searchcio-midmarket.techtarget.com/sDefinition/0,,sid183_gci212803,00.html#.

principles of object-oriented design facilitate revisions: open–closed principle, once and only once principle, dependency inversion principle, and Liskov substitution principle (Laplante, 2005).

The open–closed principle states that classes should be open to extension but at the same time closed to modification. In other words, the object should be allowed to react differently to new requirements, but at the same time, the code cannot change internally. This can be done by creating a super class, but it can represent unbounded variation by subclassing.

The once and only once principle is the idea that any portion of the software, be it algorithms, documentation, or logic, should exist only in one place. This makes maintenance and comprehension easier and isolates future changes.

The dependency principle states that high-level modules should not depend on low-level modules. Instead both should depend on abstractions, where abstractions should not depend on details, but details should depend on abstractions.

Finally, Liskov expressed the principle that "what is wanted here is something like the following substitution property: if for each object o1 of type S there is an object o2 of type T such that for all programs P defined in terms of T, the behavior of P is unchanged when o1 is substituted for o2 then S is a subtype of T" (Laplante, 2005, p. 249). This principle has led to the concept of type inheritance and is the basis for polymorphism, which was discussed earlier.

Design patterns can be defined as reusable solutions to commonly occurring problems in software design. It should be noted that a design pattern is not a finished design that can be transformed directly into code but a template for how to solve a problem. Object-oriented design patterns typically show relationships between objects without specifying the final objects involved. Indeed, developing software can be very tricky. Thus, design patterns have to be implemented such that they can solve the current problem, while the software must be general enough that it also can address future problems as well. In fact, most experienced designers know not to solve every problem from first principles but to reuse principles that they have leaned from previous designs.

Generally, a design pattern includes four main elements: a name, the problem to be solved, the solution to the problem, and the consequences of the solution. The problem to be solved describes when the design pattern should be applied in terms of specific design problems. The problem to be solved can describe class structures that indicate an inflexible design and might include conditions that have to be met first before the design pattern can be applied. The solution describes the elements that the design consists of. The solution does not describe a concrete design or implementation but provides a general arrangement of how a general arrangement of objects and classes solves the problem (Khoo, 2009).

UML is a standardized, general-purpose language that is used to construct an object-oriented software system under development, and it offers a standard way to write a system's design. Indeed, UML is sort of like a blueprint for building a house to ensure consistency and structure. UML includes concepts with a notation and rules for usage, where the notation has a set of shapes that can be combined in ways to create system diagrams. Some main types of UML diagrams include use-case diagrams, class diagrams, and implementation diagrams.[9]

4.3.2 Level-Oriented Design

There are two general approaches to level-oriented design, the top-down approach and the bottom-up approach. The top-down approach starts at a top level and breaks up the program into smaller functions. The smaller functions are more easy to analyze, easier to design, and easier to code. However, there has to be a complete understanding of the problem or system at hand when designing a system using the top-down approach. The top-down process also is dependent on decisions made in the early stages to determine structure (Khoo, 2009). Bottom-up design is an approach where a program is written in a series of layers. Each component is viewed as a tool to solve the problem. Bottom-up design is different from top-down design because the one need not know the complete problem at the outset of programming. In bottom-up design, it is important to recognize that a certain tool can solve a portion of the problem.[10]

Well-written top-down approaches have been described by Nimmer as follows:

> In practice, a programmer usually will start with a general description of the function that the program is to perform. Then, a specific outline of the approach to this problem is developed, usually by studying the needs of the end user. Next, the programmer begins to develop the outlines of the program itself, and the data structures and algorithms to be used. At this stage, flowcharts, pseudo-code, and other symbolic representations often are used to help the programmer organize the program's structure. The programmer will then break down the problem into modules or subroutines, each of which addresses a particular element of the overall programming problem, and which itself may be broken down into further modules and subroutines. Finally, the programmer writes specific source code to perform the function of each module or subroutine, as well as to coordinate the interaction between modules or subroutines (Nimmer & Nimmer, 1991).

Indeed, the top-down approach is a very modular approach to software design, where the problem is broken down into smaller, more manageable tasks. Although having a modular design has its advantages, there are drawbacks as well. For example, this approach focuses on very specific tasks that have to be done but putting little emphasis on data structures. In other words, data structures usually are only thought of after procedures have been generally defined. Moreover, any data used by several procedures usually are defined in one place and can be accessed by any module or subroutine. This may create problems if the program needs to be updated or revised because it "leads to the stack of dominoes effect familiar to anyone working in program maintenance whereby changes to one part of a software system often cause a problem in an apparently dissociated program area" (Barkan, 1993, p. 315). In other words, every time software is updated, all the procedures that rely on the old

[9] http://www.bookrags.com/research/uml-unified-modeling-language-wcs/.
[10] http://www.bookrags.com/research/bottom-up-design-wcs/.

data structure would need to be analyzed and changed accordingly. Also, top-down approaches rarely are used to solve very large, complicated programs.

Another drawback to the top-down approach is that programmers usually have to approach a program as a series of single functions. As a result, programmers are not likely to incorporate evolutionary changes in the data structures into the big picture of the overall system. Thus, the top-down approach provides few ways to reuse existing pieces of software.

In contrast, bottom-up design has the ability to be reused. Moreover, if the specifications for the program change, this impact may not be as great as it would be if a top-down approach were taken instead.[11]

4.3.3 Data Flow or Structured Design

Data flow design sometimes is referred to as the structured design approach. Structured design is the companion method to structured analysis; that is, structured analysis is functional and flat, whereas structured design is modular and hierarchal (Laplante, 2005). By using the structured design approach, emphasis is placed on the processing performed on the data, where the data are represented as a continuous flow of information that is transformed from node to node in the input–output stream (Khoo, 2009).

Structured design is characterized by the development of a structured hierarchy of modules using structure charts (SCs).[12] SCs can be used to model a group of functions defined in the specifications into modules. The SC also is used to model the hierarchical organization of the modules and the data interface between the modules. The building blocks of a SC are the module, the call, the shared data area, and the couple. The module is an independently callable unit of code. The call is an activation of a module, and the shared data represents data accessed from several modules. The couple represents an item of data or control information passed between modules.[13]

It should be noted that several significant issues are encountered when using structured analysis and structured design in modeling a real-time system. One problem with this approach is that concurrency is not depicted easily with structured design (Laplante, 2005). Also, control flows are not translated easily into code as well because they are hardware dependent.

The most troublesome part of structured design is that tracking changes can be tricky. Even more disturbing is that any change in the program requirements generally translates into significant amounts of code that will probably need to be rewritten. As a result, this approach generally is unpractical to use if significant software changes need to be made in the future. Moreover, it should be noted that none of these problem usually originate in this magnitude when using object-oriented methods (Laplante, 2005).

[11] http://www.bookrags.com/research/bottom-up-design-wcs/.
[12] http://www.cs.wvu.edu/~ammar/chapter-4.pdf.
[13] http://www.cs.wvu.edu/~ammar/chapter-4.pdf.

4.3.4 Data-Structure-Oriented Design

Last but not least, this chapter examines data-structure-oriented design. Data-structure-oriented methods focus on data structure, rather than data-flow-like structured design methods.[14] Although there are different types of data-structure-oriented methods, each having a distinct approach and notation, all have some properties in common. First, each assists in identifying key information objects and operations. Next, each assumes that the structure of information is hierarchical. Also, each provides a set of steps for mapping a hierarchical data structure into a program. Some of the main types of data-structure-oriented design methods are as follows: the Jackson Development Method, the Warnier–Orr Method, and the Logical Construction of Programs (LCP) by Warnier.[15]

The Jackson Development Method was invented in the 1970s by Michael A. Jackson and initially was used in an effort to try and make COBOL programming easier to modify and be reused.[16] However, nowadays the Jackson Development Method and be applied to all kinds of programming languages. The Jackson Development Method includes Jackson Structured Programming as well as Jackson System Development.[17] These two methods differ from other widely used methods in two main respects. First, they pay attention initially to the domain of the software and later to the software itself. Second, they focus on time-ordering that is, they focus on event sequencing rather than on static data models. Some types of Jackson System Development programs can be said to be object oriented.

Warnier–Orr diagrams are a kind of hierarchical flowchart that allows for the organization of data and procedures. Four basic constructs are used on Warnier–Orr diagrams: hierarchy, sequence, repetition, and alternation.[18] Hierarchy is the most fundamental of all of the Warnier–Orr constructs. Hierarchy can be defined as a nested group of sets and subsets as a set of nested brackets where the larger topics break down into smaller topics, which break down into even smaller topics. Sequence is the simplest structure to show and includes one level of hierarchy where the features are listed in the order in which they occur. Repetition is a kind of like a loop in programming, and happens whenever the same set of data occurs repeatedly or whenever the same group of actions is to occur repeatedly. Alternation, also known as selection, is the traditional decision process where a determination can be made to execute a process, and can be indicated as a relationship between two subsets of a set.

Last but not least is the Logical Construction of Programs, also called the Warnier Method. It is a variant of Jackson's Structured Programming, and another variant of this is the Warnier–Orr method. LCP is a data-driven program design technique and replaces the trial-and-error approach to programming with a disciplined approach, based on logical rules.[19]

[14] http://www.mhhe.com/engcs/compsci/pressman/information/olc/AltReqmets.html.

[15] http://hebb.cis.uoguelph.ca/~dave/343/Lectures/design.html#1.12.

[16] http://en.wikipedia.org/wiki/Jackson_Structured_Programming.

[17] Jackson, Michael, "The Jackson Development Methods." http://mcs.open.ac.uk/mj665/JSPDDevt.pdf.

[18] http://www.davehigginsconsulting.com/pd03.htm.

[19] http://www.wayland-informatics.com/T-LCP.htm.

4.4 ANALYSIS

The field of software engineering sometimes is criticized because it does not have the same type of rigor as other types of engineering fields. Indeed, as software design is somewhat of a creative activity, there is a tendency toward an informal approach to software design, where design and coding is done on an informal basis. However, such an informal approach actually is contrary to good software engineering techniques (Laplante, 2005). This section of this chapter will attempt to explain some factors that should be considered when evaluating a software design method, and will compare and contrast some software design methods that were discussed in the last section.

Table 4.1 is a list of basic software engineering principles that should be considered when evaluating a particular software design method.

The first principle, modularity, is the separation of concerns in software design. Specifically, it has been found that modularity is one way to divide the incremental tasks that a software designer must perform. That is, modular design involves the decomposition of software behavior into software units and, in some instances, can be done through object-oriented design (Laplante, 2005). Modularity can be achieved by grouping locally related elements together, in terms of function and responsibility.

The second principle, anticipation of change, is an extremely important topic. This is because software frequently is changed to support new features or to perform repairs, especially in industry. Indeed, according to Phillips, "a high maintainability level of the software products is one of the hallmarks of outstanding commercial software" (Laplante, 2005, p. 234). In fact, engineers often are aware that systems go through numerous changes over the life of the product, sometimes to add new features or to fix a problem in production. Real-time systems must be designed so that changes can be facilitated as easily as possible, without sacrificing other properties of the software. Moreover, it is important to ensure that when software is modified, other problems do not seem as a result of the change.

The third principle, generality, can be stated as the intent to look for a more general problem resulting from the current design concept (Laplante, 2005). That is, in other words, generality is the ability of the software to be reusable because the general idea or problem of the current software can be applied to other situations.

The last principle, consistency, allows for a user to perform a task using a familiar environment. A consistent look and feel in the software will make it easier and reduce

TABLE 4.1 Basic Software Engineering Principles

Type of Principle	Description
Modularity	Separation of concerns in software design can be achieved through modular design
Anticipation of Change	How well does the software adapt to change
Generality	The intent to look for a more general problem that can be solved
Consistency	Providing a familiar context to code

TABLE 4.2 Software Design Methods Analysis

Type of Design Method	Modularity	Anticipation of Change	Generality	Consistency
Object-Oriented	Excellent	Excellent	Excellent	Excellent
Level-Oriented Design	Excellent	Average to poor (see top-down design)	Average to Poor (see top-down design)	Good
Data Flow or Structured Design	Excellent	Poor	Poor	Good
Data-Structure Oriented Design	Good	Excellent	Excellent	Good

the time that a user takes to become familiar with the software. If a user learns the basic elements of dealing with an interface, they do not have to be relearned each time for a different software application.[20]

Table 4.2 illustrates each software design method and comments on the four factors of modularity, anticipation of change, generality, and consistency. The scale of excellent, good, average or no comment, and poor were used to compare and contrast the different software techniques and how they compare with one another.

Based on the results of this study, it seems that object-oriented design may be the best software design method, at least for some types of applications. Indeed, object-oriented programming is one of the most widely used and easiest to learn approaches. First of all, object-oriented methods are very modular, as they use black boxes known as objects that contain code. Next, one of the main benefits of using object-oriented software is that it can be reused with relative ease. Object-oriented software also includes polymorphism, which is the ability to assign different meanings to something in different contexts and allows an entity such as a variable, a function, or an object to have more than one form. Finally, tools such as design patterns and the UML make object-oriented programming user friendly and easy to use. In fact, proponents of object-oriented design argue that this type of programming is the easiest to learn and use, especially for those who are relatively inexperienced in computer programming. This is because the objects are self-contained, easily identified, and simple. However, object-oriented programming has a few drawbacks that should be noted as well. Specifically, object-oriented design takes more memory and can be slow.

Probably the next best software design method that can be used is data-structure-oriented design. Data-structure-oriented design tends to have high modularity. In fact, some types of Jackson Development Method programs can be said to be object-oriented. Data-structure-oriented design also has a high level of anticipation of change

[20] http://www.d.umn.edu/~gshute/softeng/principles.html.

and generality. In fact, the Jackson Development Method programs initially were used in an effort to try and make COBOL programming easier to modify and be reused.

Level-oriented design has some advantages as well as some drawbacks and is ranked third out of the fourth approaches. Regarding the advantages of level-oriented design, the top-down approach is a very modular approach to software design, which is an advantage. The top-down approach also is not particularly difficult to use as well. However, as discussed above, this approach focuses on very specific tasks that have to be done and puts little emphasis on data structures. In other words, data structures are usually only thought of after procedures have been defined generally. Moreover, if the program needs to updated or revised, problems may occur because changes to one part of the software system often causes problems in another portion of the software. In other words, every time software is updated, all the procedures that rely on the old data structure would need to be analyzed and changed accordingly. Programmers usually have to approach a program as a series of single functions. As a result, programmers are not likely to incorporate evolutionary changes in the data structures into the big picture of the overall system. Thus, the top-down approach provides few ways to reuse existing pieces of software.

The last ranked method is the data flow design method, also known as structured design. As discussed, this method is very modular. However, several significant issues are encountered when using structured analysis and structured design in modeling a real-time system. Probably the most troublesome part of structured design is that tracking changes can be tricky, which translates into a low level of anticipation of change. Also, any change in the program requirements generally translates into significant amounts of code that will probably need to be rewritten. As a result, this approach is unpractical to use if significant software changes need to be made in the future.

4.4.1 Future Trends

Software design is a relatively new field of engineering, especially when compared with some other engineering disciplines like mechanical or civil engineering. It is therefore important to discuss what the future may hold for software design methods.

If one were to ask any computer programmer what the future of software engineering was, there would probably be a very wide variety of answers given. However, there is a common thread among all of these answers. It is that software development continues to become more complex, and developers must work at increasingly higher levels of abstraction to cope with this complexity.[21] Indeed, if there is one issue that most software developers could agree on, it is that as software becomes more and more complicated, it is important to develop new types of methods and procedures to aid software engineers in designing a software system.

One important shift that may be occurring currently is the approach to recognize that software architecture is an important aspect of software development. Software

[21] http://www.ibm.com/developerworks/rational/library/6007.html#trends.

architecture is the integration of software development methodologies and models and is used to aid in managing the complex nature of software development.

One type of approach in particular that may be gaining some popularity recently is model-driven architecture. Model-driven architecture provides a set of guidelines for the structuring of specifications expressed as models. Model-driven architecture was launched by the Object Management Group (OMG) in 2001.[22]

Four general principles underlie model-driven architecture. First, models are expressed in a well-defined notation and are important for understanding systems for enterprise-scale solutions.[23] Second, the building of systems can be organized around a set of models by imposing a series of transformations between models. Third, describing models in a set of meta-models facilitates meaningful integration and transformation among models, which is the basis for automation through tools. Finally, acceptance and broad adoption of this model-based approach requires industry standards to provide openness to consumers and to foster competition among vendors. Indeed, model-driven architecture encourages the efficient use of system models in software development and supports the reuse of best practices when creating families of systems.

4.5 SYSTEM-LEVEL DESIGN APPROACHES

There are three *traditional* main system-level design approaches: hardware/software codesign, platform-based design, and component-based design (Cai, 2004).

- *Hardware/Software codesign* (also referred to system synthesis) is a top-down approach. It starts with system behavior and generates the architecture from the behavior. It is performed by gradually adding implementation details to the design.
- *Platform-based design:* Rather than generating the architecture from the system behavior as in codesign, platform-based design maps the system behavior to predefined system architecture. Examples of platform-based design are in (Keutzer et al., 2000), (Martin & Salefski, 1998) .
- *Component-based design:* It is a bottom-up approach. To produce the predefined platform, it assembles existing heterogeneous components by inserting wrappers between these components. An example of component-based design is described in Cesario et al., (2002).

In addition, in this book, we are adding axiomatic design[24] as a new representation method. It is presented in Chapter 13.

[22] http://en.wikipedia.org/wiki/Model-driven_architecture.

[23] http://www.ibm.com/developerworks/rational/library/3100.html.

[24] Axiomatic design is a systems design methodology using matrix methods to analyze systematically the transformation of customer needs into functional requirements, design parameters, and process variables (El-Haik, 2005).

4.5.1 Hardware/Software Codesign

Hardware/software codesign can be defined as the cooperative design of hardware[25] and software[26] to achieve system-level objectives (functionality and constraints) by exploiting the synergism of hardware and software (Niemann, 1998), (Michell & Gupta, 1997). Hardware/software codesign research focuses on presenting a unified view of hardware and software and the development of synthesis tools and simulators to address the problem of designing heterogeneous systems. Although hardware implementation provides higher performance, software implementation is more cost effective and flexible because software can be reused and modified. The choice of hardware versus software in codesign is a trade-off among various design metrics like performance, cost, flexibility and time-to-market. This trade-off represents the optimization aspect of co-design. Figure 4.1 shows the flow of a typical hardware/software codesign system.

Generally, hardware/software codesign consists of the following activities: *specification and modeling*, *design,* and *validation* (O'Nils, 1999).

4.5.2 Specification and Modeling

This is the first step in the codesign process. The system behavior at the system level is captured during the specification step (Niemann, 1998). Section 4.5.6 provides details about specification and modeling, including Models of Computation.

4.5.3 Design and Refinement

The design process follows a step-wise refinement approach using several steps to transform a specification into an implementation. Niemann (1998) and O'Nils (1999) define the following design steps:

- *Tasks assignment:* The system specification is divided into a set of tasks/basic blocks that perform the system functionality (Niemann, 1998).
- *Cost estimation*: This step estimates cost parameters for implementing the system's basic blocks (output of task assignment) in hardware or software. Examples of hardware cost parameters are as follows: gate count, chip area, and power consumption, where execution time, code size, and required code memory are examples of software cost parameters. Cost estimates are used to assist in making design decision to decrease the number of design iterations (Niemann, 1998).
- *Allocation:* This step maps functional specification into a given architecture by determining the type and number of processing components required to implement the system's functionality. To make the allocation process manageable,

[25] Hardware refers to dedicated hardware components (ASIC).

[26] Software refers to software executing on processor or ASIP (DSP, microcontroller).

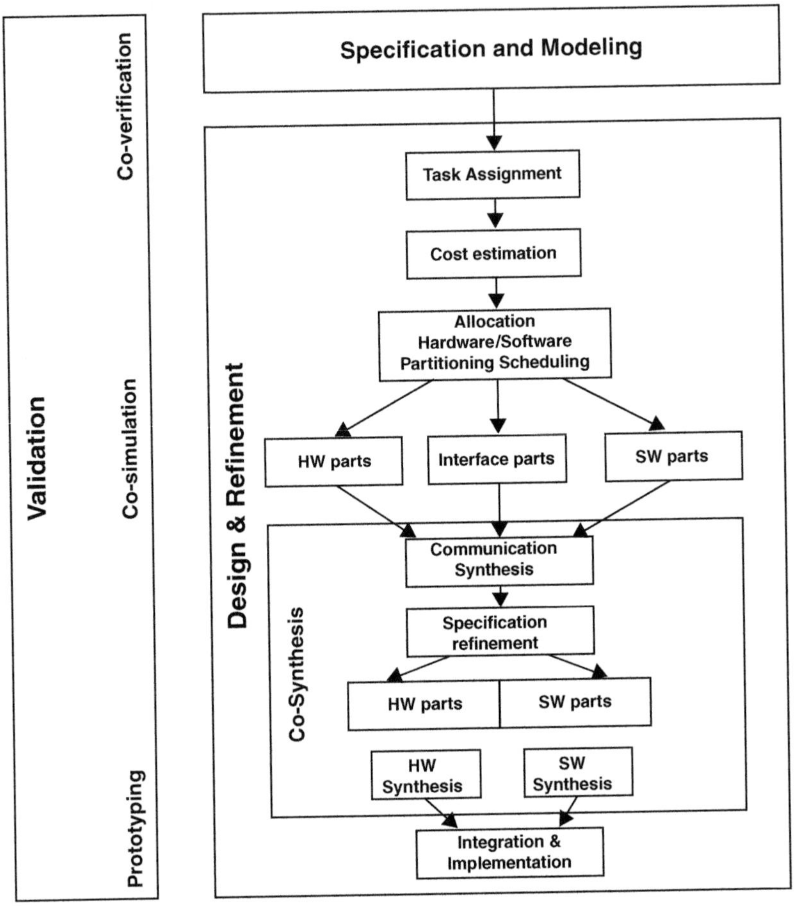

FIGURE 4.1 Flow of a typical codesign system.

codesign systems normally impose restrictions on target architectures. For example, allocation may be limited to a certain predefined components (Edwards et al., 1997).

- *Hardware/software partitioning:* This step partitions the specification into two parts: 1) a part that will be implemented in hardware and 2) a part that will be implemented in software.
- *Scheduling:* This step is concerned with scheduling the tasks assigned to processors. If tasks information (i.e., execution time, deadline, and delay) are known, scheduling is done statically at design time. Otherwise, scheduling is done dynamically at run time (i.e., using Real Time OS—RTOS). De Michell et al. (Michell & Gupta, 1997) provide an overview of techniques and algorithms to address the scheduling problem.

- *Cosynthesis*: Niemann classifies (Niemann, 1998) several design steps as part of cosynthesis:

    1. *Communication synthesis:* Implementing the partitioned system on heterogeneous target architecture requires interfacing between the ASIC components [hardware (HW)] and the processors [software (SW)] communication between the ASIC(s) and the processors. This is accomplished in communication synthesis step.
    2. *Specification refinement:* Once the system is partitioned into hardware and software, and the communication interfaces are defined (via communication synthesis), the system specification is refined into hardware specifications and software specifications, which include communication methods to allow interfacing between the hardware and software components.
    3. *Hardware synthesis*: AISC components are synthesized using behavior (high-level) synthesis and logic synthesis methods. Hardware synthesis is a mature field because of the extensive research done in this field. References (Camposano & Wolf, 1991), (Devadas et al., 1994) provide details about hardware synthesis methods.
    4. *Software synthesis*: This step is related to generating, from high-level specification, C or assembly code for the processor(s) that will be executing the software part of the heterogeneous system. Edwards et al. (1997) provides an overview of software synthesis techniques.

4.5.4 Validation

Informally, validation is defined as the process of determining that the design, at different levels of abstractions, is correct. The validation of hardware/software systems is referred to as *co-validation*. Methods for co-validations are (Edwards et al., 1997; Domer et al., XXXX).

- *Formal verification* is the process of mathematically checking that the system behavior satisfies a specific property. Formal verification can be done at the specification or the implementation level. For example, formal verification can be used to check the presence of a deadlock condition in the specification model of a system. At the implementation level, formal verification can be used to check whether a hardware component correctly implements a given finite state machine (FSM). For heterogeneous systems (i.e., composed of ASIC components and software components), formal verification is called *coverification*.
- *Simulation* validates that a system is functioning as intended by simulating a small set of inputs. Simulation of heterogeneous embedded systems requires simulating both hardware and software simultaneously, which is more complex than simulating hardware or software separately. Simulation of heterogeneous systems is referred to as *cosimulation*. A comparison of cosimulation methods is presented in Camposano and Wolf (1991).

4.5.5 Specification and Modeling

Specification is the starting point of the codesign process, where the designer specifies the systems specification without specifying the implementations. Languages are used to capture the system specifications. Modeling is the process of conceptualizing and refining the specifications. A model is different from the language used to specify the system. A model is a conceptual notation that describes the desired system behavior, whereas a language captured that concept in a concrete format. A model can be captured in a variety of languages, whereas a language can capture a variety of models (Vahid & Givargis, 2001).

To design systems that meet performance, cost, and reliability requirements, the design process need to be based on formal computational models to enable stepwise refinements from specification to implementation during the design process (Cortes et al., 2002). Codesign tools use specification languages as their input. To allow refinement during the design process, the initial specifications are transformed into intermediate forms based on the Model of Computation (MOC) (Bosman et al., 2003) used by the codesign systems. Two approaches are used for system specification, *homogeneous modeling* and *heterogeneous modeling* (Niemann, 1998), (Jerraya et al., 1999):

- *Homogeneous modeling* uses one specification language for specifying both hardware and software components of a heterogeneous system. The typical task of a codesign system using the homogeneous approach is to analyze and split the initial specification into hardware and software parts. The key challenge in this approach is the mapping of high-level concepts used in the initial specification onto low-level languages (i.e., C and VHDL) to represent hardware/software parts. To address this challenge, most co-design tools that use the homogeneous modeling approach start with a low-level specification language in order to reduce the gap between the system specification and the hardware/software models. For example, Lycos (Gajski et al., 1997) uses a C-like language called C^x and Vulcan uses another C-like language called Hardware C. Only few codesign tools start with a high-level specification language. For example, Polis (XXXX) uses Esterel (Boussinot et al., 1991) for its specification language.
- *Heterogeneous modeling* uses specific languages for hardware (e.g., VDHL) and software (e.g., C). Heterogeneous modeling allows simple mapping to hardware and software, but this approach makes validation and interfacing much more difficult. CoWare (Van Rompaey et al., 1996) is an example of a codesign methodology that uses heterogeneous modeling.

4.5.6 Models of Computation

A computational model is a conceptual formal notation that describes the system behavior (Vahid & Givargis, 2001). Ideally, a MOC should comprehend *concurrency*, *sequential behavior*, and *communication methods* (Cortes et al., 2002). Codesign systems use computational models as the underlying formal representation of a system. A variety of MOC have been developed to represent heterogeneous systems. Researchers have classified MOCs according to different criteria.

Gajski et al. (1997) classifies MOCs according to their orientation into five classes:

- *State-oriented models* use states to describe systems and events trigger transition between states.
- *Activity-oriented models* do not use states for describing systems, but instead they use data or control activities.
- *Structural-oriented models* are used to describe the physical aspects of systems. Examples are as follows: block diagrams and RT netlists.
- *Data-oriented models* describe the relations between data that are used by the systems. The entity relationship diagram (ERD) is an example of data-oriented models.
- *Heterogeneous models* merge features of different models into a heterogeneous model. Examples of heterogeneous models are program state machine (PSM) and control/data flow graphs (CDFG).

In addition to the classes described above, Bosman et al. (2003) propose a *time-oriented* class to capture the timing aspect of MOCs. Jantsch and Sander et al. (2005) group MOCs based on their timing abstractions. They define the following groups of MOCs: *continuous time models, discrete time models, synchronous models*, and *untimed models*. Continuous and discrete time models use events with a time stamp. In the case of continuous time models, time stamps correspond to a set of real numbers, whereas the time stamps correspond to a set of integer numbers in the case of discrete time models. Synchronous models are based on the *synchrony hypothesis*.[27]

Cortes et al. (2002) group MOCs based on common characteristics and the original model they are based on. The following is an overview of common MOCs based on the work by Cortes et al. (2002), and Bosman et al. (2003).

4.5.6.1 Finite State Machines (FSM). The FSM model consists of a set of states, a set of inputs, a set of outputs, an output function, and a next-state function (Gajski et al., 2000). A system is described as a set of states, and input values can trigger a transition from one state to another. FSMs commonly are used for modeling control-flow dominated systems. The main disadvantage of FSMs is the exponential growth of the number of the states as the system complexity rises because of the lack of hierarchy and concurrency. To address the limitations of the classic FSM, researcher's have proposed several derivates of the classic FSM. Some of these extensions are described as follows:

- *SOLAR* (Jerraya & O'Brien, 1995) is based on the Extended FSM Model (EFSM), which can support hierarchy and concurrency. In addtion, SOLAR supports high-level communication concepts, including channels and global

[27] Outputs are produced instantly in reaction to inputs, and no observable delay occurs in the outputs.

variables. It is used to represent high-level concepts in control-flow dominated systems, and it is mainly suited for synthesis purposes. The model provides an intermediate format that allows hardware/software designs at the system level to be synthesized.

- *Hierarchical Concurrent FSM (HCFSM)* (Niemann, 1998) solves the drawbacks of FSMs by decomposing states into a set of substates. These substates can be concurrent substates communicating via global variables. Therefore, HCFSMs supports hierarchy and concurrency. Statecharts is a graphical state machine language designed to capture the HCFSM MOC (Vahid & Givargis, 2001). The communication mechanism in statecharts is instantaneous broadcast, where the receiver proceeds immediately in response to the sender message. The HCFSM model is suitable for control-oriented/real-time systems.
- *Codesign FSM (CFSM)* (Cortes et al., 2002), (Chiodo et al., 1993) adds concurrency and hierarchy to the classic FSM and can be used to model both hardware and software. It commonly is used for modeling control-flow dominated systems. The communication primitive between CFSMs is called an event, and the behavior of the system is defined as sequences of events. CFSMs are used widely as intermediate forms in codesign systems to map high-level languages, used to capture specifications, into CFSMs. The Polis codesign system uses CFSM as its underlying MOC.

4.5.6.2 Discrete-Event Systems. In a discrete-event system, the occurrence of discrete asynchronous events triggers the transitioning from one state to another. An event is defined as an instantaneous action and has a time stamp representing when the event took place. Events are sorted globally according to their time of arrival. A signal is defined as a set of events, and it is the main method of communication between processes (Cortes et al., 2002). Discrete-event modeling often is used for hardware simulation. For example, both Verilog and VHDL use discrete-event modeling as the underlying MOC (Edwards et al., 1997). Discrete-event modeling is expensive because it requires sorting all events according to their time stamp.

4.5.6.3 Petri Nets. Petri nets are used widely for modeling systems. Petri nets consist of places, tokens, and transitions, where tokens are stored in places. Firing a transition causes tokens to be produced and consumed. Petri nets support concurrency and are asynchronous; however, they lack the ability to model hierarchy. Therefore, it can be difficult to use petri nets to model complex systems because of their lack of hierarchy. Variations of petri nets have been devised to address the lack of hierarchy. For example, the hierarchical petri nets (HPNs) were proposed by Dittrich (Agrawal, 2002). *HPNs* support hierarchy in addition to maintaining the major petri net features such as concurrency and asynchronously. HPNs use Bipartite[28] directed graphs as the underlying model. HPNs are suitable for modeling complex systems because they support both concurrency and hierarchy.

[28] A graph where the set of vertices can be divided into two disjoint sets U and V such that no edge has both end points in the same set.

4.5.6.4 Data Flow Graphs. In data flow graphs (DFGs), systems are specified using a directed graph where nodes (actors) represent inputs, outputs, and operations, and edges represent data paths between nodes (Niemann, 1998). The main usage of data flow is for modeling data flow dominated systems. Computations are executed only where the operands are available. Communications between processes is done via an unbounded FIFO buffering scheme (Cortes et al., 2002). Data flow models support hierarchy because the nodes can (Gajski et al., 1997) represent complex functions or another data flow (Niemann, 1998), (Edwards et al., 1997).

Several variations of DFGs have been proposed in the literature such as synchronous data flow (SDF) and asynchronous data flow (ADF) (Agrawal, 2002). In SDF, a fixed number of tokens is consumed, where in ADF, the number of tokens consumed is variable. Lee et al. (1995) provided an overview of data flow models and their variations.

4.5.6.5 Synchronous/Reactive Models. Synchronous modeling is based on the synchrony hypothesis, which states that outputs are produced instantly in reaction to inputs and there is no observable delay in the outputs (Watts, 1997). Synchronous models are used for modeling reactive real-time systems. Cortes et al. (2002) mention two styles for modeling reactive real-time systems: multiple clocked recurrent systems (MCRS), which are suitable for data dominated real-time systems, and state base formalisms, which are suitable for control dominated real-time systems. Synchronous languages such as Esterel (Boussinot et al., 1991) is used for capturing the synchronous/reactive MOC (Cortes et al., 2002).

4.5.6.6 Heterogeneous Models. Heterogeneous models combine features of different models of computation. Two examples of heterogeneous models are presented.

- *Programming languages* (Gajski et al., 1997) provide a heterogonous model that can support data, activity, and control modeling. Two types of programming languages are available: imperative such as C and declarative languages such as LISP and PROLOG. In imperative languages, statements are executed in the same order specified in the specification. However, execution order is not explicitly specified in declarative languages since the sequence of execution is based on a set of logic rules or functions. The main disadvantage of using programming languages for modeling is that most languages do not have special constructs to specify a system's state (Niemann, 1998).
- *PSM* is a merger between HCFSM and programming languages. A PSM model uses a programming language to capture a state's actions (Gajski et al., 1997). A PSM model supports hierarchy and concurrency inherited from HCFSM. The *Spec Charts* language, which was designed as an extension to VHDL, is capable of capturing the PSM model. The *Spec C* is another language capable of capturing the PSM model. *Spec C* was designed as an extension to C (Vahid & Givargis, 2001).

4.5.7 Comparison of Models of Computation

A comparison of various MOCs is presented by Bosman et al. (2003), and Cortes et al. (2002). Each author compares the MOCs according to certain criteria. Table 4.3 compares the MOCs discussed above based on the work done by Cortes et al., (2002), and Bosman et al., (2003).

4.6 PLATFORM-BASED DESIGN

Platform-based design was defined by Bailey et al., (2005, p. 150) as "an integration oriented design approach emphasizing systematic reuse, for developing complex products based upon platforms and compatible hardware and software virtual component, intended to reduce development risks, costs, and time to market." Platform-based design has been defined[29] as an all-encompassing intellectual framework in which scientific research, design tool development, and design practices can be embedded and justified. Platform-based design lays the foundation for developing economically feasible design flows because it is a structured methodology that theoretically limits the space of exploration, yet still achieves superior results in the fixed time constraints of the design.[30]

4.6.1 Platform-based Design Advantages

Some advantages of using the platform-based design method are as follows[31]:

- It provides a systematic method for identifying the hand-off points in the design phase.
- It eliminates costly design iterations because it fosters design reuse at all abstraction levels of a system design. This will allow the design of any product by assembling and configuring platform components in a rapid and reliable fashion.
- It provides an intellectual framework for the complete electronic design process.

4.6.2 Platform-based Design Principles

The basic principles of platform-based design are as follows:

1. Looking at the design as a meeting-in-the-middle phase, where iterative derivations of specifications phase meet with abstractions of possible implementations.
2. Identifying layers where the interface between specification and implementation phases takes place. These layers of are called platforms.[32]

[29] www1.cs.columbia.edu/~luca/research/pbdes.pdf.
[30] www1.cs.columbia.edu/~luca/research/pbdes.pdf.
[31] www1.cs.columbia.edu/~luca/research/pbdes.pdf.
[32] www1.cs.columbia.edu/~luca/research/pbdes.pdf.

TABLE 4.3 Comparison of Models of Computation[35]

MOC	Origin MOC	Main Application	Chick Mechanism	Orientation	Time	Communication Method	Hierarchy
SOLAR	FSM	Control oriented	Synchronous	State	No explicit timings	Remote procedure call	Yes
HCFSM/ Statecharts	FSM	Control oriented/ Reactive Real time	Synchronous	State	Min/Max time spent in state	Instant broadcast	Yes
CFSM	FSM	Control oriented	Asynchronous	State	Events with time stamp	Events broadcast	Yes
Discrete-Event	N/A	Real time	Synchronous	Timed	Globally sorted events with time stamp	Wired signals	No
HPN	Petri Net	Distributed	Asynchronous	Activity	No explicit timings	N/A	Yes
SDF	DFG	Signal processing	Synchronous	Activity	No explicit timing	Unbounded FIFO	Yes
ADF	DFG	Data oriented	Asynchronous	Activity	No explicit timing	Bounded FIFO	Yes

[35]In Cortes et al. (2002) and Bosman et al. (2003).

A platform is a library of components that can be assembled to generate a design for any level of abstraction. The library components are made of the following:

1. Computational units for carrying out the required computation.
2. Communication units that are used to interconnect the functional units.

A platform can be defined simply as an abstraction layer that hides the details of the several possible implementation refinements of the underlying layer.[33] Platform-based design allows designers to trade off different units of manufacturing, nonrecurring engineering and design costs, while minimally compromising design performance.

4.7 COMPONENT-BASED DESIGN

Component-based design approaches for embedded systems address in a unified way both hardware and software components. They can handle constraints on performance and dependability as well as different cost factors.[34] Component-based design is a bottom-up approach. To produce the predefined platform, it assembles existing heterogeneous components by inserting wrappers between these components. The two main design issues that component-based designs approaches need to handle are as follows:

- Presence of heterogeneous components. The components description requires concepts and languages supporting explicit behavior, time, resources, and their management because hardware components are inherently parallel, and synchronous.
- Predictability of basic properties of the designed system . The ability to describe formally the concurrent behavior of interacting components is a key aspect in component-based design.

It is necessary that theoretical results be integrated into logical component-based design flows, validated through comparison with existing industrial practice. Lately, the software engineering community has been focusing on design approaches, processes, and tools behind the concept that large software systems can be assembled from independent, reusable collections of functions (components). Some components already may be available, whereas the remaining components may need to be created. The component-based development concept is realized in technological approaches such as the Microsoft .NET platform and the Java 2 Enterprise Edition (J2EE) standards supported by products such as IBM's WebSphere and Sun's iPlanet.[35]

[33] www1.cs.columbia.edu/~luca/research/pbdes.pdf.
[34] http://www.combest.eu/home/?link=CBDforES.
[35] http://www.ibm.com/developerworks/rational/library/content/03July/2000/2169/2169.pdf.

Components are considered to be part of the starting platform for service orientation throughout software engineering, for example, Web services, and more recently, service-oriented architecture (SOA), whereby a component is converted into a service and subsequently inherits further characteristics beyond that of an ordinary component. Components can produce events or consume events and can be used for event-driven architecture.[36]

Component software is common today in traditional applications. A large software system often consists of multiple interacting components. These components can perceived as large objects with a clear and well-defined task. Different definitions of a component exist; some define objects as components, whereas others define components as large parts of coherent code, intended to be reusable and highly documented. However, all definitions have one thing in common: They focus on the functional aspect of a component. The main goal of using components is the ability to reuse them. Reuse of software currently is one of the much hyped concepts, because it enables one to build applications relatively fast.

4.8 CONCLUSIONS

This chapter has explored the past, present, and future of software design methods. Going back the 1960s and 1970s, software was developed in an unorganized fashion, leading to many safety issues. As a result, software design methods had to be developed to cope with this issue. In the early to mid-1990s, techniques such as object-oriented programming became more and more popular.

The design approaches discussed were level-oriented, data-flow-oriented, data-structure-oriented, and object-oriented. The basic software engineering principles that should be considered when evaluating a particular software design method are modularity, generality, anticipation of change, and consistency. When evaluating software design methods based on these four principles, object-oriented design is the best method available because object-oriented design is highly modular. Moreover, it can be reused with relative ease. Object-oriented software also includes polymorphism, which is the ability to assign different meanings to something in different contexts and allows an entity such as a variable, a function, or an object to have more than one form. Finally, tools such as design patterns and the UML make object-oriented programming user friendly and easy to use. In fact, proponents of object-oriented design argue that this type of programming is the easiest to learn and use, especially for those who are relatively inexperienced in computer programming.

As software programming becomes more and more complicated, software architecture may become a more important aspect of software development. Software architecture is the integration of software development methodologies and models, and it is used to aid in managing the complex nature of software development.

System-level design is considered a way to reduce the complexities and to address the challenges encountered in designing heterogeneous embedded systems. Three

[36] http://en.wikipedia.org/wiki/Component-based_software_engineering.

main approaches for system-level design are as follows: *hardware/software codesign, platform-based design*, and *component-based design*.

In this chapter, also we investigated the codesign approach of system-level design. Codesign follows a top-down design approach with a unified view of hardware and software. The approach uses step-wise refinements steps to implement from high-level specification an entire system on heterogeneous target architectures. Several codesign methodologies and tools have been developed in the research community and used in the industry. Most of them concentrate on specific aspects of the codesign process and do not cover the whole design process. Based on popularity, and literature availability, three codesign systems were studied and compared.

MOCs are used in codesign systems to specify systems using a formal representation and to allow refinement during the design process. The selection of a specific MOC is highly depended on the application intended to be modeled. As shown in Table 4.3, most MOCs support a specific application domain, whereas only one (out of the presented models) can support multiple domains.

REFERENCES

Agrawal A. (2002), "Hardware Modeling and Simulation of Embedded Applications," Master's Thesis, Vanderbilt University, 2002.

Bailey, Brian (2005), Martin, Grant, and Anderson Thomas (eds.), (2005), *Taxonomies for the Development and Verification of Digital Systems.* Springer, New York.

Barkan, David (1993), "Software litigation in the year 2000: The effect of OBJECT-ORIENTED design methodologies on traditional software jurisprudence," 7 High Technology L.J. 315.

Barkan, David M. (1992), "software litigation in the year 2000: the effect of OBJECT-ORIENTED design methodologies on traditional software jurisprudence." *Berkeley Technical Law Journal*, Fall, p. 3.

Bosman, G., Bos, I. A. M., Eussen, P. G. C., Lammel I. R. (2003), "A Survey of Co-Design Ideas and Methodologies," Master's Thesis at Vrije Universiteit, Amsterdam, The Netherlands, Oct.

Boussinot, F., de Simone, R., and Ensmp-Cma, V., (1991), "The ESTEREL language," *Proceedings of the IEEE*, Volume. 79, pp. 1293–1304.

Cai, Lukai (2004), *Estimation and Exploration Automation of System Level Design*," University of California, Irvine, CA.

Camposano, Raul and Wolf Wayne (1991), *High-Level VLSI Synthesis*, Kluwer Academic Publishers, Norwell, MA.

Cesario, Wander, Baghdadi, Ames, Gauthier, Lovis, Lyonnard, Damien, Nicolescu, Gabriela, Paviot, Yanick, Yoo, Sungjoo, Jerraya, Ahmed, and Diaz-Nava, Mario (2002), "Component-Based Design Approach for Multicore SoCs," *Proceedings of the IEEE/ACM Design Automation Conference*, Nov.

Chiodo, Massimiliano, Giusto, Paolo, Hsieh, Harry, Jurecska, Attila, Lavagno, Luciano, and Sangiovanni-Vincentelli, Alberto (1993), "A Formal Specification Model for Hardware/Software Codesign," University of California at Berkeley Berkeley, CA, USA, Technical Report: ERL-93-48.

Cortes, Luis, Eles, Petru, and Peng, Zebo (2002), *A Survey on Hardware/Software Codesign Representation Models*, Technical Report, Linköping University, Wiley, New York, 2002.

De Michell, Micheli and Gupta, Rajesh (1997), "Hardware/software co-design." *Proceedings of the IEEE*, Mar. Volume 85, pp. 349–365.

Devadas, Srinivas, Ghosh, Abhijit, and Keutzer Kurt (1994), *Logic Synthesis*, McGraw-Hill, New York.

Dömer, R., Gajski, D. and J. Zhu, "Specification and Design of Embedded Systems," *IT+ TI Magazine*, Volume *3*, #S-S, pp. 7–12.

Edwards, Stephen, Lavagno, Luciano, Lee, Edward, and Sangiovanni-Vincentelli Alberto (1997), "Design of embedded systems: Formal models, validation, and synthesis," *Proceedings of the IEEE*, Volume 85, pp. 366–390.

Gajski, Daniel, Zhu, Jainwen, and Dömer, Rainer (1997), *Essential Issues in Codesign: Information and Computer Science*, University of California, Irvine, CA.

Gajski, Danieh, Zhu, Jainwen, Dömer, Rainer, Gerstlauer, Andreas, and Zhao, Shuging (2000), *SpecC, Specification Language and [Design] Methodology* Kluwer Academic, Norwell, MA.

Gomaa, Hassan (1989), *Software Design Methods for Real Time Systems*, SEI Curriculum Module SEI-CM-22-1.0, George Mason University, Dec. 1989, p. 1.

Jantsch, Axel and Sander, Ingo (2005), Models of computation in the design process, Systemonchip: *Next Generation Electronics*, IEEE, New York.

Jerraya, Ahmed and O'Brien, Kevin (1995), "SOLAR: An intermediate format for system-level modeling and synthesis." *Computer Aided Software/Hardware Engineering*, pp. 147–175.

Jerraya, Ahmed, Romdhani, M., Le Marrec, Phillipe, Hessel, Fabino, Coste, Pascal, Valderrama, C., Marchioro, G. F., Daveau, Jean-marc, and Zergainoh, Nacer-Eddine (1999), "Multilanguage specification for system design and codesign." *System Level Synthesis*, 1999.

Keutzer, Kurt Malik, S., Newton, A. R., Rabaey, J. M., Sangiovanni-Vincentelli, Alberto (2000), " System-level design: Orthogonalization of concerns and platform-based design," *IEEE Transactions on Computer-aided Design of Integrated Circuits and Systems*, Volume 19, p. 1523.

Khoo, Benjamin Kok Swee (2009), *Software Design Methodology*, http://userpages.umbc.edu/~khoo/survey1.html.

Laplante, Phillip A. (2005), "*Real-Time Systems Design and Analysis*," 3rd Ed., IEEE Press, New York.

Lee Edward and Parks Thomas (1995), "Dataflow process networks." *Proceedings of the IEEE*, Volume. 83, pp. 773–801.

Martin, Grant and Salefski, Bill (1998), "Methodology and Technology for Design of Communications and Multimedia Products via System-Level IP Integration," *Proceedings of the DATE'98 Designers' Forum*, June, pp.11–18.

Nimmer Melville B. and Nimmer David (1991), NIMMER ON COPYRIGHT, § 13.03 [F] at 13-78.30 to .32 (1991).

Niemann, Raif (1998), *Hardware/Software Co-Design for Data Flow Dominated Embedded Systems*. Kluwer Academic Publishers, Boston, MA.

O'Nils, Mattias (1999), "Specification, Synthesis and Validation of Hardware/Software Interfaces," PhD thesis, Royal Institute of Technology, Sweden.

Polis, *A Framework for Hardware-Software Co-Design of Embedded Systems.* http://embedded.eecs.berkeley.edu/research/hsc/. Accessed August, 2008.

Software Design Consultants (2009), *What is Object-Oriented Software?* http://www.softwaredesign.com/objects.html (Last accessed on August 16, 2009).

Urlocker, Zack (1989), "Whitewater's actor: An introduction to object-oriented programming concepts." *Microsoft Systems Journal*, Volume 4, 2, p. 12.

Vahid, Frank and Givargis, Tony (2001), *Embedded System Design: A Unified Hardware/Software Introduction*. John Wiley & Sons, New York.

Van Rompaey, Karl, Verkest, Diederik, Bolsens, Ivo, De Man, Hugo, and Imec, H. (1996), "CoWare-A Design Environment for Heterogeneous Hardware/Software Systems," Design Automation Conference, 1996, with EURO-VHDL'96 and Exhibition, Proceedings EURO-DAC'96, European, pp. 252–257.

Watts, S. Humphrey (1997), *Introduction to the Personal Software Process*, Addison Wesley, Reading, MA.

CHAPTER 5

DESIGN FOR SIX SIGMA (DFSS) SOFTWARE MEASUREMENT AND METRICS[1]

> When you can measure what you are speaking about and express it in numbers, you know something about it; but when you cannot measure it, when you cannot express it in numbers, your knowledge is of a meager and unsatisfactory kind: it may be the beginnings of knowledge but you have scarcely in your thoughts advanced to the stage of Science.—Lord Kelvin (1883)

5.1 INTRODUCTION

Science,which includes software, is based on measurement. To design or redesign software, we need to understand some numerical relationships or metrics. Design for six sigma (DFSS) is no exception. Six Sigma and DFSS live and die on metrics definition, measurement, classification, optimization, and verification.

A software metric is a measure of some property of a piece of software code or its specifications. As quantitative methods have proved so powerful in other sciences, computer science practitioners and theoreticians have worked hard to bring similar measurement approaches to software development.

What is "software measurement?" The software measurement process is that portion of the DFSS software process that provides for the identification, definition, collection, and analysis of measures that are used to understand, evaluate, predict, or

[1]More on metrics are provided in Chapter 17.

Software Design for Six Sigma: A Roadmap for Excellence, By Basem El-Haik and Adnan Shaout

control software development (design/redesign) processes or products. The primary purpose of measurement is to provide insight into software processes and products so that an organization can better make decisions and manage the achievement of goals. This chapter provides a review of metrics that can be used as critical-to-quality (CTQ) with some guidelines that can help organizations integrate a measurement process with their overall DFSS software process.

What are "software metrics?" Goodman (1993) defines software metrics as "the continuous application of measurement-based techniques to the software development process and its products to supply meaningful and timely management information, together with the use of those techniques to improve that process and its products." In software organizations, measurement often is equated with collecting and reporting data and focuses on presenting the numbers. The primary purpose of this chapter is to focus measurement more on setting goals, analyzing data with respect to software development, and using the data to make decisions. The objectives of this chapter are to provide some guidelines that can be used to design and implement a process for measurement that ties measurement to software DFSS project goals and objectives; defines measurement consistently, clearly, and accurately; collects and analyzes data to measure progress toward goals; and evolves and improves as the DFSS deployment process matures. In general, measurement is for development, understanding, control, and improvement.

Modern software development practitioners likely are to point out that naive and simplistic metrics can cause more harm than good. Measurement is an essential element of software development management. There is little chance of controlling what we cannot measure. Measurement assigns numbers based on a well-defined meaning. Software metrics help avoid pitfalls such as cost overruns (most projects fail to separate design and code costs) and clarify goals. Metrics can help answer questions, such as what is the cost of each process activity? How "good" is the code being developed? How can the code under development be improved?

By aligning the measurement process with the overall software process, DFSS projects and organizations can collect and analyze data simultaneoulsy to help make decisions with respect to project goals and obtain feedback to improve the measurement process itself. Figure 5.1 presents a working definition for a software measurement process.

Measurement is related to software entities as given in Table 5.1. Input software entities include all of the resources used for software research, development, and production such as people, materials, tools, and methods. DFSS process software entities include software-related activities and events and usually are associated with a time factor. For example, activities such as developing a software system from requirements through delivery to the customer, the inspection of a piece of code, or the first months of operations after delivery, and time periods that do not necessarily correspond to specific activities. Output software entities are the products of the DFSS software process that includes all the artifacts, deliverables, and documents that are produced such as requirements documentation, design specifications, code (source, object, and executable), test documentation (plans, scripts, specifications, cases, and reports), project plans, status reports, budgets, problem reports, and software metrics.

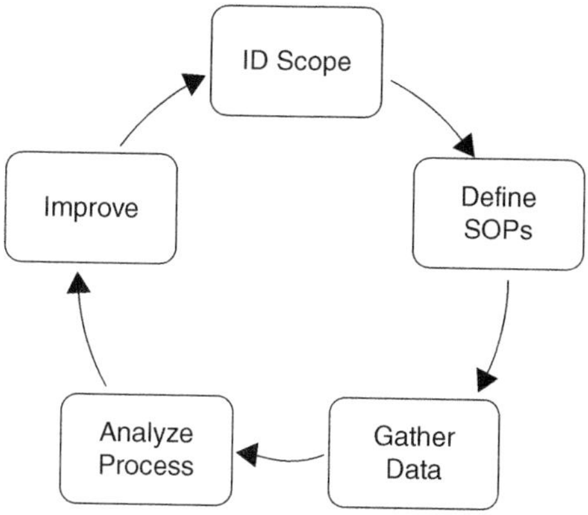

FIGURE 5.1 Software measurement cycle.

Each of these software entities has many properties or features that the DFSS team might want to measure such as computer's price, performance, or usability. In DFSS deployment, the team could look at the time or effort that it took to execute the process, the number of incidents that occurred during the development process, its cost, controllability, stability, or effectiveness. Often the complexity, size, modularity, testability, usability, reliability, or maintainability of a piece of source code can be taken as metrics.

5.2 SOFTWARE MEASUREMENT PROCESS

Software measurement process elements are constituent parts of the overall DFSS software process (Figure 11.1, Chapter 11), such as software estimating, software code, unit test, peer reviews, and measurement. Each process element covers a well-defined, bounded, closely related set of tasks (Paulk et al., 1993).

Measurements are used extensively in most areas of production and manufacturing to estimate costs, calibrate equipment, assess quality, and monitor inventories. Measurement is the process by which numbers or symbols are assigned to attributes of entities in the real world in such a way as to describe them according to clearly defined rules (Fenton, 1991).

TABLE 5.1 Examples of Entities and Metrics

Entity	Metric Measured
Software Quality	Defects discovered in design reviews
Software Design Specification	Number of modules
Software Code	Number of lines of code, number of operations
Software Development Team	Team size, average team, experience

Figure 5.1 shows the software measurement process . The process is generic in that it can be instantiated at different levels (e.g., project level, divisional level, or organizational level). This process links the measurement activities to the quantifying of software products, processes, and resources to make decisions to meet project goals.

The key principle shared by all is that projects must assess their environments so that they can link measurements with project objectives. Projects then can identify suitable measures (CTQs) and define measurement procedures that address these objectives. Once the measurement procedures are implemented, the process can evolve continuously and improve as the projects and organizations mature.

This measurement process becomes a process asset that can be made available for use by projects in developing, maintaining, and implementing the organization's standard software process (Paulk et al., 1993).

Some examples of process assets related to measurement include organizational databases and associated user documentation; cost models and associated user documentation; tools and methods for defining measures; and guidelines and criteria for tailoring the software measurement process element.

5.3 SOFTWARE PRODUCT METRICS

More and more customers are specifying software and/or quality metrics reporting as part of their contractual requirements. Industry standards like ISO 9000 and industry models like the Software Engineering Institute's (SEI) Capability Maturity Model Integration (CMMI) include measurement.

Companies are using metrics to better understand, track, control, and predict software projects, processes, and products. The term "software metrics" means different things to different people. The software metrics, as a noun, can vary from project cost and effort prediction and modeling, to defect tracking and root cause analysis, to a specific test coverage metric, to computer performance modeling. Goodman (1993) expanded software metrics to include software-related services such as installation and responding to customer issues. Software metrics can provide the information needed by engineers for technical decisions as well as information required by management.

Metrics can be obtained by direct measurement such as the number of lines of code or indirectly through derivation such as defect density = number of defects in a software product divided by the total size of product. We also can predict metrics such as the prediction of effort required to develop software from its measure of complexity. Metrics can be nominal (e.g., no ordering and simply attachment of labels), ordinal [i.e., ordered but no quantitative comparison (e.g., programmer capability: low, average, high)], interval (e.g., programmer capability: between 55th and 75th percentile of the population ability) ratio (e.g., the proposed software is twice as big as the software that has just been completed), or absolute (e.g., the software is 350,000 lines of code long).

If a metric is to provide useful information, everyone involved in selecting, designing, implementing, collecting, and using, it must understand its definition and purpose. One challenge of software metrics is that few standardized mapping systems exist. Even for seemingly simple metrics like the number of lines of code, no standard counting method has been widely accepted. Do we count physical or logical lines of code? Do we count comments or data definition statements? Do we expand macros before counting, and do we count the lines in those macros more than once? Another example is engineering hours for a project—besides the effort of software engineers, do we include the effort of testers, managers, secretaries, and other support personnel? A few metrics, which do have standardized counting criteria, include Cyclomatic Complexity (McCabe, 1976). However, the selection, definition, and consistent use of a mapping system within the organization for each selected metric are critical to a successful metrics program. A metric must obey representation condition and allow different entities to be distinguished.

Attributes, such as complexity, maintainability, readability, testability, complexity, and so on, cannot be measured directly, and indirect measures for these attributes are the goal of many metric programs. Each unit of the attribute must contribute an equivalent amount to the metric, and different entities can have the same attribute value. Software complexity is a topic that we will concentrate on going forward.

Programmers find it difficult to gauge the code complexity of an application, which makes the concept difficult to understand. The McCabe metric and Halstead's software science are two common code complexity measures. The McCabe metric determines code complexity based on the number of control paths created by the code. Although this information supplies only a portion of the complex picture, it provides an easy-to-compute, high-level measure of a program's complexity. The McCabe metric often is used for testing. Halstead bases his approach on the mathematical relationships among the number of variables, the complexity of the code, and the type of programming language statements. However, Halstead's work is criticized for its difficult computations as well as its questionable methodology for obtaining some mathematical relationships.

Software complexity deals with the overall morphology of the source code. How much fan-out do the modules exhibit? Is there an optimal amount of fan-out that reduces complexity? How cohesive are the individual modules, and does module cohesion contribute to complexity? What about the degree of coupling among modules?

Code complexity is that hard-to-define quality of software that makes it difficult to understand. A programmer might find code complex for two reasons: 1) The code does too much work. It contains many variables and generates an astronomical number of control paths. This makes the code difficult to trace. 2) The code contains language constructs unfamiliar to the programmer.

The subjective nature of code complexity cries out for some objective measures. Three common code complexity measures are the McCabe metric, Henry–Kafura Information Flow, and Halstead's software science. Each approaches the topic of code complexity from a different perspective.

These metrics can be calculated independently from the DFSS process used to produce the software and generally are concerned with the structure of source code. The most prominent metric in this category is lines of code, which can be defined as the number of "New Line" hits in the file excluding comments, blank lines, and lines with only delimiters.

5.3.1 McCabe's Cyclomatic Number

The cyclomatic complexity of a section of source code is the count of the number of linearly independent paths through the source code. For instance, if the source code contained no decision points such as IF statements or FOR loops, the complexity would be 1 because there is only a single path through the code. If the code has a single IF statement containing a single condition, then there would be two paths through the code: one path where the IF statement is evaluated as TRUE, and one path where the IF statement is evaluated as FALSE.

This is a complexity metric. The premise is that complexity is related to the control flow of the software. Using graph theory (e.g., control flow graphs), we can calculate the cyclomatic number (C) as follows:

$$C = e - n + 1 \tag{5.1}$$

where e is the number of arcs and n is the number of nodes.

McCabe uses a slightly different formula

$$C = e - n + 2p \tag{5.2}$$

where p is the number of strongly connected components (usually assumed to be 1).

In a control flow graph, each node in the graph represents a basic block (i.e., a straight-line piece of code without any jumps or jump targets; jump targets start a block, and jumps end a block). Directed edges are used to represent jumps in the control flow. There are, in most presentations, two specially designated blocks: the entry block, through which control enters into the flow graph, and the exit block, through which all control flow leaves. The control flow graph is essential to many compiler optimizations and static analysis tools.

For a single program (or subroutine or method), p is always equal to 1. Cyclomatic complexity may, however, be applied to several such programs or subprograms at the same time (e.g., to all methods in a class), and in these cases, p will be equal to the number of programs in question, as each subprogram will appear as a disconnected subset of the graph.

It can be shown that the cyclomatic complexity of any structured program with only one entrance point and one exit point is equal to the number of decision points (i.e., "if" statements or conditional loops) contained in that program plus one (Belzer et al., 1992).

Cyclomatic complexity may be extended to a program with multiple exit points; in this case, it is equal to

$$\Pi - s + 2 \tag{5.3}$$

where Π is the number of decision points in the program and s is the number of exit points.

This metric is an indication of the number of "linear" segments in a software system (i.e., sections of code with no branches) and, therefore, can be used to determine the number of tests required to obtain complete coverage. It also can be used to indicate the psychological complexity of software.

A code with no branches has a cyclomatic complexity of 1 because there is 1 arc. This number is incremented whenever a branch is encountered. In this implementation, statements that represent branching are defined as follows: "for", "while", "do", "if", "case" (optional), "catch" (optional), and the ternary operator (optional). The sum of cyclomatic complexities for software in local classes also is included in the total for a software system. Cyclomatic complexity is a procedural rather than an object-oriented metric. However, it still has meaning for object-oriented programs at the software level.

McCabe found that $C = 10$ is an acceptable threshold value when he analyzed 10 modules and modules with $C > 10$ had many maintenance difficulties and histories of error.

A popular use of the McCabe metric is for testing. McCabe himself cited software testing as a primary use for his metric. The cyclomatic complexity of code gives a lower limit for the number of test cases required for code coverage.

Other McCabe Complexity Metrics[2]:

- Actual Complexity Metric: The number of independent paths traversed during testing.
- Module Design Complexity Metric: The complexity of the design-reduced module. Reflects the complexity of the module's calling patterns to its immediate subordinate modules. This metric differentiates between modules that will seriously complicate the design of any program they are part of and modules that simply contain complex computational logic. It is the basis on which program design and integration complexities are calculated.
- Essential Complexity Metric: A measure of the degree to which a module contains unstructured constructs. This metric measures the degree of structuredness and the quality of the code. It is used to predict the maintenance effort and to help in the modularization process.
- Pathological Complexity Metric: A measure of the degree to which a module contains extremely unstructured constructs.
- Design Complexity Metric: Measures the amount of interaction between modules in a system.
- Integration Complexity Metric: Measures the amount of integration testing necessary to guard against errors.
- Object Integration Complexity Metric: Quantifies the number of tests necessary to fully integrate an object or class into an object-oriented system.

[2]http://www.mccabe.com/iq_research_metrics.htm.

- Global Data Complexity Metric: Quantifies the cyclomatic complexity of a module's structure as it relates to global/parameter data. It can be no less than one and no more than the cyclomatic complexity of the original flow graph.

McCabe Data-Related Software Metrics

- Data Complexity Metric: Quantifies the complexity of a module's structure as it relates to data-related variables. It is the number of independent paths through data logic and, therefore, a measure of the testing effort with respect to data-related variables.
- Tested Data Complexity Metric: Quantifies the complexity of a module's structure as it relates to data-related variables. It is the number of independent paths through data logic that have been tested.
- Data Reference Metric: Measures references to data-related variables independently of control flow. It is the total number of times that data-related variables are used in a module.
- Tested Data Reference Metric: The total number of tested references to data-related variables.
- Maintenance Severity Metric: Measures how difficult it is to maintain a module.
- Data Reference Severity Metric: Measures the level of data intensity within a module. It is an indicator of high levels of data-related code; therefore, a module is data intense if it contains a large number of data-related variables.
- Data Complexity Severity Metric: Measures the level of data density within a module. It is an indicator of high levels of data logic in test paths; therefore, a module is data dense if it contains data-related variables in a large proportion of its structures.
- Global Data Severity Metric: Measures the potential impact of testing data-related basis paths across modules. It is based on global data test paths.

McCabe Object-Oriented Software Metrics for ENCAPSULATION

- Percent Public Data (PCTPUB). PCTPUB is the percentage of PUBLIC and PROTECTED data within a class.
- Access to Public Data (PUBDATA). PUBDATA indicates the number of accesses to PUBLIC and PROTECTED data.

McCabe Object-Oriented Software Metrics for POLYMORPHISM

- Percent of Un-overloaded Calls (PCTCALL). PCTCALL is the number of non-overloaded calls in a system.
- Number of Roots (ROOTCNT). ROOTCNT is the total number of class hierarchy roots within a program.
- Fan-in (FANIN). FANIN is the number of classes from which a class is derived.

5.3.2 Henry–Kafura (1981) Information Flow

This is a metric to measure intermodule complexity of source code based on the in–out flow of information (e.g., parameter passing, global variables, or arguments) of a module. A count is made as follows:

I: Information count flowing in the module
O: Information count flowing out of the module
w: Weight (a measure of module size)
c: Module complexity

$$c = w(I \times O)^2 \tag{5.4}$$

For a source code of n modules, we have

$$C = \sum_{j=1}^{n} c_j = \sum_{j=1}^{n} w_j \left(I_j x O_j\right)^2 \tag{5.5}$$

5.3.3 Halstead's (1997) Software Science

Maurice Halstead's approach relied on his fundamental assumption that a program should be viewed as an expression of language. His work was based on studying the complexities of languages—both programming and written languages. Halstead found what he believed were mathematically sound relationships among the number of variables, the type of programming language statements, and the complexity of the code. He attacked part of the first and second reasons a programmer might find code complex.

Halstead derived more than a dozen formulas relating properties of code. The following is a representative sample of his work:

$$\text{Vocabulary } (\eta) = \eta_1 + \eta_2 \tag{5.6}$$

$$\text{Length } (N) \text{ as } N = N1 + N2 \tag{5.7}$$

$$\text{Volume } (V) \text{ as } V = N \log_2 \eta \text{ (the program's physical size)} \tag{5.8}$$

$$\text{Potential volume } (V^*) \text{ as } V^* = (2 + \eta_2^* \log_2(2 + \eta_2^*) \tag{5.9}$$

where η_1 is the number of distinct operators in the code, η_2 is the number of distinct operands in the code, $N1$ is the number of all operators in the code, and $N2$ is the number of all operands in the code.

V^* is the smallest possible implementation of an algorithm, where $\eta_2{}^*$ is the smallest number of operands required for the minimal implementation, which Halstead stated are the required input and output parameters.

$$\text{Program level } (L) \text{ as } L = V^*/V \tag{5.10}$$

Program level measures the program's ability to be comprehended. The closer L is to 1, the tighter the implementation. Starting with the assumption that code complexity increases as vocabulary and length increase, Halstead observed that the code complexity increases as volume increases and that code complexity increases as program level decreases. The idea is that if the team computes these variables and finds that the program level is not close to 1, the code may be too complex. The team should look for ways to "tighten" the code.

Halstead's work is sweeping, covering topics such as computing the optimal number of modules, predicting program errors, and computing the amount of time required for a programmer to implement an algorithm.

Halstead Metrics

- Program Volume: The minimum number of bits required for coding the program.
- Program Length: The total number of operator occurrences and the total number of operand occurrences.
- Program Level and Program Difficulty: Measure the program's ability to be comprehended.
- Intelligent Content: Shows the complexity of a given algorithm independent of the language used to express the algorithm.
- Programming Effort: The estimated mental effort required to develop the program.
- Error Estimate: Calculates the number of errors in a program.
- Programming Time: The estimated amount of time to implement an algorithm.
- Line Count Software Metrics
  - Lines of Code
  - Lines of Comment
  - Lines of Mixed Code and Comments
  - Lines Left Blank

A difficulty with the Halstead metrics is that they are hard to compute. How does the team easily count the distinct and total operators and operands in a program? Imagine counting these quantities every time the team makes a significant change to a program.

Code-level complexity measures have met with mixed success. Although their assumptions have an intuitively sound basis, they are not that good at predicting error rates or cost. Some studies have shown that both McCabe and Halstead do no better at predicting error rates and cost than simple lines-of-code measurements. Studies that attempt to correlate error rates with computed complexity measures show mixed results. Some studies have shown that experienced programmers provide the best prediction of error rates and software complexity.

5.4 GQM (GOAL–QUESTION–METRIC) APPROACH

Goal-oriented measurement points out that the existence of the explicitly stated goal is of the highest importance for improvement programs. GQM presents a systematic approach for integrating goals to models of the software processes, products, and quality perspectives of interest based on the specific needs of the project and the organization (Basili et al., 1994).

In other words, this means that in order to improve the process, the team has to define measurement goals, which will be, after applying the GQM method, refined into questions and consecutively into metrics that will supply all the necessary information for answering those questions. The GQM method provides a measurement plan that deals with the particular set of problems and the set of rules for obtained data interpretation. The interpretation gives us the answer if the project goals were attained.

GQM defines a measurement model on three levels: Conceptual level (goal), operational level (question), and quantitative level (metric). A goal is defined for an object for a variety of reasons, with respect to various models of quality, from various points of view, and relative to a particular environment. A set of questions is used to define the models of the object of study and then focuses on that object to characterize the assessment or achievement of a specific goal. A set of metrics, based on the models, is associated with every question in order to answer it in a measurable way. Questions are derived from goals that must be answered in order to determine whether the goals are achieved. Knowledge of the experts gained during the years of experience should be used for GQM definitions. These developers' implicit models of software process and products enable the metric definition.

Two sets of metrics now can be mutually checked for consistency and completeness. The GQM plan and the measurement plan can be developed, consecutively; data collection can be performed; and finally, the measurement results are returned to the project members for analysis, interpretation, and evaluation on the basis of the GQM plan.

The main idea is that measurement activities always should be preceded by identifying clear goals for them. To determine whether the team has met a particular goal, the team asks questions whose answers will tell them whether the goals have been achieved. Then, the team generates from each question the attributes they must measure to answer these questions.

Sometimes a goal-oriented measurement makes common sense, but there are many situations where measurement activities can be crucial even though the goals are not defined clearly. This is especially true when a small number of metrics address

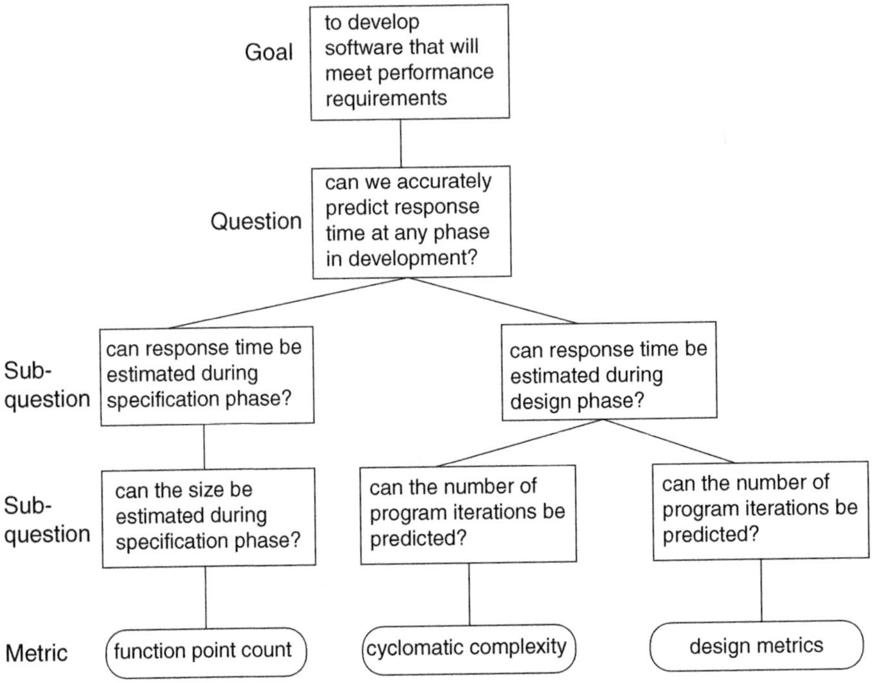

FIGURE 5.2 GQM method.[3]

different goals—in this case, it is very important to choose the most appropriate one. Figure 5.2[4] shows the GQM method.

The open literature typically describes GQM in terms of a six-step process where the first three steps are about using business goals to drive the identification of the right metrics and the last three steps are about gathering the measurement data and making effective use of the measurement results to drive decision making and improvements. Basili described his six-step GQM process as follows[5]:

1. Develop a set of corporate, division, and project business goals and associated measurement goals for productivity and quality.
2. Generate questions (based on models) that define those goals as completely as possible in a quantifiable way.
3. Specify the measures needed to be collected to answer those questions and track process and product conformance to the goals.
4. Develop mechanisms for data collection.

[3] http://www.cs.ucl.ac.uk/staff/A.Finkelstein/advmsc/11.pdf.
[4] http://www.cs.ucl.ac.uk/staff/A.Finkelstein/advmsc/11.pdf.
[5] http://en.wikipedia.org/wiki/GQM.

5. Collect, validate, and analyze the data in real time to provide feedback to projects for corrective action.
6. Analyze the data in a post mortem fashion to assess conformance to the goals and to make recommendations for future improvements.

5.5 SOFTWARE QUALITY METRICS

Software quality metrics are associated more closely with process and product metrics than with project metrics. Software quality metrics can be divided further into end-product quality metrics and into in-process quality metrics. The essence of software quality is to investigate the relationships among in-process metrics, project characteristics, and end-product quality and, based on the findings, to engineer improvements in both process and product quality.

Software quality is a multidimensional concept. It has levels of abstraction beyond even the viewpoints of the developer or user. Crosby, (1979) among many others, has defined software quality as conformance to specification. Very few end users will agree that a program that perfectly implements a flawed specification is a quality product. Of course, when we talk about software architecture, we are talking about a design stage well upstream from the program's specification. Juran and Fryna (1970) proposed a generic definition of quality. He said products must possess multiple elements of fitness for use. Two of his parameters of interest for software products were quality of design and quality of conformance. These are separate designs from implementation and may even accommodate the differing viewpoints of developer and user in each area. Moreover, we should view quality from the entire software life-cycle perspective, and in this regard, we should include metrics that measure the quality level of the maintenance process as another category of software quality metrics (Kan, 2002). Kan (2002) discussed several metrics in each of three groups of software quality metrics: product quality, in-process quality, and maintenance quality by several major software developers (HP, Motorola, and IBM) and discussed software metrics data collection. For example, by following the GQM method (Section 5.4), Motorola identified goals, formulated questions in quantifiable terms, and established metrics. For each goal, the questions to be asked and the corresponding metrics also were formulated. For example, the questions and metrics for "Improve Project Planning" goal (Daskalantonakis, 1992) are as follows:

Question 1: What was the accuracy of estimating the actual value of project schedule?
Metric 1: Schedule Estimation Accuracy (SEA)

$$SEA = \frac{\text{Actual Project Duration}}{\text{Estimated Project Duration}} \tag{5.11}$$

Question 2: What was the accuracy of estimating the actual value of project effort?
Metric 2: Effort Estimation Accuracy (EEA)

$$EEA = \frac{\text{Actual Project Effort}}{\text{Estimated Project Effort}} \tag{5.12}$$

	Capability	Usability	Performance	Reliability	Instability	Maintainability	Documentation	Availability
Capability								
Usability								
Performance	●	●						
Reliability	●	○	●					
Instability		○	○	○				
Maintainability	●	○	●	○				
Documentation	●	○				○		
Availability	●	○	○	○	○	○		

●: Conflict One Another

○: Support One Another

Blank: Not Related

FIGURE 5.3 IBM dimensions of quality.[6]

In addition to Motorola, two leading firms that have placed a great deal of importance on software quality as related to customer satisfaction are IBM and Hewlett-Packard. IBM measures user satisfaction in eight attributes for quality as well as overall user satisfaction: capability or functionality, usability, performance, reliability, installability, maintainability, documentation, and availability (see Figure 5.3).

Some of these attributes conflict with each other, and some support each other. For example, usability and performance may conflict, as may reliability and capability or performance and capability. Other computer and software vendor organizations may use more or fewer quality parameters and may even weight them differently for different kinds of software or for the same software in different vertical markets. Some organizations focus on process quality rather than on product quality. Although it is true that a flawed process is unlikely to produce a quality software product, our focus in this section is entirely on software product quality, from customer needs identification to architectural conception to verification. The developmental flaws are tackled by a robust DFSS methodology, which is the subject of this book.

5.6 SOFTWARE DEVELOPMENT PROCESS METRICS

The measurement of software development productivity is needed to control software costs, but it is discouragingly labor-intensive and expensive. Many facets of the process metrics such as yield metrics are used. For example, the application of

[6]http://www.developer.com/tech/article.php/10923_3644656_1/Software-Quality-Metrics.htm

methods and tools, the use of standards, the effectiveness of management, and the performance of development systems can be used in this category.

Productivity is another process metric and is calculated by dividing the total delivered source lines by the programmer-days attributed to the project in line of code (LOC)/programmer-day.

5.7 SOFTWARE RESOURCE METRICS[7]

These include:

- Elapsed time
- Computer resources
- Effort expended
  - On tasks within a project, classified by life-cycle phase or software function
  - On extra-project activities training

As with most projects, time and effort are estimated in software development projects. Most estimating methodologies are predicated on analogous software programs. Expert opinion is based on experience from similar programs; parametric models stratify internal databases to simulate environments from many analogous programs; engineering builds reference similar experience at the unit level; and cost-estimating relationships (like parametric models) regress algorithms from several analogous programs. Deciding which of these methodologies (or combination of methodologies) is the most appropriate for a DFSS project usually depends on availability of data, which in turn, depends on where the team is in the life cycle or project scope definition[8]:

- *Analogies*: Cost and schedule are determined based on data from completed similar efforts. When applying this method, it often is difficult to find analogous efforts at the total system level. It may be possible, however, to find analogous efforts at the subsystem or lower level computer software configuration item/computer software component/computer software unit. Furthermore, the team may be able to find completed efforts that are more or less similar in complexity. If this is the case, a scaling factor may be applied based on expert opinion. After an analogous effort has been found, associated data need to be assessed. It is preferable to use effort rather than cost data; however, if only cost data are available, these costs must be normalized to the same base year as effort using current and appropriate inflation indices. As with all methods, the quality of the estimate is directly proportional to the credibility of the data.

[7]See http://www.stsc.hill.af.mil/resources/tech_docs/gsam3/chap13.pdf for more details.
[8]http://www.stsc.hill.af.mil/resources/tech_docs/gsam3/chap13.pdf.

- *Expert opinion*: Cost and schedule are estimated by determining required effort based on input from personnel with extensive experience on similar programs. Because of the inherent subjectivity of this method, it is especially important that input from several independent sources be used. It also is important to request only effort data rather than cost data as cost estimation is usually out of the realm of engineering expertise (and probably dependent on nonsimilar contracting situations). This method, with the exception of rough orders-of-magnitude estimates, is used rarely as a primary methodology alone. Expert opinion is used to estimate low-level, low-cost pieces of a larger cost element when a labor-intensive cost estimate is not feasible.
- *Parametric models*: The most commonly used technology for software estimation is parametric models, a variety of which are available from both commercial and government sources. The estimates produced by the models are repeatable, facilitating sensitivity and domain analysis. The models generate estimates through statistical formulas that relate a dependent variable (e.g., cost, schedule, and resources) to one or more independent variables. Independent variables are called "cost drivers" because any change in their value results in a change in the cost, schedule, or resource estimate. The models also address both the development (e.g., development team skills/experience, process maturity, tools, complexity, size, and domain) and operational (how the software will be used) environments, as well as software characteristics. The environmental factors, which are used to calculate cost (manpower/effort), schedule, and resources (people, hardware, tools, etc.), often are the basis of comparison among historical programs, and they can be used to assess on-going program progress. Because environmental factors are relatively subjective, a rule of thumb when using parametric models for program estimates is to use multiple models as checks and balances against each other. Also note that parametric models are not 100 percent accurate.
- *Engineering build (grass roots or bottom-up build)*: Cost and schedule are determined by estimating effort based on the effort summation of detailed functional breakouts of tasks at the lowest feasible level of work. For software, this requires a detailed understanding of the software architecture. Analysis is performed, and associated effort is predicted based on unit-level comparisons with similar units. Often, this method is based on a notional system of government estimates of most probable cost and used in source selections before contractor solutions are known. This method is labor-intensive and usually is performed with engineering support; however, it provides better assurance than other methods that the entire development scope is captured in the resulting estimate.
- *Cost Performance Report (CPR) analysis*: Future cost and schedule estimates are based on current progress. This method may not be an optimal choice for predicting software cost and schedule because software generally is developed in three distinct phases (requirements/design, code/unit test, and integration/test) by different teams. Apparent progress in one phase may not be predictive of progress in the next phases, and lack of progress in one phase may not show up

until subsequent phases. Difficulty in implementing a poor design may occur without warning, or problems in testing may be the result of poor test planning or previously undetected coding defects. CPR analysis can be a good starting point for identifying problem areas, and problem reports included with CPRs may provide insight for risk assessments.

- *Cost-Estimating Relationships (CERs)*: Cost and schedule are estimated by determining effort based on algebraic relationships between a dependent (effort or cost) variable and independent variables. This method ranges from using a simple factor, such as cost per LOC on a similar program with similar contractors, to detailed multivariant regressions based on several similar programs with more than one causal (independent) variable. Statistical packages are available commercially for developing CERs, and if data are available from several completed similar programs (which is not often the case), this method may be a worthwhile investment for current and future cost and schedule estimating tasks. Parametric model developers incorporate a series of CERs into an automated process by which parametric inputs determine which CERs are appropriate for the program at hand.

Of these techniques, the most commonly used is parametric modeling. There is currently no list of recommended or approved models; however, the team will need to justify the appropriateness of the specific model or other technique they use. As mentioned, determining which method is most appropriate is driven by the availability of data. Regardless of which method used, a thorough understanding of software's functionality, architecture, and characteristics, and contract is necessary to accurately estimate required effort, schedule, and cost.

5.8 SOFTWARE METRIC PLAN[9]

For measurement to be effective, it must become an integral part of the team decision-making process. Insights gained from metrics should be merged with process knowledge gathered from other sources in the conduct of daily program activities. It is the entire measurement process that gives value to decision making, not just the charts and reports. Without a firm metrics plan, based on issue analysis, you can become overwhelmed by statistics, charts, graphs, and briefings to the point where the team has little time for anything other than ingestion.

Not all data are worth collecting and analyzing. Once the team development project is in-process, and your development team begins to design and produce lines-of-code, the effort involved in planning and specifying the metrics to be collected, analyzed, and reported on begins to pay dividends.

[9] http://www.stsc.hill.af.mil/resources/tech_docs/gsam3/chap13.pdf.

The ground rules for a metrics plan are as follows:

- Metrics must be understandable to be useful. For example, lines-of-code and function points are the most common, accepted measures of software size with which software engineers are most familiar.
- Metrics must be economical: Metrics must be available as a natural by-product of the work itself and integral to the software development process. Studies indicate that approximately 5% to 10% of total software development costs can be spent on metrics. The larger the software program, the more valuable the investment in metrics becomes. Therefore, the team should not waste programmer time by requiring specialty data collection that interferes with the coding task. They need to look for tools that can collect most data on an unintrusive basis.
- Metrics must be field tested: Beware of software contractors who offer metrics programs that seem to have a sound theoretical basis but have not had practical application or evaluation. The team needs to make sure proposed metrics have been successfully used on other programs or are prototyped before accepting them.
- Metrics must be highly leveraged: The team is looking for data about the software development process that permit management to make significant improvements. Metrics that show deviations of 0.005% should be relegated to the trivia bin.
- Metrics must be timely: Metrics must be available in time to effect change in the development process. If a measurement is not available until the project is in deep trouble, it has no value.
- Metrics must give proper incentives for process improvement. High-scoring teams are driven to improve performance when trends of increasing improvement and past successes are quantified. Conversely, metric data should be used very carefully during contractor performance reviews. A poor performance review, based on metrics data, can lead to negative working relationships. Metrics should not be used metrics to judge team or individual performance.
- Metrics must be spaced evenly throughout all phases of development. Effective measurement adds value to all life-cycle activities.
- Metrics must be useful at multiple levels. They must be meaningful to both management and DFSS team members for process improvement in all facets of development.

REFERENCES

Basili, V., Gianluigi, C., and Rombach, D. (1994), *The Goal Question Metric Approach.* ftp://ftp.cs.umd.edu/pub/sel/papers/gqm.pdf.

Belzer, J., Kent, A., Holzman, A.G., and Williams, J.G. (1992), *Encyclopedia of Computer Science and Technology*, CRC Press, Boca Raton, FL.

Crosby, P.B. (1979), *Quality is Free: The Art of Making Quality Certain*, McGraw-Hill, New York.

Daskalantonakis, M.K. (1992), "A practical view of software measurement and implementation experiences within Motorola (1001-1004)." *IEEE Transactions on Software Engineering*, Volume *18*, #11, pp. 998–1010.

Fenton, Norman E. (1991), *Software Metrics, A Rigorous Approach*, Chapman & Hall, London, UK.

Goodman, P. (1993), *Practical Implementation of Software Metrics*, 1st Ed., McGraw Hill, London.

Halstead, M. (1997), *Elements of Software Silence*, North Holland, New York.

Henry, S. and Kafura, D. (1981), "Software structure metrics based on information flow." *IEEE Transactions on Software Engineering*, Volume *7*, #5, pp. 510–518.

Juran, J.M. and Gryna, F.M. (1970), *Quality Planning and Analysis: From Product Development Through Use*, McGraw-Hill, New York.

Kan, S. (2002), *Metrics and Models in Software Quality Engineering*, 2nd Ed., Addison-Wesley, Upper Saddle River, NJ.

Kelvin, L. (1883), "PLA—Popular Lectures and Addresses," *Electrical Units of Measurement*, Volume *1*,

McCabe, T. (1976), "A complexity measure." *IEEE Transaction on Software Engineering*, Volume SE-*2*, #4.

Paulk, Mark C., Weber, Charles V., Garcia, Suzanne M., Chrissis, Mary Beth, and Bush, Marilyn (1993), *Key Practices of the Capabililty Maturity Model, Vwvg wx1×1'1' version 1.1 (CMU/SEI-93-TR-25)*, Software Engineering Institute, Carnegie Mellon University, Pittsburgh, PA.

CHAPTER 6

STATISTICAL TECHNIQUES IN SOFTWARE SIX SIGMA AND DESIGN FOR SIX SIGMA (DFSS)[1]

6.1 INTRODUCTION

A working knowledge of statistics is necessary to the understanding of software Six Sigma and Design for Six Sigma (DFSS). This chapter provides a very basic review of appropriate terms and statistical methods that are encountered in this book. This statistics introductory chapter is beneficial for software development professionals, including software Six Sigma and DFSS belts, measurement analysts, quality assurance personnel, process improvement specialists, technical leads, and managers.

Knowledge of statistical methods for software engineering is becoming increasingly important because of industry trends[2] as well as because of the increasing rigor adopted in empirical research. The objectives of this chapter are to introduce basic quantitative and statistical analysis techniques, to demonstrate how some of these techniques can be employed in software DFSS process, and to describe the relationship of these techniques to commonly accepted software process maturity models and standards.

Statistical analysis is becoming an increasingly important skill for software engineering practitioners and researchers. This chapter introduces the basic concepts and

[1] This chapter barely touches the surface, and we encourage the reader to consult other resources for further reference.

[2] CMMI Development Team, Capability Maturity Model—Integrated, Version 1.1, Software Engineering Institute, 2001.

Software Design for Six Sigma: A Roadmap for Excellence By Basem El-Haik and Adnan Shaout

most commonly employed techniques. These techniques involve the rigorous collection of data, development of statistical models describing that data, and application of those models to decision making by the software DFSS team. The result is better decisions with a known level of confidence.

Statistics is the science of data. It involves collecting, classifying, summarizing, organizing, analyzing, and interpreting data. The purpose is to extract information to aid decision making. Statistical methods can be categorized as *descriptive* or *inferential*. Descriptive statistics involves collecting, presenting, and characterizing data. The purpose is to describe the data graphically and numerically. Inferential statistics involves estimation and hypothesis testing to make decisions about population parameters. The statistical analysis presented here is applicable to all analytical data that involve counting or multiple measurements.

Common applications of statistics in software DFSS include developing effort and quality estimation models, stabilizing and optimizing process performance, and evaluating alternative development and testing methods. None of the techniques can be covered in sufficient detail to develop real skills in their use.[3] However, the chapter will help the practitioner to select appropriate techniques for further exploration and to understand better the results of researchers in relevant areas.

This chapter addresses basic measurement and statistical concepts. The approach presented is based on ISO/IEC Standard 15939 (Emam & Card, 2002). An effective measurement and analysis program in measurement topics include measurement scales, decision criteria, and the measurement process model provided in ISO/IEC Standard 15939. Statistical topics include descriptive statistics, common distributions, hypothesis testing, experiment design, and selection of techniques. Measurement and statistics are aids to decision making. The software DFSS team makes decisions on a daily basis with factual and systematic support. These techniques help to improve the quality of decision making. Moreover, they make it possible to estimate the uncertainty associated with a decision.

Many nonstatistical quantitative techniques help to select the appropriate statistical technique to apply to a given set of data, as well as to investigate the root causes of anomalies detected through data analysis. Root cause analysis as known today relies on seven basic tools that are the cause-and-effect diagram, check sheet, control chart (special cause vs. common cause), flowchart, histogram, Pareto chart, and scatterplot. They are captured in Figure 6.1. Other tools include check sheets (or contingency tables), Pareto charts, histograms, run charts, and scattergrams. Ishikawa's practical handbook discusses many of these.

Although many elements of the software DFSS only are implemented once or a few times in the typical project, some activities (e.g., inspections) are repeated frequently in the Verify & Validate phase. Monitoring these repeated process elements can help to stabilize the overall process elements. Many different control charts are available. The choice of techniques depends on the nature and organization of the data. Few basic statistics texts cover control charts or the more general topic of statistical process control, despite their widespread applicability in industry. Other statistical

[3]Contact www.SixSigmaPI.com for training.

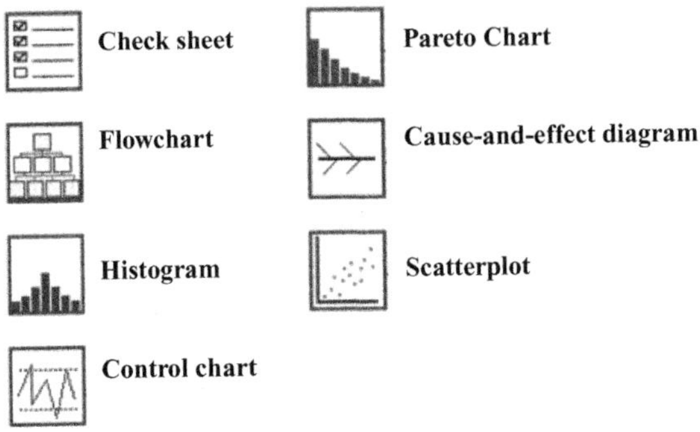

FIGURE 6.1 Seven basic quality tools.

techniques are needed when the purpose of the analysis is more complex than just monitoring the performance of a repeated process element. Regression analysis may help to optimize the performance of a process.

Development and calibration of effort, quality, and reliability estimation models often employs regression. Evaluation of alternative processes (e.g., design and inspection methods) often involves analysis of variance (ANOVA). Empirical software research also makes extensive use of ANOVA techniques. The most commonly employed regression and ANOVA techniques assume that the data under analysis follows a normal distribution. Dealing with the small samples is common in software DFSS and that assumption can be problematic. The nonparametric counterparts to the techniques based on the normal distributions should be used in these situations.

Industry use of statistical techniques is being driven by several standards and initiatives. The Capability Maturity Model Integration (CMMI) requires the "statistical management of process elements" to achieve Maturity Level 4 (Emam & Card, 2002). The latest revisions of ISO Standard 9001 have substantially increased the focus on the use of statistical methods in quality management.

6.2 COMMON PROBABILITY DISTRIBUTIONS

Table 6.1 is a description of common probability distributions.

6.3 SOFTWARE STATISTICAL METHODS

Statistical methods such as descriptive statistics, removing outliers, fitting data distributions, and others play an important role in analyzing software historical and developmental data.

The largest value added from statistical modeling is achieved by analyzing software metrics to draw statistical inferences and by optimizing the model parameters through experimental design and optimization. Statistics provide a flexible and

TABLE 6.1 Common Probability Distributions

Density Function	Graph

Bernoulli distribution:

$$p(x) = \begin{cases} 1-p, & \text{if } x = 0 \\ p, & \text{if } x = 1 \\ 0, & \text{otherwise} \end{cases}$$

Generalized random experiment two Outcomes

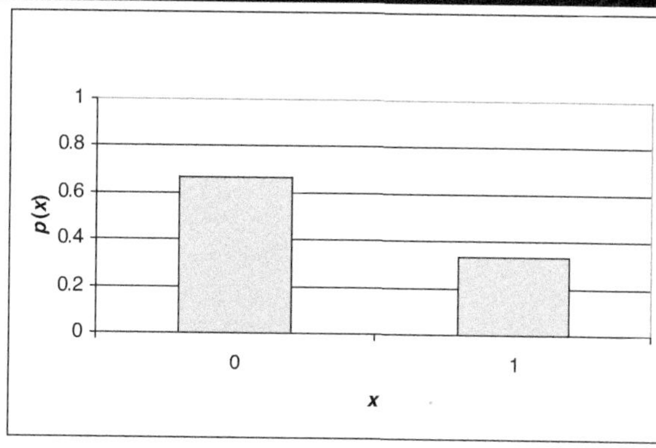

Binomial distribution:

$$p(x) = \binom{n}{x} p^{x}(1-p)^{n-x}$$

Number of successes in n experiments (number of defective items in a batch)

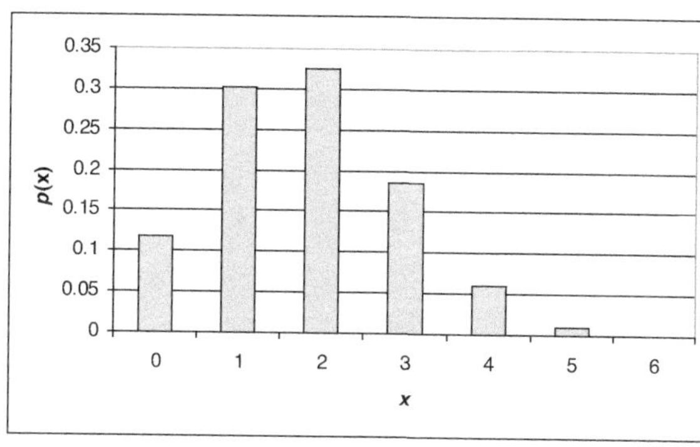

Poisson distribution:

$$p(x) = \frac{e^{-\lambda}\lambda^{x}}{x!}, \; x = 0, 1, \ldots$$

Stochastic arrival processes

λ: average number of arrivals per time unit

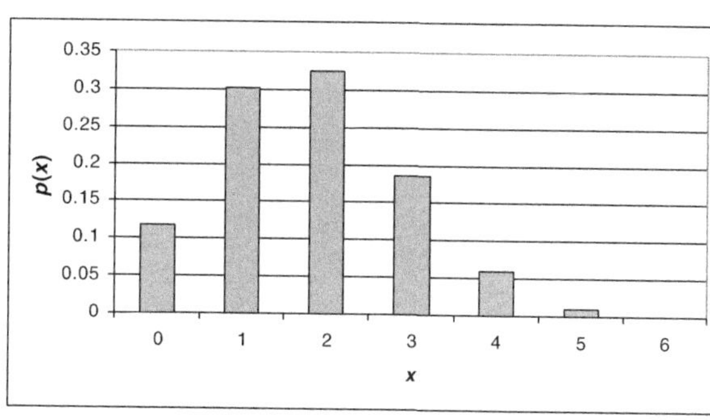

(Continued)

TABLE 6.1 Common Probability Distributions *(Continued)*

Geometric distribution:

$$p(x) = p(1-p)^x$$

Number of failures before success in a series of independent Bernoulli trials

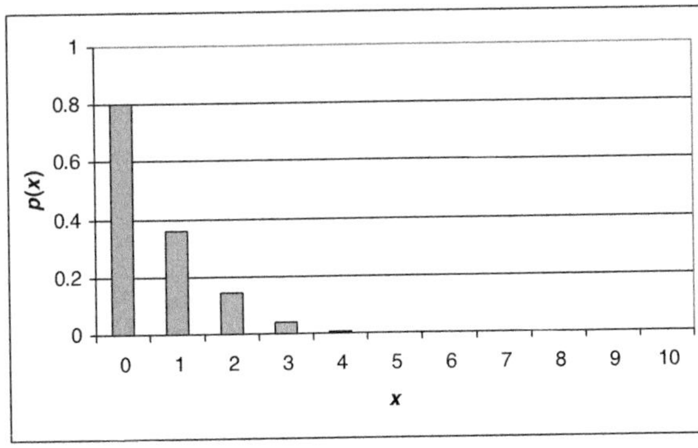

Uniform distribution:

$$f_U(x) = \frac{1}{b-a}, \quad a \le x \le b$$

Random number generation (RNG)

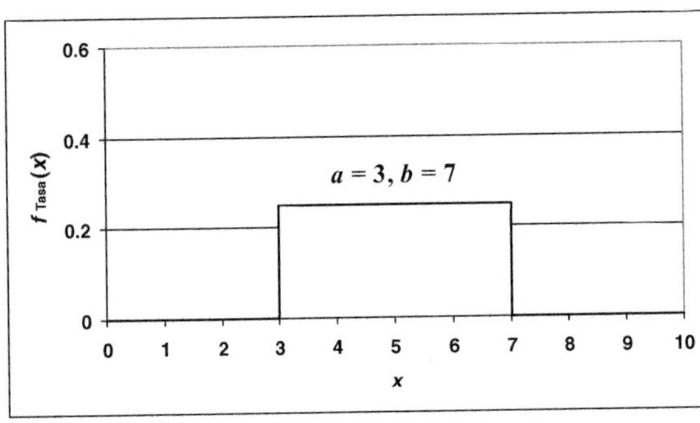

Normal distribution:

$$f_N(x) = \frac{1}{\sqrt{2\pi}\sigma} \exp\left[-\frac{(x-\mu)^2}{2\sigma^2}\right]$$

Natural phenomena of large population size

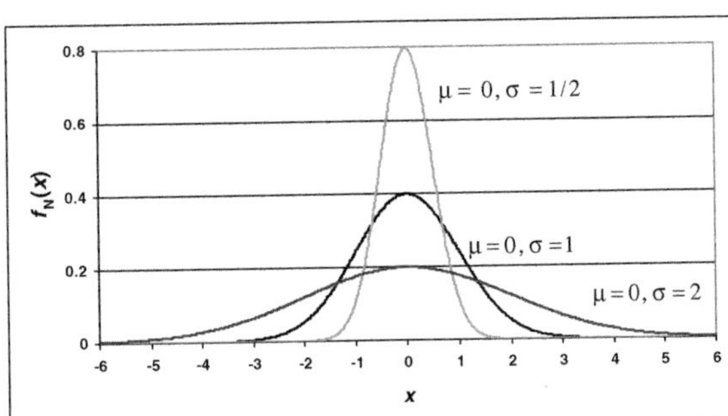

Exponential distribution:

$$f_{Exp}(x) = \lambda e^{-\lambda x}$$

Reliability models:

Lifetime of a component

Service time

Time between arrivals

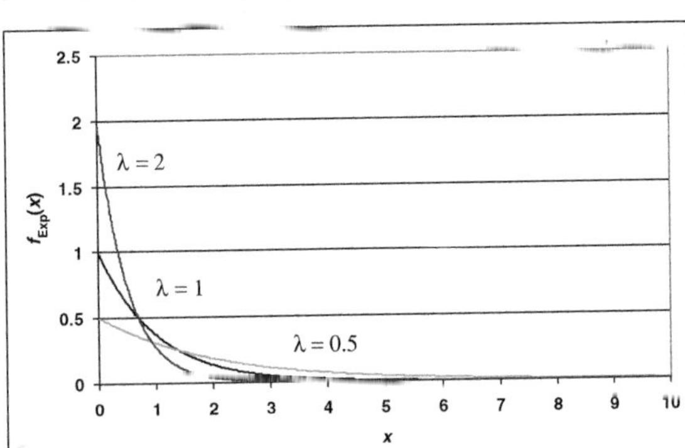

TABLE 6.1 Common Probability Distributions *(Continued)*

Triangular distribution:

$$f_{\text{Tria}}(x) = \begin{cases} \dfrac{2(x-a)}{(b-a)(c-a)}, & \text{if } a \leq x \leq c \\ \dfrac{2(b-x)}{(b-a)(b-c)}, & \text{if } c < x \leq b \end{cases}$$

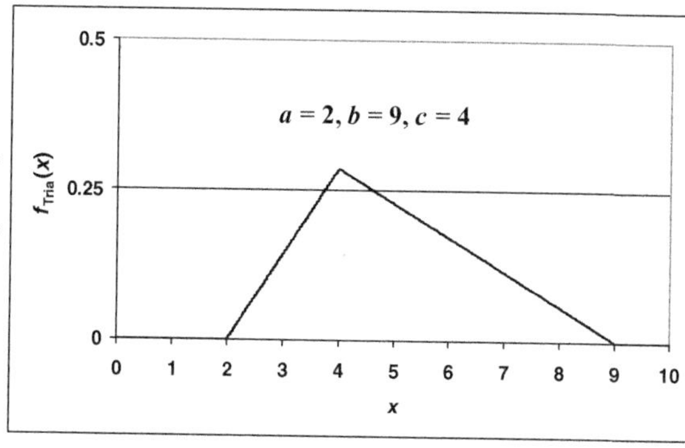

Gamma distribution:

$$f_{\text{Gamma}}(x) = \frac{\lambda}{\Gamma(\lambda)} \lambda x^{k-1} e^{-\lambda x}$$

Failure from repetitive disturbances

Duration of a multiphase task

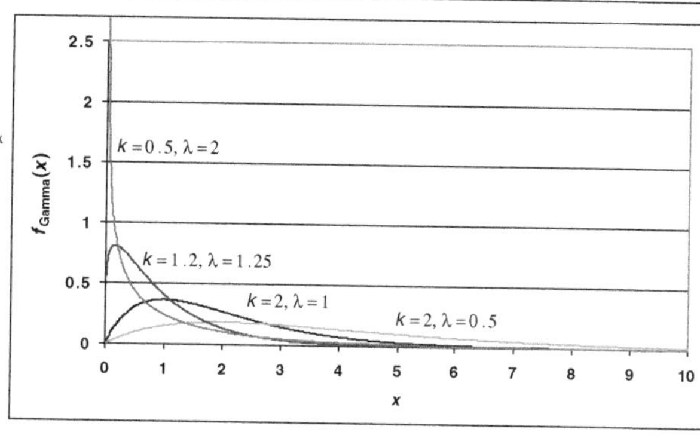

cost-effective platform for running experimental design, what-if analysis, and optimization methods. Using the results obtained, software design teams can draw better inferences about the code behavior, compare multiple design alternatives, and optimize the metric performance.

Along with statistical and analytical methods, a practical sense of the underlying assumptions can assist greatly the analysis activity. Statistical techniques often lead to arriving at accurate analysis and clear conclusions. Several statistical methods skills are coupled together to facilitate the analysis of software developmental and operational metrics.

This chapter provides a survey of basic quantitative and statistical techniques that have demonstrated wide applicability in software design. The chapter includes examples of actual applications of these techniques. Table 6.2 summarizes the statistical methods and the modeling skills that are essential at each one of the major statistical modeling activities.

Statistical analysis in design focuses on measuring and analyzing certain metric output variables. A variable, or in DFSS terminology, a critical-to-quality (CTQ) characteristic, is any measured characteristic or attribute that differs from one code to another or from one application to another.

TABLE 6.2 Modeling and Statistical Methods

Statistical Modeling	Statistical Methods	Modeling Skills
Software metrics input modeling	– Sampling techniques – Probability models – Histograms – Theoretical distributions – Parameter estimation – Goodness-of-fit – Empirical distributions	– Data collection – Random generation – Data classification – Fitting distributions – Modeling variability – Conformance test – Using actual data
Software metrics output analysis	– Graphical tools – Descriptive statistics – Inferential statistics – Experimental design – Optimization search – Transfer function – Scorecard	– Output representation – Results summary – Drawing inferences – Design alternatives – Optimum design

For example, the extracted biohuman material purity from one software to another and the yield of a software varies over multiple collection times. A CTQ can be cascaded at lower software design levels (system, subsystem, or component) where measurement is possible and feasible to functional requirements (FRs). At the software level, the CTQs can be derived from all customer segment wants, needs, and delights, which are then cascaded to functional requirements, the outputs at the various hierarchical levels.

Software variables can be quantitative or qualitative. Quantitative variables are measured numerically in a discrete or a continuous manner, whereas qualitative variables are measured in a descriptive manner. For example, the memory size of software is a quantitative variable, wherease the ease of use can be looked at as a qualitative variable. Variables also are dependent and independent. Variables such as passed arguments of a called function are independent variables, whereas function-calculated outcomes are dependent variables. Finally, variables are either continuous or discrete. A continuous variable is one for which any value is possible within the limits of the variable ranges. For example, the time spent on developing a DFSS project (in man-hours) is a continuous variable because it can take real values between an acceptable minimum and 100%. The variable "Six Sigma Project ID" is a discrete variable because it only can take countable integer values such as 1, 2, 3. . ., etc. It is clear that statistics computed from continuous variables have many more possible values than the discrete variables themselves.

The word "statistics" is used in several different senses. In the broadest sense, "statistics" refers to a range of techniques and procedures for analyzing data,

TABLE 6.3 Examples of Parameters and Statistics

Measure	Parameter	Statistics
Mean	μ	$\overline{X}$
Standard deviation	Σ	s
Proportion	Π	p
Correlation	ρ	r

interpreting data, displaying data, and making decisions based on data. The term "statistic" refers to the numerical quantity calculated from a sample of size n. Such statistics are used for parameter estimation.

In analyzing outputs, it also is essential to distinguish between statistics and parameters. Although statistics are measured from data samples of limited size (n), a parameter is a numerical quantity that measures some aspect of the data population. Population consists of an entire set of objects, observations, or scores that have something in common. The distribution of a population can be described by several parameters such as the mean and the standard deviation. Estimates of these parameters taken from a sample are called statistics. A sample is, therefore, a subset of a population. As it usually is impractical to test every member of a population (e.g., 100% execution of all feasible verification test scenarios), a sample from the population is typically the best approach available. For example, the mean time between failures (MTBF) in 10 months of run time is a "statistics," whereas the MTBF mean over the software life cycle is a parameter. Population parameters rarely are known and usually are estimated by statistics computed using samples. Certain statistical requirements are, however, necessary to estimate the population parameters using computed statistics. Table 6.3 shows examples of selected parameters and statistics.

6.3.1 Descriptive Statistics

One important use of statistics is to summarize a collection of data in a clear and understandable way. Data can be summarized numerically and graphically. In numerical approach, a set of descriptive statistics are computed using a set of formulas. These statistics convey information about the data's central tendency measures (mean, median, and mode) and dispersion measures (range, interquartiles, variance, and standard deviation). Using the descriptive statistics, data central and dispersion tendencies are represented graphically (such as dot plots, histograms, probability density functions, steam and leaf, and box plot).

For example, a sample of an operating system CPU usage (in %) is depicted in Table 6.4 for some time. The changing usage reflects the variability of this variable that typically is caused by elements of randomness in current running processes, services, and background code of the operating system performance.

The graphical representations of usage as an output help to understand the distribution and the behavior of such a variable. For example, a histogram representation can be established by drawing the intervals of data points versus each interval's frequency

TABLE 6.4 CPU Usage (in %)

55	52	55	52	50	55	52	49	55	52
48	45	42	39	36	48	45	48	48	45
65	62	59	56	53	50	47	44	41	38
49	46	43	40	37	34	31	28	25	22
64	61	64	61	64	64	61	64	64	61
63	60	63	58	63	63	60	66	63	63
60	57	54	51	60	44	41	60	63	50
65	62	65	62	65	65	62	65	66	65
46	43	46	43	46	46	43	46	63	46
56	53	56	53	56	56	53	56	60	66

of occurrence. The probability density function (pdf) curve can be constructed and added to the graph by connecting the centers of data intervals. Histograms help in selecting the proper distribution that represents simulation data. Figure 6.2 shows the histogram and normal curve of the data in Table 6.4 as obtained from Minitab (Minitab Inc., PA, USA). Figure 6.4 also displays some useful statistics about the *central tendency*, *skewness*, *dispersion (variation)*, and *distribution fitness to normality*.

Several other types of graphical representation can be used to summarize and represent the distribution of a certain variable. For example, Figures 6.3 and 6.4 show another two types of graphical representation of the yield requirement design output using the box plot and dot plot, respectively.

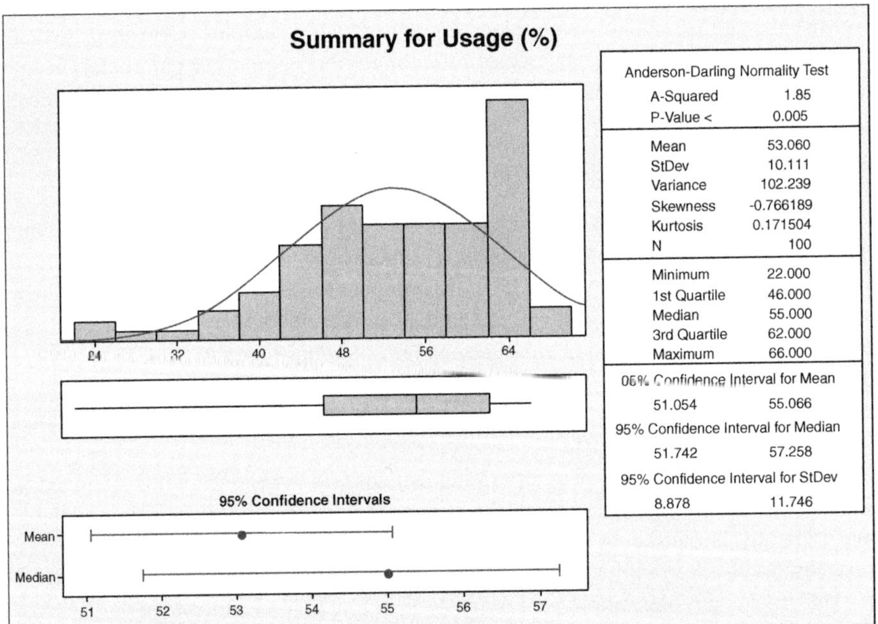

FIGURE 6.2 Histogram and normal curve of data in Table 6.4.

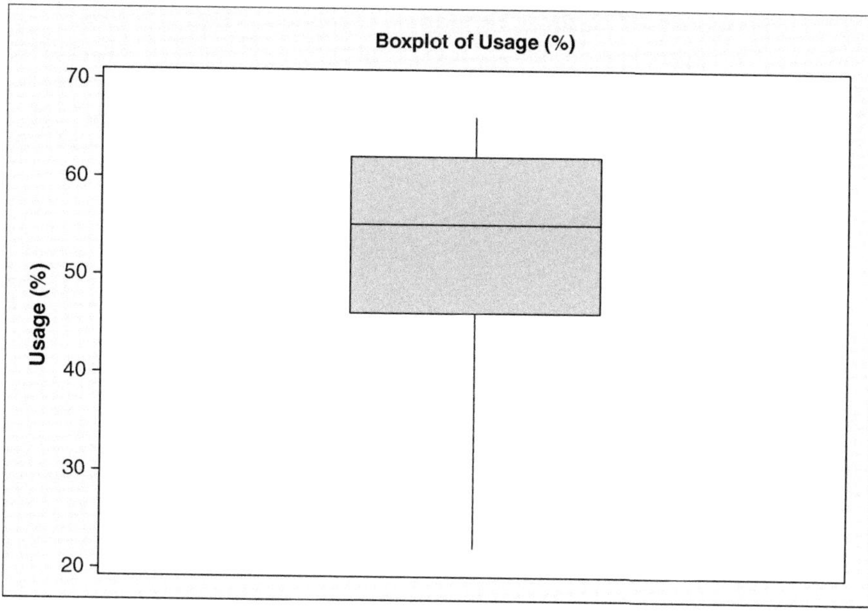

FIGURE 6.3 Box plot of usage data in Table 6.4.

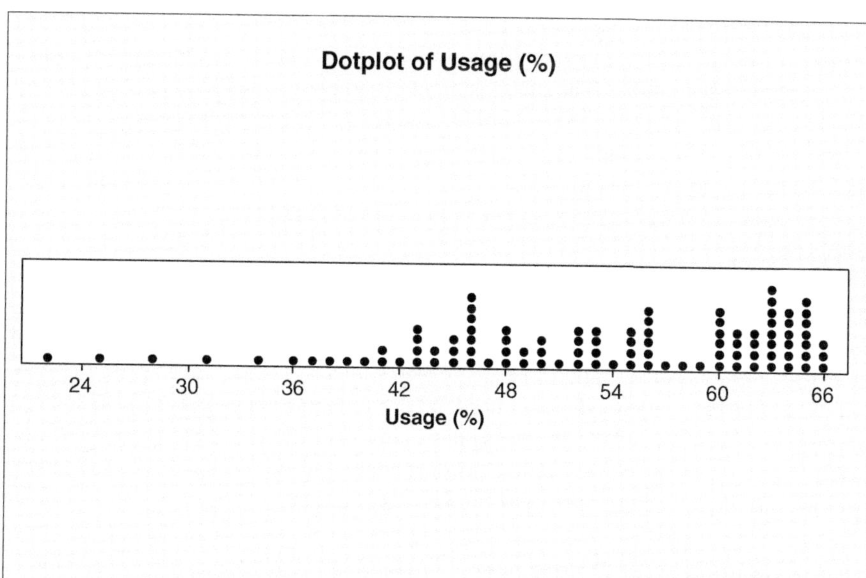

FIGURE 6.4 Dot plot of usage data in Table 6.4.

6.3.1.1 *Measures of Central Tendency.* Measures of central tendency are measures of the location of the middle or the center of a distribution of a functional requirement variable (denoted as y). The mean is the most commonly used measure of central tendency. The arithmetic mean is what is commonly called the average. The mean is the sum of all the observation divided by the number of observations in a sample or in a population:

The mean of a population is expressed mathematically as:

$$\mu_y = \frac{\sum_{i=1}^{N} y_i}{N}$$

where N is the number of population observations.

The average of a sample is expressed mathematically as:

$$\bar{y} = \frac{\sum_{i=1}^{n} y_i}{n}$$

where n is the sample size.

The mean is a good measure of central tendency for roughly *symmetric* distributions but can be misleading in *skewed* distributions because it can be influenced greatly by extreme observations. Therefore, other statistics such as the median and mode may be more informative for skewed distributions. The mean, median, and mode are equal in symmetric distributions. The mean is higher than the median in *positively skewed* distributions and lower than the median in *negatively skewed* distributions.

The median is the middle of a distribution where half the scores are above the median and half are below the median. The median is less sensitive to extreme scores than the mean, and this makes it a better measure than the mean for highly skewed distributions.

The mode is the most frequently occurring score in a distribution. The advantage of the mode as a measure of central tendency is that it has an obvious meaning. Furthermore, it is the only measure of central tendency that can be used with nominal data (it is not computed). The mode is greatly subject to sample fluctuation and is, therefore, not recommended to be used as the only measure of central tendency. Another disadvantage of the mode is that many distributions have more than one mode. These distributions are called "multimodal." Figure 6.5 illustrates the mean, median, and mode in symmetric and skewed distributions.

6.3.1.2 *Measures of Dispersion.* A functional requirement (FR $= y$) dispersion is the degree to which scores on the FR variable differ from each other. "Variability" and "spread" are synonyms for dispersion. There are many measures of spread. The range (R) is the simplest measure of dispersion. It is equal to the difference between the largest and the smallest values. The range can be a useful measure of

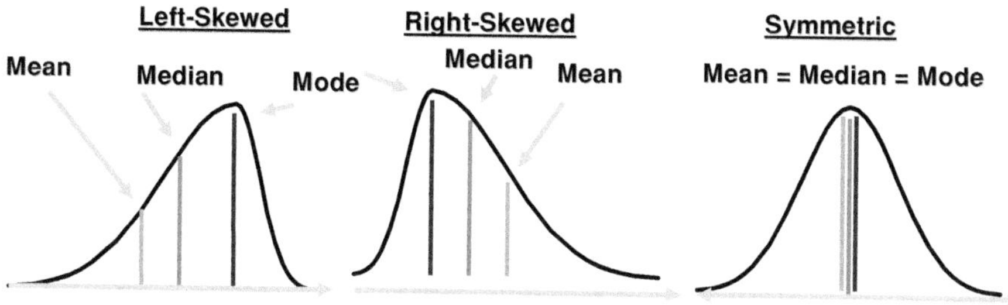

FIGURE 6.5 Symmetric and skewed distribution.

spread because it is understood so easily. However, it is very sensitive to extreme scores because it is based on only two values. The range should almost never be used as the only measure of spread, but it can be informative if used as a supplement to other measures of spread such as the standard deviation and interquartile range. For example, the range is determined for the following "y" sample as follows:

$$[10, 12, 4, 6, 13, 15, 19, 16]$$

$$\begin{aligned} Ry &= \text{Max}[10, 12, 4, 6, 13, 15, 19, 16] - \text{Min}[10, 12, 4, 6, 13, 15, 19, 16] \\ &= 19 - 4 = 15 \end{aligned} \tag{6.1}$$

The range is a useful statistic to know but not as a stand-alone dispersion measure because it takes into account only two scores.

The variance is a measure of the spreading out of a distribution. It is computed as the average squared deviation of each number from its mean. Formulas for the variance are as follows.

For a population:

$$\sigma_y^2 = \frac{\sum_{i=1}^{N} (y_i - \mu_y)^2}{N} \tag{6.2}$$

where N is the number of population observations

For a sample:

$$s_y^2 = \frac{\sum_{i=1}^{n} (y_i - \bar{y})^2}{n - 1} \tag{6.3}$$

where n is the sample size

The standard deviation is the measure of dispersion most commonly used. The formula for the standard deviation is the square root of the variance. An important

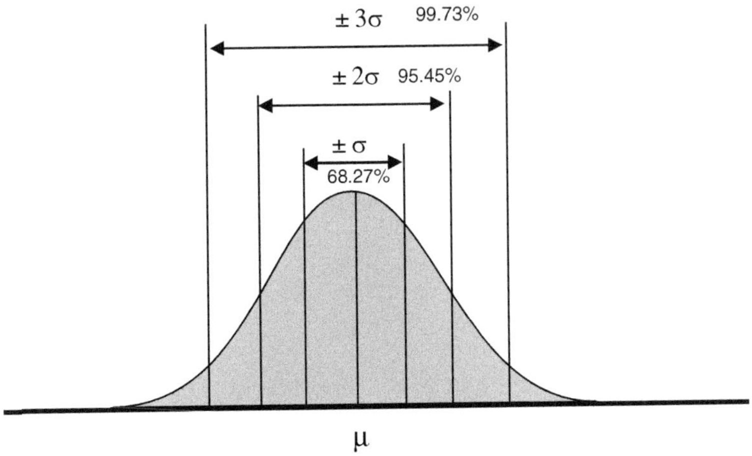

FIGURE 6.6 Normal distribution curve.

attribute of the standard deviation is that if the mean and standard deviation of a normal distribution are known, it is possible to compute the percentile rank associated with any given observation. For example, the empirical rule states that in a normal distribution, approximately 68.27% of the data points are within 1 standard deviation of the mean, approximately 95.45% of the data points are within 2 standard deviations of the mean, and approximately 99.73% of the data points are within 3 standard deviations of the mean. Figure 6.6 illustrates the normal distribution curve percentage data points contained within several standard deviations from the mean.

The standard deviation often is not considered a good measure of spread in highly skewed distributions and should be supplemented in those cases by the interquartile range (Q_3–Q_1). The interquartile range rarely is used as a measure of spread because it is not very mathematically tractable. However, it is less sensitive to extreme data points than the standard deviation, and subsequently, it is less subject to sampling fluctuations in highly skewed distributions.

For the data set shown in Table 6.4, a set of descriptive statistics, shown in Table 6.5, is computed using a Microsoft Excel (Microsoft Corporation, Redmond, WA) sheet to summarize the behavior of y = "Usage" data in Table 6.4.

6.4 INFERENTIAL STATISTICS

Inferential statistics are used to draw inferences about a population from a sample on n observations. Inferential statistics generally require that sampling be both random and representative. Observations are selected by randomly choosing the sample that resembles the population's functional requirement. This can be obtained as follows:

1. A sample is random if the method for obtaining the sample meets the criterion of randomness (each item or element of the population having an equal chance of

TABLE 6.5 Descriptive Statistics Summary for Data in Table 6.4 (%)

Mean	53.06
Standard error	1.01
Median	55
Mode	63
Standard deviation	10.11
Sample variance	102.24
Range	44
Minimum	22
Maximum	66
First quartile (IQ_1)	46
Third quartile (IQ_3)	62
Interquartile range	16
Count	100
Sum	5306

A typical Minitab descriptive statistics command will produce the following:

Descriptive Statistics: Usage (%)

```
Variable    N     N*  Mean   SE Mean  StDev  Minimum  Q1     Median  Q3
Usage(%)    100   0   53.06  1.01     10.11  22.00    46.00  55.00   62.00

Variable  Maximum
Usage(%)    66.00
```

being chosen). Hence, random numbers typically are generated from a uniform distribution $U\,[a, b]$.[4]

2. Samples are drawn independently with no sequence, correlation, or autocorrelation between consecutive observations.
3. The sample size is large enough to be representative, usually $n \geq 30$.

The two main methods used in inferential statistics are parameter estimation and hypothesis testing.

6.4.1 Parameter Estimation

In estimation, a sample is used to estimate a parameter and to construct a confidence interval around the estimated parameter. Point estimates are used to estimate the parameter of interest. The mean (μ_y) and standard deviation (σ_y) are the most common point estimates. As discussed, the population mean (μ_y) and standard deviation (σ_y) are estimated using sample average ($\bar{y}$) and standard deviation (s_y), respectively.

[4]The continuous uniform distribution is a family of probability distributions such that for each member of the family, all intervals of the same length on the distribution's support are equally probable. The support is defined by the two parameters, a and b, which are its minimum and maximum values. The distribution is often abbreviated $U[a, b]$.

A point estimate, by itself, does not provide enough information regarding variability encompassed into the simulation response (output measure). This variability represents the differences between the point estimates and the population parameters. Hence, an interval estimate in terms of a confidence interval is constructed using the estimated average ($\bar{y}$) and standard deviation (s_y). A confidence interval is a range of values that has a high probability of containing the parameter being estimated. For example, the 95% confidence interval is constructed in such a way that the probability that the estimated parameter is contained with the lower and upper limits of the interval is of 95%. Similarly, 99% is the probability that the 99% confidence interval contains the parameter.

The confidence interval is symmetric about the sample mean $\bar{y}$. If the parameter being estimated is μ_y, for example, the 95% confidence interval (CI) constructed around an average of $\bar{y} = 28.0\%$ is expressed as follows:

$$25.5\% \leq \mu_y \leq 30.5\%$$

this means that we can be 95% confident that the unknown performance mean (μ_y) falls within the interval [25.5%, 30.5%].

Three statistical assumptions must be met in a sample of data to be used in constructing the confidence interval. That is, the data points should be normally, independent, and identically distributed. The following formula typically is used to compute the CI for a given significance level (α):

$$\bar{y} - t_{\alpha/2,\, n-1}\, s/\sqrt{n} \leq \mu \leq \bar{y} + t_{\alpha/2,\, n-1}\, s/\sqrt{n} \tag{6.4}$$

where $\bar{y}$ is the average of multiple data points, t_{n-1}, $\alpha/2$ is a value from the Student t distribution[5] for an α level of significance.

For example, using the data in Table 6.4, Figure 6.2 shows a summary of both graphical and descriptive statistics along with the computed 95% CI for the mean, median, and standard deviation. The graph is created with Minitab statistical software.

The normality assumption can be met by increasing the sample size (n) so that the central limit theorem (CLT) is applied. Each average performance $\bar{y}$ (average "Usage," for example) is determined by summing together individual performance values (y_1, y_2, ..., y_n) and by dividing them by n. The CLT states that the variable representing the sum of several independent and identically distributed random values tends to be normally distributed. Because (y_1, y_2, . . ., y_n) are not independent and identically distributed, the CLT for correlated data suggests that the average performance ($\bar{y}$) will be approximately normal if the sample size (n) used to compute $\bar{y}$ is large, $n \geq 30$. The 100%(1 − α) confidence interval on the true population mean is expressed

[5]A probability distribution that originates in the problem of estimating the mean of a normally distributed population when the sample size is small. It is the basis of the popular Students t tests for the statistical significance of the difference between two sample means, and for confidence intervals for the difference between two population means.

as follows:

$$\bar{y} - Z_{\alpha/2}\,\sigma/\sqrt{n} \leq \mu \leq \bar{y} + Z_{\alpha/2}\sigma/\sqrt{n} \tag{6.5}$$

6.4.1.1 Hypothesis Testing. Hypothesis testing is a method of inferential statistics that is aimed at testing the viability of a null hypothesis about a certain population parameter based on some experimental data. It is common to put forward the null hypothesis and to determine whether the available data are strong enough to reject it. The null hypothesis is rejected when the sample data are very different from what would be expected under a true null hypothesis assumption. It should be noticed, however, that failure to reject the null hypothesis is not the same thing as accepting the null hypothesis.

In Six Sigma, hypothesis testing primarily is used for making comparisons. Two or more software packages can be compared with the goal of identifying the superior design alternative relative to some functional requirement performance. In testing a hypothesis, the null hypothesis often is defined to be the reverse of what the team actually believes about the performance. Thus, the collected data are used to contradict the null hypothesis, which may result in its rejection. For example, if the design team has proposed a new design alternative, team members would be interested in testing experimentally whether the proposed design works better than the current baseline. To this end, the team would design an experiment comparing the two packages. The Usage of both software packages could be collected and used as data for testing the viability of the null hypothesis. The null hypothesis would be, for example, that there is no difference between the CPU usage of the two packages (i.e., the usage population means of the two population μ_1 and μ_2 are identical). In such a case, the software DFSS team would be hoping to reject the null hypothesis and conclude that the new proposed software developed is a better one.

The symbol H_0 is used to indicate the null hypothesis, where "null" refers to the hypothesis of no difference. This is expressed as follows:

$$H_0\colon \mu_1 - \mu_2 = 0 \text{ or } H_0\colon \mu_1 = \mu_2$$

The alternative hypothesis (H_1 or H_a) simply is set to state that the mean usage (%) of the proposed package (μ_1) is higher than that of the current baseline (μ_2). That is:

$$H_a\colon \mu_1 - \mu_2 > 0 \text{ or } H_a\colon \mu_1 > \mu_2$$

Although H_0 is called the "null hypothesis," there are occasions when the parameter of interest is not hypothesized to be 0. For instance, it is possible for the null hypothesis to be that the difference (d) between population means is of a particular value ($H_0\colon \mu_1 - \mu_2 = d$). Or, the null hypothesis could be that the population mean is of a certain value ($H_0\colon \mu = \mu_0$).

The used test statistics in hypothesis testing depends on the hypothesized parameter and the data collected. In practical comparison studies, most tests involve comparisons of a mean performance with a certain value or with another software mean. When the variance (σ^2) is known, which rarely is the case in real-world applications, Z_0 is used as a test statistic for the null hypothesis H_0: $\mu = \mu_0$, assuming that the observed population is normal or the sample size is large enough so that the CLT applies. Z_0 is computed as follows:

$$Z_0 = \frac{\bar{y} - \mu_0}{\sigma/\sqrt{n}} \tag{6.6}$$

The null hypothesis H_0: $\mu = \mu_0$ would be rejected if $|Z_0| > Z_{\alpha/2}$ when H_a: $\mu \neq \mu_0$, $Z_0 < -Z_\alpha$ when H_a: $\mu < \mu_0$, and $Z_0 > Z_\alpha$ when H_a: $\mu > \mu_0$.

Depending on the test situation, several test statistics, distributions, and comparison methods also can be used at several hypothesis tests. Let us look at some examples.

For the null hypothesis, H_0: $\mu_1 = \mu_2$, Z_0 is computed as follows:

$$Z_0 = \frac{\bar{y}_1 - \bar{y}_2}{\sqrt{\dfrac{\sigma 1^2}{n_1} + \dfrac{\sigma 2^2}{n_2}}} \tag{6.7}$$

The null hypothesis H_0: $\mu_1 = \mu_2$ would be rejected if $|Z_0| > Z_{\alpha/2}$ when H_a: $\mu_1 \neq \mu_2$, $Z_0 < -Z_\alpha$ when H_a: $\mu_1 < \mu_2$, and $Z_0 > Z_\alpha$ when H_a: $\mu_1 > \mu_2$.

When the variance (σ^2) is unknown, which is typically the case in real-world applications, t_0 is used as a test statistic for the null hypothesis H_0: $\mu = \mu_0$ and t_0 is computed as follows:

$$t_0 = \frac{\bar{y} - \mu_0}{s/\sqrt{n}} \tag{6.8}$$

The null hypothesis H_0: $\mu = \mu_0$ would be rejected if $|t_0| > t_{\alpha/2,\, n-1}$ when H_a: $\mu \neq \mu_0$, $t_0 < -t_{\alpha,\, n-1}$ when H_a: $\mu < \mu_0$, and $t_0 > t_{\alpha,\, n-1}$ when H_a: $\mu > \mu_0$.

For the null hypothesis H_0: $\mu_1 = \mu_2$, t_0 is computed as:

$$t_0 = \frac{\bar{y}_1 - \bar{y}_2}{\sqrt{\dfrac{s1^2}{n_1} + \dfrac{s2^2}{n_2}}} \tag{6.9}$$

Similarly, the null hypothesis H_0: $\mu_1 = \mu_2$ would be rejected if $|t_0| > t_{\alpha/2,\, v}$ when H_a: $\mu_1 \neq \mu_2$, $t_0 < -t_{\alpha,\, v}$ when H_a: $\mu_1 < \mu_2$, and $t_0 > t_{\alpha, v}$ when H_a: $\mu_1 > \mu_2$, where $v = n_l + n_2 - 2$.

The discussed examples of null hypotheses involved the testing of hypotheses about one or more population means. Null hypotheses also can involve other parameters such as an experiment investigating the variance (σ^2) of two populations, the proportion (π), and the correlation (ρ) between two variables. For example, the correlation between project size and design effort on the job would test the null hypothesis that the population correlation (ρ) is 0. Symbolically, H_0: $\rho = 0$.

Sometimes it is required for the design team to compare more than two alternatives for a system design or an improvement plan with respect to a given performance measure. Most practical studies tackle this challenge by conducting multiple paired-comparisons using several paired-*t* confidence intervals, as discussed. Bonferroni's approach is another statistical approach for comparing more than two alternative software packages in some performance metric or a functional requirement. This approach also is based on computing confidence intervals to determine whether the true mean performance of a functional requirement of one system (μ_i) is significantly different from the true mean performance of another system (μ_i') in the same requirement. ANOVA is another advanced statistical method that often is used for comparing multiple alternative software systems. ANOVA's multiple comparison tests are used widely in experimental designs.

To draw the inference that the hypothesized value of the parameter is not the true value, a significance test is performed to determine whether an observed value of a statistic is sufficiently different from a hypothesized value of a parameter (null hypothesis). The significance test consists of calculating the probability of obtaining a sample statistic that differs from the null hypothesis value (given that the null hypothesis is correct). This probability is referred to as a *p* value. If this probability is sufficiently low, then the difference between the parameter and the statistic is considered to be "statistically significant." The probability of a Type I error (α) is called the significance level and is set by the experimenter. The significance level (α) commonly is set to 0.05 and 0.01. The significance level is used in hypothesis testing to:

- Determine the difference between the results of the statistical experiment and the null hypothesis.
- Assume that the null hypothesis is true.
- Compute the probability (*p* value) of the difference between the statistic of the experimental results and the null hypothesis.
- Compare the *p* value with the significance level (α). If the probability is less than or equal to the significance level, then the null hypothesis is rejected and the outcome is said to be statistically significant.

The lower the significance level, therefore, the more the data must diverge from the null hypothesis to be significant. Therefore, the 0.01 significance level is more conservative because it requires a stronger evidence to reject the null hypothesis then that of the 0.05 level.

Two kinds of errors can be made in significance testing: Type I error (α), where a true null hypothesis can be rejected, incorrectly and Type II error (β), where a false null hypothesis can be accepted incorrectly. A Type II error is only an error in the

TABLE 6.6 The Two Types of Test Errors

Statistical	True state of null hypothesis (H_0)	
Decision	H_0 is true	H_0 is false
Reject H_0	Type I error (α)	Correct
Accept H_0	Correct	Type II error (β)

sense that an opportunity to reject the null hypothesis correctly was lost. It is not an error in the sense that an incorrect conclusion was drawn because no conclusion is drawn when the null hypothesis is accepted. Table 6.6 summarized the two types of test errors.

A type I error generally is considered more serious than a Type II error because it results in drawing a conclusion that the null hypothesis is false when, in fact, it is true. The experimenter often makes a tradeoff between Type I and Type II errors. A software DFSS team protects itself against Type I errors by choosing a stringent significance level. This, however, increases the chance of a Type II error. Requiring very strong evidence to reject the null hypothesis makes it very unlikely that a true null hypothesis will be rejected. However, it increases the chance that a false null hypothesis will be accepted, thus lowering the hypothesis *test power*. Test power is the probability of correctly rejecting a false null hypothesis. Power is, therefore, defined as: $1 - \beta$, where β is the Type II error probability. If the power of an experiment is low, then there is a good chance that the experiment will be inconclusive. There are several methods for estimating the test power of an experiment. For example, to increase the test power, the team can be redesigned by changing one factor that determines the power, such as the sample size, the standard deviation (σ), and the size of difference between the means of the tested software packages.

6.4.2 Experimental Design

In practical Six Sigma projects, experimental design usually is a main objective for building the *transfer function model*. Transfer functions models are fundamentally built with an extensive effort spent on data collection, verification, and validation to provide a flexible platform for optimization and tradeoffs. Experimentation can be done in hardware and software environments.

Software experimental testing is any activity aimed at evaluating an attribute or capability of a program or system and at determining that it meets its required results. The difficulty in software testing stems from the complexity of software. Software experimental testing is more than just debugging. The purpose of testing can be quality assurance, verification and validation, or reliability estimation. Testing can be used as a generic metric as well. Correctness testing and reliability testing are two major areas of testing. Software testing is a tradeoff among budget, time, and quality.

Experimenting in a software environment is a typical practice for estimating performance under various running conditions, conducting "what-if" analysis, testing hypothesis, comparing alternatives, factorial design, and optimization. The results of such experiments and methods of analysis provide the DFSS team with insight, data, and necessary information for making decisions, allocating resources, and setting optimization strategies.

An experimental design is a plan that is based on a systematic and efficient application of certain treatments to an experimental unit or subject, an object, or a source code. Being a flexible and efficient experimenting platform, the experimentation environment (hardware or software) represents the subject of experimentation at which different treatments (factorial combinations) are applied systematically and efficiently. The planned treatments may include both structural and parametric changes applied to the software. Structural changes include altering the type and configuration of hardware elements, the logic and flow of software entities, and the structure of the software configuration. Examples include adding a new object-oriented component, changing the sequence of software operation, changing the concentration or the flow, and so on. Parametric changes, however, include making adjustments to software size, complexity, arguments passed to functions or calculated from such functions, and so on.

In many applications, parameter design is more common in software experimental design than that of structural experimental design. In practical applications, DFSS teams often adopt a certain concept structure and then use the experimentation to optimize its functional requirement (FR) performance. Hence, in most designed experiments, design parameters are defined as decision variables and the experiment is set to receive and run at different levels of these decision variables in order to study their impact on certain software functionality, an FR. Partial or full factorial design is used for two purposes:

- Finding those design parameters (variables) of greatest significance on the system performance.
- Determining the levels of parameter settings at which the *best* performance level is obtained. Direction of goodness (i.e., best) performance can be maximizing, minimizing, or meeting a preset target of a functional requirement.

The success of experimental design techniques is highly dependent on providing an efficient experiment setup. This includes the appropriate selection of design parameters, functional requirements, experimentation levels of the parameters, and number of experimental runs required. To avoid conducting a large number of experiments, especially when the number of parameters (a.k.a. factors in design of experiment terminology) is large, certain experimental design techniques can be used. An example of such handling includes using screening runs to designate insignificant design parameters while optimizing the software system.

Experimental design, when coupled with software available testing tools and techniques, is very insightful. An abundance of software testing tools exist. The correctness testing tools often are specialized to certain systems and have limited ability and generality. Robustness and stress testing tools are more likely to be made generic. Mothora (DeMillo, 1991) is an automated mutation testing tool set developed at Purdue University. Using Mothora, the tester can create and execute test cases, measure test case adequacy, determine input–output transfer function correctness, locate and remove faults or bugs, and control and document the test. For run-time checking and debugging aids, you can use NuMega's Boundschecker[6] or Rational's Purify.[7] Both can both check and protect against memory leaks and pointer problems. Ballista COTS Software Robustness Testing Harness[8] is a full-scale automated robustness testing tool. The first version supports testing up to 233 POSIX[9] function calls in UNIX operating systems. The second version also supports testing of user functions provided that the data types are recognized by the testing server. The Ballista testing harness gives quantitative measures of robustness comparisons across operating systems. The goal is to test automatically and to harden commercial off-the-shelf (COTS) software against robustness failures.

In experimental design, decision variables are referred to as *factors* and the output measures are referred to as *response, software metric* (e.g., complexity), or *functional requirement* (e.g., GUI). Factors often are classified into *control* and *noise* factors. Control factors are within the control of the design team, whereas noise factors are imposed by operating conditions and other internal or external uncontrollable factors. The objective of software experiments usually is to determine settings to the software control factors so that software response is optimized and system random (noise) factors have the least impact on system response. You will read more about the setup and analysis of designed experiments in the following chapters.

6.5 A NOTE ON NORMAL DISTRIBUTION AND NORMALITY ASSUMPTION

Normal distribution is used in different domains of knowledge, and as such, it is standardized to avoid the taxing effort of generating specialized statistical tables. A standard normal has a mean of 0 and a standard deviation of 1, and functional requirement, y, values are converted into Z-scores or Sigma levels using $Z_i = \frac{(y_i - \mu)}{\sigma}$ transformation. A property of the normal distribution is that 68% of all of its observations fall within a range of ± 1 standard deviation from the mean, and a range of ± 2 standard deviations includes 95% of the scores. In other words, in a normal distribution, observations that have a Z-score (or Sigma value) of less than -2 or more than $+2$ have a relative frequency of 5% or less. A Z-core value means that a value is expressed in terms of its difference from the mean, divided by the standard deviation. If you have access to statistical software such as Minitab, you can explore the exact values of probability associated with different values in the normal distribution using the Probability Calculator tool; for example, if you enter the Z value (i.e., standardized value) of 4, the associated probability computed will be less than

[6] http://www.numega.com/devcenter/bc.shtml.
[7] http://www.rational.com/products/purify_unix/index.jtmpl.
[8] http://www.cs.cmu.edu/afs/cs/project/edrc-ballista/www/.
[9] POSIX (pronounced/pvziks/) or "Portable Operating System Interface [for Unix]".

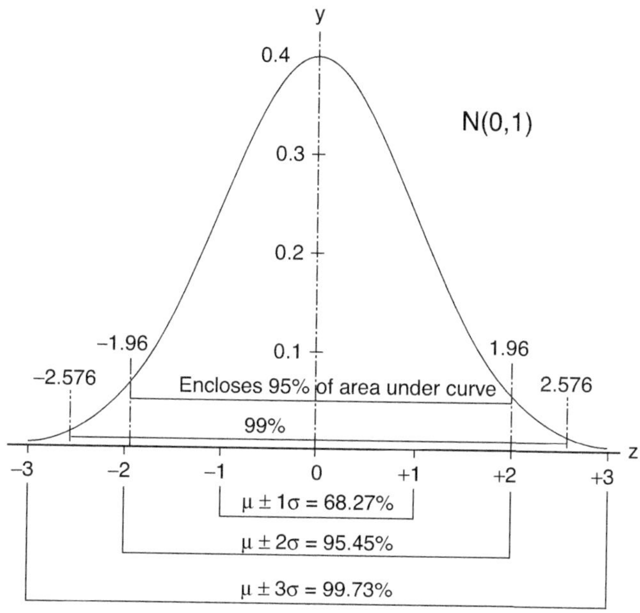

FIGURE 6.7 The standardized normal distribution $N(0,1)$ and its properties.

0.0001, because in the normal distribution, almost all observations (i.e., more than 99.99%) fall within the range of ±4 standard deviations. A population of measurements with normal or Gaussian distribution will have 68.3% of the population within $\pm 1\sigma$, 95.4% within $\pm 2\sigma$, 99.7% within $\pm 3\sigma$, and 99.9% within $\pm 4\sigma$ (Figure 6.7).

The normal distribution is used extensively in statistical reasoning (induction), the so-called inferential statistics. If the sample size is large enough, the results of randomly selecting sample candidates and measuring a response or FR of interest is "normally distributed," and thus knowing the shape of the normal curve, we can calculate precisely the probability of obtaining "by chance" FR outcomes representing various levels of deviation from the hypothetical population mean of zero.

In hypothesis testing, if such a calculated probability is so low that it meets the previously accepted criterion of statistical significance, then we only have one choice: conclude that our result gives a better approximation of what is going on in the population than the "null hypothesis." Note that this entire reasoning is based on the assumption that the shape of the distribution of those "data points" (technically, the "sampling distribution") is normal.

Are all test statistics normally distributed? Not all, but most of them are either based on the normal distribution directly or on distributions that are related to, and can be derived from, normal, such as Students t, Fishers F, or chi-square. Typically, those tests require that the variables analyzed are normally distributed in the population; that is, they meet the so-called "normality assumption." Many observed variables actually are normally distributed, which is another reason why the normal distribution represents a "general feature" of empirical reality. The problem may occur when one tries to use a normal-distribution-based test to analyze data from variables that are not normally distributed. In such cases, we have two general choices. First, we can use some alternative "nonparametric" test (a.k.a. "distribution-free test"), but this often is inconvenient because such tests typically are less powerful and less flexible in terms of types of conclusions that they can provide. Alternatively, in many cases we can still use the normal-distribution-based test if we only make sure that the size of our samples is large enough. The latter option is based on an extremely important principle, which is largely responsible for the popularity of tests that are based on the normal function. Namely, as the sample size increases, the shape of the sampling distribution (i.e., distribution of a statistic from the sample; this term was first used by Fisher, 1928) approaches normal shape, even if the distribution of the variable in question is not normal.

However, as the sample size (of samples used to create the sampling distribution of the mean) increases, the shape of the sampling distribution becomes normal. Note that for $n = 30$, the shape of that distribution is "almost" perfectly normal. This principle is called the central limit theorem (this term was first used by Pólya in 1920).

6.5.1 Violating the Normality Assumption

How do we know the consequences of violating the normality assumption? Although many statements made in the preceding paragraphs can be proven mathematically, some of them do not have theoretical proofs and can be demonstrated only empirically, via so-called Monte Carlo experiments. In these experiments, large numbers of samples are generated by a computer following predesigned specifications, and the results from such samples are analyzed using a variety of tests. This way we can evaluate empirically the type and magnitude of errors or biases to which we are exposed when certain theoretical assumptions of the tests we are using are not met by our data. Specifically, Monte Carlo studies were used extensively with normal-distribution-based tests to determine how sensitive they are to violations of the assumption of normal distribution of the analyzed variables in the population. The general conclusion from these studies is that the consequences of such violations are less severe than previously thought. Although these conclusions should not entirely discourage anyone from being concerned about the normality assumption, they have increased the overall popularity of the distribution-dependent statistical tests in many areas.

6.6 SUMMARY

In this chapter, we have given a very basic review of appropriate statistical terms and methods that are encountered in this book. We reviewed collection, classification, summarization, organization, analysis, and interpretation of data. We covered with examples both descriptive and inferential statistics. A practical view of common probability distributions, modeling, and statistical methods was discussed in the chapter.

We expressed the criticality of understanding hypothesis testing and discussed examples of null hypotheses involving testing of hypotheses about one or more population means. Next we moved into an explanation of ANOVA and types of test errors, Type I and Type II errors.

Experimental design and its objective in building the transfer function model were explained. Normal distribution and normality assumption were explained, and an answer to how we know the consequences of violating the normality assumption was discussed.

REFERENCES

CMMI Development Team (2001), Capability Maturity Model—Integrated, Version 1.1, Software Engineering Institute, Pittsburgh, PA.

Demillo, R.A. (1991), "Progress Toward Automated Software Testing," Proceedings of the 13th International Conference on Software Engineering, p. 180.

Emam, K. and Card, D. (Eds.) (2002), ISO/IEC Std 15939, Software Measurement Process.

CHAPTER 7

SIX SIGMA FUNDAMENTALS

7.1 INTRODUCTION

Through out the evolution of quality there has always been on manufacturing industry (the production of hardware parts). In recent years, more application has focused on process in general; however, the application of a full suite of tools to nonmanufacturing industries is rare and still considered risky or challenging. Only companies that have mature Six Sigma deployment programs see the application of Design for Six Sigma (DFSS) to information technology (IT) applications and software development as an investment rather than as a needless expense. Even those companies that embark on DFSS seem to struggle with confusion over the DFSS "process" and the process being designed.

Multiple business processes can benefit from DFSS. Some of these are listed in Table 7.1.

If properly measured, we would find that few if any of these processes perform at Six Sigma performance levels. The cost, timeliness, or quality (accuracy and completeness) are never where they should be and hardly world class from customer perspectives.

Customers may be internal or external; if it is external, the term "consumer" (or end user) will be used for clarification purposes. Six Sigma is process oriented, and a short review of process and transaction may be beneficial at this stage. Some processes (e.g., dry cleaning) consist of a single process, whereas many services consist of several processes linked together. At each process, transactions occur. A

Software Design for Six Sigma: A Roadmap for Excellence, By Basem El-Haik and Adnan Shaout

TABLE 7.1 Examples of Organizational Functions

Marketing	Sales	HR	Design
• Brand Management • Prospect	• Discovery • Account Management	• Staffing • Training	• Change Control • New Product
Production Control	**Sourcing**	**Information Technology**	**Finance**
• Inventory Control • Scheduling	• Commodity • Purchasing	• Help Desk • Training	• Accounts Payable • Accounts Receivable

transaction is the simplest process step and typically consists of an input, procedures, resources, and a resulting output. The resources can be people or machines, and the procedures can be written, learned, or even digitized in software code. It is important to understand that some processes are enablers to other processes, whereas some provide their output to the end customer. For example, the transaction centered around the principal activities of an order-entry environment include transactions such as entering and delivering orders, recording payments, checking the status of orders, and monitoring the stock levels at the warehouse. Processes may involve a mixture of concurrent transactions of different types and complexity either executed online or queued for deferred execution. In a real-time operating system, real-time transactions in memory management, peripheral communication [input/output(I/O)], task management and so on. are transactions within their repective processes and processors.

We experience processes, which spans the range from ad hoc to designed.[1] Our experience indicates that most processes are ad hoc and have no metrics associated with them and that many consist solely of a person with a goal and objectives. These processes have large variation in their perceived quality and are very difficult to improve. It is akin to building a house on a poor foundation.

Processes affect almost every aspect of our life. There are restaurant, health-care, financial, transportation, software, entertainment, and hospitality, processes, and they all have the same elements in common. Processes can be modeled, analyzed, and improved using simulation and other IT applications.

In this chapter we will cover an overview of Six Sigma and its development as well as the traditional deployment for process/product improvement called DMAIC and its components. The DMAIC platform also is referenced in several forthcoming chapters. The focus in this chapter is on the details of Six Sigma DMAIC methodology, value stream mapping (VSM) and lean manufacturing techniques, and the synergy and benefits of implementing a Lean Six Sigma (LSS) system.

[1] See software development classification in Section 2.1.1.

Because of similarity between software development and transaction-based applications, we will start introducing concepts in transaction-based Six Sigma as an introduction to software Six Sigma and software Design for Six Sigma in what follows. Where we see fit, we start merging concepts and define interfaces between transaction-based and software Six Sigma applications.

7.2 WHY SIX SIGMA?

Typically, the answer is purely and simply economic. Customers are demanding it. They want components and systems that work the first time and every time. A company that cannot provide ever increasing levels of quality, along with competitive pricing, is headed out of business. There are two ways to get quality in a product. One is to test exhaustively every product headed for the shipping dock, 100% inspection. Those that do not pass are sent back for rework, retest, or scrap. And rework can introduce new faults, which only sends product back through the rework loop once again. Make no mistake, much of this test, and all of the rework, are overhead. They cost money but do not contribute to the overall productivity. The other approach to quality is to build every product perfectly the first time and provide only a minimal test, if any at all. This would drive the reject rate so low that those units not meeting specification are treated as disposable scrap. It does involve cost in training, in process equipment, and in developing partnerships with customers *and* suppliers. But in the long run, the investments here will pay off, eliminating excessive test and the entire rework infrastructure releases resources for truly productive tasks. Overhead goes down, productivity goes up, costs come down, and pricing stays competitive.

Before diving into Six Sigma terminology, a main enemy threatening any development process should be agreed upon: *Variation*. The main target of Six Sigma is to minimize variation because it is somehow impossible to eliminate it totally. Sigma (σ), as shown in Figure 7.1, in the statisical field is a metric used to represent the

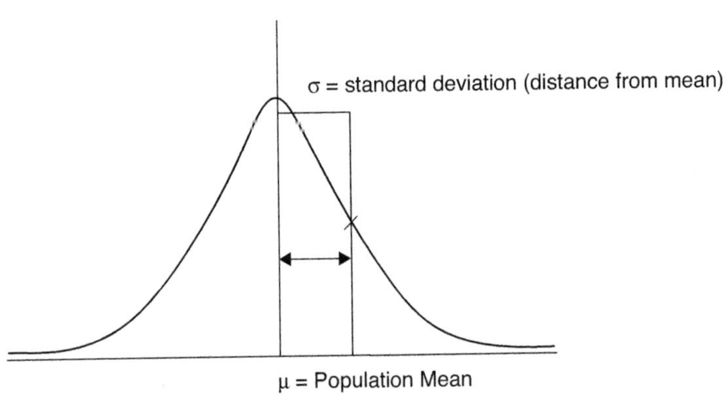

FIGURE 7.1 Standard deviation and population mean.

TABLE 7.2 Sigma Scale

Sigma	DPMO	Efficiency (%)
1	691,462	30.9
2	308,538	69.1
3	66,807	93.3
4	6,210	99.4
5	233	99.98
6	3.4	99.9999966

distance in standard deviation units from the mean to a specific limit. Six Sigma is a representation of 6 standard deviations from the distribution mean. But what does this mean? What is the diffence between 6 sigma and 4 sigma or 3 sigma? Six Sigma is almost defect free: "If a process is described as within six sigma, the term quantitatively means that the process produces fewer than 3.4 defects per million opportunities (DPMO). That represents an error rate of 0.0003%; conversely, that is a defect-free rate of 99.9999966% (Wikipedia Contributors, 2009; Section: Holistic Overview, para 5)." However, Four Sigma is 99.4% good or 6,210 DPMO (Siviy et al., 2007). This does not sound like a big difference; however, those are defects that will be encountered and noticed by the customers and will reduce their satisfaction. So to point out briefly why a Six Sigma quality level is important is simple; this company will definitely be saving money, unlike most companies who operate at a lower sigma level and bear a considerable amount of losses resulting from the cost of poor quality, known as COPQ. Table 7.2 shows how exponential the sigma scale is between levels 1 and 6.

7.3 WHAT IS SIX SIGMA?

We all use services and interact with processes each day. When was the last time you remember feeling really good about a transaction or a service you experienced? What about the last poor service you received? It usually is easier for us to remember the painful and dissatisfying experiences than it is to remember the good ones. One of the authors recalls sending a first-class registered letter, and after eight business days, he still could not see that the letter was received so he called the postal service provider's toll-free number and had a very professional and caring experience. It is a shame they could not perform the same level of service at delivering a simple letter. It turns out that the letter was delivered, but their system failed to track it. So how do we measure quality for a process? For a software performance? For an IT application?

In a traditional manufacturing environment, conformance to specification and delivery are the common quality items that are measured and tracked. Often, lots are rejected because they do not have the correct documentation supporting them. Quality in manufacturing then is conforming product, delivered on time, and having all of the supporting documentation. With software, quality is measured as conformance to expectations, availability, experience of the process, and people interacting with the software or the IT application.[2]

If we look at Figure 7.2, we can observe the customer's experience through three aspects: (1) The specific product or service has attributes such as availability, "it's what I wanted, it works;" (2) the process through which the product (including software) is delivered can be ease of use or value added; and (3) the people (or system) should be knowledgeable and friendly. To fulfill these needs, there is a life cycle to which we apply a quality operating system.

Six Sigma is a philosophy, measure, and methodology that provides businesses with perspective and tools to achieve new levels of performance in both services and products. In Six Sigma, the focus is on process improvement to increase capability and reduce variation. The vital few inputs are chosen from the entire system of controllable and noise variables, and the focus of improvement is on controlling these vital few inputs.

Six Sigma as a philosophy helps companies achieve very low defects per million opportunities over long-term exposure. Six Sigma as a measure gives us a statistical scale to measure our progress and to benchmark other companies, processes, or products. The defect per million opportunities measurement scale ranges from 0 to 1,000,000, whereas the realistic sigma scale ranges from 0 to 6. The methodologies used in Six Sigma build on all of the tools that have evolved to date but put them into a data-driven framework. This framework of tools allows companies to achieve the lowest defects per million opportunities possible.

The simplest definition of a defect is that a defect is anything that causes customer disatisfaction. This may be a product that does not work, an incorrect component inserted on the manufacturing line, a delivery that is not on time, a software that takes too long to produce results, or a quotation with an arithmetic error. Specifically for a product, a defect is any variation in a required characteristic that prevents meeting the customer's requirements. An opportunity is defined as any operation that may introduce an error (defect). With those definitions in hand, one might think that it is straightforward, although perhaps tedious, to count defects and opportunities. Consider the case of writing a specification. An obvious defect would be any wrong value. What about typographical errors? Should a misspelled word be counted as a defect? Yes, but what is the unit of opportunity? Is it pages, words, or letters? If the unit is pages, and a ten-page specification has three errors, then the defect rate is 300,000 per million. If the unit is characters, then the defect rate is approximately 85 per million—a value much more likely to impress management. What if the unit of opportunity is each word or numerical value? The defect rate is then approximately 500 per million, a factor of 100 away from Six Sigma.

Reduction of defects in a product is a key requirement in manufacturing for which six sigma techniques are widely used. DMAIC (Define opportunity, Measure performance, Analyze opportunity, Improve performance, and Control performance) is a Six Sigma methodology often used in effecting incremental changes to product or

[2]See Chapter 1.

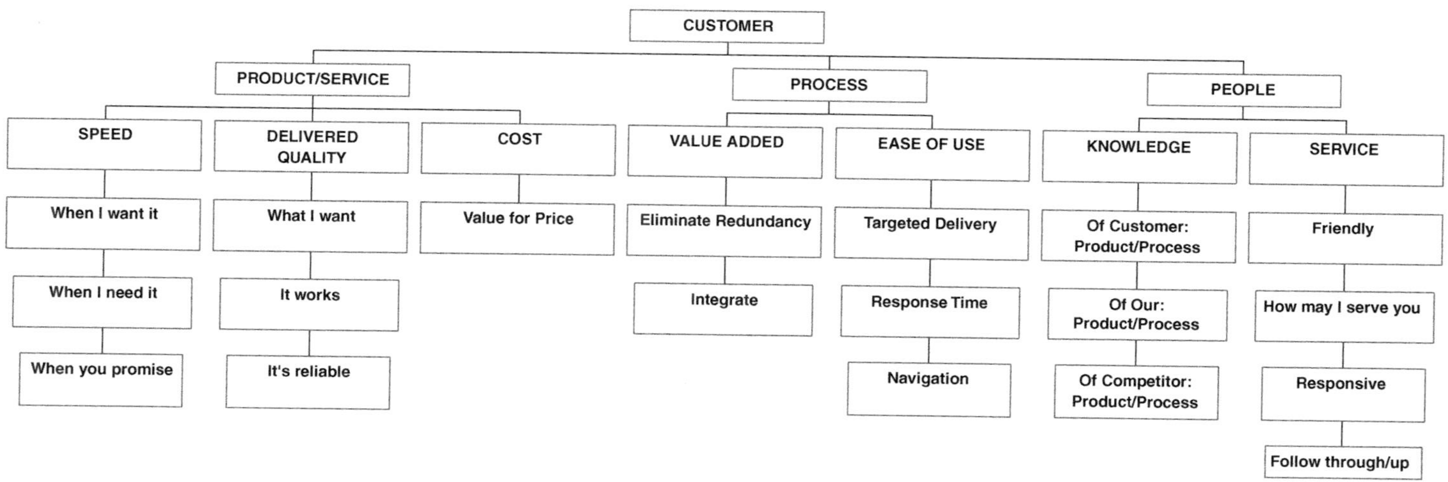

FIGURE 7.2 Customer experience channels.

service offerings focusing on the reduction of defects. DFSS (Design for Six Sigma), however, is used in the design of new products with a view to improving overall initial quality.

Six Sigma evolved from the early total quality management (TQM) efforts as discussed in El-Haik and Roy (2005). Motorola initiated the movement and then it spread to Asea Brown Boveri, Texas Instruments Missile Division, and Allied Signal. It was at this juncture that Jack Welch became aware from Larry Bossidy of the power of Six Sigma and in the nature of a fast follower committed GE to embracing the movement. It was GE who bridged the gap between just manufacturing process and product focus and took it to what was first called transactional processes and later changed to commercial processes. One reason that Jack was so interested in this program was that an employee survey had just been completed, and it had revealed that the top-level managers of the company believed that GE had invented quality, after all Armand Feigenbaum worked at GE; however, the vast majority of employees did not think GE could spell quality. Six Sigma has turned out to be the methodology to accomplish Crosby's goal of zero defects. Understanding what the key process input variables are and that variation and shift can occur we can create controls that maintain Six Sigma, or 6σ for short, performance on any product or service and in any process. The Greek letter σ is used by statisticians to indicate standard deviation, a statistical parameter, of the population of interest.

Six Sigma is process oriented, and a generic process with inputs and outputs can be modeled. We can understand clearly the process inputs and outputs if we understand process modeling.

7.4 INTRODUCTION TO SIX SIGMA PROCESS MODELING

Six Sigma is a process-focused approach to achieving new levels of performance throughout any business or organization. We need to focus on a process as a system of inputs, activities, and output(s) in order to provide a holistic approach to all the factors and the way they interact together to create value or waste. Many products (including software) and services, when used in a productive manner, also are processes. An ATM machine takes your account information, personal identification number, energy, and money and processes a transaction that dispenses funds or an account rebalance. A computer can take keystroke inputs, energy, and software to process bits into a word document.

At the simplest level, the process model can be represented by a process diagram, often called an IPO diagram for input–process–output (Figure 7.3).

If we take the IPO concept and extend the ends to include the suppliers of the inputs and the customers of the outputs, then we have the SIPOC, which stands for supplier–input–process–output–customer (Figure 7.4). This is a very effective tool in gathering information and modeling any process. A SIPOC tool can take the form of a column per each category in the name.

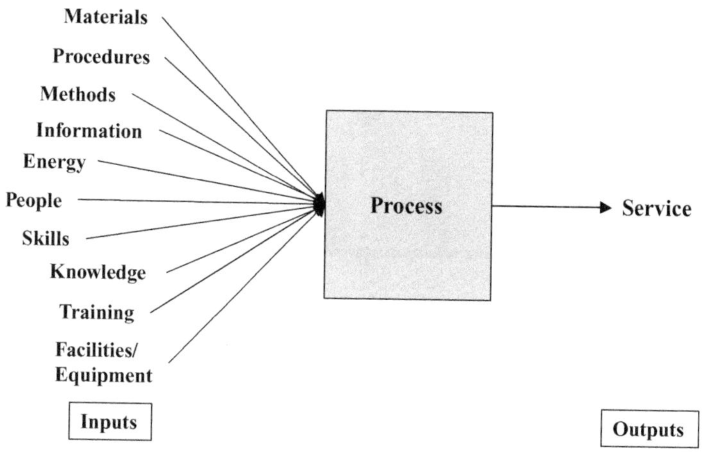

FIGURE 7.3 The IPO diagram.

7.4.1 Process Mapping

Where the SIPOC is a linear flow of steps, process mapping is a means of displaying the relationship between process steps and allows for the display of various aspects of the process, including delays, decisions, measurements, and rework and decision loops.

Process mapping builds on the SIPOC information by using standard symbols to depict varying aspects of the processes flow linked together with lines with arrows demonstrating the direction of flow.

Suppliers	Inputs	Inputs Characteristic	Process	Outputs	Output Characteristic	Customers
7. Who are the suppliers of the inputs?	6. What are the inputs of the process?	8. What are the characteri-stics of the inputs?	2a. What is the start of the process? 1. What is the process? 2b. What is the end of the process?	3. What are the outputs of the process?	5. What are the characteri-stics of the outputs?	4. Who are the customers of the outputs?

FIGURE 7.4 SIPOC table.

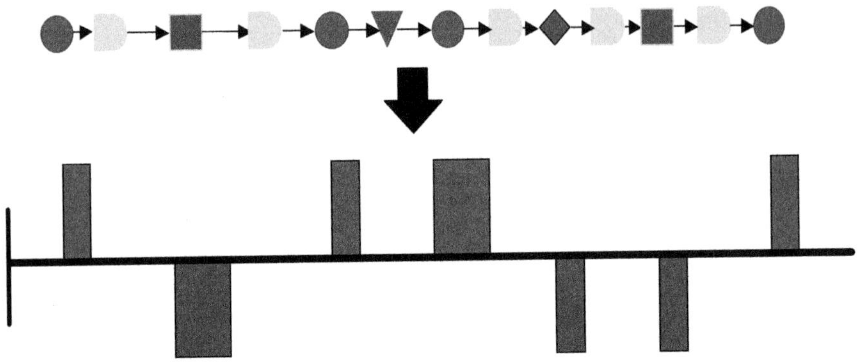

FIGURE 7.5 Process map transition to value stream map.

7.4.2 Value Stream Mapping

Process mapping can be used to develop a value stream map to understand how well a process is performing in terms of value and flow. Value stream maps can be performed at two levels. One can be applied directly to the process map by evaluating each step of the process map as value added or non-value added (see Figures 7.5 and 7.6). This type of analysis has been in existence since at least the early 1980s, but a good reference is the book, *Hunter's and the Hunted* (Swartz, 1996). This is effective if the design team is operating at a local level. However, if the design team is at more of an enterprise level and needs to be concerned about the flow of information as well as the flow of product or service, then the higher level value stream map is needed (see Figure 7.7). This methodology is best described in Rother and Shook (2003), *Learning to See*.

7.5 INTRODUCTION TO BUSINESS PROCESS MANAGEMENT

Most processes are ad hoc or allow great flexibility to the individuals operating them. This, coupled with the lack of measurements of efficiency and effectiveness, result in the variation to which we have all become accustomed. In this case, we

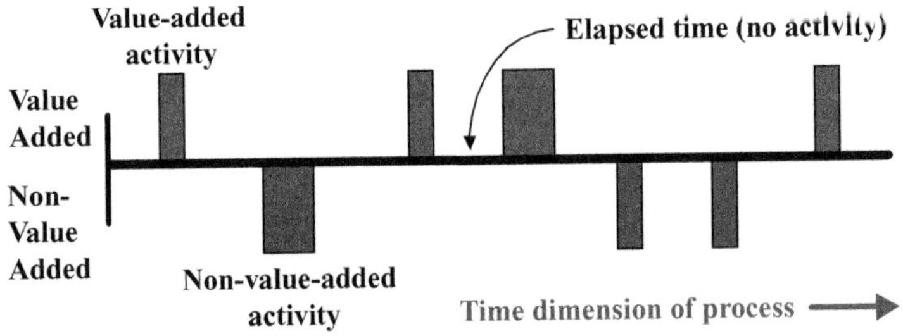

FIGURE 7.6 Value stream map definitions.

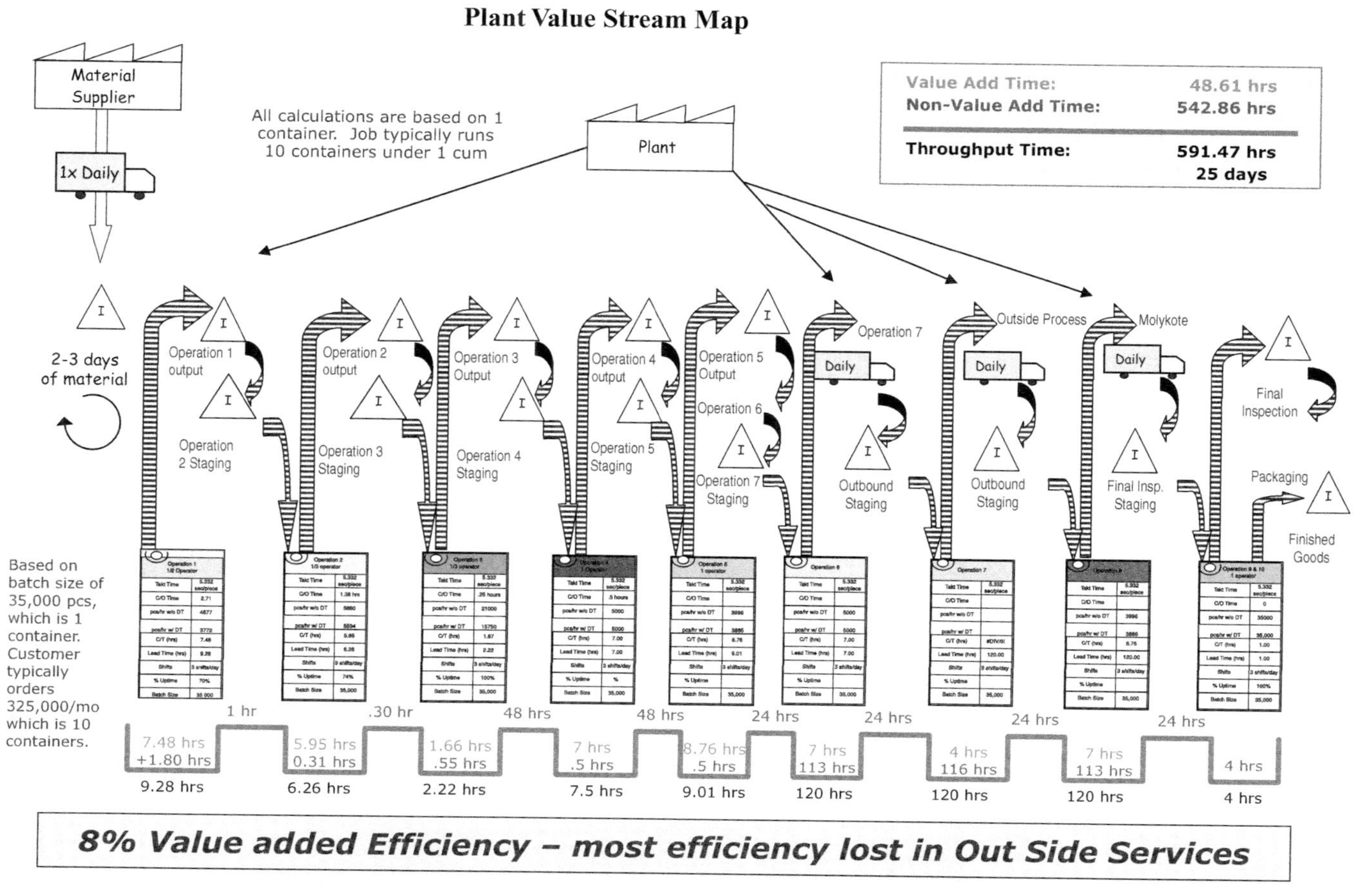

FIGURE 7.7 High-level value stream map example.

use the term "efficiency" for the within process step performance (often called the voice of the process, VOP), whereas effectiveness is how all of the process steps interact to perform as a system (often called the voice of the customer, VOC). This variation we have become accustomed to is difficult to address because of the lack of measures that allow traceability to the root cause. Businesses that have embarked on Six Sigma programs have learned that they have to develop process management systems and implement them in order to establish baselines from which to improve. The deployment of a business process management system (BPMS) often results in a marked improvement in performance as viewed by the customer and associates involved in the process. The benefits of implementing BPMS are magnified in cross-functional processes.

7.6 SIX SIGMA MEASUREMENT SYSTEMS ANALYSIS

Now that we have some form of documented process from the choices ranging from IPO, SIPOC, process map, value stream map, or BPMS, we can begin our analysis of what to fix, what to enhance, and what to design. Before we can focus on what to improve and how much to improve it, we must be certain of our measurement system. Measurements can start at benchmarking through to operationalization. We must answer how accurate and precise is the measurement system versus a known standard? How repeatable is the measurement? How reproducible? Many process measures are the results of calculations; when performed manually, the reproducibility and repeatability can astonish you if you take the time to perform the measurement system analysis (MSA).

For example, in supply chain, we might be interested in promises kept, such as on-time delivery, order completeness, deflation, lead time, and acquisition cost. Many of these measures require an operational definition in order to provide for repeatable and reproducible measures. The software measurement is discussed in Chapter 5.

Referring to Figure 7.8, is on-time delivery the same as on-time shipment? Many companies do not have visibility as to when a client takes delivery or processes a receipt transaction, so how do we measure these? Is it when the item arrives, when the paperwork is complete, or when the customer actually can use the item?

We have seen a customer drop a supplier for a 0.5% lower cost component only to discover that the new multiyear contract that they signed did not include transportation

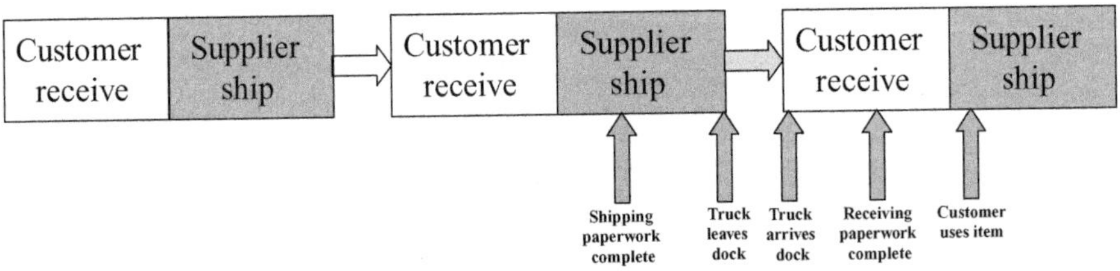

FIGURE 7.8 Supplier-to-customer cycle.

and they ended up paying 4.5% higher price for three years. The majority of measures in a service or process will focus on:

- Speed
- Cost
- Quality
- Efficiency as defined as the first-pass yield of a process step.
- Effectiveness as defined as the rolled throughput yield of all process steps.

All of these can be made robust at a Six Sigma level by creating operational definitions, defining the start and stop, and determining sound methodologies for assessing. It should come as no surprise that "If you can't measure it, you can't improve it" is a statement worth remembering and ensuring that adequate measurement sytems are available throughout the project life cycle. Software is no exception.

Software measurement is a big subject, and in the next section, we barely touch the surface. We have several objectives in this introduction. We need to provide some guidelines that can be used to design and implement a process for measurement that ties measurement to software DFSS project goals and objectives; defines measurement consistently, clearly, and accurately; collects and analyzes data to measure progress toward goals; and evolves and improves as the DFSS deployment process matures.

Some examples of process assets related to measurement include organizational databases and associated user documentation; cost models and associated user documentation; tools and methods for defining measures; and guidelines and criteria for tailoring the software measurement process element. We discussed the software CTQs or metrics and software measurement in Chapter 5.

7.7 PROCESS CAPABILITY AND SIX SIGMA PROCESS PERFORMANCE

Process capability is when we measure a process's performance and compare it with the customer's needs (specifications). Process performance may not be constant and usually exhibits some form of variability. For example, we may have an Accounts Payable (A/P) process that has measures accuracy and timeliness (same can be said about CPU utilization, memory mangemnt metrics, etc.) For the first two months of the quarter, the process has few errors and is timely, but at the quarter point, the demand goes up and the A/P process exhibits more delays and errors.

If the process performance is measurable in real numbers (continous) rather than pass or fail (discrete) categories, then the process variability can be modeled with a normal distribution. The normal distribution usually is used because of its robustness in modeling many real-world performance, random variables. The normal distribution has two parameters quantifying the central tendency and variation. The center is the average (mean) performance, and the degree of variation is expressed by the standard

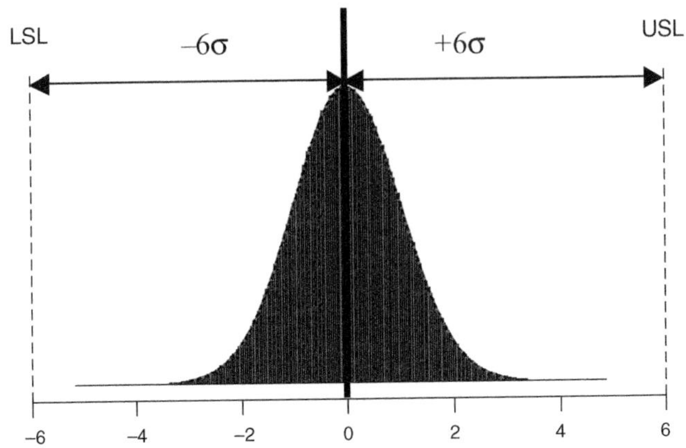

FIGURE 7.9 Highly capable pocess.

deviation. If the process cannot be measured in real numbers, then we convert the pass/fail, good/bad (discrete) into a yield and then convert the yield into a sigma value. Several transformations from discrete distributions to continuous distribution can be borrowed from mathematical statistics.

If the process follows a normal probability distribution, 99.73 % of the values will fall between the $\pm 3\sigma$ limits, where σ is the standard deviation, and only 0.27 % will be outside of the $\pm 3\sigma$ limits. Because the process limits extend from -3σ to $+3\sigma$, the total spread amounts to 6σ total variation. This total spread is the process spread and is used to measure the range of process variability.

For any process performance metrics, usually there are some performance specification limits. These limits may be single sided or two sided. For the A/P process, the specification limit may be no less than 95 % accuracy. For receipt of material into a plant, it may be two days early and zero days late. For a call center, we may want the phone conversation to take between two minutes and four minutes. For each of the last two double-sided specifications, they also can be stated as a target and as a tolerance. The material receipt could be one-day early ± 1 day, and for the phone conversation, it could be three minutes ± 1 minute.

If we compare the process spread with the specification spread, we can usually observe three conditions:

- Condition I: Highly Capable Process (see Figure 7.9). The process spread is well within the specification spread.

$$6\sigma < (USL - LSL)$$

The process is capable because it is extremely unlikely that it will yield unacceptable performance.

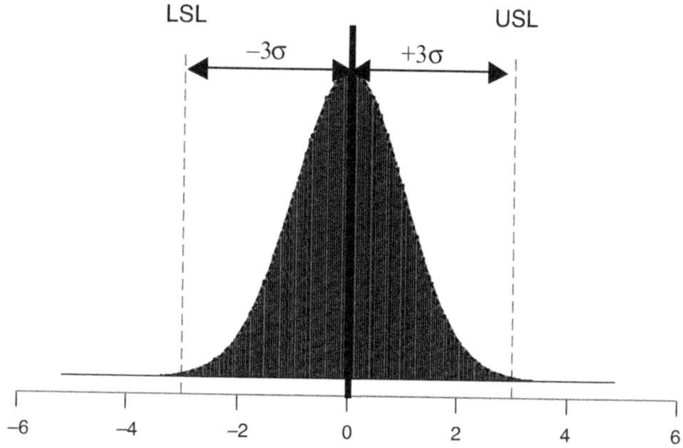

FIGURE 7.10 Marginally capable pocess.

- Condition II: Marginally Capable Process (see Figure 7.10). The process spread is approximately equal to the specification spread.

$$6\sigma = (\text{USL} - \text{LSL})$$

When a process spread is nearly equal to the specification spread, the process is capable of meeting the specifications. If we remember that the process center is likely to shift from one side to the other, then a significant amount of the output will fall outside of the specification limit and will yield unacceptable performance.

- Condition III: Incapable Process (see Figure 7.11). The process spread is greater than the specification spread.

$$6\sigma > (\text{USL} - \text{LSL})$$

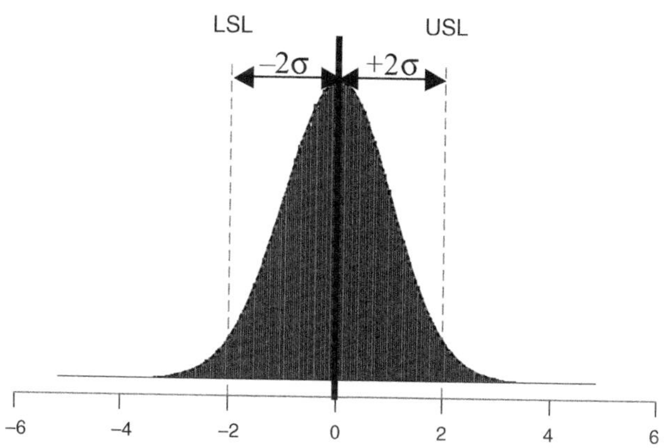

FIGURE 7.11 Incapable process.

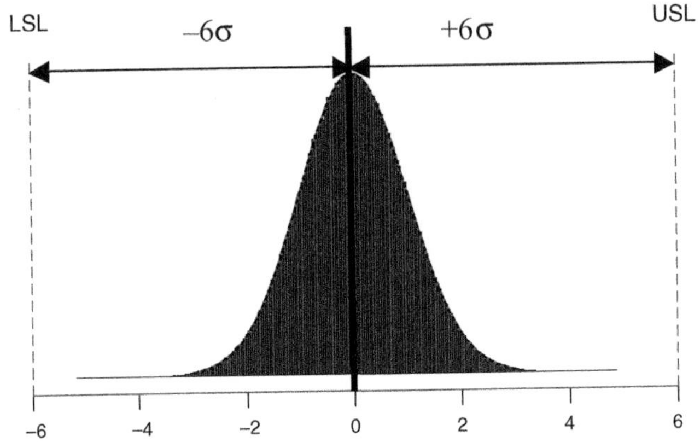

FIGURE 7.12 Six Sigma capable process (short term).

When a process spread is greater than the specification spread, the process is incapable of meeting the specifications and a significant amount of the output will fall outside of the specification limit and will yield unacceptable performance. The sigma level is also know as the Z value (assuming normal distribution) and for a certain CTQ is given by

$$Z = \frac{\text{USL} - \text{mean}}{\sigma} \text{ or } \frac{\text{mean} - \text{LSL}}{\sigma} \tag{7.1}$$

where USL is the upper specification limit and LSL is the lower specification limit.

7.7.1 Motorola's Six Sigma Quality

In 1986, the Motorola Corporation won the Malcolm Baldrige National Quality Award. Motorola based its success in quality on its Six Sigma program. The goal of the program was to reduce the variation in every process such that a spread of 12σ (6σ on each side of the average) fits within the process specification limits (see Figure 7.12).

Motorola accounted for the process average to shift side to side over time. In this situation, one side shrinks to a 4.5σ gap, and the other side grows to 7.5σ (see Figure 7.13). This shift accounts for 3.4 parts per million (ppm) on the small gap and a fraction of parts per billion on the large gap. So over the long term, a 6σ process will generate only 3.4 ppm defect.

To achieve Six Sigma capability, it is desirable to have the process average centered within the specification window and to have the process spread approximately one half of the specification window.

There are two approaches to accomplish Six Sigma levels of performance . When dealing with an existing process, there is the process improvement method also known

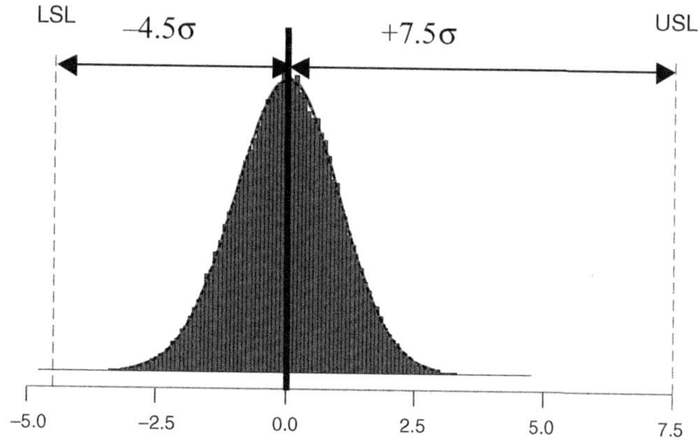

FIGURE 7.13 Six Sigma capable process (long term).

as DMAIC, and if there is a need for a new process, then it is Design For Six Sigma (DFSS). Both of these will be discussed in the following sections.

7.8 OVERVIEW OF SIX SIGMA IMPROVEMENT (DMAIC)

Applying Six Sigma methodology to improve an existing process or product follows a five-phase process of:

- Define: Define the opportunity and customer requirements
- Measure: Ensure adequate measures, process stability, and initial capability
- Analyze: Analyze the data and discover the critical inputs and other factors
- Improve: Improve the process based on the new knowledge
- Control: Implement adequate controls to sustain the gain

This five-phase process often is referred to as DMAIC, and each phase is described briefly below.

7.8.1 Phase 1: Define

First we create the project definition that includes the problem/opportunity statement, the objective of the project, the expected benefits, what items are in scope and what items are out of scope, the team structure, and the project timeline. The scope will include details such as resources, boundaries, customer segments, and timing.

The next step is to determine and define the customer requirements. Customers can be both external consumers or internal stakeholders. At the end of this step you should have a clear operational definition of the project metrics (called Big Y's,

CTQs, or the outputs)[3] and their linkage to critical business levers as well as the goal for improving the metrics. Business levers, for example, can consist of return on invested capital, profit, customer satisfaction, and responsiveness.

The last step in this phase is to define the process boundaries and high-level inputs and outputs using the SIPOC as a framework and to define the data collection plan.

7.8.2 Phase 2: Measure

The first step is to make sure that we have good measures of our Y's through validation or measurement system analysis.

Next we verify that the metric is stable over time and then determine what our baseline process capability is using the method discussed earlier. If the metric is varying wildly over time, then we must first address the special causes creating the instability before attempting to improve the process. Many times the result of stabilizing the performance provides all of the improvement desired.

Lastly, in the Measure phase, we define all of the possible factors that affect the performance and use qualitative methods of Pareto, cause-and-effect diagrams, cause-and-effect matrices, failure modes and their effects, and detailed process mapping to narrow down to the potential influential (significant) factors (denoted as the x's).

7.8.3 Phase 3: Analyze

In the Analyze phase, we first use graphical analysis to search out relationships between the input factors (x's) and the outputs (Y's).

Next we follow this up with a suite of statistical analysis (Chapter 6) including various forms of hypothesis testing, confidence intervals, or screening design of experiments to determine the statistical and practical significance of the factors on the project Y's. A factor may prove to be statistically significant; that is, with a certain confidence, the effect is true and there is only a small chance it could have been by mistake. The statistically significant factor is not always practical in that it may only account for a small percentage of the effect on the Y's; in which case, controlling this factor would not provide much improvement. The transfer function $Y = f(x)$ for every Y measure usually represents the regression of several influential factors on the project outputs. There may be more than one project metric (output), hence, the Y's.

7.8.4 Phase 4: Improve

In the Improve phase, we first identify potential solutions through team meetings and brainstorming or through the use of TRIZ in product and service concepts, which are covered in El-Haik and Roy (2005) and El-Haik and Mekki (2008). It is important at this point to have completed a measurement system analysis on the key factors (x's) and possibly to have performed some confirmation design of experiments.

[3]See Chapter 5 for software metrics.

The next step is to validate the solution(s) identified through a pilot run or through optimization design of experiments.

After confirmation of the improvement, then a detail project plan and cost benefit analysis should be completed.

The last step in this phase is to implement the improvement. This is a point where change management tools can prove to be beneficial.

7.8.5 Phase 5: Control

The Control phase consist of four steps, In the first step, we determine the control strategy based on the new process map, failure mode and effects, and a detailed control plan. The control plan should balance between the output metric and the critical few input variables.

The second step involves implementing the controls identified in the control plan. This typically is a blend of poka yoke's and control charts as well as of clear roles and responsibilities and operator instructions depicted in operational method sheets.

Third, we determine what the final capability of the process is with all of the improvements and controls in place.

The final step is to perform the ongoing monitoring of the process based on the frequency defined in the control plan. The DMAIC methodology has allowed businesses to achieve lasting breakthrough improvements that break the paradigm of reacting to the causes rather than the symptoms. This method allows design teams to make fact-based decisions using statistics as a compass and to implement lasting improvements that satisfy the external and internal customers.

7.9 DMAIC SIX SIGMA TOOLS

The DMAIC is a defined process that involves a sequence of five phases (define, measure, analyze, improve, and control). Each phase has a set of tasks that get accomplished using a subset of tools. Figure 7.14 (Pan et al., 2007) provides an overview of the tools/techniques that are used in DMAIC.

Most of the tools specified in Figure 7.14 above are common across Six Sigma projects and tend to be used in DMAIC-and DFSS-based projects. Some additional ones are used and will be explored in Chapters 10 and 11. Many statistical needs (e.g., control charts and process capability) specified in the tools section are available through Minitab (Minitab Inc., State College, PA).

The DMAIC methodology is an acronym of the process steps. Although rigorous, it provides value in optimizing repeatable processes by way of reducing waste and making incremental changes. However, with increasing competition and the human resources needed to rework a product, there is a greater need to bring out products that work correctly the first time around (i.e., the focus of new product development is to prevent defects rather than fixing defects). Hence, a DFSS approach that is the next evolution of the Six Sigma methodology often is used in new product initiatives

DMAIC Phase Steps	Tools Used
• **D - Define Phase:** Define the project goals and customer (internal and external) deliverables.	
• Define Customers and Requirements (CTQs) • Develop Problem Statement, Goals, and Benefits • Identify Champion, Process Owner, and Team • Define Resources • Evaluate Key Organizational Support • Develop Project Plan and Milestones • Develop High-Level Process Map	• Project Charter • Process Flowchart • SIPOC Diagram • Stakeholder Analysis • DMAIC Work Breakdown Structure • CTQ Definitions • Voice of the Customer Gathering
• **M - Measure Phase:** Measure the process to determine current performance; quantify the problem.	
• Define Defect, Opportunity, Unit, and Metrics • Detailed Process Map of Appropriate Areas • Develop Data Collection Plan • Validate the Measurement System • Collect the Data • Begin Developing Y = f(x) Relationship • Determine Process Capability and Sigma Baseline	• Process Flowchart • Data Collection Plan/Example • Benchmarking • Measurement System Analysis/Gage R&R • Voice of the Customer Gathering • Process Sigma Calculation
• **A - Analyze Phase:** Analyze and determine the root cause(s) of the defects.	
• Define Performance Objectives • Identify Value/Non-Value-Added Process Steps • Identify Sources of Variation • Determine Root Cause(s) • Determine Vital Few x's, Y = f(x) Relationship	• Histogram • Pareto Chart • Time Series/Run Chart • Scatter Plot • Regression Analysis • Cause-and-Effect/Fishbone Diagram • 5 Whys • Process Map Review and Analysis • Statistical Analysis • Hypothesis Testing (Continuous and Discrete) • Non-Normal Data Analysis
• **I - Improve Phase:** Improve the process by eliminating defects.	
• Perform Design of Experiments • Develop Potential Solutions • Define Operating Tolerances of Potential System • Assess Failure Modes of Potential Solutions • Validate Potential Improvement by Pilot Studies • Correct/Re-Evaluate Potential Solution	• Brainstorming • Mistake Proofing • Design of Experiments • Pugh Matrix • House of Quality • Failure Modes and Effects Analysis (FMEA) • Simulation Software
• **C - Control Phase:** Control future process performance.	
• Define and Validate Monitoring and Control System • Develop Standards and Procedures • Implement Statistical Process Control • Determine Process Capability • Develop Transfer Plan, Handoff to Process Owner • Verify Benefits, Cost Savings/Avoidance, Profit Growth • Close Project, Finalize Documentation • Communicate to Business, Celebrate	• Process Sigma Calculation • Control Charts (Variable and Attribute) • Cost-Savings Calculations • Control Plan

FIGURE 7.14 DMAIC steps and tools.

Differences between SIx Sigma and Design For Six Sigma	
Six Sigma	Design for Six Sigma
• DMAIC: Define, Measure, Analyze, Improve, and Control.	• DMADV: Define, Measure, Analyze, Design, and Verify. • DMADOV: Define, Measure, Analyze, Design, Optimize, and Verify.
Looks at existing processes and fixes problems.	Focuses on the upfront design of the product and process.
More reactive.	More proactive.
Dollar benefits obtained from Six Sigma can be quantified rather quickly.	Benefits are more difficult to quantify and tend to be more long term. It can take 6 to 12 months after the launch of the new product before you will obtain proper accounting on the impact.

FIGURE 7.15 DMAIC versus DFSS comparison.[4]

today. The differences between the two approaches are captured in Figure 7.15. In addition to ICOV, DMADV and DMADOV are used as depicted in Figure 7.15.

Unlike different models where the team members on a project need to figure out the way and technique to obtain the data they need, Six Sigma provides a set of tools making the process clear and structured and therefore easier to proceed through in order to save both time and effort and get to the final goal sooner. Table 7.3 shows a list of some of these tools and their use.

7.10 SOFTWARE SIX SIGMA

Jeannine Siviy and Eileen Forrester (Siviy & Forrester, 2004) suggest "Line of sight" or alignment to business needs should be consistently clear and quantitative in the Six Sigma process. "The ability of Six Sigma's focus on" should also be critical to quality factors and to bottom-line performance to provide resolution among peers with a similar rating and to provide visibility into (or characterization of) the specific performance strengths of each. As an example, with Six Sigma, an organization might be enabled to reliably make a statement such as, "We can deliver this project in ±2% cost, and we have the capacity for five more projects in this technology domain. If we switch technologies, our risk factor is "xyz" and we may not be able to meet cost or may not be able to accommodate the same number of additional projects."

7.10.1 Six Sigma Usage in Software Industry

The earliest attempts to use Six Sigma methodology in development were considered part of electronic design where mapping the Six Sigma process steps to the

[4]http://www.plm.automation.siemens.com/en_us/Images/wp_nx_six_sigma_tcm1023-23275.pdf.

TABLE 7.3 A Sample List of Some Six Sigma Tools and Their Usage

Six Sigma Tool	Use
Kano model, Benchmark	To support product specification and discussion through better development team understanding.
GQM	"Goal, Question, Metric," is an approach to software metrics.
Data collection methods	A process of preparing and collecting data. It provides both a baseline from which to measure from and in certain cases a target on what to improve.
Measurement system evaluation	It is a specially designed experiment that seeks to identify the components of variation in the measurement.
Failure modes and effects analysis (FMEA)	It is a procedure for analysis of potential failure modes within a system for classification by severity or determination of the effect of failures on the system.
Statistical interference	To estimate the probability of failure or the frequency of failure.
Reliability analysis	It is to test the ability of a system or component to perform its required functions under stated conditions for a specified period of time.
Root cause analysis	It is a class of problem-solving methods aimed at identifying the root causes of problems or events.
Hypothesis test	Deciding whether experimental results contain enough information to cast doubt on conventional wisdom.
Design of experiments	Often the experimenter is interested in the effect of some process or intervention (the "treatment") on some objects (the "experimental units"), which may be people.
Analysis of vari-ance(ANOVA)	It is a collection of statistical models, and their associated procedures, in which the observed variance is partitioned into components resulting from different explanatory variables. It is used to test for differences among two or more independent groups.
Decision and risk analysis	It should be performed as part of the risk management process for each project. The data of which would be based on risk discussion workshops to identify potential issues and risks ahead of time before these were to pose cost and/ or schedule negative impacts.
Platform-specific model (PSM)	It is a model of a software or business system that is linked to a specific technological platform.
Control charts	It is a tool that is used to determine whether a manufacturing or business process is in a state of statistical control. If the process is in control, all points will plot within the control limits. Any observations outside the limits, or systematic patterns within, suggest the introduction of a new (and likely unanticipated) source of variation, known as a special-cause variation. Because increased variation means increased quality costs, a control chart "signaling" the presence of a special cause requires immediate investigation.
Time-series methods	It is the use of a model to forecast future events based on known past events: to forecast future data points before they are measured.

(*Continued*)

TABLE 7.3 (*Continued*)

Six Sigma Tool	Use
Procedural adherence	It is the process of systematic examination of a quality system carried out by an internal or external quality auditor or an audit team.
Performance management	It is the process of assessing progress toward achieving predetermined goals. It involves building on that process, adding the relevant communication and action on the progress achieved against these predetermined goals.
Preventive measure	To use risk prevention to safeguard the quality of the product.
Histogram	It is a graphical display of tabulated frequencies, shown as bars. It shows what proportion of cases fall into each of several categories.
Scatterplot	It is a type of display using Cartesian coordinates to display values for two variables for a set of data.
Run chart	It is a graph that displays observed data in a time sequence. Run charts are analyzed to find anomalies in data that suggest shifts in a process over time or special factors that may be influencing the variability of a process.
Flowchart	Flowcharts are used in analyzing, designing, documenting, or managing a process or a program in various fields.
Brainstorming	It is a group creativity technique designed to generate a large number of ideas for the solution of a problem.
Pareto chart	It is a special type of bar chart where the values being plotted are arranged in descending order, and it is used in quality assurance.
Cause-and-effect diagram	It is a diagram that shows the causes of a certain event. A common use of it is in product design, to identify potential factors causing an overall effect.
Baselining, surveying methods	They are used to collect quantitative information about items in a population. A survey may focus on opinions or factual information depending on its purpose, and many surveys involve administering questions to individuals.
Fault tree analysis (FTA)	It is basically composed of logic diagrams that display the state of the system and is constructed using graphical design techniques. Fault tree analysis is a logical, structured process that can help identify potential causes of system failure before the failures actually occur. Fault trees are powerful design tools that can help ensure that product performance objectives are met.

manufacture of an electronic overcurrent detection circuit were presented (White, 1992). An optimal design from the standpoint of a predictable defect rate is attempted by studying $Y = f(x_1, x_2, x_3, \ldots x_n)$, where Y is the current threshold of the detector circuit and $x_1, x_2, x_3, \ldots x_n$ are the circuit components that go into the detection circuit. Recording Y and error (deviation from Y) by changing parameter(s) one at a time using a Monte Carlo simulation technique results in a histogram or forecast chart that shows the range of possible outcomes and probability of the occurrence of the outcome. This helps with identification of the critical x(s) causing predominant variation.

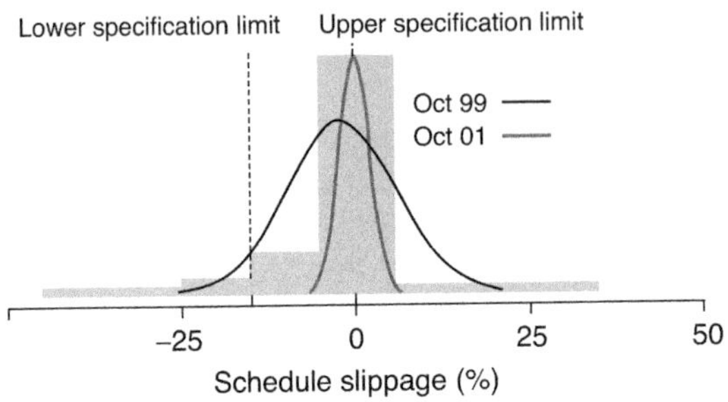

FIGURE 7.16 Process capability analysis for schedule slippage (Muruguppan & Keeni, 2003).

Monitoring of software project schedules as part of the software development cycle is another aspect where Six Sigma methodology has been used, as shown in Figure 7.16. During a two-year period, the company claims to have reduced the variation (sigma) associated with slippage on project schedules making its customer commitments more consistent. This claim could be a murky one because the study does not indicate how many projects were delivered during the timeframe and how many projects were similar. These factors could alter the conclusion as Six-Sigma-based statistics requires a sufficient sample size for results to be meaningful.

In addition, this there are other instances where the Six Sigma technology has been applied effectively to the software development cycle. Although Six Sigma continued to be practiced in manufacturing as a way to optimize processes, its use in the software development cycle, particularly in the area of problem solving, seems to have gained traction since the late 1990s.

7.11 SIX SIGMA GOES UPSTREAM—DESIGN FOR SIX SIGMA

The Six Sigma DMAIC[5] (Define-Measure-Analyze-Improve-Control) methodology is excellent when dealing with an existing process in which reaching the *entitled level of performance* will provide all of the benefit required. *Entitlement* is the best the process or product (including software) is capable of performing with adequate control. Reviewing historical data it is often evident as the best performance point. But what do we do if reaching entitlement is not enough or there is a need for an innovative solution never before deployed? We could continue with the typical code it–build it–fix it cycle, as some of the traditional software development processes promote in this chapter, or we can use the most powerful tools and methods available for developing an optimized, robust, derisked software design. These tools and methods can be aligned with an existing new software development process or used in a stand-alone manner.

[5]http://www.plm.automation.siemens.com/en_us/Images/wp_nx_six_sigma_tcm1023-23275.pdf.

DFSS is a disciplined methodology with a collection of tools to ensure that products and processes are developed systematically to provide reliable results that exceed customer requirements. A key function of DFSS is to understand and prioritize the needs, wants, and desires of customers and to translate those requirements into products and processes that will consistently meet those needs. The DFSS tool set can be used in support of major new software product development initiatives, or in stand-alone situations to ensure that proper decisions are made. DFSS is a highly discipline approach to embedding the principle of Six Sigma as possible in the design and development process. When a problem is not discovered until well into the product life cycle, the costs to make a change, not to mention the intangible costs, such as customer dissatisfaction, are considerable (Figure 1.2).

The rest of this book is devoted to explaining and demonstrating the DFSS tools and methodology. Chapter 8 is the introductory chapter for DFSS, giving the overview for DFSS theory; DFSS-gated process, and DFSS application. Chapter 9 provides a detailed description about how to deploy DFSS in a software development organization, covering the training, organization support, financial management, and deployment strategy. Chapter 11 provides a very detailed "road map" of the whole software DFSS project execution, which includes an in-depth description of the DFSS stages, task management, scorecards, and how to integrate all DFSS methods into developmental stages. Chapters 12 through 19 provide detailed descriptions with examples on all of the major methods and tools used in DFSS.

7.12 SUMMARY

The term "Six Sigma" is heard often today. Suppliers offer Six Sigma as an incentive to buy; customers demand Six Sigma compliance to remain on authorized vendor lists. It was known that it has to do with quality, and obviously something to do with statistics, but what exactly is it? Six Sigma is a lot of things: a methodology, a philosophy, an exercise in statistics, and a way of doing business, a tool for improving quality. Six Sigma is only one of several tools and processes that an organization needs to use to achieve world-class quality. Six Sigma places an emphasis on identifying and eliminating defects from one's products, sales quotations, and proposals to a customer or a paper presented at a conference. The goal is to improve one's processes by eliminating waste and opportunity for waste so much that mistakes are nearly impossible. The goal of a process that is Six Sigma good is a defect rate of only a few parts per million. Not 99% good, not even 99.9% good, but 99.999996% good.

In this chapter, we have explained what 6σ is and how it has evolved over time. We explained how it is a process-based methodology and introduced the reader to process modeling with a high-level overview of IPO, process mapping, value stream mapping and value analysis, as well as BPMS. we discussed the criticality of understanding the measurements of the process or system and how this is accomplished with measurement systems analysis (MSA). Once we understand the goodness of our measures, we can evaluate the capability of the process to meet customer requirements and can demonstrate what is 6σ capability. Next we moved into an explanation of the DMAIC methodology and how it incorporates these concepts into a road-map method. Finally we covered how 6σ moves upstream to the design environment with the application of DFSS. In Chapter 8, we will introduce the reader to the software DFSS process.

REFERENCES

El-Haik, Basem, S. and Mekki, K. (2008). *Medical Device Design for Six Sigma: A Road Map for Safety and Effectiveness*, 1st Ed., Wiley-Interscience, New York.

El-Haik, Basem, S. and Roy, D. (2005). *Service Design for Six Sigma: A Roadmap for Excellence*, Wiley-Interscience, New York.

Muruguppan, M and Keeni, G. (2003), "Blending CMMM and Six Sigma to Meet Business Goals." *IEEE Software*, Volume 20, #2, pp. 42–48.

Pan, Z., Park, H., Baik, J., and Choi, H. (2007), "A Six Sigma Framework for Software Process Improvement and Its Implementation," IEEE, Proc. of the 14th Asia Pacific Software Engineering Conference.

Shook, J., Womack, J., and Jones, D. (1999). *Learning to See: Value Stream Mapping to Add Value and Eliminate MUDA*, Lean Enterprise Institute, Cambridge, MA.

Sivi, J. M., Penn, M. L., and Stoddard, R. W. (2007). *CMMI and Six Sigma: Partners in Process Improvement*, 1st Ed., Addison-Wesley Professional, Upper Saddle River, NJ.

Siviy, Jeannine and Forrester, Eileen (2004), "Enabling Technology Transition Using Six Sigma," Oct, http://www.sei.cmu.edu/library/abstracts/reports/04tr018.cfm.

Swartz, James B. (1996). *The Hunters and the Hunted: A Non-Linear Solution for Reengineering the Workplace*, 1st Ed., Productivity Press, New York.

White, R.V. (1992), "An Introduction to Six Sigma with a Design Example," APEC '92 Seventh Annual Applied Power Electronics Conference and Exposition, Feb, pp. 28–35.

Wikipedia Contributors, *Six Sigma*. http://en.wikipedia.org/w/index.php?title=Six_Sigma&oldid=228104747. Accessed August, 2009.

CHAPTER 8

INTRODUCTION TO SOFTWARE DESIGN FOR SIX SIGMA (DFSS)[1]

8.1 INTRODUCTION

The objective of this chapter is to introduce the software Design for Six Sigma (DFSS) process and theory as well as to lay the foundations for the subsequent chapters of this book. DFSS combines design analysis (e.g., requirements cascading) with design synthesis (e.g., process engineering) within the framework of the deploying company's software (product) development systems. Emphasis is placed on Critical-To-Satisfaction (CTS) requirements (a.k.a Big Y's), identification, optimization, and verification using the transfer function and scorecard vehicles. A transfer function in its simplest form is a mathematical relationship between the CTSs and/or their cascaded functional requirements (FRs) and the critical influential factors (called the X's). Scorecards help predict risks to the achievement of CTSs or FRs by monitoring and recording their mean shifts and variability performance.

DFSS is a disciplined and rigorous approach to software, process, and product design by ensuring that new designs meet customer requirements at launch. It is a design approach that ensures complete understanding of process steps, capabilities, and performance measurements by using scorecards, transfer functions, and tollgate

[1]The word "Sigma" refers to the Greek letter, σ, that has been used by statisticians to measure variability. As the numerical levels of Sigma or (σ) increase, the number of defects in a process falls exponentially. Six Sigma design is the ultimate goal since it means if the same task performed one million times, there will be only 3.4 defects assuming normality. The DMAIC Six Sigma approach was introduced in Chapter 7.

Software Design for Six Sigma: A Roadmap for Excellence, By Basem El-Haik and Adnan Shaout

reviews to ensure accountability of all the design team members, Black Belt, Project Champions, and Deployment Champions[2] as well as the rest of the organizations.

The software DFSS objective is to attack the design vulnerabilities in both the conceptual and the operational phase by deriving and integrating tools and methods for their elimination and reduction. Unlike the DMAIC methodology, the phases or steps of DFSS are not defined universally as evidenced by the many customized training curriculum available in the market. Many times the deployment companies will implement the version of DFSS used by their choice of the vendor assisting in the deployment. However, a company will implement DFSS to suit its business, industry, and culture, creating its own version. However, all approaches share common themes, objectives, and tools.

DFSS is used to design or redesign a service, physical product, or software generally called "product" in the respective industries. The expected process Sigma level for a DFSS product is at least 4.5,[3] but it can be Six Sigma or higher depending on the designed product. The production of such a low defect level from product or software launch means that customer expectations and needs must be understood completely before a design can be operationalized. That is, quality is defined by the customer.

The material presented herein is intended to give the reader a high-level understanding of software DFSS, its uses, and its benefits. Following this chapter, readers should be able to assess how it could be used in relation to their jobs and identify their needs for further learning.

DFSS as defined in this book has a two-track deployment and application. By deployment, we mean the strategy adopted by the deploying company to launch the Six Sigma initiative. It includes putting into action the deployment infrastructure, strategy, and plan for initiative execution (Chapter 9). In what follows, we are assuming that the deployment strategy is in place as a prerequisite for application and project execution. The DFSS tools are laid on top of four phases as detailed in Chapter 11 in what we will be calling the software DFSS project road map.

There are two distinct tracks within the term "Six Sigma" initiative as discussed in previous chapters. The retroactive Six Sigma DMAIC[4] approach takes problem solving as an objective, whereas the proactive DFSS approach targets redesign and new software introductions on both development and production (process) arenas.

DFSS is different than the Six Sigma DMAIC approach in being a proactive prevention approach to design.

The software DFSS approach can be phased into **I**dentify, **C**onceptualize, **O**ptimize, and **V**erify/**V**alidate or **ICOV**, for short. These are defined as follows:

Identify customer and design requirements. Prescribe the CTSs, design parameters and corresponding process variables.

[2]We will explore the roles and responsibilities of these Six Sigma operatives and others in Chapter 9.

[3]No more than approximately 1 defect per thousand opportunities.

[4]**D**efine: project goals and customer deliverables. **M**easure: the process and determine baseline. **A**nalyze:determine rooat causes. **I**mprove: the process by optimization (i.e., eliminating/reducing defects). **C**ontrol: sustain the optimized solution.

Conceptualize the concepts, specifications, and technical and project risks.

Optimize the design transfer functions and mitigate risks.

Verify that the optimized design meets intent (customer, regulatory, and deploying software function).

In this book, both ICOV and DFSS acronyms will be used interchangeably.

8.2 WHY SOFTWARE DESIGN FOR SIX SIGMA?

Generally, the customer-oriented design is a development process of transforming customers' wants into design software solutions that are useful to the customer. This process is carried over several development stages starting at the conceptual stage. In this stage, conceiving, evaluating, and selecting good design solutions are difficult tasks with enormous consequences. It usually is the case that organizations operate in two modes "proactive" (i.e., conceiving feasible and healthy conceptual entities) and "retroactive" (i.e., problem solving such that the design entity can live to its committed potentials). Unfortunately, the latter mode consumes the largest portion of the organization's human and nonhuman resources. The Design for Six Sigma approach highlighted in this book is designed to target both modes of operations.

DFSS is a premier approach to process design that can embrace and improve developed homegrown supportive processes (e.g., sales and marketing) within its development system. This advantage will enable the deploying company to build on current foundations while enabling them to reach unprecedented levels of achievement that exceed the set targets.

The link of the Six Sigma initiative and DFSS to the company vision and annual objectives should be direct, clear, and crisp. DFSS have to be the crucial mechanism to develop and improve the business performance and to drive up the customer satisfaction and quality metrics. Significant improvements in all health metrics are the fundamental source of DMAIC and DFSS projects that will, in turn, transform culture one project at a time. Achieving a Six Sigma culture is very essential for the future well-being of the deploying company and represents the biggest return on investment beyond the obvious financial benefits. Six Sigma initiatives apply to all elements of a company's strategy, in all areas of the business if massive impact is really the objective.

The objective of this book is to present the software Design for Six Sigma approach, concepts, and tools that eliminate or reduce both the conceptual and operational types of vulnerabilities of software entities and releases such entities at Six Sigma quality levels in all of their requirements.

Operational vulnerabilities take variability reduction and mean adjustment of the critical-to-quality, critical-to-cost, critical-to-delivery requirements, the CTSs, as an objective and have been the subject of many knowledge fields such as parameter design, DMAIC Six Sigma, and tolerance design/tolerancing techniques. On the contrary, the conceptual vulnerabilities usually are overlooked because of the lack of a compatible systemic approach to find ideal solutions, the ignorance of the designer, the pressure of the deadlines, and budget limitations. This can be attributed, in part, to the fact that traditional quality methods can be characterized as after-the-fact practices because they use lagging information for developmental activities such as bench tests and field data. Unfortunately, this practice drives design toward endless cycles of design–test–fix–retest, creating what broadly is known as the "fire fighting" mode of the design process (i.e., the creation of design-hidden factories). Companies who follow these practices usually suffer from high development costs, longer time-to-market, lower quality levels, and marginal competitive edge. In addition, corrective actions to improve the conceptual vulnerabilities via operational vulnerabilities improvement means are marginally effective if at all useful. Typically, these corrections are costly and hard to implement as the software project progresses in the development process. Therefore, implementing DFSS in the conceptual stage is a goal, which can be achieved when systematic design methods are integrated with quality concepts and methods upfront. Specifically, on the technical side, we developed an approach to DFSS by borrowing from the following fundamental knowledge arenas: process engineering, quality engineering, axiomatic design (Suh, 1990), and theories of probability and statistics. At the same time, there are several venues in our DFSS approach that enable transformation to a data-driven and customer-centric culture such as concurrent design teams, deployment strategy, and plan.

In general, most current design methods are empirical in nature. They represent the best thinking of the design community that, unfortunately, lacks the design scientific base while relying on subjective judgment. When the company suffers in detrimental behavior in customer satisfaction, judgment and experience may not be sufficient to obtain an optimal Six Sigma solution, which is another motivation to devise a software DFSS method to address such needs.

Attention starts shifting from improving the performance during the later stages of the software design life cycle to the front-end stages where design development takes place at a higher level of abstraction. This shift also is motivated by the fact that the design decisions made during the early stages of the software design life cycle have the largest impact on the total cost and quality of the system. It often is claimed that up to 80% of the total cost is committed in the concept development stage (Fredrikson, 1994). The research area of design currently is receiving increasing focus to address industry efforts to shorten lead times, cut development and manufacturing costs, lower total life-cycle cost, and improve the quality of the design entities in the form of software products. It is the experience of the authors that at least 80% of the design quality also is committed in the early stages as depicted in Figure 8.1 (El-Haik & Roy, 2005). The "potential" in the figure is defined as the difference between the impact (influence) of the design activity at a certain design stage and the total development cost up to that stage. The *potential* is positive but decreasing as design progresses implying reduced design freedom over time. As financial resources are committed (e.g., buying process equipment and facilities and hiring staff), the *potential* starts changing sign, going from positive to negative. For the cunsumer, the potential is negative and the cost overcomes the impact tremendously. At this stage, design changes for corrective actions only can be achieved at a high cost, including

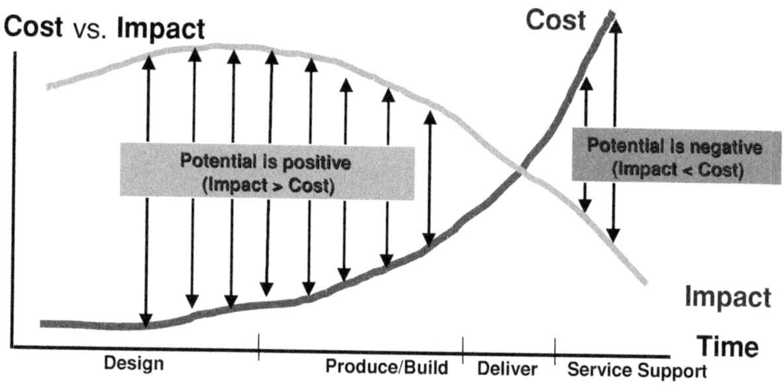

FIGURE 8.1 Effect of design stages on life cycle.

customer dissatisfaction, warranty, marketing promotions, and in many cases under the scrutiny of the government (e.g., recall costs).

8.3 WHAT IS SOFTWARE DESIGN FOR SIX SIGMA?

Software DFSS is a structured, data-driven approach to design in all aspects of software functions (e.g., human resources, marketing, sales, and IT) where deployment is launched, to eliminate the defects induced by the design process and to improve customer satisfaction, sales, and revenue. To deliver on these benefits, DFSS applies design methods like software methods, axiomatic design,[5] creativity methods, and statistical techniques to all levels of design decision making in every corner of the business to identify and optimize the critical design factors (the X's) and validate all design decisions in the use (or surrogate) environment of the end user.

DFSS is not an add-on but represents a cultural change within different functions and organizations where deployment is launched. It provides the means to tackle weak or new processes, driving customer and employee satisfaction. DFSS and Six Sigma should be linked to the deploying company's annual objectives, vision, and mission statements. It should not be viewed as another short-lived initiative. It is a vital, permanent component to achieve leadership in design, customer satisfaction, and cultural transformation. From marketing and sales, to development, operations, and finance, each business function needs to be headed by a deployment leader or a deployment champion. This local deployment team will be responsible for delivering dramatic change thereby removing the number of customer issues and internal problems and expediting growth. The deployment team can deliver on their objective through Six Sigma operatives called Black Belts and Green Belts who will be executing scoped projects that are in alignment with the objectives of the

[5] A perspective design method that employs two design axioms: the independence axioms and the information axiom. See Chapter 11 for more details.

company. Project Champions are responsible for scoping projects from within their realm of control and handing project charters (contracts) over to the Six Sigma resource. The Project Champion will select projects consistent with corporate goals and remove barriers. Six Sigma resources will complete successful projects using Six Sigma methodology and will train and mentor the local organization on Six Sigma. The deployment leader, the highest initiative operative, sets meaningful goals and objectives for the deployment in his or her function and drives the implementation of Six Sigma publicly.

Six Sigma resources are full-time Six Sigma operatives on the contrary to Green Belts who should be completing smaller projects of their own, as well as assisting Black Belts. They play a key role in raising the competency of the company as they drive the initiative into day-to-day operations.

Black Belts are the driving force of software DFSS deployment. They are project leaders that are removed from day-to-day assignments for a period of time (usually two years) to focus exclusively on design and improvement projects with intensive training in Six Sigma tools, design techniques, problem solving, and team leadership. The Black Belts are trained by Master Black Belts who initially are hired if not homegrown.

A Black Belt should possess process and organization knowledge, have some basic design theory and statistical skills, and be eager to learn new tools. A Black Belt is a "change agent" to drive the initiative into his or her teams, staff function, and across the company. In doing so, their communication and leadership skills are vital. Black Belts need effective intervention skills. They must understand why some team members may resist the Six Sigma cultural transformation. Some soft training on leadership training should be embedded within their training curriculum. Soft-skills training may target deployment maturity analysis, team development, business acumen, and individual leadership. In training, it is wise to share several initiative maturity indicators that are being tracked in the deployment scorecard, for example, alignment of the project to company objectives in its own scorecard (the Big Y's), readiness of project's mentoring structure, preliminary budget, team member identification, and scoped project charter.

DFSS Black Belt training is intended to be delivered in tandem with a training project for hands-on application. The training project should be well scoped with ample opportunity for tool application and should have cleared Tollgate "0" prior to training class. Usually, project presentations will be weaved into each training session. More details are given in Chapter 9.

While handling projects, the role of the Black Belts spans several functions, such as learning, mentoring, teaching, and coaching. As a mentor, the Black Belt cultivates a network of experts in the project on hand, working with the process operators, design owners, and all levels of management. To become self-sustained, the deployment team may need to task their Black Belts with providing formal training to Green Belts and team members.

Software DFSS is a disciplined methodology that applies the transfer function [CTSs $= f(X)$] to ensure customer expectations are met, embeds customer expectations into the design, predicts design performance prior to pilot, builds performance measurement systems (Scorecards) into the design to ensure effective ongoing process management, and leverages a common language for design within a design tollgate process.

DFSS projects can be categorized as design or redesign of an entity whether it is a product, process, or software. "Creative design" is the term that we will be using to indicate new software design, design from scratch, and "incremental design" to indicate the redesign case or design from a datum (e.g., next-generation Micrsoft Office suite). In the latter case, some data can be used to baseline current performance. The degree of deviation of the redesign from datum is the key factor on deciding on the usefulness of relative existing data. Software DFSS projects can come from historical sources (e.g., software redesign from customer issues) or from proactive sources like growth and innovation (new software introduction). In either case, the software DFSS project requires greater emphasis on:

- Voice of the customer collection scheme
- Addressing all (multiple) CTS's as cascaded by the customer
- Assessing and mitigating technical failure modes and project risks in their own environments as they linked to the tollgate process reviews
- Project management with some communication plan to all affected parties and budget management
- Detailed project change management process

8.4 SOFTWARE DFSS: THE ICOV PROCESS

As mentioned in Section 8.1, Design for Six Sigma has four phases over seven development stages. They are as follows: Identify, Conceptualize, Optimize, and Verify. The acronym ICOV is used to denote these four phases. The software life cycle is depicted in Figure 8.2. Notice the position of the software ICOV phases of a design project.

Naturally, the process of software design begins when there is a need, an impetus. People create the need whether it is a problem to be solved (e.g., if a functionality or use interface is not user friendly, then the GUI needs to be redesigned) or a new invention. Design objective and scope are critical in the impetus stage. A design project charter should describe simply and clearly what is to be designed. It cannot be vague. Writing a clearly stated design charter is just one step. In stage 2, the design team must write down all the information they may need, in particular the voice of the customer (VOC) and the voice of the business (VOB). With the help of the quality function deployment (QFD) process, such consideration will lead the definition of the software design functional requirements to be later grouped into programs and routine codes. A functional requirement must contribute to an innovation or to a solution of the objective described in the design charter. Another question that should be on the minds of the team members relates to how the end result will look. The simplicity, comprehensiveness, and interfaces should make the software attractive. What options

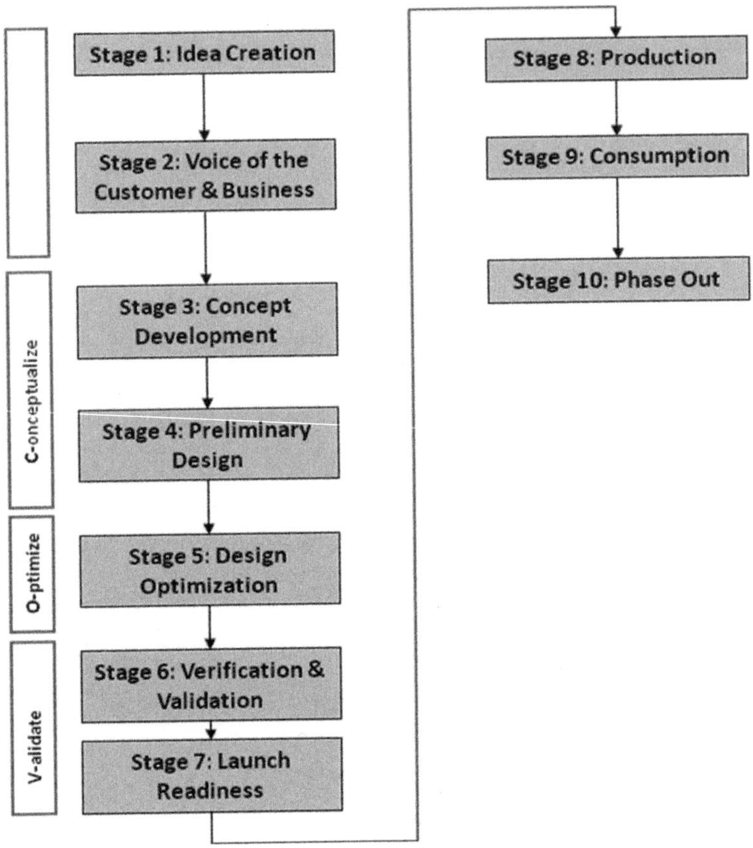

FIGURE 8.2 The software life cycle.

are available to the team? And at what cost? Do they have the right attributes, such as completeness, language, and reliability? Will it be difficult to operate and maintain? What methods will they need to process, store, and deliver the software?

In stage 3, the design team should produce several solutions. It is very important that they write or draw every idea on paper as it occurs to them. This will help them remember and describe them more clearly. It also is easier to discuss them with other people if drawings are available. These first drawings do not have to be very detailed or accurate. Sketches will suffice and should be made quickly. The important thing is to record all ideas and develop solutions in the preliminary design stage (stage 4). The design team may find that they like several solutions. Eventually, the design team must choose one. Usually, careful comparison with the original design charter will help them to select the best subject to the constraints of cost, technology, and skills available. Deciding among the several possible solutions is not always easy. It helps to summarize the design requirements and solutions and to put the summary in a matrix called the morphological matrix.[6] An overall design alternative set is synthesized from this matrix that is conceptually high-potential and feasible solutions. Which solution should they choose? The Pugh matrix, a concept selection tool named after Stuart Pugh, can be used. The selected solution will be subjected to a thorough design optimization stage (stage 5). This optimization could be deterministic and/or statistical in nature. On the statistical front, the design solution will be made insensitive to uncontrollable factors (called the noise factors) that may affect its performance. Factors like customer usage profile and use environment should be considered as noise. To assist on this noise insensitivity task, we rely on the transfer function as an appropriate vehicle. In stage 5, the team needs to make detailed documentation of the optimized solution. This documentation must include all of the information needed to produce the software. Consideration for design documentation, process maps, operational instructions, software code, communication, marketing, and so on should be put in place. In stage 6, the team can make a model assuming the availability of the transfer functions and later a prototype or they can go directly to making a prototype or a pilot. A model is a full-size or small-scale simulation. Architects, engineers, and most designers use models. Models are one more step in communicating the functionality of the solution. A scale model is used when design scope is very large. A prototype is the first working version of the team's solution. Design verification and validation, stage 6, also includes testing and evaluation, which is basically an effort to answer these very basic questions: Does it work? (Does it meet the design charter? If failures are discovered, will modifications improve the solution?) These questions have to be answered. After having satisfactory answers, the team can move to the next development and design stage.

In stage 7, the team needs to prepare the production facilities where the software will be produced for launch. At this stage, they should ensure that the software is marketable and that no competitors beat them to the market. The team together with the project stakeholders must decide how many to make. Similar to products, software may be mass-produced in low volume or high volume. The task of making the software is divided into jobs. Each worker trains to do his or her assigned job. As workers complete their special jobs, the software product takes shape. Post stage 7, the mass production saves time and other resources. Because workers train to do a certain job, each becomes skilled in that job.

8.5 SOFTWARE DFSS: THE ICOV PROCESS IN SOFTWARE DEVELOPMENT

Because software DFSS integrates well with a software life-cycle system, it is an event-driven process, in particular, the development (design) stage. In this stage, milestones occur when the entrance criteria (inputs) are satisfied. At these milestones, the stakeholders including the project champion, design owner, and deployment

[6]A morphological matrix is a way to show all functions and corresponding possible design parameters (solutions).

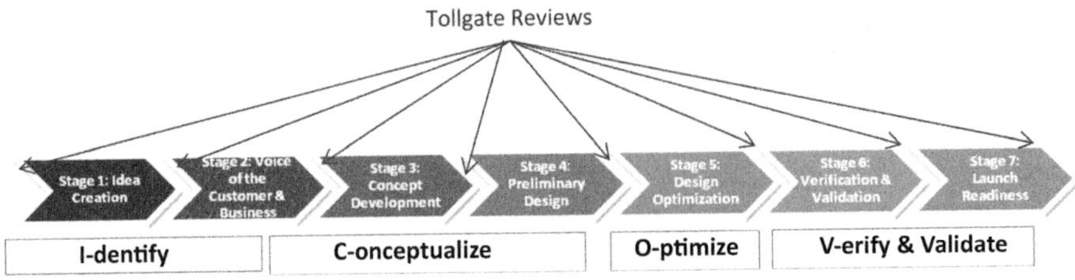

FIGURE 8.3 The ICOV DFSS process.

champion (if necessary) conduct reviews called "tollgate" reviews. A development stage has some *thickness*, that is, entrance criteria and exit criteria for the bounding tollgates. The ICOV DFSS phases as well as the seven stages of the development process are depicted in Figure 8.3. In these reviews, a decision should be made whether to proceed to the next phase of development, recycle back for further clarification on certain decisions, or cancel the project altogether. Cancellation of problematic projects, as early as possible, is a good thing. It stops nonconforming projects from progressing further while consuming resources and frustrating people. In any case, the Black Belt should quantify the size of the benefits of the design project in language that will have an impact on upper management, identify major opportunities for improving customer dissatisfaction and associated threats to salability, and stimulate improvements through publication of DFSS approach.

In tollgate reviews, work proceeds when the exit criteria (required decisions) are made. As a DFSS deployment side bonus, a standard measure of development progress across the deploying company using a common development terminology is achieved. Consistent exit criteria from each tollgate with both software DFSS own deliverables from the application of the approach itself, and the business unit or function-specific deliverables. The detailed entrance and exit criteria by stage are presented in Chapter 11.

8.6 DFSS VERSUS DMAIC

Although the terminology is misleading, allowing us to assume that DFSS and Six Sigma are interrelated somehow, DFSS is in its roots a distinguishable methodology very different than the Six Sigma DMAIC because it is not intended to improve but to *innovate*. Moreover, in opposition to DMAIC, the DFSS spectrum does not have a main methodology to be applied as is the case for Six Sigma but has multiple different processes and templates.[7] The one we adopt is ICOV as discussed earlier. However, the objective is the same: a newly designed product with higher quality level—a Six Sigma level of quality. The ICOV DFSS approach can be used for designing of products (Yang & El-Haik, 2003), services, or processes (El-Haik & Yang, 2005)

[7] See Section 8.7.

from scratch. It also can be used for the redesign of existing products, services, and processes where the defects are so numerous that it is more efficient to redesign it from the beginning using DFSS than to try to improve it using the traditional Six Sigma methodology. Although Christine Tayntor (2002) states simply that the DFSS "helps companies build in quality from the beginning," Yang and El-Haik (2008) presents it in a more detailed statement saying that "instead of simply plugging leak after leak, the idea is to figure out why it is leaking and where and attack the problem at its source."

Organizations usually realize their design shortcomings and reserve a certain budget for warranty, recalls, and other design defects. Planning for rework is a fundamental negative behavior that resides in most process developments. This is where DFSS comes in to change this mentality toward a new trend of thinking that focuses on minimizing rework and later corrections by spending extra efforts on the design of the product to make it the best possible upfront. The goal is to replace as many inspectors as possible and put producers in their place. From that point, we already can make a clear distinction between Six Sigma and Design for Six Sigma giving an implicit subjective preference to the DFSS approach. It is important to point out that DFSS is indeed the best remedy but sometimes not the fastest, especially for those companies already in business having urgent defects to fix. Changing a whole process from scratch is neither simple nor cost free. It is a hard task to decide whether the innovative approach is better than the improving one, and it is up to the company's resources, goals, situation, and motivations to decide whether they are really ready for starting the innovation adventure with DFSS. But on the other side, actually some specific situations will force a company to innovate by using the DFSS. Some motivations that are common to any industry could be:

- They face some technical problem that cannot be fixed anymore and need a breakthrough changes.
- They might have a commercial product that needs a business differentiator feature to be added to overcome its competitors.
- The development process or the product itself became too complex to be improved.
- High risks are associated with the current design.

Six Sigma is a process improvement philosophy and methodology, whereas DFSS is centered on designing new products and services. The main differences are that Six Sigma focuses on one or two CTQ metrics, looks at processes, and aims to improve the CTQ performance. In contrast, DFSS focuses on every single CTQ that matters to the customer, looks at products and services as well as the processes by which they are delivered, and aims to bring forth a new product/service with a performance of about 4.5 sigma (long terms) or better. Other differences are that DFSS projects often are much larger and take longer and often are based on a long-term business need for new products, rather than a short-term need to fix a customer problem.

In practicality the divide between a formal DFSS project and a "simple" Six Sigma project can be indistinct—at times there is a need for a Six Sigma project to improve radically the capability (rather than, or as well as, performance) of a broken or nonexistent process using design or redesign.

DFSS brings about a huge change of roles in an organization. The DFSS team is cross-functional, as the key factor is covering all aspects for the product from market research to process launch. DFSS provides tools to get the improvement process done efficiently and effectively. It proves to be powerful management technique for projects. It optimizes the design process so as to achieve the level of Six Sigma for the product being designed.

The DFSS methodology should be used when a product or process is not in existence at your company and one needs to be developed or when the product or process exists and has been optimized and reached their entitlement (using either DMAIC, for example) and still does not meet the level of customer specification or Six Sigma level.

It is very important to have practical experience of Six Sigma, as DFSS builds on the concepts and tools from a typical DMAIC approach. Becuase DFSS works with products and services rather than with processes and because design and creativity are important, a few new tools are common to any DFSS methodology. Strong emphasis is placed on customer analysis, an the transition of customer needs and requirements down to process requirements, and on error and failure proofing. Because the product/service often is very new, modeling and simulation tools are important, particularly for measuring and evaluating in advance the anticipated performance of the new process.

If DFSS is to work successfully, it is important that it covers the full life cycle of any new software product. This begins when the organization formally agrees with the requirement for something new, and ends when the new software is in full commercial delivery.

The DFSS tools are used along the entire life cycle of product. Many tools are used in each phase. Phases like (DOE), which are used to collect data, assess impact, predict performance, design for robustness, and validate performance. Table 8.1 classifies DFSS tools used by design activity. In the next section, we will discuss the DFSS tool usage by ICOV phase.

8.7 A REVIEW OF SAMPLE DFSS TOOLS BY ICOV PHASE

The origin of DFSS seems to have its beginnings with NASA and the U.S. Department of Defense. In the late 1990s, early 2000s, GE Medical systems was among the forerunners in using DFSS for new product development with its use in the design of the light speed computed tomography (CT) system.

DFSS provides a systematic integration of tools, methods, processes and team members throughout product and process design. Initiatives vary dramatically from company to company but typically start with a charter (linked to the organization's strategic plan), an assessment of customer needs, functional analysis, identification

TABLE 8.1 Sample DFSS Tools by Development Activity (Pan, 2007)

Define and Manage Requirement	Voice of customer (VOC) Contextual inquiry Quality function deployment House of quality HOQ Analytic hierarchy process (AHP)
Prioritize/Narrow Focus	Kano's model Normal group technique CTQ tree Pugh concept selection Pareto chart Pugh concept selection Pareto chart Pugh concept selection Axiomatic design (El-Haik, 2005)
Generation and Select Design Concept	Axiomatic design TRIZ (El-Haik and Roy, 2005)
Perform Functional Analysis	Capability analysis
Predict Performance	Histograms Modeling and simulation Simulation DFSS scorecard Control plans Failure mode and effect analysis (FMEA)
Evaluate and Mitigate Risk	Probability distribution[8] Axiomatic design Gap analysis
Evaluate/Assess/Improve Design	Design for X-ability (DFX) Statistical process control (SPC) Design of experiment (DOE) Monte Carlo simulation
Design for Robustness Evaluate Robustness to Noise	Correlation (disambiguation) Regression analysis Robust design[9] Design of experiment (DOE) CE diagram
Validate Performance	FMEA[10] High throughput testing (HTT) Capability analysis

[8]See Chapter 6.
[9]See Chapter 18.
[10]See Chapter 16.

of critical to quality characteristics, concept selection, detailed design of products and processes, and control plans.[11]

To achieve this, most DFSS methodologies tend to use advanced design tools (quality function deployment, failure modes and effects analysis, benchmarking, axiomatic design, simulation, design of experiments, simulation, statistical optimization, error proofing, cause- and effect-matrix, Kano analysis, Pugh matrix, and so on). Some of these techniques are discussed in here. We selected a critical one to cover in dedicated chapters (Chapters 12–19).

8.7.1 Sample Identify Phase DFSS Tools

Design should begin with the customer. DFSS focuses on determining what customers require and value through a range of tools, including customer voices analysis, Affinity diagramming, quality function deployment (QFD),[12] house of quality (HOQ), Kano model, voice of the customer table, and analytic hierarchy process.

The VOC is a process used to capture the requirements and feedback from the customer (internal or external) to provide the customers with the best-in-class product (or service) quality. This process is all about being proactive and constantly innovative to capture the changing requirements of the customers with time. Within any organization, there are multiple customer voices: the procuring unit, the user, and the supporting maintenance unit. Within any of those units, there also may be multiple customer voices. The "voice of the customer" is the term used to describe the stated and unstated needs or requirements of the customer. The voice of the customer can be captured in a variety of ways: direct discussion or interviews, surveys, focus groups, customer specifications, observation, warranty data, field reports, complaint logs, and so on. These data are used to identify the quality attributes needed for a supplied component or material to incorporate in the process or product.

VOC is methodology that allows a project team to record information about customer needs in a way that captures the context of those needs to enable the team to better understand an explicit and implicit customer requirement. For each customer statement, the team identifies the demographic information and information about software use. The information is categorized in terms of basic questions—what, where, when, why, and how—that provide a context for analyzing and understanding the customer statement.

HOQ the major matrix in QFD helps the software DFSS team member structure his or her thinking, reach a consensus about where to focus the attention of the organization, and communicate this information throughout the organization. This tool helps ensure that they do not leave anything out where they identify CTQs that are the source of customer satisfaction, at the system level, subsystem level, and component level.

QFD is a systematic process for motivating a team to focus on their customers to identify and resolve issues involved in providing software products, processes,

[11] http://www.plm.automation.siemens.com/en_us/Images/wp_nx_six_sigma_tcm1023-23275.pdf
[12] See Chapter 12.

services, and strategies that will more than satisfy their customers is a structured approach. Defining customer needs or requirements and translating them into specific plans to produce products to meet those needs are major QFD activities. It is effective for focusing and aligning the project team very early in the identify phase of software DFSS, identifying gaps and targets, and planning and organizing requirements at all levels of the design. QFD can be used in all phases of DFSS (ICOV).

Survey analysis is a popular technique to collect VOC. This survey is used to gather information from a sample of individuals, usually a fraction of the population being studied. In a bona fide survey, the sample is scientifically chosen so that each person in the population will have a measurable chance of being selected. Survey can be conducted in various ways, including over the telephone, by mail, and in person. Focus groups and one-on-one interviews are popular types of VOC collection techniques. Without surveying the customers adequately, it is difficult to know which features of a product or a service will contribute to its success or failure or to understand why. Surveys are useful in some situations, but there are weak in terms of getting the types of data necessary for new design.

Kano analysis[13] is a tool that can be used to classify and prioritize customer needs. This is useful because customer needs are not all of the same kind, not all have the same importance, and are different for different populations. The results can be used to prioritize the team effort in satisfying different customers. The Kano model divides the customer requirement into three categories (basic CTQs, satisfier CTQs, and delighter CTQs).

Analytic hierarchy process (AHP) is a tool for multicriteria analysis that enables the software DFSS team to rank explicitly an intangible factor against each other in order to establish priorities. The first step is to decide on the relative importance of the criteria, comparing each one against each other. Then, a simple calculation determines the weight that will be assigned to each criterion: This weight will be a value between 0 and 1, and the sum of weight for all criteria will be 8. This tool for multicriteria analysis has another benefit for software DFSS project teams. By breaking down the steps in the selection process, AHP reveals the extent to which team members understand and can evaluate factors and criteria. The team leaders can use it to simulate discussion of alternatives.

Pareto chart[14] provides facts needed for setting priorities. Typically, it organizes and displays information to show the relative importance of various problems or causes of problems. In DFSS, it can be used to prioritize CTQs in the QFD from importance perspectives. It is a form of a vertical bar chart that puts items in order (from the highest to the lowest) relative to some measurable CTQ importance. The chart is based on the Pareto principle, which states that when several factors (or requirements) affect a situation, a few factors will account for most of the impact. The Pareto principle describes a phenomenon in which 80% of variation observed in everyday processes can be explained by a mere 20% of the causes of that variation. Placing the items in descending order of frequency makes it easy to discern those

[13]See Chapter 12.

[14]See Chapter 1.

problems that are of greatest importance or those causes that seem to account for most of the variation. Thus, a Pareto chart helps teams to focus their efforts where they can have the greatest potential impact. Pareto charts help teams focus on the small number of really important problems or their causes. They are useful for establishing priorities by showing which are the most critical CTQs to be tackled or causes to be addressed. Comparing Pareto charts of a given situation over time also can determine whether an implemented solution reduced the relative frequency or cost of that problem or cause.

A CTQ tree is used to decompose broad customer requirements into more easily quantified requirements. CTQ trees often are used in the Six Sigma DMAIC methodology. CTQs are derived from customer needs. Customer delight may be an add-on while deriving CTQ parameters. For cost considerations, one may remain focused an customer needs at the initial stage. CTQs are the key measurable characteristics of a product or process whose performance standards or specification limits must be met in order to satisfy the customer. They align improvement or design efforts with customer requirements. CTQs represent the product or service characteristics that are defined by the customer (internal or external). They may include the upper and lower specification limits or any other factors related to the product or service. A CTQ usually must be interpreted from a qualitative customer statement to an actionable, quantitative business specification.

Pugh concept selection is a method, an iterative evaluation, that tests the completeness and understanding of requirements and quickly identifies the strongest software concept. The method is most effective if each member of the DFSS team performs it independently. The results of the comparison usually will lead to repetition of the method, with iteration continued until the team reaches a consensus. Pugh concept selection refers to a matrix that helps determine which potential conceptual solutions are *best*.[15] It is to be done after you capture VOC and before design, which means after product-planning QFD. It is a scoring matrix used for concept selection, in which options are assigned scores relative to criteria. The selection is made based on the consolidated scores. Before you start your detailed design, you must have many options so that you can choose the best from among them.

The Pugh matrix is a tool used to facilitate a disciplined, team-based process for concept generation and selection. Several concepts are evaluated according to their strengths and weaknesses against a reference concept called the datum (base concept). The Pugh matrix allows the DFSS team to compare differ concepts, create strong alternative concepts from weaker concepts, and arrive at a conceptually best (optimum) concept that may be a hybrid or variant of the best of other concepts

The Pugh matrix encourages comparison of several different concepts against a base concept, creating stronger concepts and eliminating weaker ones until an optimal concept finally is reached. Also, the Pugh matrix is useful because it does not require a great amount of quantitative data on the design concepts, which generally is not available at this point in the process.

[15]El-Haik formulated the Concept Selection Problem as an integer program in El-Haik (2005).

8.7.2 Sample Conceptualize Phase DFSS Tools

Axiomatic design (AD)[16] is a perspective design methodology using matrix formulation to analyze systematically the transformation of customer needs into functional requirements, design parameters, and process variables. Axiomatic design is a general methodology that helps software DFSS teams to structure and understand design projects, thereby facilitating the synthesis and analysis of suitable design requirements, solutions, and processes. This approach also provides a consistent framework from which some metrics of design alternatives (e.g., coupling) can be quantified. The basic premise of the axiomatic approach to design is that there are basic *axioms* that govern decision making in design, just as the laws of nature govern the physics and chemistry of nature. Two basic principles, independence axiom and information axiom, are derived from the generation of good design practices. The corollaries and theorems, which are direct consequences or are derived from the axioms, tend to have the flavor of design rules. Axiomatic design pays much attention to the functional, physical, and process hierarchies in the design of a system. At each layer of the hierarchy, two axioms are used to assess design solutions.

A key aspect of axiomatic design is the separation between what a system has to achieve (functional requirements) and the design choices involved in how to achieve it (design parameters). Our preemptive software DFSS technology focuses on the effectiveness of the earliest phases of the solution development process: requirements analysis and solution synthesis. Therefore AD is more than appropriate way in this way.

TRIZ[17] offers a wide-ranging series of tools to help designers and inventors to avoid the trial-and-error approach during the design process and to solve problems in creative and powerful ways. For the most part, TRIZ tools were created by means of careful research of the world patent database (mainly in Russian), so they have been evolved independent and separate from many of the design strategies developed outside of Russia. TRIZ abstracts the design problem as either the contradiction, or the Su-field model, or the required function realization. Then corresponding knowledge base tools are applied once the problem is analyzed and modeled. Although approaches to the solutions are of some differences, many design rules in AD and problem-solving tools in TRIZ are related and share the same ideas in essence (El-Haik, 2005).

Capability Analysis[18] is a statistical tool that visually or mathematically compares actual process performance with the performance standards established by the customer, the specification limits. To analyze (plot or calculate) capability you need the mean and standard deviation associated with the required attribute in a sample of the software product, and customer requirements associated with that software metric of interest, the CTQ.

[16] See Chapter 13. Also El-Haik (2005).

[17] Theory of Inventive Problem Solving (TIPS). TRIZ in Russian. See El-Haik and Roy (2005).

[18] See Chapter 4.

Histograms[19] are graphs of a distribution of data designed to show the centering, dispersion (spread), and shape (relative frequency) of the data. Histograms can provide a visual display of large amounts of data that are difficult to understand in a tabular, or spreadsheet, form. They are used to understand how the output of a process relates to customer expectations (targets and specifications) and to help answer the question: "Is the process capable of meeting customer requirements?" In other words, how the voice of the process (VOP) measures up to the voice of the customer (VOC). Histograms are used to plot the density of data and often for density estimation: estimating the probability density function of the underlying variable. The total area of a histogram always equals 1. If the lengths of the intervals on the x-axis are all 1, then a histogram is identical to a relative frequency plot. An alternative to the histogram is kernel density estimation, which uses a kernel to smooth samples.

DFSS scorecard (El-Haik & Yang, 2003) is the repository for all managed CTQ information. At the top level, the scorecard predicts the defect level for each CTQ. The input sheets record the process capability for each key input. The scorecard calculates short-term Z scores and long-term DPMO (see Chapter 7). By layering scorecards, they become a systems integration tool for the project team and manager.

If a model can be created to predict the team's designs performance with respect to a critical requirement, and if this model can be computed relatively quickly, then powerful statistical analyses become available that allow the software DFSS team to reap the full benefits of DFSS. They can predict the probability of the software design meeting the requirement given environmental variation and usage variation using statistical analysis techniques (see Chapter 6). If this probability is not sufficiently large, then the team can determine the maximum allowable variation on the model inputs to achieve the desired output probability using statistical allocation techniques. And if the input variation cannot be controlled, they can explore new input parameter values that may improve their design's statistical performance with respect to multiple requirements simultaneously using optimization techniques (see Chapters 17 and 18).

Risk is a natural part of the business landscape. The software industry is no difference. If left unmanaged, the uncertainty can spread like weeds. If managed effectively, losses can be avoided and benefits can be obtained. Too often, software risk (risk related to the use of software) is overlooked. Other business risks, such as market risks, credit risk and operational risks have long been incorporated into the corporate decision-making processes. *Risk Management*[20] is a methodology based on a set of guiding principles for effective management of software risk.

Failure Mode and Effect Analysis (FMEA)[21] is a proactive tool, technique, and quality method that enables the identification and prevention of process or software product errors before they occur. As a tool embedded within DFSS methodology, FMEA can help identify and eliminate concerns early in the development of a process or new service delivery. It is a systematic way to examine a process prospectively for possible ways in which failure can occur, and then to redesign the product so

[19] See Chapter 5.
[20] See Chapter 15.
[21] See Chapter 16.

that the new model eliminates the possibility of failure. Properly executed, FMEA can assist in improving overall satisfaction and safety levels. There are many ways to evaluate the safety and quality of software products and developmental processes, but when trying to design safe entities, a proactive approach is far preferable to a reactive approach.

Probability distribution: Having one prototype that works under controlled conditions does not prove that the design will perform well under other conditions or over time. Instead a statistical analysis is used to assess the performance of the software design across the complete range of variation. From this analysis, an estimate of the probability of the design performing acceptably can be determined. There are two ways in which this analysis can be performed: 1) Build many samples and test and measure their performance, or 2) predict the design's performance mathematically. We can predict the probability of the design meeting the requirement given sources of variation experienced by a software product. If this probability is not sufficiently large, then the team can determine the maximum allowable variation on the model's inputs to achieve the desired output probability. And if the input variation cannot be controlled, the team can explore new input parameter values that may improve their design's statistical performance with respect to multiple requirements simultaneously.

The control chart, also known as the Stewart chart or process-behavior chart, in statistical process control is a tool used to determine whether a process is in a state of statistical control. If the chart indicates that the process is currently under control, then it can be used with confidence to predict the future performance of the process. If the chart indicates that the process being monitored is not in control, the pattern it reveals can help determine the source of variation to be eliminated to bring the process back into control. A control chart is a specific kind of run chart that allows significant change to be differentiated from the natural variability of the process. This is the key to effective process control and improvement. On a practical level, the control chart can be considered part of an objective disciplined approach that facilitates the decision as to whether process (e.g., a Chapter 2 software development process) performance warrants attention.

We ultimately can expect the technique to penetrate the software industry. Although a few pioneers have attempted to use statistical process control in software-engineering applications, the opinion of many academics and practitioners is that it simply does not fit in the software world. These objections probably stem from unfamiliarity with the technique and how to use it to best advantage. Many tend to dismiss it simply on the grounds that software can not be measured, but properly applied, statistical process control can flag potential process problems, even though it cannot supply absolute scores or goodness ratings.

8.7.3 Sample Optimize Phase DFSS Tools

Axiomatic design implementation in software DFSS is a systematic process, architecture generator, and disciplined problem-prevention approach to achieve excellence. Robust design is the heart of the software DFSS optimize phase. To ensure the success of robust parameter design, one should start with good design concepts. Axiomatic design, a fundamental set of principles that determine good design practice, can help to facilitate a project team to accelerate the generation of good design concepts. Axiomatic design holds that *uncoupled* designs are to be preferred over *coupled* designs. Although uncoupled designs are not always possible, application of axiomatic design principles in DFSS presents an approach to help the DFSS team focus on functional requirements to achieve software design intents and maximize product reliability. As a result of the application of axiomatic design followed by parameter design, a robust design technique, the DFSS team achieved design robustness and reliability.

Design for X-ability (DFX)[22] is the value-added service of using best practices in the design stage to improve X where X is one of the members of the growing software DFX family (e.g., reliability, usability, and testability). DFX focuses on a vital software element of concurrent engineering, maximizing the use of limited recourses available to the DFSS teams. DFX tools collect and present facts about both the software design entity and its production processes, analyze all relationships between them, and measure the CTQ of performance as depicted by the conceptual architectures. The DFX family generates alternatives by combining strength and avoiding vulnerabilities, provides a redesign recommended for improvement, provides an if–then scenario, and does all that in many iterations.

A gap analysis identifies the difference between the optimized allocation and integration of the input and the current level of allocation. This helps provide the team with insight into areas that could be improved. The gap analysis process involves determining, documenting, and approving the variance between project requirements and current capabilities. Gap analysis naturally flows from benchmarking and other assessments. Once the general expectation of performance in the industry is understood, it is possible to compare that expectation with the current level of performance. This comparison becomes the gap analysis. Such analysis can be performed at the strategic or operational level of an organization.

Robust Design[23] variation reduction is recognized universally as a key to reliability and productivity improvement. There are many approaches to reducing the variability, each one having its place in the product development cycle. By addressing variation reduction at a particular stage in a product's life cycle, one can prevent failures in the downstream stages. The Six Sigma approach has made tremendous gains in cost reduction by finding problems that occur in operations and fixing the immediate causes. The robustness strategy of the CTQs is to prevent problems through optimizing software product designs and their production operations.

Regression is a powerful method for predicting and measuring CTQ responses. Unfortunately, simple linear regression is abused easily by not having sufficient understanding of when to—and when not to—use it. Regression is a technique that investigates and models the relationship between a dependent variable (Y) and its independent predictors (Xs). It can be used for hypothesis testing, modeling causal relationships ($Y = f(x)$), or a prediction model. However, it is important to make sure that the underlying model assumptions are not violated. One of the key outputs in a

[22] See Chapter 14.
[23] See Chapter 18.

regression analysis is the regression equation and correlation coefficients. The model parameters are estimated from the data using the method of least squares. The model also should be checked for adequacy by reviewing the quality of the fit and checking residuals.

8.7.4 Sample Verify and Validate Phase DFSS Tools

FMEA can provide an analytical approach when dealing with potential failure modes and their associated causes. When considering possible failures in a software design—like safety, cost, performance, quality, and reliability—a team can get a lot of information about how to alter the development and production process, in order to avoid these failures. FMEA provides an easy tool to determine which risk has the greatest concern, and therefore, an action is needed to prevent a problem before it develops. The development of these specifications will ensure the product will meet the defined requirements.

Capability analysis is about determining how well a process meets a set of specification limits, based on a sample of data taken from a process. It can be used to establish a baseline for the process and measure the future state performance of the process for comparison.

It graphically illustrates the relationship between a given outcome and all the factors that influence the outcome. This type of diagram is sometimes called an Ishikawa diagram (a.k.a. Fishbone or cause-and–effect). A cause-and-effect diagram is a tool that is useful for identifying and organizing the known or possible causes of quality, or the lack of it. The structure provided by the diagram helps team members think in a very systematic way. Some of the benefits of constructing a cause-and-effect diagram are as follows:

- Helps determine the root causes of a problem or a CTQ using a structured approach
- Encourages group participation and uses group knowledge of the process
- Uses an orderly, easy-to-read format to diagram cause-and-effect relationships
- Increases knowledge of the development process by helping everyone to learn more about the factors at work and how they relate
- Identifies areas where data should be collected for further study

For many engineered systems, it is necessary to predict measures such as the system's reliability (the probability that a component will perform its required function over a specified time period) and availability (the probability that a component or system is performing its required function at any given time). For some engineered systems (e.g., processing plants and transportation systems), these measures directly impact the system's throughput: the rate at which material (e.g., rocks, chemicals, and products) move through the system. Reliability models are used frequently to compare design alternatives on the basis of metrics such as warranty and maintenance costs. Throughput models typically are used to compare design alternatives in order to optimize throughput and/or minimize processing costs. Software design for reliability is discussed in Chapter 14.

When it is used for software testing, there is a large amount of savings in testing time and cost. Design of experiments has been proven to be one of the best known methods for validating and discovering relationships between CTQs (Y's) and factors (x's).

8.8 OTHER DFSS APPROACHES

DFSS can be accomplished using any one of many other methodologies besides the one presented in this book. IDOV[24] is one popular methodology for designing products to meet Six Sigma standards. It is a four-phase process that consists of Identify, Design, Optimize, and Verify. These four phases parallel the four phases of the ICOV process presented in this book.

- Identify phase: It begins the process with a formal tie of design to VOC. This phase involves developing a team and a team charter, gathering VOC, performing competitive analysis, and developing CTSs.
- Design phase: This phase emphasizes CTSs and consists of identifying functional requirements, developing alternative concepts, evaluating alternatives, selecting a best-fit concept, deploying CTSs, and predicting sigma capability.
- Optimize phase: The Optimize phase requires use of process capability information and a statistical approach to tolerancing. Developing detailed design elements, predicting performance, and optimizing design take place within this phase.
- Validate phase: The Validate phase consists of testing and validating the design. As increased testing using formal tools occurs, feedback of requirements should be shared with production operations and sourcing, and future operations and design improvements should be noted.

Another popular Design for Six Sigma methodology is called DMADV, and it retains the same number of letters, number of phases, and general feel as the DMAIC acronym. The five phases of DMADV are:

- Define: Define the project goals and customer (internal and external) requirements.
- Measure: Measure and determine customer needs and specifications; benchmark competitors and industry.
- Analyze: Analyze the process options to meet the customer's needs.
- Design: Design (detailed) the process to meet the customer's needs.
- Verify: Verify the design performance and ability to meet the customer's needs.

[24] See Dr. David Woodford's article at http://www.isixsigma.com/library/content/c020819a.asp.

Another flavor of the DMADV methodology is DMADOV, that is, Design, Measure, Analyze, Design, Optimize, and Verify. Other modified versions include DCCDI and DMEDI. DCCDI is being pushed by Geoff Tennant and is defined as Define, Customer Concept, Design, and Implement, which is a replica of the DMADV phases. DMEDI is being taught by PriceWaterhouseCoopers and stands for Define, Measure, Explore, Develop, and Implement. The fact is that all of these DFSS methodologies use almost the same tools (quality function deployment, failure mode and effects analysis, benchmarking, design of experiments, simulation, statistical optimization, error proofing, robust design, etc.) and provide little difficulty in alternating using them. On top of these common elements, the ICOV offers a thread through a road map with overlaid tools that is based on nontraditional tools such as design mappings, design axioms, creativity tools, as well as cultural treatments.

A DFSS approach can be mapped closely to the software development cycle as illustrated in the development of a DVD player (Shenvi, 2008) from Philips, where a reduction in cost of non quality (CONQ) is attempted using a DFSS approach. The case study is summarized in Appendix 8.A.

8.9 SUMMARY

Software DFSS offers a robust set of tools and processes that address many of today's complex business design problems. The DFSS approach helps design teams frame their project based on a process with financial, cultural, and strategic implications to the business. The software DFSS comprehensive tools and methods described in this book allow teams to assess software issues quickly and identify financial and operational improvements that reduce costs, optimize investments, and maximize returns. Software DFSS leverages a flexible and nimble organization and maintains low development costs allowing deploying companies to pass these benefits on to their customers. Software DFSS employs a unique gated process that allows teams to build tailor-made approaches (i.e., not all the tools need to be used in each project). Therefore, it can be designed to accommodate the specific needs of the project charter. Project by project, the competency level of the design teams will be enhanced leading to deeper knowledge and broader experience.

In this book, we formed and integrated several strategic and tactical and methodologies that produce synergies to enhance software DFSS capabilities to deliver a broad set of optimized solutions. The method presented in this book has a widespread application to help design teams and the belt population in different project portfolios (e.g., staffing and other human resources functions; finance, operations, and supply chain functions; organizational development; financial software; training; technology; and tools and methods)

Software DFSS provides a unique commitment to the project customers by guaranteeing agreed upon financial and other results. Each project must have measurable outcomes, and the design team is responsible for defining and achieving those outcomes. Software DFSS ensures these outcomes through risk identification and mitigation plans, variable (DFSS tools that are used over many stages) and fixed

(DFSS tool that is used once) tool structures and advanced conceptual tools. The DFSS principles and structure should motivate design teams to provide business and customers with a substantial return on their design investment.

8.A.1 APPENDIX 8.A (Shenvi, 2008)

8.A.1.1 Design of DivX DVD Player Using DIDOVM Process

New product or service introduction in the software arena, be it embedded or otherwise, is characterized by an increasing need to get designs right the first time. In areas such as consumer electronics (DVD players, iPhones, cell phones, etc.) or household appliances (microwave ovens, refrigerators, etc.), the margin on a product often is low, but the sale quantity often is in the order of thousands, if not millions. Hence, it is all the more important to get the desired product quality out the very first time because the cost of recalls and re-work if at all possible often ends up being a losing proposition.

The number of research papers in the public domain on the benefits of software Six Sigma and software DFSS as practiced by industry is limited as companies continue to view Six Sigma as a differentiator in the marketplace. In addition, companies often use Six Sigma in conjunction with Lean practices and do not wish to divulge specifics for competition reasons. The DivX DVD DFSS player case study is an example.

The case study outlines in the following discussion illustrates at a high level the application of DFSS to the DivX DVD player. The intent here is not to make the reader an expert but to provide a flavor and to pave the way for subsequent chapters. The case follows DIDOVM: Define–Identify–Design–Optimize–Verify–Monitor methodology.

8.A.2 DIDOVM PHASE: DEFINE

This phase is characterized by the definition of the problem (CONQ reduction), as shown in Figure 8.4. Discovery of the needs of the customer constitutes the prime focus in this phase where both the development and product management community folks are involved. From a software development cycle standpoint, VOC information typically is a part of the requirement specifications and includes information based on marketing intelligence, customer interviews, and surveys.

Software artifacts to this phase include competitive advances and technology road maps. Tools such as the cause-and-effect matrix, QFD, risk–benefit matrix, and Kano analysis are used to provide shape to "fuzzy" requirements that are translated and prioritized into critical-to-quality (CTQ) characteristics to aid further design.

QFD (a.k.a. house of quality) is among the most often used tool in most DFSS strategies. Quite often project teams use the Kano model to start with and proceed

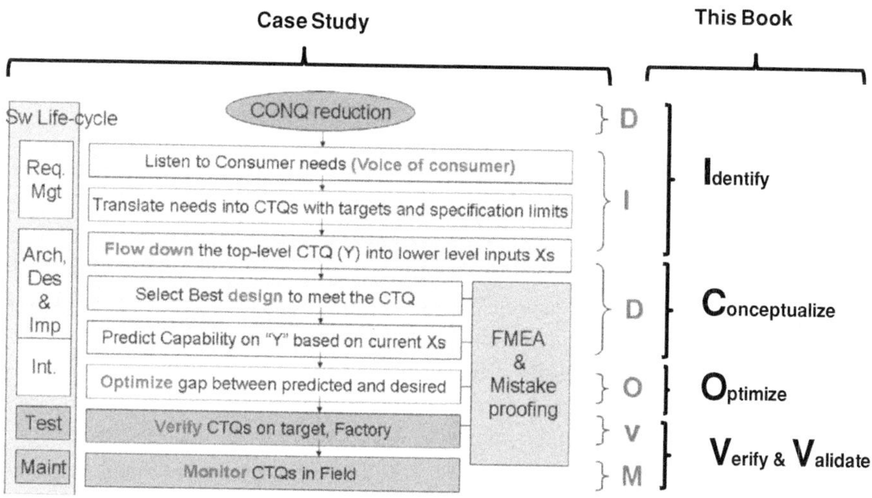

FIGURE 8.4 DFSS software development cycle mapping (Shenvi, 2008).

to the voice of the customer table and subsequently to the house of quality when identifying the CTQ characteristics.

Kano analysis helps categorize requirements and in turn the VOC into essential and differentiating attributes by simple ranking them into one of several buckets. Figure 8.A.1 shows an example involving the design of the DVD player. The team has three buckets that are must have's (essential customer needs), satisfiers (aspects that increase customer satisfaction), and delighters (good to have, "WOW" factor).

Voice of Customer
Voice of Business

Must Have's	Satisfiers	Delighters
Pause Live TV	Recording (Faster + more)	Best AV experience
Robustness	Usability - intuitiveness	On-line Help (Help Menu)
Hard Disk Functionality	Better - Archiving	DivX (multiple titles in single file)
Installation and Connectivity		
Digital Terrestrial Tuner (DTT)		
DivX - Playability		
UI Responsiveness (fast)		

FIGURE 8.A.1 Kano analysis of DVD player (Lowe, 2000).

Classification in this manner aids CTQ definition and paves the way for development of the QFD that includes several components besides the customer CTQs, as shown in Figure 8.A.2.

The HOQ is built with the following rooms (Chapter 12):

- *Customer needs (Room 1):* What is needed for the house gets specified here with each row representing a VOC (need, want, or delight).
- *Characteristic measured (Room 3)*: Identify the CTQs that are captured as a technical requirement and are assigned a column in the house. There may be a need to dive deeper into each of the How(s) until such time the factor becomes a measurable quantity. This results in the HOQ extending beyond one level.
- *Correlation (Room 4):* Reflects the impact of each CTQ on the customer requirement. The impact is color coded as strong, medium, or weak. Empty spaces indicate that there is no interaction.
- *Competitive customer rating (Room 2):* Top product or technical requirements based on customer needs are identified by assigning an influence factor on a scale of 1...10, where 1 implies least impact, which is used to find effects.
- *Conflicts (Room 8):* Provides correlation information in terms of how meeting the technical requirement impacts the product design. This information typically is updated during the design phase and is used in design tradeoffs.
- *Targets and limits (Room 7):* Get incorporated into the QFD as part of the Measure phase.
- *Customer importance (Room 1):* Ranking of the VOC on a scale of 1...5, where 5 is the most important.

8.A.3 DIDOVM PHASE: IDENTIFY

In this phase, other aspects that are a focus of this phase include the creation of a project charter that identifies the various stakeholders, the project team.

The identification of stakeholders as in Figure 8.A.3 ensures that linkages are established to the various levels (technical, commercial, sales, finance, etc.) to obtain necessary buy in and involvement from all concerned. This is of great importance in ensuring that bottlenecks get resolved in the best possible way and that change management requests are getting the appropriate attention.

The CTQ(s) identified in the Define phase are referred as the Y(s). Each Y can be either continuous or discrete. For each Y, the measurement method, target, and specification limits are identified as a part of the Measure phase.

If the CTQ is a continuous output, typical measurements and specifications relate to the performance of the CTQ or to a time-specific response (e.g., DVD playback time after insertion of a DVD and selection of the play button). Discrete CTQ(s) could pose challenges in terms of what constitutes a specification and what is a measure of fulfillment. It may be necessary to identify the critical factors associated

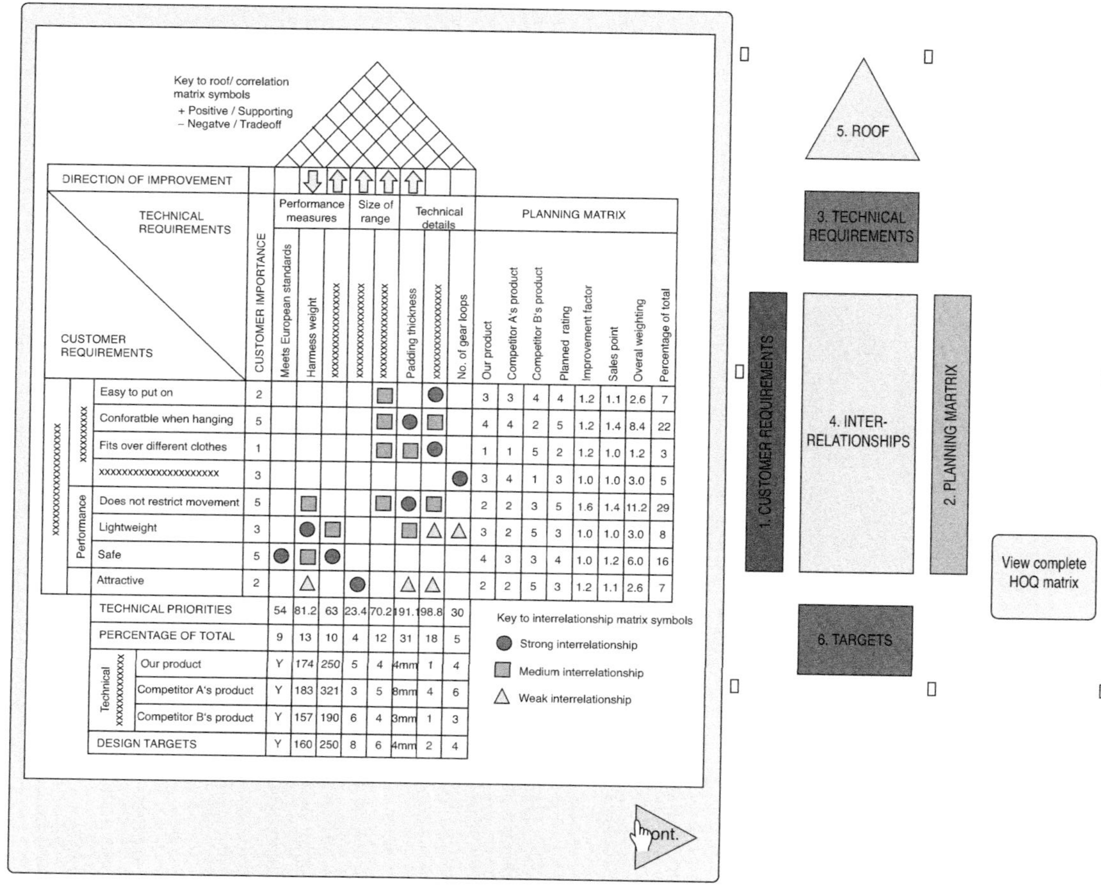

FIGURE 8.A.2 QFD/house-of-quality components.

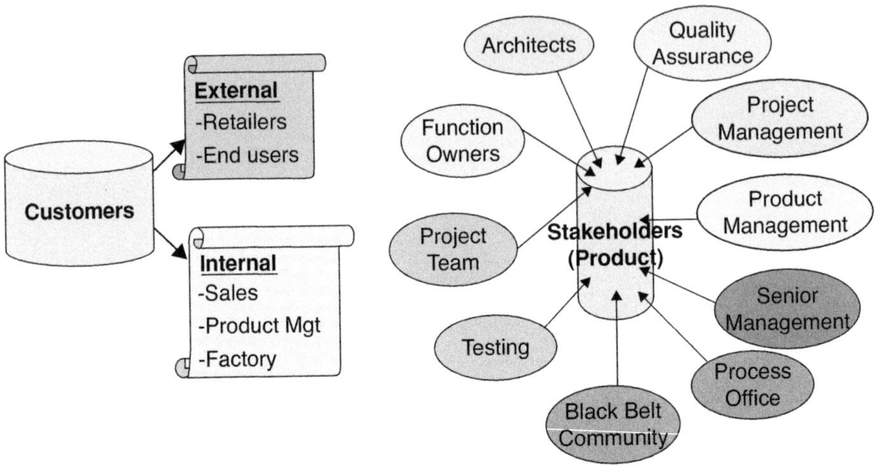

FIGURE 8.A.3 Customers and stakeholders.

with the discrete CTQ and use indirect measures to make these quantifiable. One such challenge in the case of the DVD player was the CTQ–DivX Playability feature (Yes/No). This is discrete but made quantifiable by the team as follows:

> DivX playability was an interesting case. An end user would typically want everything that is called as DivX content to play on his device. This is a free content available on the Internet and it is humanly impossible to test all. To add to the problems, users can also create text files and associate with a DivX content as "external subtitles". Defining a measurement mechanism for this CTQ was becoming very tricky and setting target even trickier. So we again had a brainstorming with product management and development team, searched the Internet for all patterns of DivX content available, and created a repository of some 500 audio video files. This repository had the complete spectrum of all possible combinations of DivX content from best case to worst case and would address at least 90% of use cases. The measurement method then was to play all these 500 files and the target defined was at least 90% of them should play successfully. So DivX playability then became our discrete CTQ (Shenvi, 2008, p. 99).

Artifacts in the software development cycle needed for this phase include the requirement specifications. A general rule of thumb governing the definition of upper and lower specification limits is the measure of success on a requirement, and hence, the tolerance on the specification often is tighter than the customer measure of success. If $Y = f(X_1, X_2, X_3, \ldots X_n)$, X2... X*n*), then the variation of Y is determined by the variation of the independent variables x(*s*). The aim of the measure phase is to define specifications for the individual X's that influence the Y such that the design is both accurate (on target) and precise (small variation). By addressing the aspect of target and variation in this phase, the DFSS ensures that the design would fully meet customer requirements.

8.A.4 DIDOVM PHASE: DESIGN

The goal of the design phase is twofold:

- Select the best design.
- Decompose CTQ(s) into actionable, low-level factors—X(s) referred to as CTQ flow-down.

Decomposition of CTQ(s) helps to identify correlations and aids in the creation of transfer functions that can be used to model system behavior and can be used in prediction of output performance. However, transfer functions may not be derivable at all times. In such cases, it often is very important to identify the critical factors X(s), the inputs that are constant or fixed and the items that are noise. For instance, in designing the DVD player, the DivX transfer function gets represented as shown in Figure 8.A.4 and helps establish the critical X factors to be controlled for achieving predictability on the Y(s). This is referred to as CTQ flow-down.

Predicting product performance, also known as capability flow-up, on CTQ(s) is another key aspect in this phase. It often is difficult to predict performance during the early stages of product development for CTQ(s) in the absence of a clear set of correlations. In some cases, this may, however, be possible. For example, in the case of the DVD player, the CTQ startup time (‘Y’) and each of the X’s 1, 2, and 3 that contribute to it can be quantified as:

$$\text{Startup time (Y)} = \text{drive initialization (X1)} + \text{software initialization (X2)} \\ + \text{diagnostic check time (X3)}$$

The measurable aspect of the startup time makes it a candidate that will be examined during the Unit-Testing phase. In CTQ flow-down, the average value of Y and the desired variation we want in the Y’s are used to derive the needed value of X’s,

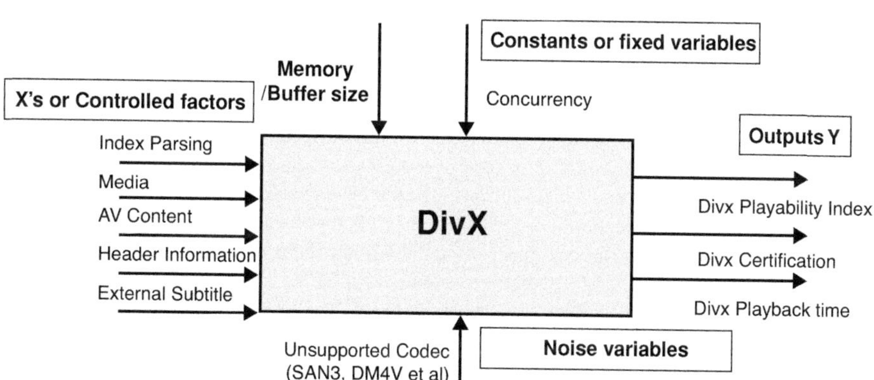

FIGURE 8.A.4 DivX feature transfer function.

whereas in CTQ flow-up, data obtained via simulation or empirical methods of the various X's is used to predict the final performance on Y.

Predicting design behavior also brings to the fore another critical DFSS methodology component: process variation, part variation, and measurement variation. For instance, change in the value of a factor (X1) may impact outputs (Y1 and Y2) of interest in opposite ways. How do we study the effect of these interactions in a software design? The Main effects plot and interaction plots available through Minitab (Minitab Inc., State College, PA)—the most widely used Six Sigma analysis tool—often are used to study the nature of interaction.

FMEA often is carried out during this phase to identify potential failure aspects of the design and plans to overcome failure. FMEA involves computation of a risk priority number (RPN) for every cause that is a source of variation in the process. For each cause severity, correction is rated on a scale of 1...10, with 1 being the best and 10 the worst. The detection aspect for each cause also is rated on a scale of 1...10, but here a rating of 10 is most desirable, whereas 1 is least desirable.

- *Severity*—How significant is the impact of the cause on the output?
- *Occurrence*—How likely is it that the cause of the failure mode will occur?
- *Detection*—How likely is it that the current design will be able to detect the cause or mode of failure should

$$\text{Risk Priority Number} = \text{Severity} \times \text{Occurrence} \times \text{Detection}$$

If data from an earlier design were available, regression is a possible option, whereas design of experiments (DOE), inputs from domain experts, factorial design, simulation, or a combination often is adopted when past data are not available. Businesses also could use techniques such as ATAM (Kazman et al., 2000) that place emphasis on performance, modifiability, and availability characteristics to determine the viability of a software design from an architectural standpoint. This offers a structured framework to evaluate designs with a view to determining the design tradeoffs and is an aspect that makes for interesting study.

> Each quality attribute characterization is divided into three categories: *external stimuli, architectural decisions,* and *responses*. *External stimuli* (or just *stimuli* for short) are the events that cause the architecture to respond or change. To analyze architecture for adherence to quality requirements, those requirements need to be expressed in terms that are concrete and measurable or observable. These measurable/observable quantities are described in the *responses* section of the attribute characterization. *Architectural decisions* are those aspects of an architecture i.e. components, connectors, and their properties—that have a direct impact on achieving attribute responses. For example, the external stimuli for performance are events such as messages, interrupts, or user keystrokes that result in computation being initiated. Performance architectural decisions include processor and network arbitration mechanisms; concurrency structures including processes, threads, and processors; and properties including process priorities and execution times. Responses are characterized by measurable quantities such as latency and throughput. For modifiability, the external stimuli are change requests to the

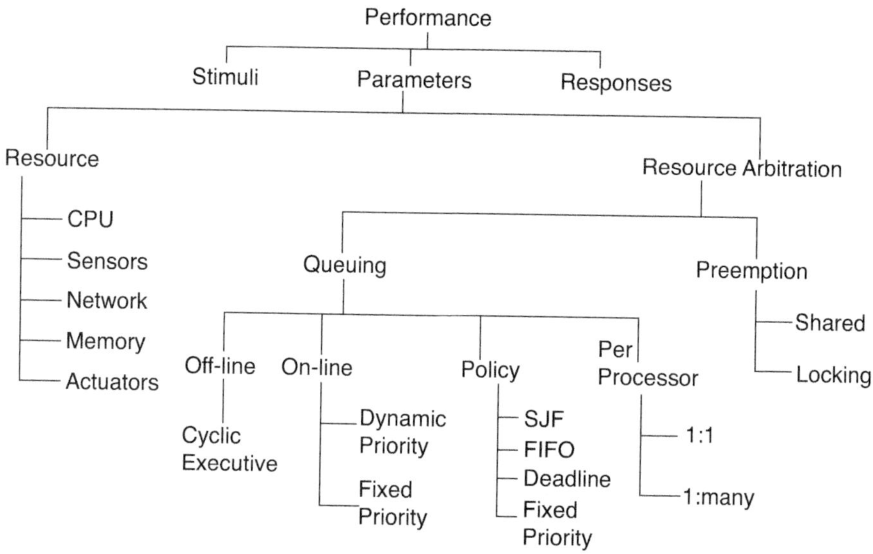

FIGURE 8.A.5 ATAM—performance characterization architectural methods.

system's software. Architectural decisions include encapsulation and indirection mechanisms, and the response is measured in terms of the number of affected components, connectors, and interfaces and the amount of effort involved in changing these affected elements. Characterizations for performance, availability, and modifiability are given below in Figures: 8.A.5–8.A.9 (Kazman et al., 2000, p. 100).

Figures 8.A.5–8.A.9 outline the aspects to consider when issues of software robustness and quality are to be addressed from a design perspective. These are not

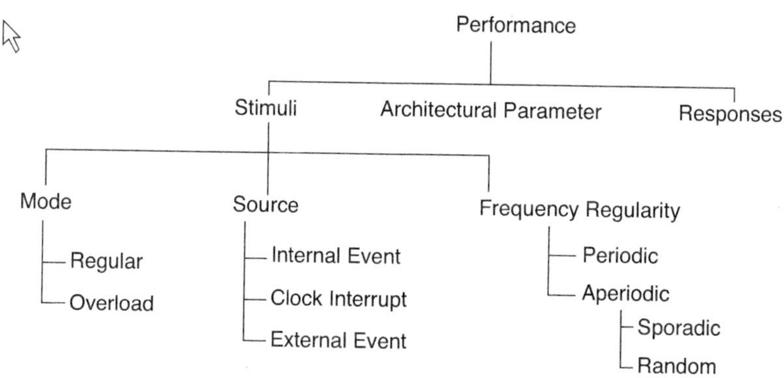

FIGURE 8.A.6 ATAM—performance characterization stimuli.

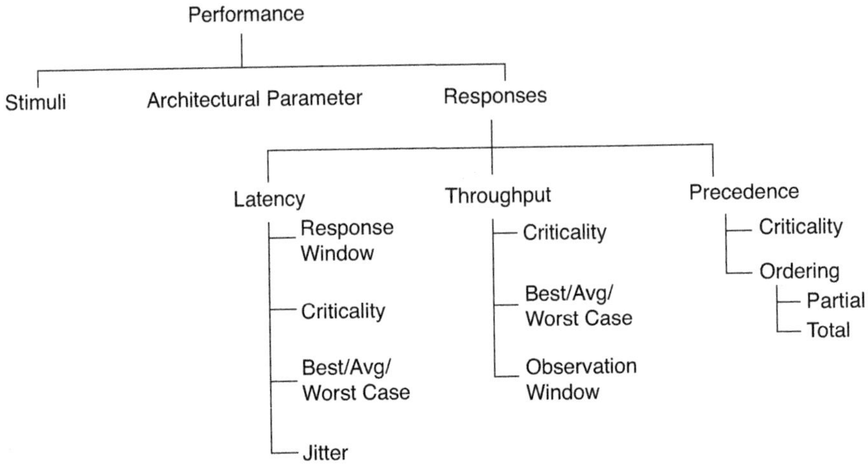

FIGURE 8.A.7 ATAM—performance characterization response to stimuli.

discussed as a part of this chapter but are intended to provide an idea of the factors that the software design should address for it to be robust.

The Design phase maps to the Design and Implementation phase of the software development cycle. The software architecture road map, design requirements, and use cases are among the artifacts that are used in this phase.

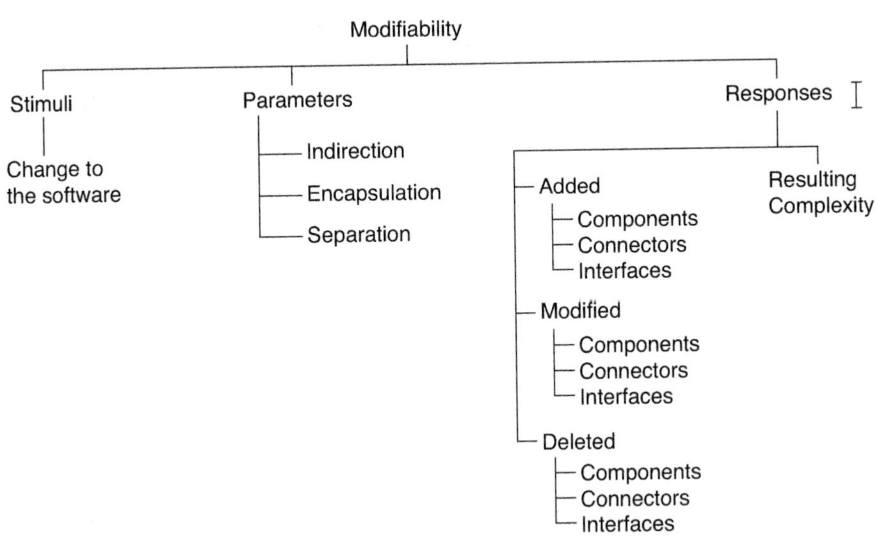

FIGURE 8.A.8 ATAM—modifiability characterization.

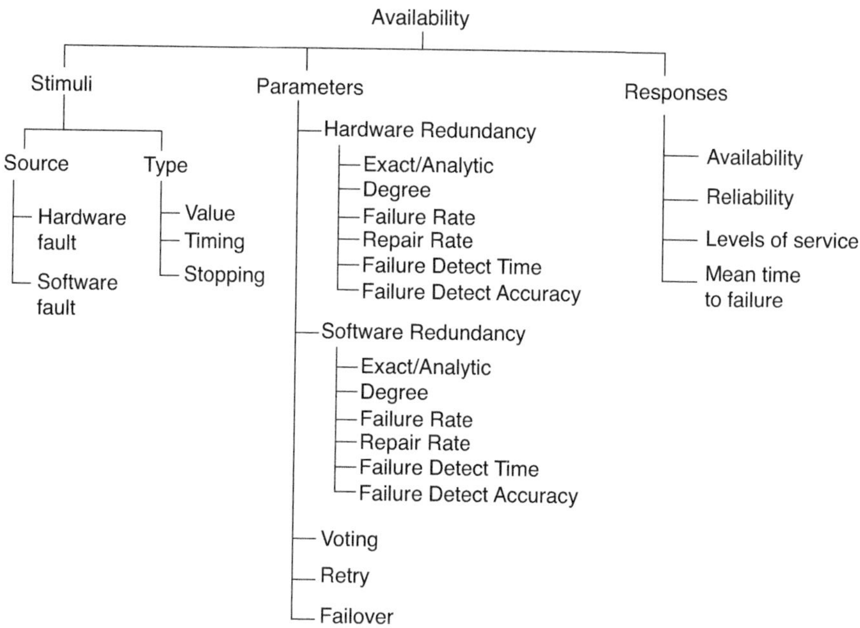

FIGURE 8.A.9 ATAM—availability characterization.

8.A.5 DIDOVM PHASE: OPTIMIZE

Optimizing the design typically involves one or more of the following:

- Statistical analysis of variance drivers
- Robustness
- Error proofing

One way to address robustness from a coding standpoint discussed in the DVD player case study is to treat this as a CTQ, determine the X factors, and look at effective methods to address the risks associated with such causes.

Robustness $= f$(Null pointers, Memory leaks, CPU load, Exceptions, Coding errors)

Error-proofing aspects typically manifest as opportunities originating from the FMEA study, performed as part of the design. There are six mistake-proofing principles[25] or methods that can be applied to the software design. Table 8.A.1 shows the details of the error-proofing methods.

[25]Crow, K. @ http://www.npd-solutions.com/mistake.html—Error Proofing and Design.

TABLE 8.A.1 Error-Proofing Methods

Method	Explanation	Example
Elimination	Redesign product to avoid usage of the component.	Redesign code to avoid use of GOTO statements.
Replacement	Substitute with a more reliable process.	Replace multiple "If Then Else" statements with a "Case" statement.
Prevention	Design the product such that it is impossible to make a mistake.	Use polarized connectors on electronic circuit boards.
Facilitation	Combine steps to simplify the design.	Reduce number of user interfaces for data entry.
Detection	Identify error before processing.	Validate data type when processing user data.
Mitigation	Minimize effect of errors.	Provide graceful exit and error recovery in code.

From a software development cycle, this phase may be treated as an extension of the Design phase.

8.A.6 DIDOVM PHASE: VERIFY

The Verify phase is akin to the Testing phase of a software development cycle. Tools like Minitab are used extensively in this phase where statistical tests and Z scores are computed and control charts are used extensively to determine how well the CTQ(s) are met. When performing response time or other performance related tests, it is important that the measurement system is calibrated and that errors in the measurement system are avoided. One technique used to avoid measurement system errors is to use instruments from the same manufacturer so that testers can avoid device-related errors from creeping in.

The example in Figure 8.A.10 relates to the DVD player example where the "content feedback time" CTQ performance was verified. Notice that the score for Z is very high, indicating that the extent of variation in the measured metric is very low.

One aspect to be kept in mind when it comes to software verification is the aspect of repeatability. Because software results often are repeatable, the Z scores often tend to be high but the results can be skewed when tests are run in conjunction with the hardware and the environment in which the system will operate in an integrated fashion.

8.A.7 DIDOVM PHASE: MONITOR

It is in this phase that the product becomes a reality and hence the customer response becomes all the more important. A high spate of service calls after a new product

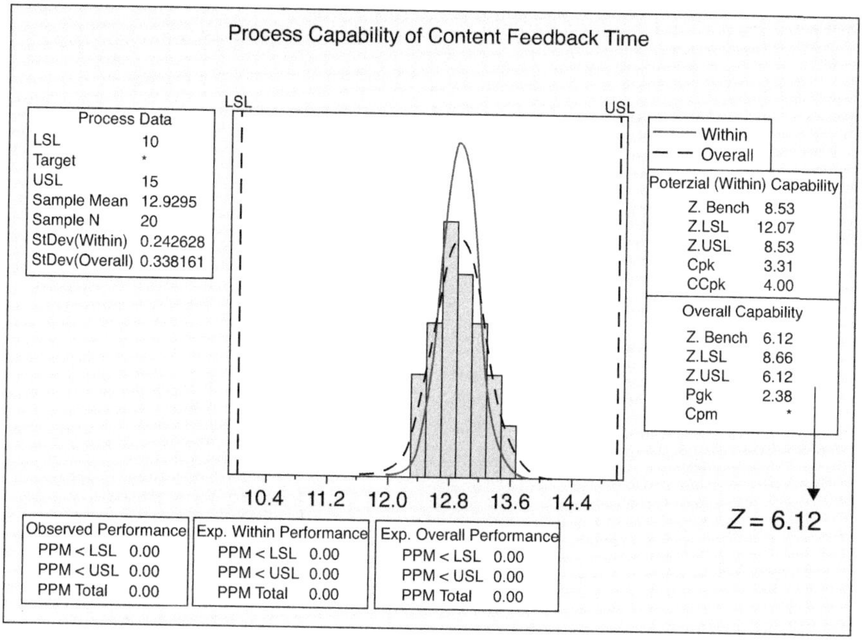

FIGURE 8.A.10 Process capability—content feedback time (CTQ).

launch could indicate a problem. However, it often is difficult to get a good feel for how good the product is, until we start seeing the impact in terms of service calls and warranty claims for at least a three-month period. The goal of the DFSS is to minimize the extent of effort needed in terms of both resources and time during this phase, but this would largely depend on how well the product is designed and fulfills customer expectations. Information captured during this phase typically is used in subsequent designs as part of continual improvement initiatives.

REFERENCES

El-Haik, Basem, S. (2005), *Engineering*, 1st Ed., Wiley-Interscience, New York.

El-Haik, Basem S. and Roy, D. (2005), *Service Design for Six Sigma: A Roadmap for Excellence*, Wiley-Interscience, New York.

Fredrikson, B. (1994), Holostic Systems Engineering in Product Development, The Saab-Scania Griffin, Saab-Scania, AB, Linkoping, Sweden.

Kazman, R., Klein, M., and Clements, P. (2000), *ATAM: Method for Architecture Evaluation* (CMU/SEI-2000-TR-004, ADA382629). Software Engineering Institute, Pittsburgh, PA.

Pan, Z., Park, H., Baik, J., and Choi, H. (2007), "A Six Sigma Framework for Software Process Improvement and Its Implementation," IEEE, Proc. of the 14th Asia Pacific Software Engineering Conference.

Shenvi, A.A. (2008), "Design for Six Sigma: Software Product Quality," IEEE, Proc. of the 1st Conference on India Software Engineering Conference: ACM, pp. 97–106.

Suh, N.P. (1990), *The Principles of Design*, Oxford University Press, New York.

Tayntor, C. (2002), *Six Sigma Software Development*, 1st Ed., Auerbach Publications, Boca Raton, FL.

Yang and El-Haik, Basem, S. (2003).

Yang, K. and El-Haik, Basem, S. (2008), Design for Six Sigma: A Roadmap for Product Development, 2nd Ed., McGraw-Hill Professional, New York.

CHAPTER 9

SOFTWARE DESIGN FOR SIX SIGMA (DFSS): A PRACTICAL GUIDE FOR SUCCESSFUL DEPLOYMENT

9.1 INTRODUCTION

Software Design for Six Sigma (DFSS) is a disciplined methodology that embeds customer expectations into the design, applies the transfer function approach to ensure customer expectations are met, predicts design performance prior to pilot, builds performance measurement systems (scorecards) into the design to ensure effective ongoing process management, leverages a common language for design, and uses tollgate reviews to ensure accountability

This chapter takes the support of a software DFSS deployment team that will launch the Six Sigma program as an objective. A deployment team includes different levels of the deploying company leadership, including initiative senior leaders, project champions, and other deployment sponsors. As such, the material of this chapter should be used as deployment guidelines with ample room for customization. It provides the considerations and general aspects required for a smooth and successful initial deployment experience.

The extent to which software DFSS produces the desired results is a function of the adopted deployment plan. Historically, we can observe that many sound initiatives become successful when commitment is secured from involved people at all levels. At the end, an initiative is successful when crowned as the new norm in the respective functions. Software Six Sigma and DFSS are no exception. A successful DFSS deployment is people dependent, and as such, almost every level, function, and division involved with the design process should participate including the customer.

Software Design for Six Sigma: A Roadmap for Excellence, By Basem El-Haik and Adnan Shaout

9.2 SOFTWARE SIX SIGMA DEPLOYMENT

The extent to which a software Six Sigma program produces results is directly affected by the plan with which it is deployed. This section presents a high-level perspective of a sound plan by outlining the critical elements of successful deployment. We must point out up front that a successful Six Sigma initiative is the result of key contributions from people at all levels and functions of the company. In short, successful Six Sigma initiatives require buy-in, commitment, and support from officers, executives, and management staff before and while employees execute design and continuous improvement projects.

This top-down approach is critical to the success of a software Six Sigma program. Although Black Belts are the focal point for executing projects and generating cash from process improvements, their success is linked inextricably to the way leaders and managers establish the Six Sigma culture, create motivation, allocate goals, institute plans, set procedures, initialize systems, select projects, control resources, and maintain an ongoing recognition and reward system.

Several scales of deployment may be used (e.g., across the board, by function, or by product); however, maximum entitlement of benefits only can be achieved when all affected functions are engaged. A full-scale, company-wide deployment program requires senior leadership to install the proper culture of change before embarking on their support for training, logistics, and other resources required. People empowerment is the key as well as leadership by example.

Benchmarking the DMAIC Six Sigma program in several successful deployments, we can conclude that a top-down deployment approach will work for software DFSS deployment as well. This conclusion reflects the critical importance of securing and cascading the buy-in from the top leadership level. The Black Belts and the Green Belts are the focused force of deployment under the guidance of the Master Black Belts and champions. Success is measured by an increase in revenue and customer satisfaction as well as by generated cash flow in both the long and short terms (soft and hard), one a project at a time. Belted projects should, diligently, be scoped and aligned to the company's objectives with some prioritization scheme. Six Sigma program benefits cannot be harvested without a sound strategy with the long-term vision of establishing the Six Sigma culture. In the short term, deployment success is dependent on motivation, management commitment, project selection and scoping, an institutionalized reward and recognition system, and optimized resources allocation. This chapter is organized into the following sections, containing the information for use by the deployment team.

9.3 SOFTWARE DFSS DEPLOYMENT PHASES

We are categorizing the deployment process, in term of evolution time, into three phases:

- The Predeployment phase to build the infrastructure

- The Deployment phase where most activities will happen
- The Postdeployment phase where sustainment needs to be accomplished

9.3.1 Predeployment

Predeployment is a phase representing the period of time when a leadership team lays the groundwork and prepares the company for software Six Sigma design implementation, ensures the alignment of its individual deployment plans, and creates synergy and heightened performance.

The first step in an effective software DFSS deployment starts with the top leadership of the deployment company. It is at this level that the team tasked with deployment works with the senior executives in developing a strategy and plan for deployment that is designed for success. Six Sigma initiative marketing and culture selling should come from the top. Our observation is that senior leadership benchmark themselves across corporate America in terms of results, management style, and company aspirations. Six Sigma, in particular DFSS, is no exception. The process usually starts with a senior leader or a pioneer who begins to research and learn about Six Sigma and the benefits/results it brings to the culture. The pioneer starts the deployment one step at a time and begins shaking old paradigms. The old paradigm guards become defensive. The defense mechanisms begin to fall one after another based on the undisputable results from several benchmarked deploying companies (GE, 3M, Motorola, Textron, Allied Signal, Bank of America, etc.). Momentum builds, and a team is formed to be tasked with deployment. As a first step, it is advisable that select senior leadership as a team meet jointly with the assigned deployment team offsite (with limited distractions) that entails a balanced mixture of strategic thinking, Six Sigma high-level education, interaction, and hands-on planning. On the education side, overviews of Six Sigma concepts, presentation of successful deployment benchmarking, and demonstration of Six Sigma statistical methods, improvement measures, and management controls are very useful. Specifically, the following should be a minimum set of objectives of this launch meeting:

- Understand the philosophy and techniques of software DFSS and Six Sigma, in general.
- Experience the application of some tools during the meeting.
- Brainstorm a deployment strategy and a corresponding deployment plan with high first-time-through capability.
- Understand the organizational infrastructure requirements for deployment.
- Set financial and cultural goals, targets, and limits for the initiative.
- Discuss project pipeline and Black Belt resources in all phases of deployment.
- Put a mechanism in place to mitigate deployment risks and failure modes. Failure modes like the following are indicative of a problematic strategy: training Black Belts before champions; deploying DFSS without multigenerational software plans and software technology road maps; validing data and

measurement systems; leadership development; compensation plan; or change management process.

- Design a mechanism for tracking the progress of the initiative. Establish a robust "financial" management and reporting system for the initiative.

Once this initial joint meeting has been held, the deployment team could replicate to other additional tiers of leadership whose buy-in is deemed necessary to *push* the initiative through the different functions of the company. A software Six Sigma *pull* system needs to be created and sustained in the Deployment and Postdeployment phases. Sustainment indicates the establishment of bottom-up pulling power. Software Six Sigma, including DFSS, has revolutionized many companies in the last 20 years. On the software side, companies of various industries can be found implementing software DFSS as a vehicle to plan growth, improve software products and design process quality, delivery performance, and reduce cost. In parallel, many deploying companies also find themselves reaping the benefits of increased employee satisfaction through the true empowerment Six Sigma provides. Factual study of several successful deployments indicates that push and pull strategies need to be adopted based on needs and differ strategically by objective and phase of deployment. A push strategy is needed in the Predeployment and Deployment phases to jump-start and operationalize deployment efforts. A pull system is needed in the Postdeployment phase once sustainment is accomplished to improve deployment process performance on a continuous basis. In any case, top and medium management should be on board with deployment; otherwise, the DFSS initiative will fade away eventually.

9.3.2 Predeployment Considerations

The impact of a DFSS initiative depends on the effectiveness of deployment (i.e., how well the Six Sigma design principles and tools are practiced by the DFSS project teams). Intensity and constancy of purpose beyond the norm are required to improve deployment constantly. Rapid deployment of DFSS plus commitment, training, and practice characterize winning deploying companies.

In the Predeployment phase, the deployment leadership should create a compelling business case for initiating, deploying, and sustaining DFSS as an effort. They need to raise general awareness about what DFSS is, why the company is pursuing it, what is expected of various people, and how it will benefit the company. Building the commitment and alignment among executives and deployment champions to support and drive deployment aggressively throughout the designated functions of the company is a continuous activity. Empowerment of leaders and DFSS operatives to carry out effectively their respective roles and responsibilities is a key to success. A successful DFSS deployment requires the following prerequisites in addition to the senior leadership commitment previously discussed.

9.3.2.1 *Deployment Structure Established (Yang and El-Haik, 2008).* The first step taken by the senior deployment leader is to establish a deployment team to develop strategies and oversee deployment. With the help of the deployment team, the leader is responsible for designing, managing, and delivering successful deployment of the initiative throughout the company, locally and globally. He or she needs to work with Human Resources to develop a policy to ensure that the initiative becomes integrated into the culture, which may include integration with internal leadership development programs, career planning for Belts and deployment champions, a reward and recognition program, and progress reporting to the senior leadership team. In addition, the deployment leader needs to provide training, communication (as a single point of contact to the initiative), and infrastructure support to ensure consistent deployment.

The critical importance of the team overseeing the deployment cannot be overemphasized to ensure the smooth and efficient rollout. This team sets a DFSS deployment effort in the path to success whereby the proper individuals are positioned and support infrastructures are established. The deployment team is on the deployment forward edge assuming the responsibility for implementation. In this role, team members perform a company assessment of deployment maturity, conduct a detailed gap analysis, create an operational vision, and develop a cross-functional Six Sigma deployment plan that spans human resources, information technology (IT), finance, and other key functions. Conviction about the initiative must be expressed at all times, even though in the early stages there is no physical proof for the company's specifics. They also accept and embody the following deployment aspects:

- Visibility of the top-down leadership commitment to the initiative (indicating a push system).
- Development and qualification of a measurement system with defined metrics to track the deployment progress. The objective here is to provide a tangible picture of deployment efforts. Later a new set of metrics that target effectiveness and sustainment needs to be developed in maturity stages (end of Deployment phase).
- Stretch-goal setting process in order to focus culture on changing the process by which work gets done rather than on adjusting current processes, leading to quantum rates of improvement.
- Strict adherence to the devised strategy and deployment plan.
- Clear communication of success stories that demonstrate how DFSS methods, technologies, and tools have been applied to achieve dramatic operational and financial improvements.
- Provide a system that will recognize and reward those who achieve success.

The deployment structure is not only limited to the deployment team overseeing deployment both strategically and tactically, but also it includes project champions, functional areas, deployment champions, process and design owners who will implement the solution, and Master Black Belts (MBBs) who mentor and coach the Black Belts. All should have very crisp roles and responsibilities with defined objectives. A premier deployment objective can be that the Black Belts are used as a task force to improve customer satisfaction, company image and other strategic long-term objectives of the deploying company. To achieve such objectives, the deploying division should establish a deployment structure formed from deployment directors, centralized deployment team overseeing deployment, and Master Black Belts (MBBs) with defined roles and responsibilities as well as long- and short-term planning. The structure can take the form of a council with a definite recurring schedule. We suggest using software DFSS to design the DFSS deployment process and strategy. The deployment team should:

- Develop a Green Belt structure of support to the Black Belts in every department.
- Cluster the Green Belts (GBs) as a network around the Black Belts for synergy and to increase the velocity of deployment.
- Ensure that the scopes of the projects are within control, that the project selection criteria are focused on the company's objective like quality, cost, customer satisfiers, delivery drivers, and so on.
- Handing-off (matching) the right scoped projects to Black Belts.
- Support projects with key up-front documentation like charters or contracts with financial analysis highlighting savings and other benefits, efficiency improvements, customer impact, project rationale, and so on. Such documentation will be reviewed and agreed to by the primary stakeholders (deployment champions, design owners, Black Belts, and Finance Leaders),
- Allocate the Black Belt resources optimally across many divisions of the company targeting first high-impact projects as related to deployment plan and business strategy, and create a long-term allocation mechanism to target a mixture of DMAIC versus DFSS to be revisited periodically. In a healthy deployment, the number of DFSS projects should grow, whereas the number of DMAIC[1] projects should decay over time. However, this growth in the number of DFSS projects should be managed. A growth model, an S-curve, can be modeled over time to depict this deployment performance. The initiating condition of how many and where DFSS projects will be targeted is a significant growth control factor. This is very critical aspect of deployment, in particular, when the deploying company chooses not to separate the training track of the Black Belts to DMAIC and DFSS and to train the Black Belt on both methodologies.
- Use available external resources as leverage when advantageous, to obtain and provide the required technical support.
- Promote and foster work synergy through the different departments involved in the DFSS projects.
- Maximize the utilization of the continually growing DFSS community by successfully closing most of the matured projects approaching the targeted completion dates.

[1]Chapter 7

- Keep leveraging significant projects that address the company's objectives, in particular, the customer satisfaction targets.
- Maximize Black Belt certification turnover (set target based on maturity).
- Achieve and maintain working relationships with all parties involved in DFSS projects that promotes an atmosphere of cooperation, trust, and confidence between them.

9.3.2.2 Other Deployment Operatives. Several key people in the company are responsible for jump-starting the company for successful deployment. The same people also are responsible for creating the momentum, establishing the culture, and driving DFSS through the company during the Predeployment and Deployment phases. This section describes who these people are in terms of their roles and responsibilities. The purpose is to establish clarity about what is expected of each deployment team member and to minimize the ambiguity that so often characterizes change initiatives usually tagged as the flavor-of-the-month.

9.3.2.2.1 Deployment Champions. In the deployment structure, the deployment champion role is a key one. This position usually is held by an executive-ranked vice president assigned to various functions within the company (e.g., marketing, IT, communication, or sales). His or her task as a part of the deployment team is to remove barriers within their functional area and to make things happen, review DFSS projects periodically to ensure that project champions are supporting their Black Belts' progress toward goals, assist with project selection, and serve as "change agents."

Deployment champions are full time into this assignment and should be at a level to execute the top-down approach, the push system, in both the Predeployment and Deployment phases. They provide key individuals with the managerial and technical knowledge required to create the focus and facilitate the leadership, implementation, and deployment of DFSS in designated areas of their respective organizations. In software DFSS deployment, they are tasked with recruiting, coaching, and developing (not training, but mentoring) Black Belts; identifying and prioritizing projects; leading software programs and design owners; removing barriers; providing the drum beat for results; and expanding project benefits across boundaries via a mechanism of replication. Champions should develop a big-picture understanding of DFSS, deliverables, tools to the appropriate level, and how DFSS fits within the software life cycle.

The deployment champion will lead his or her respective function's total quality efforts toward improving growth opportunities, quality of operations, and operating margins among others using software DFSS. This leader will have a blend of business acumen and management experience, as well as process improvement passion. The deployment champions need to develop and grow a Master Black Belt training program for the purpose of certifying and deploying homegrown future Master Back Belts throughout deployment. In summary, the deployment champion is responsible for broad-based deployment, common language, and culture transformation by weaving Six Sigma into the company DNA as an elevator speech, a consistent, teachable point of view of their own.

9.3.2.2.2 Project Champions. The project champions are accountable for the performance of Belts and the results of projects; for selection, scoping, and successful completion of Belt projects; for removal of roadblocks for Belts within their span of control; and for ensuring timely completion of projects. The following considerations should be the focus of the deployment team relative to project champions as they lay down their strategy relative to the champion role in deployment:

- What does a DFSS champion need to know to be effective?
- How should the champion monitor impact and progress projects?
- What are the expectations from senior leadership, the Black Belt population, and others?
- How are the expectations relative to the timeline for full adoption of DFSS into the development process?
- What is the playbook (reference) for the champions?
- What are the "must have" versus the "nice to have" tools (e.g., Lean DFSS project application)?
- How should the champion be used as a "change agent?"
- Which failure mode and effects analysis (FMEA) exercise will the champion complete—identifying deployment failure modes, ranking, or corrective actions? The FMEA will focus on potential failure modes in project execution.
- How will the champion plan for DFSS implementation: timely deployment plan within his or her span of control, project selection, project resources, and project pipeline?
- Will the champion develop guidelines, references, and checklists (cheat sheets) to help him or her understand (force) compliance with software DFSS project deliverables?

The roles and responsibilities of a champion in project execution are a vital dimension of successful deployment that needs to be iterated in the deployment communication plan. Champions should develop their teachable point of view, elevator speech, or resonant message.

A suggested deployment structure is presented in Figure 9.1.

9.3.2.2.3 Design Owner. This population of operative is the owner of the software development program or software design where the DFSS project results and conclusion will be implemented. As owner of the design entity and resources, his or her buy-in is critical and he or she has to be engaged early on. In the Predeployment phase, design owners are overwhelmed with the initiative and wondering why a Belt was assigned to fix their design. They need to be educated, consulted on project selection, and responsible for the implementation of project findings. They are tasked with project gains sustainment by tracking project success metrics after full implementation. Typically, they should serve as a team member on the project, participate in reviews, and push the team to find permanent innovative solutions

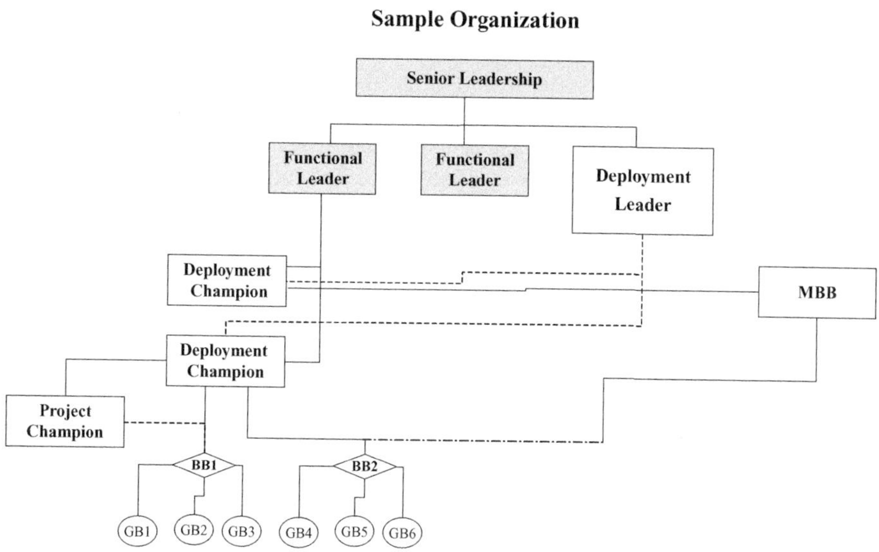

FIGURE 9.1 Suggested deployment structure.

In the Deployment and Postdeployment phases, design owners should be the first in line to staff their projects with the Belts.

9.3.2.2.4 Master Black Belt (MBB). A software Master Black Belt should possess expert knowledge of the full Six Sigma tool kit, including proven experience with DFSS. As a full-time assignment, he or she also will have experience in training, mentoring, and coaching Black Belts, Green Belts, champions, and leadership. Master Black Belts are ambassadors for the business and the DFSS initiative, someone who will be able to go to work in a variety of business environments and with varying scales of Six Sigma penetration. A Master Black Belt is a leader with good command of statistics as well as of the practical ability to apply Six Sigma in an optimal manner for the company. Knowledge of Lean also is required to move the needle on the initiative very fast. The MBB should be adaptable to the Deployment phase requirement.

Some businesses trust them with the management of large projects relative to deployment and objective achievements. MBBs also need to get involved with project champions relative to project scoping and coach the senior teams at each key function.

9.3.2.2.5 Black Belt (BB).[2] Black Belts are the critical resource of deployment as they initiate projects, apply software DFSS tools and principles, and close them with tremendous benefits. Being selected for technical proficiency, interpersonal skills,

[2]Although Black Belts are deployment portative individuals and can be under the previous section, we chose to separate them in one separate section because of their significant deployment role.

and leadership ability, a Black Belt is an individual who solves difficult business issues for the last time. Typically, the Black Belts have a couple of years on software life during the Deployment phase. Nevertheless, their effect as a disciple of software DFSS when they finish their software life (postdeployment for them) and move on as the next-generation leaders cannot be trivialized. It is recommended that a fixed population of Black Belts (usually computed as a percentage of affected functions masses where software DFSS is deployed) be kept in the pool during the designated deployment plan. This population is not static; however, it is kept replenished every year by new blood. Repatriated Black Belts, in turn, replenish the disciple population and the cycle continues until sustainment is achieved. Software DFSS becomes the way of doing design business.

Black Belts will learn and understand software DFSS methodologies and principles and find application opportunities within the project, cultivate a network of experts, train and assist others (e.g., Green Belts) in new strategies and tools, leverage surface business opportunities through partnerships, and drive concepts and methodology into the way of doing work.

The deployment of Black Belts is a subprocess with the deployment process itself with the following steps: 1) Black Belt identification, 2) Black Belt project scoping, 3) Black Belt training, 4) Black Belt deployment during the software life, and 5) Black Belt repatriation into the mainstream.

The deployment team prepares designated training waves or classes of software Black Belts to apply DFSS and associated technologies, methods, and tools on scoped projects. Black Belts are developed by project execution, training in statistics and design principles with on-the-project application, and mentored reviews. Typically, with a targeted quick cycle time, a Black Belt should be able to close a set number of projects a year. Our observations indicate that Black Belt productivity, on the average, increases after his/her training projects. After their training focused descoped project, the Black Belt projects can get more complex and evolve into cross-function, supply-chain, and customer projects.

The Black Belts are the leaders of the future. Their visibility should be apparent to the rest of the organization, and they should be cherry-picked to join the software DFSS program with the "leader of the future" stature. Armed with the right tools, processes, and DFSS principles, Black Belts are the change agent network the deploying company should use to achieve its vision and mission statements. They need to be motivated and recognized for their good effort while mentored at both the technical and leadership fronts by the Master Black Belt and the project champions. Oral and written presentation skills are crucial for their success. To increase the effectiveness of the Black Belts, we suggest building a Black Belt collaboration mechanism for the purpose of maintaining structures and environments to foster individual and collective learning of initiative and DFSS knowledge, including initiative direction, vision, and prior history. In addition, the collaboration mechanism, whether virtual or physical, could serve as a focus for Black Belt activities to foster team building, growth, and inter- and intra-function communication and collaboration. Another important reason for establishing such a mechanism is to ensure that the deployment team gets its information accurate and timely to prevent and mitigate failure modes

TABLE 9.1 Deployment Operative Roles Summary

• Project Champions	• Manage projects across company • Approve the resources • Remove the barriers • Create vision
• Master Black Belts	• Review project status • Teach tools and methodology • Assist the champion • Develop local deployment plans
• Black Belts	• Train their teams • Apply the methodology and lead projects • Drive projects to completion
• Green Belts	• Same as Black Belts (but done in conjunction with other full-time job responsibilities)
• Project Teams	• Implement process improvements • Gather data

downstream of Deployment and Postdeployment phases. Historical knowledge might include lessons learned, best-practices sharing, and deployment benchmarking data.

In summary, Table 9.1 summarizes the roles and responsibilities of the deployment operatives presented in this section.

In addition, Figure 9.2 depicts the growth curve of the Six Sigma deployment operatives. It is the responsibility of the deployment team to shape the duration and slopes of these growth curves subject to the deployment plan. The pool of Black Belts

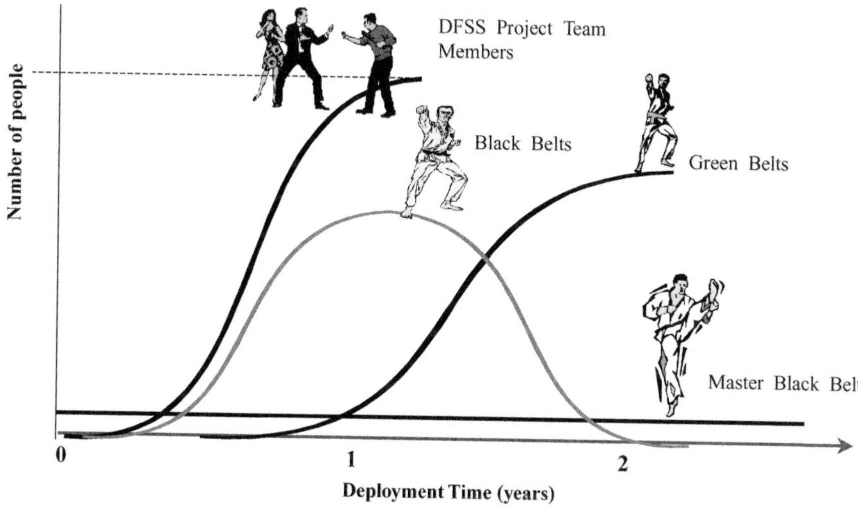

FIGURE 9.2 Deployment operative growth curves.

is replenished periodically. The 1% role (i.e., 1 Black Belt per 100 employees), has been adopted by several successful deployments. The number of MBBs is a fixed percentage of the Black Belt population. Current practice ranges from 10 to 20 Black Belts per MBB.

9.3.2.2.6 Green Belt. A Green Belt is an employee of the deploying company who has been trained on Six Sigma and will participate on project teams as part of their full-time job. The Green Belt penetration of knowledge and Six Sigma skills is less than that of a Black Belt. The Green Belt business knowledge in their company is a necessity to ensure the success of their improvement task. The Green Belt employee plays an important role in executing the Six Sigma process on day-to-day operations by completing smaller scope projects. Black Belts should be networked around Green Belts to support and coach Green Belts. Green Belt training is not for awareness. The deployment plan should enforce certification while tracking their project status as control mechanisms over deployment. Green Belts, like Black Belts, should be closing projects as well.

In summary, Green Belts are employees trained in Six Sigma methodologies that are conducting or contributing to a project that requires Six Sigma application. After successful completion of training, Green Belts will be able to participate in larger projects being conducted by a Black Belt, lead small projects, and apply Six Sigma tools and concepts to daily work.

9.3.2.3 Communication Plan. To ensure the success of software DFSS, the deployment team should develop a communication plan that highlights the key steps as software DFSS is being deployed. In doing so, they should target the audiences that will receive necessary communication at various points in the deployment process with identifiable possible mediums of communication deemed most effective by the company. The deployment team should outline the overriding communication objectives at each major phase of software DFSS deployment and provide a high-level, recommended communications plan for each of the identified communicators during company DFSS initialization.

As software DFSS is deployed in a company, we recommend that various people communicate certain messages at certain relative times. For example, at the outset of deployment, the CEO should send a strong message to the executive population that the corporation is adopting software DFSS, why it is necessary, who will be leading the effort both at leadership and deployment team levels, why their commitment and involvement is absolutely required, as well as other important items. The CEO also sends, among other communiqués to other audiences, a message to the deployment champions, explaining why they have been chosen, what is expected of them, and how they are empowered to enact their respective roles and responsibilities.

Several key people will need to communicate key messages to key audiences as DFSS is initialized, deployed, and sustained. For example, the training and development leader, finance leader, human resources (HR) leader, IT leader, project champions, deployment champions (functional leaders), managers and supervisors, Black Belts, and Green Belts, to name a few. Every leader involved in DFSS processes must have conviction in the cause to mitigate derailment. Leaders as communicators must have total belief to assist in this enabler of cultural evolution driven by DFSS. Every leader must seek out information from the deployment team to validate his or her conviction to the process.

To assist in effective communications, the leader and others responsible for communicating DFSS deployment should delineate who delivers messages to whom during the predeployment. It is obvious that certain people have primary communication responsibility during the initial stages of Six Sigma deployment, specifically the CEO, software DFSS deployment leader, deployment champions, and so on. The company communications leader plays a role in supporting the CEO, deployment leader, and other leaders as they formulate and deliver their communiqués in support of predeployment. The communication plan should include the following minimum communiqués:

- A discussion of why the company is deploying DFSS, along with several key points about how Six Sigma supports and is integrated with company's vision, including other business initiatives.
- A set of financial targets, operational goals, and metrics that will be providing structure and guidance to DFSS deployment effort. To be done with discretion of the targeted audience.
- A breakdown of where DFSS will be focused in the company; a rollout sequence by function, geography, product, or other scheme; a general timeframe for how quickly and aggressively DFSS will be deployed.
- A firmly established and supported long-term commitment to the DFSS philosophy, methodology, and anticipated results.
- Specific managerial guidelines to control the scope and depth of deployment for a corporation or function.
- A review and interrogation of key performance metrics to ensure the progressive utilization and deployment of DFSS.
- A commitment from the part-time and full-time deployment champion, full-time project champion, and full-time Black Belt resources.

9.3.2.4 Software DFSS Project Sources. The successful deployment of the DFSS initiative within a company is tied to projects derived from the company breakthrough objectives; multigeneration planning, growth, and innovation strategy; and chronic pressing redesign issues. Such software DFSS project sources can be categorized as retroactive and as proactive sources. In either case, an active measurement system should be in place for both internal and external critical-to-satisfaction (CTS's) metrics, sometimes called the "Big Y's." The measurement system should pass a Gage R&R study in all Big Y metrics. We discussed software process and product metrics in Chapter 5. So how do we define Big Y's? This question underscores why we need to decide early who is the primary customer (internal and external) of our potential DFSS project. What is the Big Y (CTS) in customer terms? It does us no good, for example, to develop a delivery system to shorten delivery processes if

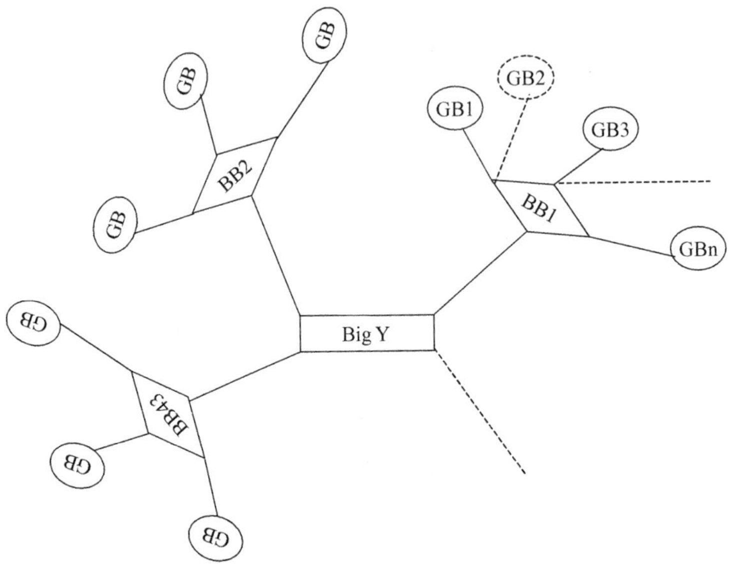

FIGURE 9.3 Green Belt (GB) and Black Belt (BB) clustering scheme.

the customer is mainly upset with quality and reliability. Likewise, it does us no good to develop a project to reduce tool breakage if the customer is actually upset with inventory cycle losses. It pays dividends to later project success to know the Big Y. No Big Y (CTS), simply means no project! Potential projects with hazy Big Y definitions are setups for Black Belt failure. Again, it is unacceptable to not know the Big Y's of top problems (retroactive project sources) or those of proactive project sources aligned with the annual objectives, growth and innovation strategy, benchmarking, and multigeneration software planning and technology road maps.

On the proactive side, Black Belts will be claiming projects from a multigenerational software plan or from the Big Y's replenished prioritized project pipeline. Green Belts should be clustered around these key projects for the deploying function or business operations and tasked with assisting the Black Belts as suggested by Figure 9.3.

We need some useful measure of Big Y's, in *variable terms*,[3] to establish the transfer function, Y = f(y). The transfer function is the means for dialing customer satisfaction, or other Big Y's, and can be identified by a combination of design mapping and design of experiment (if transfer functions are not available or cannot be derived). A transfer function is a mathematical relationship, in the concerned mapping, linking controllable and uncontrollable factors.

Sometimes we find that measurement of the Big Y opens windows to the mind with insights powerful enough to solve the problem immediately. It is not rare to find customer complaints that are very subjective, unmeasured. The Black Belt needs

[3]The transfer function will be weak and questionable without it.

to find the best measure available to his/her project Big Y to help you describe the variation faced and to support Y = f(x) analysis. The Black Belt may have to develop a measuring system for the project to be true to the customer and Big Y definition!

We need measurements of the Big Y that we trust. Studying problems with false measurements leads to frustration and defeat. With variable measurements, the issue is handled as a straightforward Gage R&R question. With attribute or other subjective measures, it is an attribute measurement system analysis (MSA) issue. It is tempting to ignore the MSA of the Big Y. This is not a safe practice. More than 50% of the Black Belts we coached encounter MSA problems in their projects. This issue in the Big Y measurement is probably worse because little thought is conventionally given to MSA at the customer level. The Black Belts should make every effort to ensure themselves that their Big Y's measurement is error minimized. We need to be able to establish a distribution of Y from which to model or draw samples for Y = f(x) study. The better the measurement of the Big Y, the better the Black Belt can see the distribution contrasts needed to yield or confirm Y = f(x).

What is the value to the customer? This should be a mute point if the project is a top issue. The value decisions are made already. Value is a relative term with numerous meanings. It may be cost, appearance, or status, but the currency of value must be decided. In Six Sigma, it is common practice to ask that each project generate average benefits greater than $250,000. This is seldom a problem in top projects that are aligned to business issues and opportunities.

The Black Belt together with the finance individual assigned to the project should decide a value standard and do a final check for potential project value greater than the minimum. High-value projects are not necessarily harder than low-value projects. Projects usually hide their level of complexity until solved. Many low-value projects are just as difficult to complete as high-value projects, so the deployment champions should leverage their effort by value.

Deployment management, including the local Master Black Belt, has the lead in identifying redesign problems and opportunities as good potential projects. The task, however, of going from potential to assigned Six Sigma project belongs to the project champion. The deployment champion selects a project champion who then carries out the next phases. The champion is responsible for the project scope, Black Belt assignment, ongoing project review, and, ultimately, the success of the project and Black Belt assigned. This is an important and responsible position and must be taken very seriously. A suggested project initiation process is depicted in Figure 9.4.

It is a significant piece of work to develop a good project, but Black Belts, particularly those already certified, have a unique perspective that can be of great assistance to the project champions. Green Belts, as well, should be taught fundamental skills useful in developing a project scope. Black Belt and Green Belt engagement is the key to helping champions fill the project pipeline, investigate potential projects, prioritize them, and develop achievable project scopes, however, with stretched targets. It is the observation of many skilled problem solvers that adequately defining the problem and setting up a solution strategy consumes the most time on the path to a

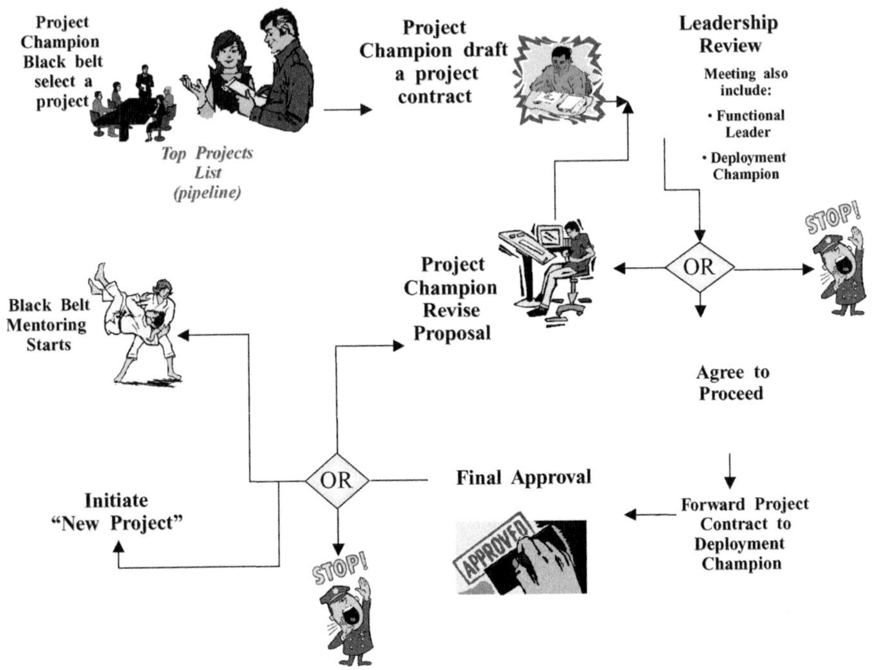

FIGURE 9.4 Software DFSS project initiation process.

successful project. The better we define and scope a project, the faster the deploying company and its customer base benefit from the solution! That is the primary Six Sigma objective.

It is the responsibility of management, deployment and project champions, with the help of the design owner, to identify both retroactive and proactive sources of DFSS projects that are important enough to assign the company's limited, valuable resources to find a Six Sigma solution. Management is the caretaker of the business objectives and goals. They set policy, allocate funds and resources, and provide the personnel necessary to carry out the business of the company. Individual Black Belts may contribute to the building of a project pipeline, but it is entirely management's list.

It is expected that an actual list of projects will always exist and be replenished frequently as new information or policy directions emerge. Sources of information from which to populate the list include all retroactive sources, support systems such as a warranty system, internal production systems related to problematic metrics such as scrap and rejects, customer repairs/complaints database, and many others. In short, the information comes from the strategic vision and annual objectives; multigeneration software plans; the voice of the customer surveys or other engagement methods; and the daily business of deployment champions, and it is their responsibility to approve what gets into the project pipeline and what does not. In general, software

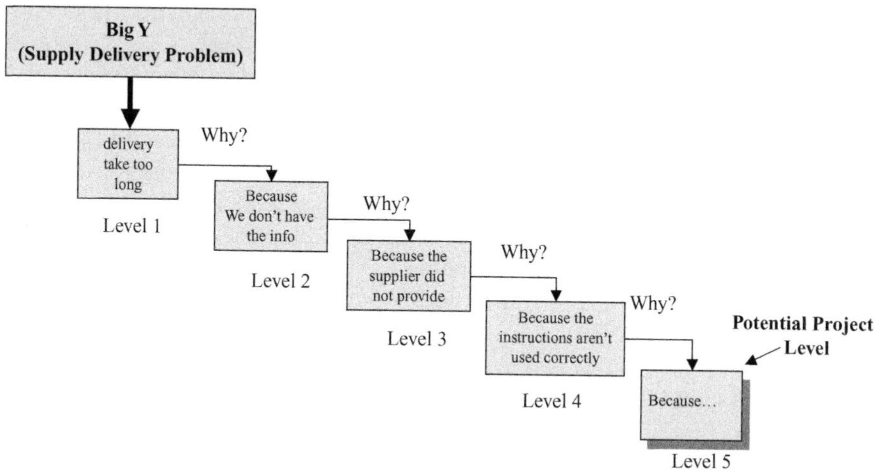

FIGURE 9.5 The "five Why" scoping technique.

DFSS projects usually come from processes that reached their ultimate capability (entitlement) and are still problematic or those targeting a new process design because of their nonexistence.

In the case of retroactive sources, projects derive from problems that champions agree need a solution. Project levels can be reached by applying the "five why" technique (see Figure 9.5) to dig into root causes prior to the assignment of the Black Belt.

A scoped project will always give the Black Belt a good starting ground and reduce the Identify phase cycle time within the ICOV DFSS approach. They must prioritize because the process of going from potential project to a properly scoped Black Belt project requires significant work and commitment. There is no business advantage in spending valuable time and resources on something with a low priority? Usually, a typical company scorecard may include metrics relative to safety, quality, delivery, cost, and environment. We accept these as big sources (buckets); yet each category has a myriad of its own problems and opportunities to drain resources quickly if champions do not prioritize. Fortunately, the Pareto principle applies so we can find leverage in the significant few. It is important to assess each of the buckets to the 80–20 principles of Pareto. In this way, the many are reduced to a significant few that still control more than 80% of the problem in question. These need review and renewal by management routinely as the business year unfolds. The top project list emerges from this as a living document.

From the individual bucket Pareto lists, champions again must give us their business insight to plan an effective attack on the top issues. Given key business objectives, they must look across the several Pareto diagrams, using the 80–20 principle, and sift again until we have few top issues on the list with the biggest impact on the business. If the champions identify their biggest problem elements well, based on management business objectives and the Pareto principle, then how could any manager or supervisor in their right mind refuse to commit resources to achieve a solution? Solving any problems but these gives only marginal improvement.

Resource planning for Black Belts, Green Belts, and other personnel is visible and simplified when they are assigned to top projects on the list. Opportunities to assign other personnel such as project team members are clear in this context. The local deployment champion and/or Master Black Belt needs to manage the list. Always remember, a project focused on the top problems is worth a lot to the business. All possible effort must be exerted to scope problems and opportunities into projects that Black Belts can drive to a Six Sigma solution.

The following process steps help us turn a problem into a scoped project (Figure 9.6).

A critical step in the process is to define the customer. This is not a question that can be taken lightly! How do we satisfy customers, either internal or external to the business, if the Black Belt is not sure who they are? The Black Belt and his team must know customers to understand their needs, delights, and satisfiers. Never guess or assume what your customers need, ask them. Several customer interaction methods will be referenced in the next chapters. For example, the customer of a software project on improving the company image is the buyer of the software, the consumer. However, if the potential project is to reduce tool breakage in a manufacturing process, then the buyer is too far removed to be the primary customer. Here the customer is more likely the design owner or other business unit manager. Certainly, if we reduce tool breakage, then we gain efficiency that may translate to cost or availability satisfaction, but this is of little help in planning a good project to reduce tool breakage.

No customer, no project! Know your customer. It is unacceptable, however, to not know your customer in the top project pipeline. These projects are too important to allow this kind of lapse.

9.3.2.5 Proactive DFSS Project Sources: MultiGeneration Planning. A multigeneration plan is concerned with developing a timely design evolution of software products and of finding optimal resource allocation. An acceptable plan must be capable of dealing with the uncertainty about future markets and the availability of software products when demanded by the customer. The incorporation of uncertainty into a resource-planning model of a software multigeneration plan is essential. For example, on the personal financial side, it was not all that long ago that a family was only three generations deep—grandparent, parent, and child. But as life expectancies increase, four generations are common and five generations are no longer unheard of. The financial impact of this demographic change has been dramatic. Instead of a family focused only on its own finances, it may have to deal with financial issues that cross generations. Where once people lived only a few years into retirement, now they live 30 years or more. If the parents cannot take care of themselves, or they cannot afford to pay for high-cost, long-term care either at home or in a facility, their children may need to step forward. A host of financial issues are involved such as passing on the estate, business succession, college versus retirement, life insurance, and loaning money. These are only a smattering of the many multigenerational financial issues that may originate.

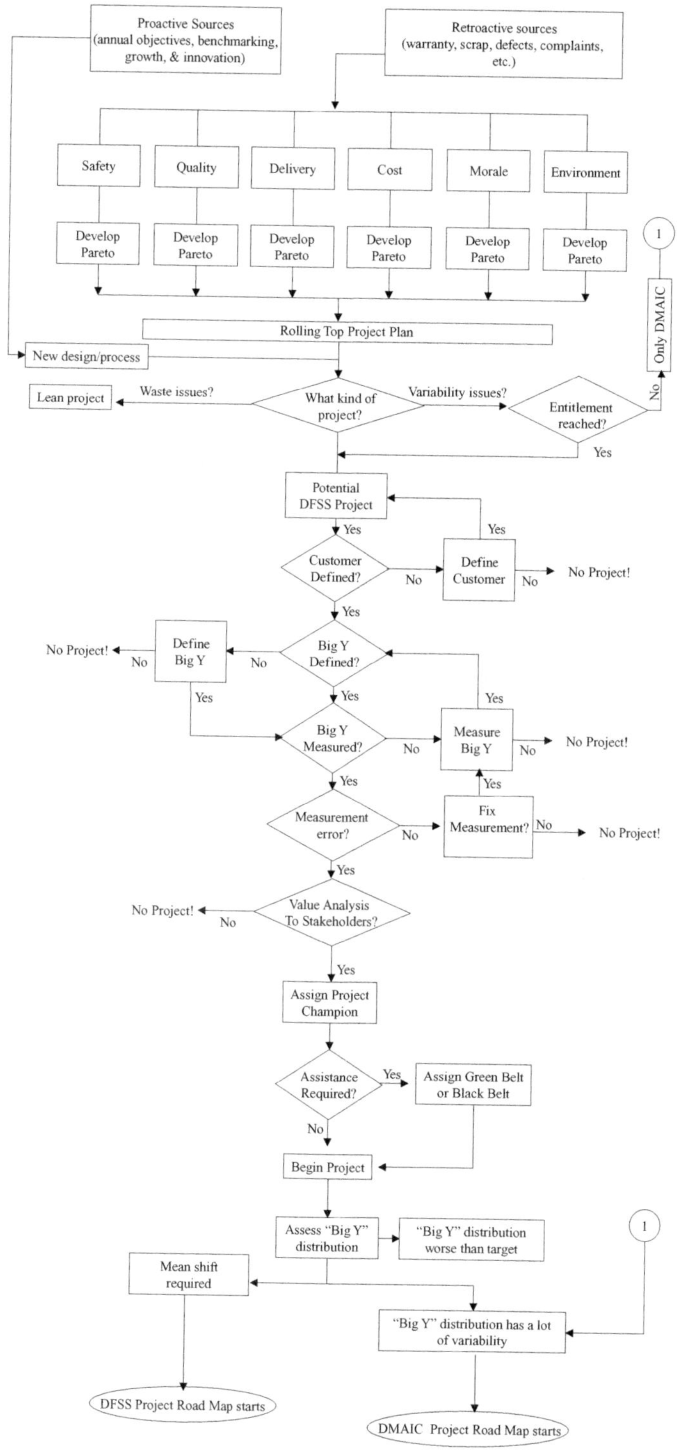

FIGURE 9.6 Six Sigma project identification and scoping process.

Software design requires a multigeneration planning that takes into consideration demand growth and the level of coordination in planning, and resource allocation among functions within a company. The plan should take into consideration uncertainties in demand and technology and other factors by means of defining strategic design generations, which reflect gradual and realistic possible evolutions of the software of interest. The decision analysis framework needs to be incorporated to quantify and minimize risks for all design generations. Advantages associated with generational design in mitigating risks, financial support, economies of scale, and reductions of operating costs are key incentives for growth and innovation.

The main step is to produce generation plans for software design CTSs and functional requirements or other metrics with an assessment of uncertainties around achieving them. One key aspect for defining the generation is to split the plan into periods where flexible generations can be decided. The beginning of generational periods may coincide with milestones or relevant events. For each period, a generational plan gives an assessment of how each generation should performs against an adopted set of metrics. For example, a company generational plan for its SAP[4] system may be depicted in Figure 9.7 where a multigenerational plan lays out the key metrics and the enabling technologies and processes by time horizon.

9.3.2.6 Training. To jump start the deployment process, DFSS training is usually outsourced in the first year or two into deployment (www.SixSigmaPI.com).[5] The deployment team needs to devise a qualifying scheme for training vendors once their strategy is finalized and approved by the senior leadership of the company. Specific training session content for executives leadership, champions, and Black Belts should be planned with strong participation by the selected vendor. This facilitates a coordinated effort, allowing better management of the training schedule and more prompt software. In this section, simple guidelines for training deployment champions, project champions, and any other individual whose scope of responsibility intersects with the training function needs to be discussed. Attendance is required for each day of training. To get the full benefit of the training course, each attendee needs to be present for all material that is presented. Each training course should be developed carefully and condensed into the shortest possible period by the vendor. Missing any part of a course will result in a diminished understanding of the covered topics and, as a result, may severely delay the progression of projects.

9.3.2.7 Existence of a Software Program Development Management System. Our experience is that a project road map, a design algorithm, is required

[4]SAP stands for "Systems, Applications, Products" (German: Systeme, Anwendungen, Produkte). SAP AG, headquartered in Walldorf, Germany, is the third-largest software company in the world and the world's largest inter-enterprise software company, providing integrated inter-enterprise software solutions as well as collaborative e-business solutions for all types of industries and for every major market.

[5]Six Sigma Professionals, Inc. (www.SixSigmaPI.com) has a portfolio of software Six Sigma and DFSS programs tiered at executive leadership, deployment champions, project champions, Green Belts, Black Belts, and Master Black Belts in addition to associated deployment expertise.

	Gen 0 "As Is"	Gen 1 120 days	Gen 2 6–12 months
VISION		Use DFSS to create Standard process with scalable features that provide a framework to migrate to future state	Evolve process into SAP environment and drive 20% productivity improvement
METRICS			
Touch Time	unknown	L – 40 hrs M – 20 hrs S – 10 hrs	Same
Cycle Time	Manual 1–20 weeks	Manual 3–10 days	Automated
Win Rate	Unknown	Measured	Automated
Accuracy	Unknown	Accuracy	Accuracy
Completeness	Unknown	Completeness	Completeness
Win Rate	Unknown	Win Rate	Win Rate
Compliance	hope	planned	Mistake proofed
Auditable/ Traceable	hope	planned	Mistake proofed
SCOPE		Service 1	Service 2

FIGURE 9.7 SAP software design multigeneration plan.

for successful DFSS deployment. The algorithm works as a compass leading Black Belts to closure by laying out the full picture of the DFSS project. We would like to think of this algorithm as a recipe that can be tailored to the customized application within the company's program management system that spans the software design life cycle.[6] Usually, the DFSS deployment team encounters two venues at this point: 1) Develop a new program management system (PMS) to include the proposed DFSS algorithm. The algorithm is best fit after the research and development and prior to the customer-use era. It is the experience of the authors that many companies lack such universal discipline from a practical sense. This venue is suitable for such companies and those practicing a variety of PMS hoping that alignment will evolve. 2) Integrate with the current PMS by laying this algorithm over and synchronizing when and where needed.

In either case, the DFSS project will be paced at the speed of the leading program from which the project was derived in the PMS. Initially, high-leverage projects should target subsystems to which the business and the customer are sensitive. A sort of requirement flow-down, a cascading method should be adopted to identify these

[6]The design life cycle spans the research and development, development, production and release, customer, and post-customer (e.g., software and after market).

subsystems. Later, when DFSS becomes the way of doing business, system-level DFSS deployment becomes the norm and the issue of synchronization with PMS will diminish eventually. Actually, the PMS will be crafted to reflect the DFSS learning experience that the company gained during the years of experience.

9.3.3 Deployment

This phase is the period of time when champions are trained and when they select initial Black Belt projects, as well as when the initial wave of Black Belts are trained and when they complete projects that yield significant operational benefit both soft and hard. The training encompasses most of the deployment activities in this phase, and it is discussed in the following section. Additionally, this deployment phase includes the following assignment of the deployment team:

- Reiterate to key personnel their responsibilities at critical points in the deployment process.
- Reinforce the commitment among project champions and Black Belts to execute selected improvement projects aggressively. Mobilize and empower both populations to carry out effectively their respective roles and responsibilities.
- Recognize exemplary performance in execution and in culture with the project champion and Black Belt levels.
- Inform the general employee population about the tenets of Six Sigma and the deployment process.
- Build information packets for project champions and Black Belts that contain administrative, logistical, and other information they need to execute their responsibilities at given points in time.
- Document and publicize successful projects and the positive consequences for the company and its employees.
- Document and distribute project-savings data by business unit, product, or other appropriate area of focus.
- Hold Six Sigma events or meetings with all employees at given locations where leadership is present and involved and where such topics are covered.

9.3.3.1 Training. The critical steps in DFSS training are 1) determining the content and outline, 2) developing the materials, and 3) deploying the training classes. In doing so, the deployment team and its training vendor of choice should be very cautious about cultural aspects and to weave into the soft side of the initiative the culture change into training. Training is the significant mechanism within deployment that, in addition to equiping trainees with the right tools, concepts, and methods, will expedite deployment and help shape a data-driven culture. This section will present a high-level perspective of the training recipients and what type of training they should receive. They are arranged as follows by the level of complexity.

9.3.3.1.1 Senior Leadership. Training for senior leadership should include an overview, business and financial benefits of implementation, benchmarking of successful deployments, and specific training on tools to ensure successful implementation.

9.3.3.1.2 Deployment Champions. Training for Deployment Champions is more detailed than that provided to senior leadership. Topics would include the DFSS concept, methodology, and "must-have" tools and processes to ensure successful deployment within their function. A class focused on how to be an effective champion as well as on their roles and responsibilities often is beneficial.

9.3.3.1.3 Master Black Belts. Initially, experienced Master Black Belts are hired from the outside to jump start the system. Additional homegrown MBBs may need to go to additional training beyond their Black Belt training.[7] Training for Master Black Belts must be rigorous about the concept, methodology, and tools, as well as provide detailed statistics training, computer analysis, and other tool applications. Their training should include soft and hard skills to get them to a level of proficiency compatible with their roles. On the soft side, topics include strategy, deployment lesson learned, their roles and responsibilities, presentation and writing skills, leadership and resource management, and critical success factors benchmarking history and outside deployment. On the hard side, a typical training may go into the theory of topics like DOE and ANOVA, axiomatic design, hypothesis testing of discrete random variables, and Lean tools.

9.3.3.1.4 Black Belts. The Black Belts as project leaders will implement the DFSS methodology and tools within a function on projects aligned with the business objectives. They lead projects, institutionalize a timely project plan, determine appropriate tool use, perform analyses, and act as the central point of contact for their projects. Training for Black Belts includes detailed information about the concept, methodology, and tools. Depending on the curriculum, the duration usually is between three to six weeks on a monthly schedule. Black Belts will come with a training focused descoped project that has an ample opportunity for tool application to foster learning while delivering to deployment objectives. The weeks between the training sessions will be spent on gathering data, forming and training their teams, and applying concepts and tools where necessary. DFSS concepts and tools flavored by some soft skills are the core of the curriculum. Of course, DFSS training and deployment will be in synch with the software development process already adopted by the deploying company. We are providing in Chapter 11 of this book a suggested software DFSS project road map serving as a design algorithm for the Six Sigma team. The algorithm will work as a compass leading Black Belts to closure by laying out the full picture of a typical DFSS project.

[7] See www.SixSigmaPI.com training programs.

9.3.3.1.5 Green Belts. The Green Belts may also take training courses developed specifically for Black Belts where there needs to be more focus. Short-circuiting theory and complex tools to meet the allocated short training time (usually less than 50% of Black Belt training period) may dilute many subjects. Green Belts can resort to their Black Belt network for help on complex subjects and for coaching and mentoring.

9.3.3.2 Six Sigma Project Financial. In general, DFSS project financials can be categorized as ***hard*** or ***soft*** savings and are mutually calculated or assessed by the Black Belt and the assigned financial analyst to the project. The financial analyst assigned to a DFSS team should act as the lead in quantifying the financials related to the project "actions" at the initiation and closure phases, assist in identification of "hidden factory" savings, support the Black Belt on an ongoing basis, and if financial information is required from areas outside his/her area of expertise, he/she needs to direct the Black Belt to the appropriate contacts, follow up, and ensure the Black Belt receives the appropriate data. The analyst, at project closure, also should ensure that the appropriate stakeholders concur with the savings. This primarily affects processing costs, design expense, and nonrevenue items for rejects not directly led by Black Belts from those organizations. In essence, the analyst needs to provide more than an audit function.

The financial analyst should work with the Black Belt to assess the projected annual financial savings based on the information available at that time (e.g., scope or expected outcome). This is not a detailed review but a rough order of magnitude approval. These estimates are expected to be revised as the project progresses and more accurate data become available. The project should have the potential to achieve an annual preset target. The analyst confirms the business rationale for the project where necessary.

El-Haik in Yang and El-Haik (2008) developed a scenario of Black Belt target cascading that can be customized to different applications. It is based on project cycle time, number of projects handled simultaneously by the Black Belt, and their importance to the organization.

9.3.4 Postdeployment Phase

This phase spans the period of time when subsequent waves of Black Belts are trained, when the synergy and scale of Six Sigma build to critical mass, and when additional elements of DFSS deployment are implemented and integrated.

In what follows, we are presenting some thoughts and observations that were gained through our deployment experience of Six Sigma and, in particular, DFSS. The purpose is to determine factors toward keeping and expanding the momentum of DFSS deployment to be sustainable.

This book presents the software DFSS methodology that exhibits the merging of many tools at both the conceptual and analytical levels and penetrates dimensions like conceptualization, optimization, and validation by integrating tools, principles, and concepts. This vision of DFSS is a core competency in a company's overall technology strategy to accomplish its goals. An evolutionary strategy that moves the deployment of the DFSS method toward the ideal culture is discussed. In the strategy, we have identified the critical elements, needed decisions, and deployment concerns.

The literature suggests that more innovative methods fail immediately after initial deployment than at any stage. Useful innovation attempts that are challenged by cultural change are not terminated directly but allowed to fade slowly and silently. A major reason for the failure of technically viable innovations is the inability of leadership to commit to integrated, effective, cost justified, and the evolutionary program for sustainability, which is consistent with the company's mission. The DFSS deployment parallels in many aspects the technical innovation challenges from a cultural perspective. The DFSS initiatives are particularly vulnerable if they are too narrowly conceived, built on only one major success mechanism, or lack fit to the larger organizational objectives. The tentative top-down deployment approach has been working where the top leadership support should be the significant driver. However, this approach can be strengthened when built around mechanisms like the superiority of DFSS as a design approach, and the attractiveness of the methodologies to designers who want to become more proficient on their jobs.

Although there are needs to customize a deployment strategy, it should not be rigid. The strategy should be flexible enough to meet unexpected challenges. The deployment strategy itself should be DFSS driven and robust to anticipated changes. It should be insensitive to expected swings in the financial health of a company and should be attuned to the company's objectives on a continuous basis.

The strategy should consistently build coherent linkages between DFSS and daily software development and design business. For example, engineers and architectures need to see how all of the principles and tools fit together, complement one another, and build toward a coherent whole process. DFSS needs to be perceived, initially, as an important part, if not the central core, of an overall effort to increase technical flexibility.

9.3.4.1 DFSS Sustainability Factors. In our view, DFSS possesses many inherent sustaining characteristics that are not offered by current software development practices. Many deign methods, some called best practices, are effective if the design is at a low level and need to satisfy a minimum number of functional requirements. As the number of the software product requirements increases (design becomes more complex), the efficiency of these methods decreases. In addition, these methods are hinged on heuristics and developed algorithms limiting their application across the different development phases.

The process of design can be improved by constant deployment of DFSS, which begins from different premises, namely, the principle of design. The design axioms and principles are central to the conception part of DFSS. As will be defined in Chapter 13, axioms are general principles or truths that cannot be derived except there are no counterexamples or exceptions. Axioms are fundamental to many engineering disciplines such as thermodynamics laws, Newton's laws, the concepts of force and energy, and so on. Axiomatic design provides the principles to develop a good software design systematically and can overcome the need for customized approaches.

In a sustainability strategy, the following attributes would be persistent and pervasive features:

- A deployment measurement system that tracks the critical-to-deployment requirements and failure modes as well as implements corrective actions
- Continued improvement in the effectiveness of DFSS deployment by benchmarking other successful deployment elsewhere
- Enhanced control (over time) over the company's objectives via selected DFSS projects that really move the needle
- Extended involvement of all levels and functions
- DFSS embedded into the everyday operations of the company

The prospectus for sustaining success will improve if the strategy yields a consistent day-to-day emphasis of recognizing that DFSS represents a cultural change and a paradigm shift and allows the necessary time for a project's success. Several deployments found it very useful to extend their DFSS initiative to key suppliers and to extend these beyond the component level to subsystem and system-level projects. Some call these projects intra-projects when they span different areas, functions, and business domains. This ultimately will lead to integrating the DFSS philosophy as a superior design approach within the program management system (PMS) and to aligning the issues of funding, timing, and reviews to the embedded philosophy. As a side bonus of the deployment, conformance to narrow design protocols will start fading away. In all cases, sustaining leadership and managerial commitment to adopting appropriate, consistent, relevant, and continuing reward and recognition mechanism for Black Belts and Green Belts is critical to the overall sustainment of the initiative.

The vision is that DFSS as a consistent, complete, fully justified, and usable process should be expanded to other new company-wide initiatives. The deployment team should keep an eye on the changes that are needed to accommodate altering a Black Belt tasks from individualized projects to broader scope, intra-team assignments. A prioritizing mechanism for future projects of this kind that target the location, size, complexity, involvement of other units, type of knowledge to be gained, and potential for fit within the strategic plan should be developed.

Another sustaining factor lies in providing relevant, on-time training and opportunities for competency enhancement of the Black Belt and Green Belt. The capacity to continue learning and alignment of rewards with competency and experience must be fostered. Instituting an accompanying accounting and financial evaluation that enlarges the scope of consideration of the impact of the project on both fronts' hard and soft savings is a lesson learned. Finance and other resources should be moving upfront toward the beginning of the design cycle in order to accommodate DFSS methodology.

If the DFSS approach is to become pervasive as a central culture underlying a development strategy, it must be linked to larger company objectives. In general, the DFSS methodology should be linked to:

1. The societal contribution of the company in terms of developing more reliable, efficient, environmentally friendly software products
2. The goals of the company, including profitability and sustainability in local and global markets
3. The explicit goals of management embodied in company mission statements, including characteristics such as greater design effectiveness, efficiency, cycle time reduction, responsiveness to customers, and the like
4. A greater capacity for the deploying company to adjust and respond to customers and competitive conditions
5. The satisfaction of managers, supervisors, and designers

A deployment strategy is needed to sustain the momentum achieved in the deployment phase. The strategy should show how DFSS allows Black Belts and their teams to respond to a wide variety of externally induced challenges and that complete deployment of DFSS will fundamentally increase the yield of company operations and its ability to provide a wide variety of design responses. DFSS deployment should be a core competency of a company. DFSS will enhance the variety of quality of software entity and design processes. These two themes should, continuously, be stressed in strategy presentations to more senior leadership. As deployment proceeds, the structures and processes used to support deployment also will need to evolve. Several factors need to be considered to build into the overall sustainability strategy. For example, the future strategy and plan for sustaining DFSS needs to incorporate a more modern learning theory on the usefulness of the technique for Green Belts and other members at the time they need the information. On the sustainment of DFSS deployment, we suggest that the DFSS community (Black Belts, Green Belts, Master Black Belts, champions, and deployment directors) will commit to the following:

- Support their company image and mission as a highly motivated producer of choice of world-class, innovative complete software solutions that lead in quality and technology and exceed customer expectations in satisfaction and value.
- Take pride in their work and in the contribution they make internally and externally.
- Constantly pursue "Do It Right the First Time" as a means of reducing the cost to their customers and company.
- Strive to be recognized as a resource, vital to both current and future development programs and management of operations.
- Establish and foster a partnership with subject matter experts, the technical community in their company.

- Treat DFSS lessons learned as a corporate source of returns and savings through replicating solutions and processes to other relevant entities.
- Promote the use of DFSS principles, tools, and concepts where possible at both project and day-to-day operations and promote the data-driven decision culture, the crest of the Six-Sigma culture.

9.4 BLACK BELT AND DFSS TEAM: CULTURAL CHANGE

We are adopting the Team Software Process (TSP) and Personal Software Process (PSP) as a technical framework for team operations. This is discussed in Chapter 10. In here, the soft aspects of cultural changes are discussed.

The first step is to create an environment of teamwork. One thing the Black Belt eventually will learn is that team members have very different abilities, motivations, and personalities. For example, there will be some team members that are pioneers and others who will want to vanish. If Black Belts allow the latter behavior, they become dead weight and a source of frustration. The Black Belt must not let this happen. When team members vanish, it is not entirely their fault. Take someone who is introverted. They find it stressful to talk in a group. They like to think things through before they start talking. They consider others' feelings and do not find a way to participate. It is the extroverts' responsibility to consciously include the introvert, to not talk over them, to not take the floor away from them. If the Black Belt wants the team to succeed, he or she has to accept that you must actively manage others. One of the first things the Black Belt should do as a team is make sure every member knows every other member beyond name introduction. It is important to get an idea about what each person is good at and about what resources they can bring to the project.

One thing to realize is that when teams are new, each individual is wondering about their identity within the team. Identity is a combination of personality, competencies, behavior, and position in an organization chart. The Black Belt needs to push for another dimension of identity, that is, the belonging to the same team with the DFSS project as task on hand. Vision is of course a key. Besides the explicit DFSS project phased activities, what are the real project goals? A useful exercise, a deliverable, is to create a project charter, with a vision statement, among themselves and with the project stakeholders. The charter is basically a contract that says what the team is about, what their objectives are, what they are ultimately trying to accomplish, where to get resources, and what kind of benefits will be gained as a return on their investment on closing the project. The best charters usually are those that synthesize from each member's input. A vision statement also may be useful. Each member should separately figure out what they think the team should accomplish, and then together see whether there are any common elements out of which they can build a single, coherent vision to which each person can commit. The reason why it is helpful to use common elements of members' input is to capitalize on the common direction and to motivate the team going forward.

It is a critical step, in a DFSS project endeavor, to establish and maintain a DFSS project team that has a shared vision. Teamwork fosters the Six Sigma transformation and instills the culture of execution and pride. It is difficult for teams to succeed without a leader, the Belt, who should be equipped with several leadership qualities acquired by experience and through training as the leader. It is a fact that there will be team functions that need to be performed, and he or she can do all of them, or split up the job among pioneer thinkers within his team. One key function is that of facilitator. The Black Belt will call meetings, keeps members on track, and pay attention to team dynamics. As a facilitator, the Black Belt makes sure that the team focuses on the project, engages participation from all members, prevents personal attacks, suggests alternative procedures when the team is stalled, and summarizes and clarifies the team's decisions. In doing so, the Black Belt should stay neutral until the data starts speaking, stop meetings from running too long, even if it is going well or people will try to avoid coming next time. Another key function is that of liaison. The Black Belt will serve as liaison between the team and the project stakeholders for most of the work-in-progress. Finally, there is the project management function. As a manager of the DFSS project, the Black Belt organizes the project plan and sees that it is implemented. He or she needs to be able to take a whole project task and break it down into scoped and bounded activities with crisp deliverables to be handed out to team members as assignments. The Black Belt has to be able to budget time and resources and get members to execute their assignments at the right time.

Team meetings can be very useful if done right. One simple thing that helps a lot is having an updated agenda. Having a written agenda, the Black Belt will make it useful for the team to steer things back to the project activities and assignments, the compass.

There will be many situations in which the Black Belt needs to give feedback to other team members. It is extremely important to avoid any negative comment that would seem to be about the member, rather than about the work or the behavior. It is very important that teams assess their performance from time to time. Most teams have good starts and then drift away from their original goals and eventually collapse. This is much less likely to happen if, from time to time, the Black Belt asks everyone how they are feeling about the team, and does a performance pulse of the team against the project charter. It is just as important to the Black Belt to *maintain* the team to improve its performance. This function, therefore, is an ongoing effort throughout the project's full cycle.

The DFSS teams emerge and grow through systematic efforts to foster continuous learning, shared direction, interrelationships, and a balance between intrinsic motivators (a desire that comes from within) and extrinsic motivators (a desire stimulated by external actions). Winning is usually contagious. Successful DFSS teams foster other teams. Growing synergy originates from ever-increasing numbers of motivated teams and accelerates improvement throughout the deploying company. The payback for small, up-front investments in team performance can be enormous.

DFSS deployment will shake many guarded and old paradigms. People's reaction to change varies from denial to pioneering passing through many stages. On this

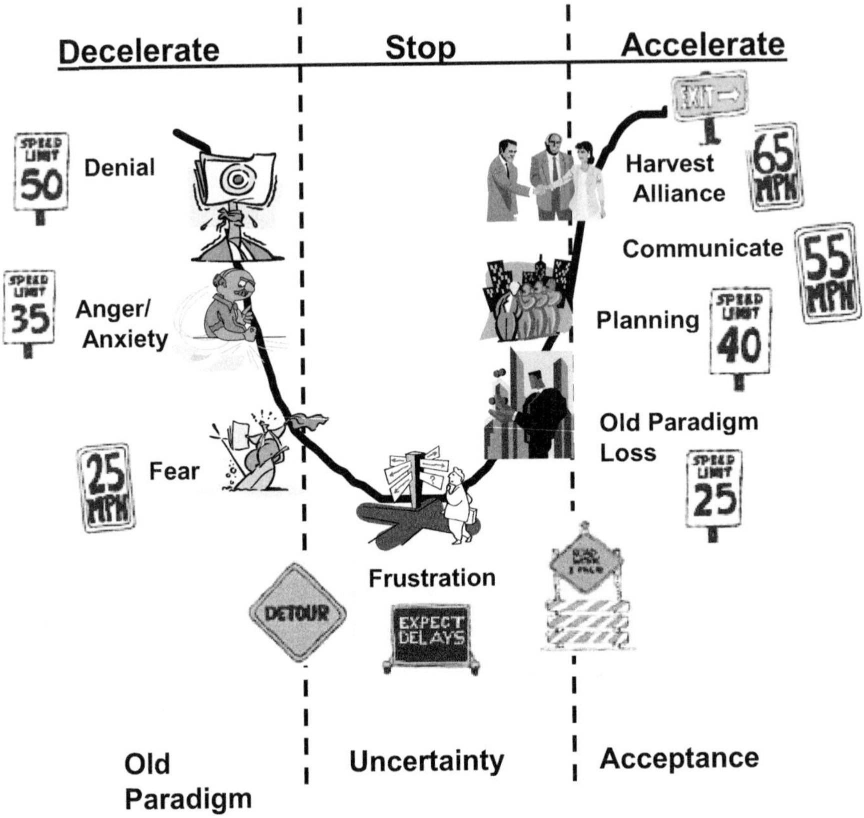

FIGURE 9.8 The "frustration curve."

venue, the objective of the Black Belt is to develop alliances for his efforts as he or she progresses. El-Haik and Roy (2005) depict the different stages of change in Figure 9.8. The Six Sigma change stages are linked by what is called the "frustration curves." We suggest that the Black Belt draw such a curve periodically for each team member and use some or all of the strategies listed to move his or her team members to the positive side, the "recommitting" phase.

What about Six Sigma culture? What we are finding powerful in cultural transformation is the premise that the company results wanted is the culture wanted. Leadership must first identify objectives that the company must achieve. These objectives must be defined carefully so that the other elements such as employee's beliefs, behaviors, and actions support them. A company has certain initiatives and actions that it must maintain in order to achieve the new results. But to achieve Six Sigma results, certain things must be stopped while others must be started (e.g., deployment). These changes will cause a behavioral shift the people must make in order for the Six Sigma cultural transition to evolve. True behavior change will not occur, let alone last, unless there is an accompanying change in leadership and deployment team belief. Beliefs are powerful in that they dictate action plans that produce desired results. Successful deployment benchmarking (initially) and experiences (later) determine the beliefs, and beliefs motivate actions, so ultimately leaders must create experiences that foster beliefs in people. The bottom line is that for a Six Sigma data-driven culture to be achieved, the company cannot operate with the old set of actions, beliefs, and experiences; otherwise the results it gets are those results that it is currently having. Experiences, beliefs, and actions—these have to change.

The biggest impact on the culture of a company is the initiative founders themselves, starting from the top. The new culture is just maintained by the employees once transition is complete. They keep it alive. Leadership set up structures (deployment team) and processes (deployment plan) that consciously perpetuate the culture. New culture means new identity and new direction, the Six Sigma way.

Implementing large-scale change through Six Sigma deployment, the effort enables the company to identify and understand the key characteristics of the current culture. Leadership together with the deployment team then develops the Six Sigma culture characteristics and the deployment plan of "how to get there." Companies with great internal conflicts or with accelerated changes in business strategy are advised to move with more caution in their deployment.

Several topics that are vital to deployment success should be considered from a cultural standpoint such as:

- Elements of cultural change in the deployment plan
- Assessment of resistance
- Ways to handle change resistance relative to culture
- Types of leaders and leadership needed at different points in the deployment effort
- How to communicate effectively when very little is certain initially
- Change readiness and maturity measurement or assessment

A common agreement between the senior leadership and deployment team should be achieved on major deployment priorities and timing relative to cultural transformation, and those areas where further work is needed to reach consensus.

At the team level, there are several strategies a Black Belt could use to his or her advantage in order to deal with team change in the context of Figure 9.7. To help reconcile, the Black Belt needs to listen with empathy, acknowledge difficulties, and define what is out of scope and what is not. To help stop the old paradigm and reorient the team to the DFSS paradigm, the Black Belt should encourage redefinition, use management to provide structure and strength, rebuild a sense of identity, gain a sense of control and influence, and encourage opportunities for creativity. To help recommit the team in the new paradigm, he or she should reinforce the new beginning, provide a clear purpose, develop a detailed plan, be consistent in the spirit of Six Sigma, and celebrate success.

REFERENCES

El-Haik, Basem S. and Roy, D. (2005), *Service Design for Six Sigma: A Roadmap for Excellence*, Wiley-Interscience, New York.

Yang, K. and El-Haik, Basem. (2008), *Design for Six Sigma: A Roadmap for Product Development*, 2nd Ed., McGraw-Hill Professional, New York.

CHAPTER 10

DESIGN FOR SIX SIGMA (DFSS) TEAM AND TEAM SOFTWARE PROCESS (TSP)

10.1 INTRODUCTION

In this chapter we discuss the operational and technical aspect of a software DFSS team. The soft aspects were discussed in Chapter 9. We are adopting the *Team Software Process* (TSP) along with the *Personal Software Process* (PSP) as an operational DFSS team framework. Software DFSS teams can use the TSP to apply integrated team concepts to the development of software systems within the DFSS project road map (Chapter 11). The PSP shows DFSS belts how to manage the quality of their projects, make commitments they can meet, improve estimating and planning, and reduce defects in their products. The PSP can be used by belts as a guide to a disciplined and structured approach to developing software. The PSP is a prerequisite for an organization planning to introduce the TSP. PSP can be applied to small-program development, requirement definition, document writing, and systems tests and systems maintenance.

A launch process walks teams and their managers through producing a team plan, assessing development risks, establishing goals, and defining team roles and responsibilities. TSP ensures quality software products, creates secure software products, and improves the DFSS process management. The process provides a defined process framework for managing, tracking, and reporting the team's progress.

Using TSP, a software company can build self-directed teams that plan and track their work, establish goals, and own their processes and plans. TSP will help a company establish a mature and disciplined engineering practice that produces secure, reliable software.

Software Design for Six Sigma: A Roadmap for Excellence, By Basem El-Haik and Adnan Shaout

In this chapter we will explore further the Personal Software Process and the Team Software Process highlighting interfaces with DFSS practices and exploring areas where DFSS can add value through a deployment example.

10.2 THE PERSONAL SOFTWARE PROCESS (PSP)

DFSS teams can use the TSP to apply integrated team concepts to the development of software-intensive systems. The PSP is the building block of TSP. The PSP is a personal process for developing software or for doing any other defined activity. The PSP includes defined steps, forms, and standards. It provides a measurement and analysis framework for characterizing and managing a software professional's personal work. It also is defined as a procedure that helps to improve personal performance (Humphrey, 1997). A stable, mature PSP allows teams to estimate and plan work, meet commitments, and resist unreasonable commitment pressures. Using the PSP process, the current performance of an individual could be understood and could be equipped better to improve the capability (Humphrey, 1997).

The PSP process is designed for individual use. It is based on scaled-down industrial software practice. The PSP process demonstrates the value of using a defined and measured process. It helps the individual and the organization meet the increasing demands for high quality and timely delivery. It is based on the following principles (Humphrey, 1997):

- **PSP Principles 1:** The quality of a software system is determined by the quality of its worst developed component. The quality of a software component is governed by the quality of the process used to develop it. The key to quality is the individual developer's skill, commitment, and personal process discipline.
- **PSP Principles 2:** As a software professional, one is responsible for one's personal process. And should measure, track, and analyze one's work. Lessons learned from the performance variations should be incorporated into the personal practices.

The PSP is summarized in the following phases:

- **PSP0: Process Flow** PSP0 should be the process that is used to write software. If there is no regular process, then PSP0 should be used to design, code, compile, and test phases done in whatever way one feels is most appropriate. Figure 10.1 shows the PSP0 process flow.

  The first step in the PSP0 is to establish a baseline that includes some basic measurements and a reporting format. The baseline provides a consistent basis for measuring progress and a defined foundation on which to improve.

  PSP0 critical-to-satisfaction measures include:

- The time spent per phase—Time Recording Log
- The defects found per phase—Defect Recording Log

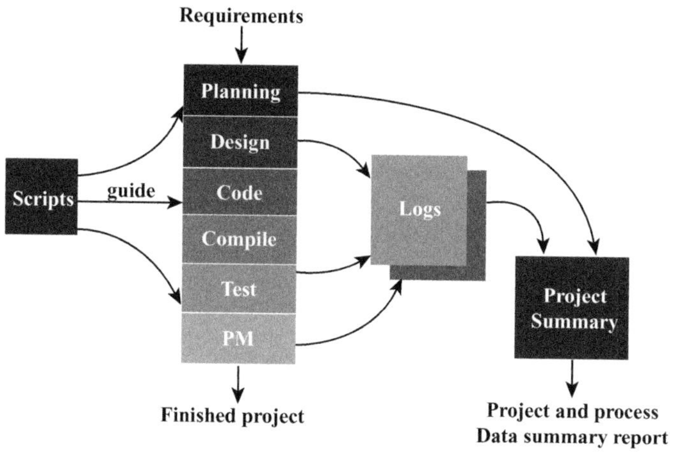

FIGURE 10.1 The PSP0 process flow (Humphrey, 1999).

- **PSP1: Personal Planning Process** PSP1 adds planning steps to PSP0 as shown in Figure 10.2. The initial increment adds test report, size, and resource estimation. In PSP1, task and schedule planning are introduced.

  The intention of PSP1 is to help understand the relation between the size of the software and the required time to develop it, which can help the software professional make reasonable commitments. Additionally, PSP1 gives an orderly plan for doing the work and gives a framework for determining the status of the software project (Humphrey, 1997).

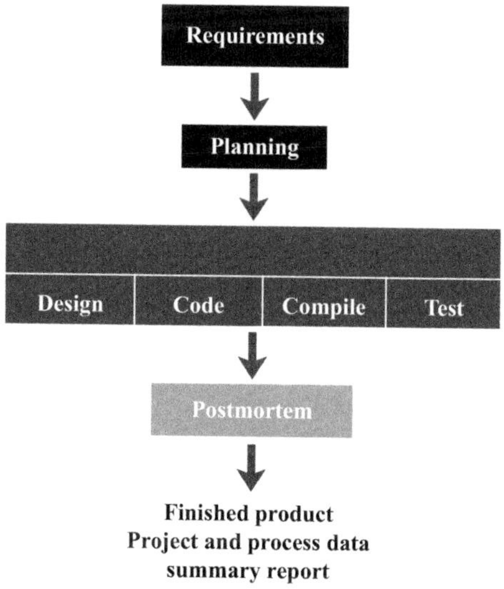

FIGURE 10.2 PSP1—Personal planning process (Humphrey, 1997).

- **PSP2: Personal Quality Management Process** PSP2 adds review techniques to PSP1 to help the software professional find defects early when they are least expensive to fix. It comprises gathering and analyzing the defects found in compile and test of the software professional's earlier programs. With these data, one can establish review checklists and make one's own process quality assessments. PSP2 addresses the design process in a nontraditional way. Here, PSP does not tell a software professional how to design but rather how to complete a design. PSP2 establishes design completeness criteria and examines various design verification and consistency techniques.
- **PSP3: A Cyclic Personal Process** There are times when a program gets bigger [e.g., a program of 10,000 lines of code (LOCs)]. This type of a program is too big to write, debug, and review using PSP2. In that case, instead of PSP2, use the abstraction principle embodied in PSP3. The PSP3 is an example of a large-scale personal process. The strategy is to subdivide a larger program into PSP2-sized pieces (a few thousand LOCs, KLOCs). The first build is a base module or kernel that is enhanced in iterative cycles. In each cycle, a complete PSP2 is performed, including design, code, compile, and test. Each enhancement builds on the previously completed increments, so PSP3 is suitable for programs of up to several thousand LOCs (Humphrey, 1997). Its strategy is to use a cyclic process. Each cycle is progressively unit tested and integrated, and at the end, you have the integrated, complete program ready for system integration or system test (Kristinn et al., 2004).
- **PSP3: A Cyclical Personal Process** PSP3 starts with a requirements and planning step that produces a conceptual design for the overall system, estimates its size, and plans the development work (Kristinn et al., 2004). In the high-level design, the product's natural division is identified and a cyclic strategy is devised. After a high-level design review, then cyclic development takes place. A good rule of thumb is to keep each cycle between 100 and 300 lines of new and changed source code (Kristinn et al., 2004). In cyclic development, the specifications for the current cycle are established. Each cycle essentially is a PSP2 process that produces a part of the product. Because each cycle is the foundation for the next, the review and tests within a cycle must be as complete as possible. Scalability is preserved as long as incremental development (cycle) is self-contained and defect free. Thus, thorough design reviews and comprehensive tests are essential parts of the cyclic development process (Kristinn et al., 2004). In cyclic testing, the first test will generally start with a clean slate. Each subsequent cycle then adds functions and progressively integrates them into the previously tested product. After the final cycle, the entire program would be completely unit and integration tests. This PSP3 designed software is now ready for system test or for integration into a larger system. Figure 10.3 shows the evolution of PSP processes from PSP0 to PSP3, whereas Figure 10.4 shows evolution within each of the PSP stages and its final evolution to PSP3.

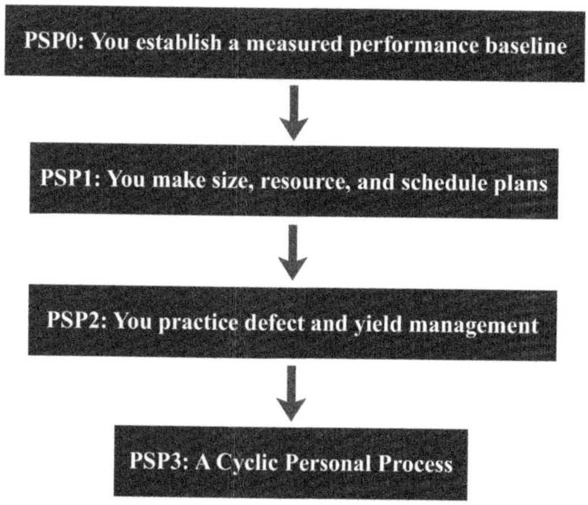

FIGURE 10.3 PSP3 evolution (Kristinn et al., 2004).

10.3 THE TEAM SOFTWARE PROCESS (TSP)

Using PSP3, programs can be built with more than 10 KLOCs. However, there are two problems: First, as the size grows so does the time and effort required, and second, most engineers have trouble visualizing all the important facets of even moderately sized programs. There are so many details and interrelationships that they may

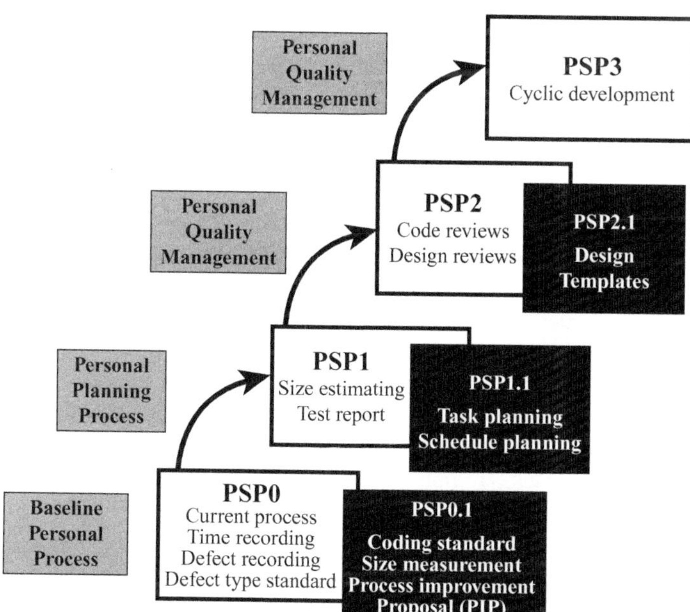

FIGURE 10.4 PSP evolution (Kristinn et al., 2004).

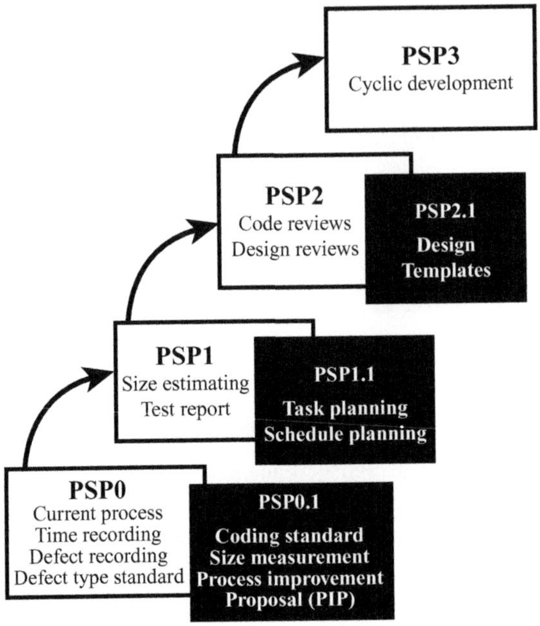

FIGURE 10.5 PSP3 to TSP evolution (Humphrey, 2005).

overlook some logical dependencies, timing interactions, or exception conditions. This may cause missing obvious mistakes because the problem is compounded by habituation, or self-hypnosis (Humphrey, 1997).

One of the most powerful software processes, however, is the Team Software Process (TSP) where the support of peers is called and asked for. When several people cooperate on a common project, they can finish it sooner, and a habituation problem can be addressed by reviewing each other's work. This review is only partially effective because teams too can suffer from excessive habituation. This can be countered by periodically including an outsider in the design reviews. The outsider's role is to ask "dumb" questions. A surprising percentage of these "dumb" questions will identify fundamental issues (Humphrey, 1997).

A defined and structured process can improve working efficiency. Defined personal processes should conveniently fit the individual skills and preferences of each software engineer. For professionals to be comfortable with a defined process, they should be involved in its definition. As the professionals' skills and abilities evolve, their processes should evolve too. Continuous process improvement is enhanced by rapid and explicit feedback (Humphrey, 1997, 2005). An evolution from PSP3 to TSP is shown in Figure 10.5.

10.3.1 Evolving the Process

The software industry is rapidly evolving. The functionality and characteristics of software products are changing at the same rate. The software development task also is evolving as fast or faster. Consequently, software belts can expect their jobs to become more challenging every year. Software Six Sigma belt skills and abilities thus must evolve with their jobs. If their processes do not evolve in response to these challenges, those developmental processes will cease to be useful. As a result, their processes may not be used (Humphrey, 1997).

10.4 PSP AND TSP DEPLOYMENT EXAMPLE

In this section, PSP and TSP processes will be used for three real-world applications in the automotive embedded controls industry while working on a hybrid vehicle using the Spiral Model, which is defined in Section 2.2, mapped to PSP and TSP as shown in Figure 10.6. The Spiral Model was chosen as a base model over other models because of its effectiveness for embedded applications with prototype iterations. To evaluate these processes thoroughly, simple and small (S&S) software with a size of 1 KLOC, moderately complex and medium (M&M) software with a size of

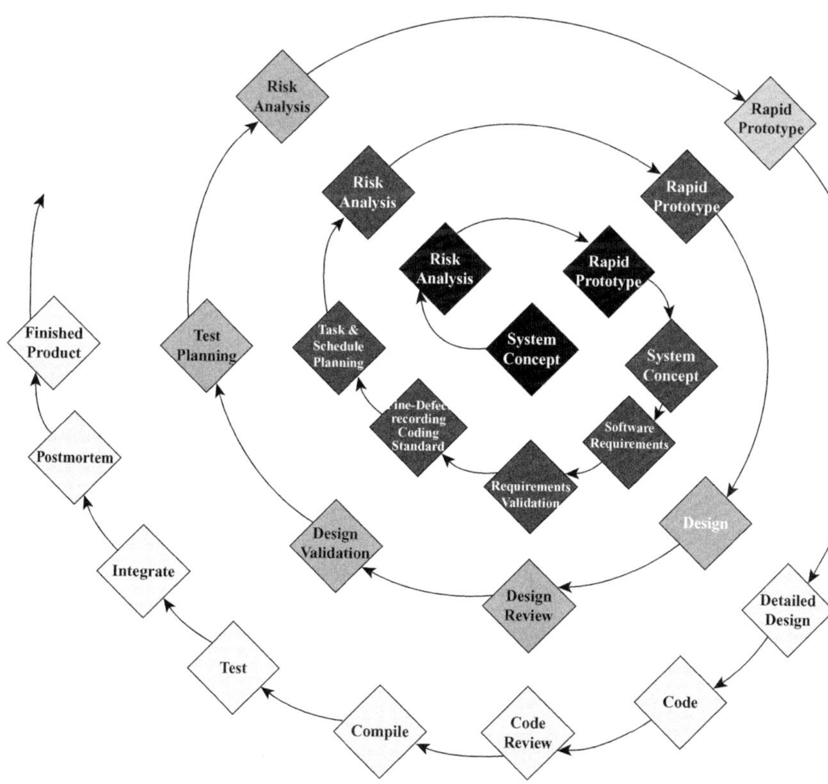

FIGURE 10.6 Practicing PSP using the Spiral Model.

10 KLOCs, and finally complex and large (C&L) software with a size of 90 KLOCs were chosen.

Here an S&S application was started after an M&M application, and fault tree analysis[1] (FTA) was conducted during the execution of the applications. FTA is a logical, structured process that can help identify potential causes of system failure before the failures actually occur. FTAs are powerful design tools that can help ensure that product performance objectives are met. FTA has many benefits such as identify possible system reliability or safety problems at design time, assess system reliability or safety during operation, improve understanding of the system, identify components that may need testing or more rigorous quality assurance scrutiny, and identify root causes of equipment failures (Humphrey, 1995). It was required for these applications to understand various human factors to have engineers with different educations backgrounds, years of experience, and level of exposure to these systems and to have personal quality standards. However, in this case, to simplify error calculations, all of these engineers were considered to be at the same level. An accurate log was maintained during the execution of the various application trials as well as available scripts in a UNIX environment to calculate the compilation, parse and build time, error count, and so on. There was one more factor where server compilation speed was changing day-to-day depending on the number of users trying to compile their software on a given day and time. For these reasons, time was averaged out for a day to reduce time calculation discrepancies. The errors also were logged systematically and flushed per the software build requirements.

10.4.1 Simple and Small-Size Project

Figure 10.6 shows a working software model using both PSP and Spiral Model software processes (Shaout and Chhaya, 2008, 2009; Chhaya, 2008). The model will be applied to an engine control subsystem with approximately 10 input and output interfaces and a relatively easy algorithm with approximately 1 KLOC.

10.4.1.1 Deployment Example: Start–Stop Module for a Hybrid Engine Controls Subsystem. *DFSS Identify Phase*—While working on various modules within engine controls, a start–stop module with approximately 1 KLOC was chosen. This involved gathering software interface and control requirements from internal departments of the organization. The time line was determined to be two persons for approximately four weeks of time. The following were the software variable requirements:

- Hybrid_Selection_Calibration
- Hybrid_ Mode
- Engine_Start_Not_Inhibit
- Over_Current_Fault_Not_Active

[1] See Chapter 15.

TABLE 10.1 Example Pseudo-Code

```
Engine Start Stop ()
// ***** Check all the conditions for Start and Stop ***** //

   Hybrid_Selection_Calibration
&& Hybrid_ Mode
&& Engine_Start_Not_Inhibit
&& Over_Current_Fault_Not_Active
&& Motor_Fault_Not_Active
&& High_Voltage_Interlock_Close
&& Alternative_Energy_Diagnostic_Fault_Not_Active
&& High_Voltage_Greater_Than_HV_Crank_Min
// *****Engine Start Stop ( ) - Continued***** //

Stop Engine if
{
   Engine_Stop_Request = True
OR Vehicle_Speed = Zero_for_ CAL_ Seconds
}
// ***** If any of the conditions below is true then start engine ***** //

Start if
{
      Immediate_Hybrid_Engine_ Start
   OR Accelerator_Pedal_Position > CAL_Minimum_Value
```

- Motor_Fault_Not_Active
- High_Voltage_Interlock_Close
- Alternative_Energy_Diagnostic_Fault_Not_Active
- High_Voltage_Greater_Than_HV_Crank_Min
- Engine_Stop_Request = True
- Vehicle_Speed
- Immediate_Hybrid_Engine_ Start
- Accelerator_Pedal_Position

DFSS Conceptualize Phase—A pseudo-code was constructed based on the requirements of the application (Table 10.1). Figure 10.7 shows the State Flow Diagram for the start–stop control algorithm module.

DFSS Optimize and Verify Phases—After understanding the requirements, design, and going through the current algorithm, it was determined that a new strategy was required to design such a vehicle because a temporary fix could not work in this case and the existence of unknown issues was generated during the operation of the vehicle.

Design discussions were held between cross-functional teams and a concept was finalized as shown in Figure 10.7. Initially hand coding was done to prototype the

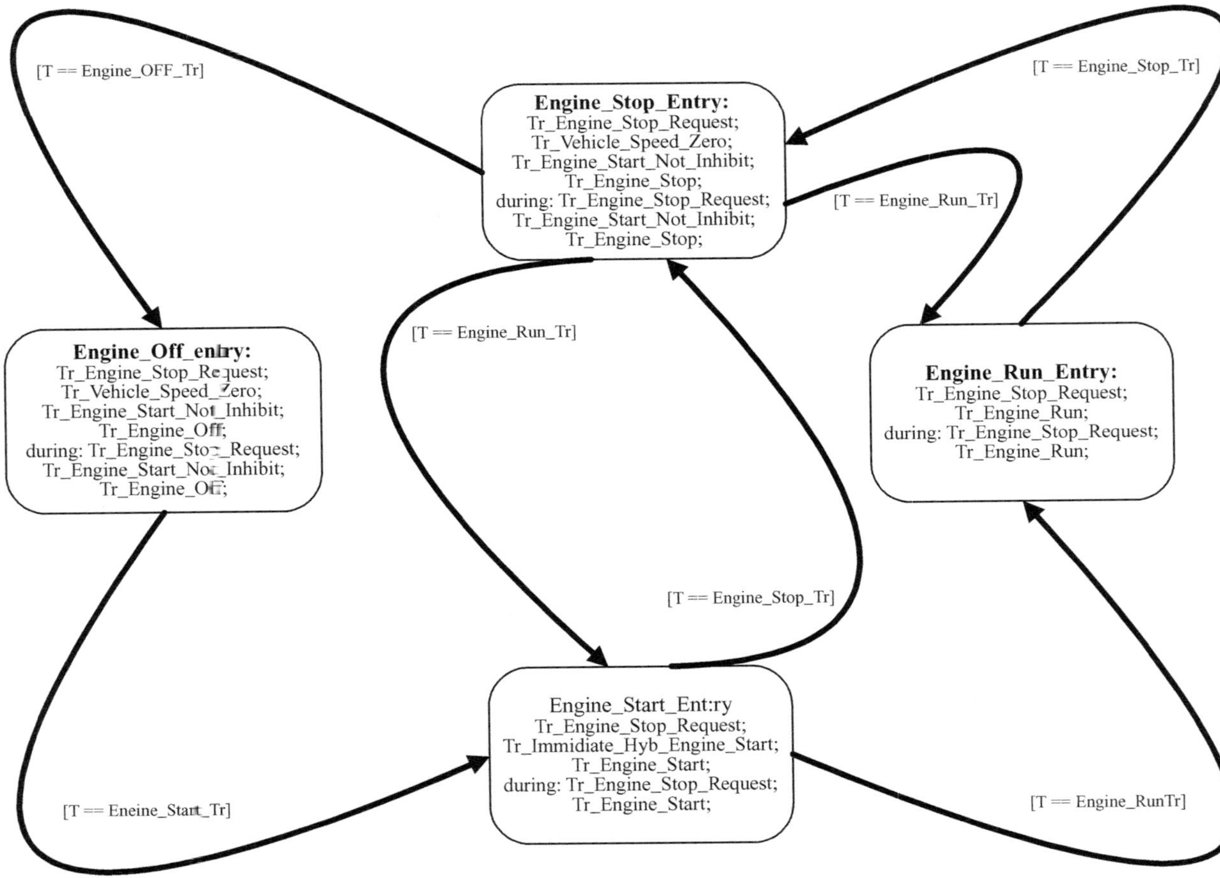

FIGURE 10.7 State flow diagram for start–stop.

algorithm. Some extra efforts were required during the compilation phase because various parameters were required to parse while compiling a single module.

The implementation and integration with the main code was done, and a vehicle test was conducted to verify the functionality of the vehicle, because it involved some mechanical nuances to check and finalize the calibration values.

Time Recording Log, Defect Recording Log, and PSP Project Plan Summary were used to determine Plan, Actual, To Date, and To Date% PSP process parameters during this program. In this case, PSP processes for two persons were used and the combined results related to time, defects injected, and defects removed are logged in Table 10.2, which shows the Simple and Small-Size PSP Project Plan Summary. During the bench test, software defects were injected to observe the proper functionality and response to errors and its diagnostics. No operating issue with software was found during this time. However, during the integration with the rest of the software modules at the vehicle level, a mismatch in software variable name (typo) defect was found that was caught as a result of an improper system response. The templates for Tables 10.2–10.7 were provided in a package downloaded from the SEI Web site "PSP-for-Engineers-Public-Student-V4.1.zip" after the necessary registering procedure. For Table 10.2 and Table 10.3 calculations, please refer to Appendix 10.A1, 10.A2, and 10.A3.

Although this example project is discussed here first, it was actually conducted after the 'M&M' project. Also it was decided to apply FTA to understand fault modes while designing the S&S project.

In conclusion, PSP processes provided a methodical and yet very lean approach to practice software processes while working on the 'S&S' project. The deviation in the achievement could be a result of a few constraints like newness of the process, the size of software project, number of people involved, and finally taking into consideration the individual software development person's personal software quality standard. The final summary results for the S&S project are shown in Table 10.3.

10.4.2 Moderate and Medium-Size Project

In this case, an M&M software project in the range of 10 KLOCs was chosen to understand the effectiveness of PSP and TSP while using the Spiral Model as shown in Figure 10.6 (Shaout and Chhaya, 2008).

10.4.2.1 Deployment Example: Electrical Power Steering Subsystem (Chhaya, 2008). *DFSS Identify Phase*—Discussions were held with the vehicle system team and steering system team to identify the high-level requirements. Next, the system requirements were interfaced to the vehicle, design guidelines, vehicle standards (SAE and ISO), safety standards application implementation, and integration environment, and team interfaces were discussed and agreed to during this phase. After jotting down the rough requirements, each of the requirements was discussed thoroughly with internal and external interfaces. The following were the requirements:

TABLE 10.2 Simple and Small-Size PSP Project Plan Summary

	Simple and Small Size Project PSP Project Plan Summary			
Program Size (LOC):	Plan	Actual	To Date	
Base(B)	0 (Measured)	0 (Measured)		
Deleted (D)	0 (Estimated)	0 (Counted)		
Modified (M)	200 (Estimated)	190 (Counted)		
Added (A)	800 (N−M)	900 (Counted)		
Reused (R)	0 (Estimated)	0 (Counted)		
Total New & Changed (N)	0 (Estimated)	1090 (A+M)	0	
Total LOC (T)	1000 (N+B−M−D+R)	1090 (Measured)	1090	
Total New Reused	0	0	0	
Time in Phase (minute)	**Plan**	**Actual**	**To Date**	**To Date %**
Planning	480	480	480	3.33
Design	3600	2400	2400	16.67
Design review	480	480	480	3.33
Code	3600	2400	2400	16.67
Code review	1200	1200	1200	8.33
Compile	960	960	960	6.67
Test	7920	6000	6000	41.67
Postmortem	960	480	480	3.33
Total	19200	14400	14400	100.00
Defects Injected	**Plan**	**Actual**	**To Date**	**To Date %**
Planning	0	0	0	0.00
Design	2	2	2	6.25
Design review	0	0	0	0.00
Code	20	27	27	84.38
Code review	0	0	0	0.00
Compile	0	0	2	6.25
Test	0	0	1	3.13
Total Development	22	29	32	100.00
Defects Removed	**Plan**	**Actual**	**To Date**	**To Date %**
Planning	0	0	0	0.00
Design	0	0	0	0.00
Design review	0	0	0	0.00
Code	0	0	1	25.00
Code review	0	5	0	0.00
Compile	4	0	2	50.00
Test	2	1	1	25.00
Total Development	6	6	4	100.00
After Development	0	0	0	

TABLE 10.3 Simple and Small-Size Project Result

Results using PSP Simple and Small Size Project		
(i) Project	**Plan**	**Actual**
Size (LOC)	**1000**	**1090**
Effort (People)	**2**	**2**
Schedule (Weeks)	**4**	**3**
Project Quality (Defect/KLOC removed in phase) Simple and Small Size Project		
Integration	**0.001 Defect/KLOC**	**0.001 Defect/KLOC**
(ii) System Test	**0.001 Defect/KLOC**	**0.000 Defect/KLOC**
Field Trial	**0.000 Defect/KLOC**	**0.000 Defect/KLOC**
Operation	**0.000 Defect/KLOC**	**0.000 Defect/KLOC**

- Electronic Control Unit and Sensor Interfaces: This section details requirements, related to interfacing of position sensors, temperature sensors, and current sensors with an electronic control unit.
- Position Sensors: Two encoders were used in this application to sense the position of the steering control. A resolver was used to sense motor rotation direction and determine the revolution per minute for controls.
  - Encoder—type, operating range, resolution, supply, number of sensors required, interface, placement, and enclosure requirements.
  - Resolver—for motor position—type, operating range, resolution, supply, and number of sensors required, interface, placement, and enclosure requirements.
- Temperature Sensor:
  - Motor temperature—type, operating range, resolution, supply voltages, required number of sensors, interface, placement, and enclosure requirements.
  - Inverter temperature—type, operating range, resolution, supply voltages, number of sensors required, interface, placement, and enclosure requirements.
- Current Sensor:
  - Motor Current Measurement—type, operating range, resolution, supply voltages, number of sensors required, interface, placement, and enclosure requirements.
- Motor Information (not part of interfaces): To provide a general idea of the type of motor used in this application, typical motor specifications also were provided, which were not directly required for hardware interface purpose. Only a software variable to sense the current and voltages of the three phases of the motor as well as the output required voltage and current to drive the motor were required to be calculated and to be sent to the Motor Control Unit.

- Motor:
  - Motor type
  - Size—KW/HP
  - RPM min–max, range, resolution
  - Supply voltage range, min–max, tolerance
  - Temperature range, min–max, tolerance
  - Torque range
  - Current range, min–max, tolerance
  - Connections
  - Wiring harness (control and high voltage)
- Electronic Control Unit (ECU)—Software—The detailed software interface requirements document was prepared for software variables related to sensor(s) measurement, resolution, accuracy, error diagnostics, and for local/global information handling. Also, a detailed algorithm and controls document was prepared for controls-related local and global software variables, error diagnostics, and software interfaces with other software modules.

The following high-level software variables were further detailed in either the sensor interface or the algorithm and controls requirements document.

- Communication protocols and diagnostics requirements
- Control voltage—Low voltage (align with h/w constraint)
- Motor power—High voltage (align with h/w constraint)
- Resolver interface
- Motor angle range
- Motor angle measurement
- Encoder interface
- Steering angle range
- Steering angle measurement
- Steering angle min–max limits
- Temperature sensor interface
- Temperature range
- Temperature measurement
- Temperature resolution
- Temperature min–max
- Motor frequency range
- Motor frequency measurement
- Motor frequency resolution
- Motor frequency min–max
- Motor voltage measurement

- Motor voltage range
- Motor voltage resolution
- Motor current range
- Motor current measurement
- Motor current min–max
- Size—KW/HP (h/w constraint)
- Torque limits—minimum and maximum
- Diagnostics conditions
  - Resolver interface diagnostics
  - Resolver out-of-range diagnostics
  - Encoder interface diagnostics
  - Encoder out-of-range diagnostics
  - Temperature interface diagnostics
  - Temperature out-of-range diagnostics
  - Safety interlocks
  - Sensor failures
  - Input/output failures
  - Module overvoltage
  - Module overcurrent
  - Module overtemperature
  - Module overtemperature
  - Module overtemperature
  - Short to GND
  - Short to VSS
  - Loss of high-voltage isolation detection
  - Torque limits
  - Supply voltage fault
  - Micro-controller fault
  - Power-On RAM diagnostics
  - Power-On EEPROM diagnostics
  - Hardware watchdog timeout and reset
  - Software watchdog timeout and reset
- Electronic Control Unit (ECU)—Power—These are general ECU hardware requirements related to power, sleep current, wake-up, efficiency, hardware input/output, cold crank operation, and EMC.
- Onboard—3.3-V and 5-V supply (sensor and controls)
- Vehicle—12-V supply (sensor and controls)
- PWM (control)
- Low- and high-side drivers

- Efficiency of power supply
- EMC compliance
- Module operational temperature range
- Cold-crank operation
- Wake-up
- Sleep current

DFSS Conceptualize Phase—Understanding the requirements in detail for a "Program Plan" consisting of a time line, the deliverables at each milestone, and the final buy-off plan were prepared. Before the requirements discussions, roughly eight personnel were chosen to handle a different task to finish the system in eight weeks based on the engineering judgment, as no past data were available. During the initial stage, it was decided to reduce the head count to six, which included three software engineers, two hardware engineers, and one integration and test engineer because the design was based heavily on the previously exercised concept.

DFSS Optimize Phase—With the detailed requirements, understanding the design was based on a previous concept that required adopting the proven architecture to the new vehicle with minimal changes at the architecture level. The addition to the previous architecture was adding the Measurement Validity Algorithm to ensure the sensor measurement. The Spiral Model was used for this embedded controls example project. Figure 10.8 shows the Electrical Steering Control Unit Design Architecture. Encoder 1, Encoder 2, Resolver, Motor Temperature, and Inverter Temperature were interfaced with the "sensor measurement" block in Figure 10.8. The sensor diagnostics block was designed to perform a power-on sensor health and a periodic sensor health check and to report sensor errors upon detection. If sensors were determined to be good, and no hybrid and motor safety interlock fault or ECU health check faults were set, then a "NO FAULT" flag was SET. The measurement validity algorithm block was designed to determine the validity of the sensor measurement. Vehicle parameters such as torque, RPM, speed, acceleration, deceleration, motor phase R, S, and T voltage, and current were fed to the motor control algorithm block in addition to the measurements from the sensor measurement block. Finally this block determined the required amount of steering angle by determining the required motor voltage and current for the R, S, and T phases of the motor.

DFSS Verify and Validate Phase—Here the scope is not to discuss the software implementation because the intention is to evaluate the software process and its effectiveness on the software product quality and reliability. After going through the requirements and suitability of previous software architecture, new functional requirements were discussed with the internal project team and the external teams.

Together it was decided that the previous architecture could be used with some modification for a new functionality with a portability of available code. This was done to ensure that only necessary changes were made in order to reduce errors during various phases to provide maximum quality with the highest reliability with minimal effort and cost. Also, in this particular software development, no operating system or lower layer software development was carried out. The changes in the

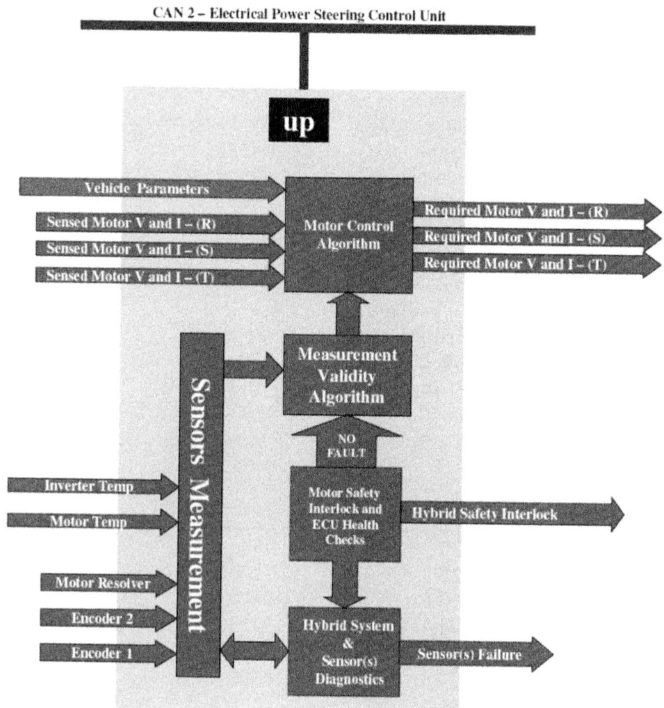

FIGURE 10.8 Electrical Steering Control Unit Design Architecture.

Application Layer were hand coded in C++. It was decided that during a later stage, it would be transferred to the Matlab (The MathWorks, Inc., Natick, MA) environment. The modular coding approach was taken, and each module was checked with its corresponding functional requirements by a coder. After approximately four weeks, core modules were made available and the integration phase was started.

Test cases for white box testing and black box testing with hardware-in-loop were written jointly by the test engineer and coder and reviewed by different teams. Time Recording Log, Defect Recording Log, and PSP Project Plan Summary were used to determine Planned, Actual, To Date, and To Date% PSP process parameters during this project.

In this case, PSP process results for six persons who had worked for eight weeks and their combined efforts in terms of time, defects injected, and defects removed were logged in Table 10.4. Also, defects related to Code errors, Compile errors, and Testing errors were identified and removed as detailed in Table 10.4 and fixed before final delivery of the software product for vehicle-level subsystems' integration and testing. For Table 10.4 and Table 10.5 calculations, please refer to Appendix A1, A2, and A3.

An error was corrected caused by a communication issue that was found, identified, notified, and resolved during the test phase. Also, there were approximately four

TABLE 10.4 Moderately Complex and Medium-Size PSP Project Plan Summary

Program Size (LOC):	Plan	Actual	To Date	
Base(B)	15000 (Measured)	15000 (Measured)		
Deleted (D)	12500 (Estimated)	12600 (Counted)		
Modified (M)	2500 (Estimated)	3100 (Counted)		
Added (A)	7600 (N-M)	7100 (T-B+D-R)		
Reused (R)	0 (Estimated)	0 (Counted)	0	
Total New & Changed (N)	10000 (Estimated)	10200 (A+M)	0	
Total LOC (T)	10000 (N+B-M-D+R)	9500 (Measured)	9500	
Total New Reused	0	0	0	
Time in Phase (minute)	**Plan**	**Actual**	**To Date**	**To Date %**
Planning	480	480	480	0.42
Design	7200	7200	7200	6.25
Design review	3300	2400	2400	2.08
Code	90420	57120	57120	49.58
Code review	3300	2400	2400	2.08
Compile	12900	9600	9600	8.33
Test	34560	34560	34560	30.00
Postmortem	1440	1440	1440	1.25
Total	153600	115200	115200	100.00
Defects Injected	**Plan**	**Actual**	**To Date**	**To Date %**
Planning	0	0	0	0.00
Design	10	10	12	2.38
Design review	0	0	0	0.00
Code	0	12	15	2.97
Code review	0	78	90	17.82
Compile	200	340	378	74.85
Test	0	0	10	1.98
Total Development	210	440	505	100.00
Defects Removed	**Plan**	**Actual**	**To Date**	**To Date %**
Planning	0	0	0	0.00
Design	0	0	0	0.00
Design review	0	0	0	0.00
Code	2	0	0	0.00
Code review	3	5	5	55.56
Compile	0	0	0	0.00
Test	3	4	4	44.44
Total Development	8	9	9	100.00
After Development	0	0	0	

TABLE 10.5 Moderately Complex and Medium-Size Project Result

Results using PSP and TSP Moderately complex & Medium Size Project		
(i) Project	**Plan**	**Actual**
Size (LOC)	**10000**	**9500**
Effort (People)	**8**	**6**
Schedule (Weeks)	**8**	**8**
Project Quality (Defect/KLOC removed in phase) Moderately Complex & Medium Size Project		
Integration	**0.001 Defect/KLOC**	**0.06 Defect/KLOC**
(ii) System Test	**0.001 Defect/KLOC**	**0.003 Defect/KLOC**
Field Trial	**0.000 Defects/KLOC**	**0.001 Defect/KLOC**
Operation	**0 Defects/KLOC**	**0.001 Defect/KLOC**

changes that were required in the diagnostics and interfaces to match the vehicle requirements because of new safety standards adoption by vehicle architecture after lengthy discussions with different program teams working for the same vehicle. Overall the example project was integrated successfully with the rest of the vehicle subsystems. Different teams carried out vehicle-level integration and then the final vehicle testing that was not the scope of this chapter. Table 10.5 shows that the results were near the estimated but not that encouraging while comparing them with Six Sigma. Looking at these results and system performance issues, in the later stage, it was determined that the current embedded controls design and its implementation does not provide industry-required reliability and quality, and thus, more efforts were asked to be put in by management.

10.4.3 Complex and Large Project

In this case, a complex and large-size embedded controls project with a software size in the range of 100 KLOCs was chosen to evaluate the efficiency of PSP and TSP (Shaout & Chhaya, 2008), (Chhaya, 2008). While following these processes, the Spiral Model was used during the entire life cycle of this embedded controls project as shown in Figure 10.9.

10.4.3.1 Deployment Example: Alternative Energy Controls and Torque. Arbitration Controls. The scope of this example application was to design alternative energy controls and hybrid controls for a hybrid system of a vehicle to store and provide alternative power to the internal combustion engine (ICE) and to do arbitration of torque for the vehicle.

DFSS Identify Phase—During the early start of this project, several discussions were held between various personnel from the internal combustion controls team, electrical motor controls team, high-voltage electrical team, vehicle system controls

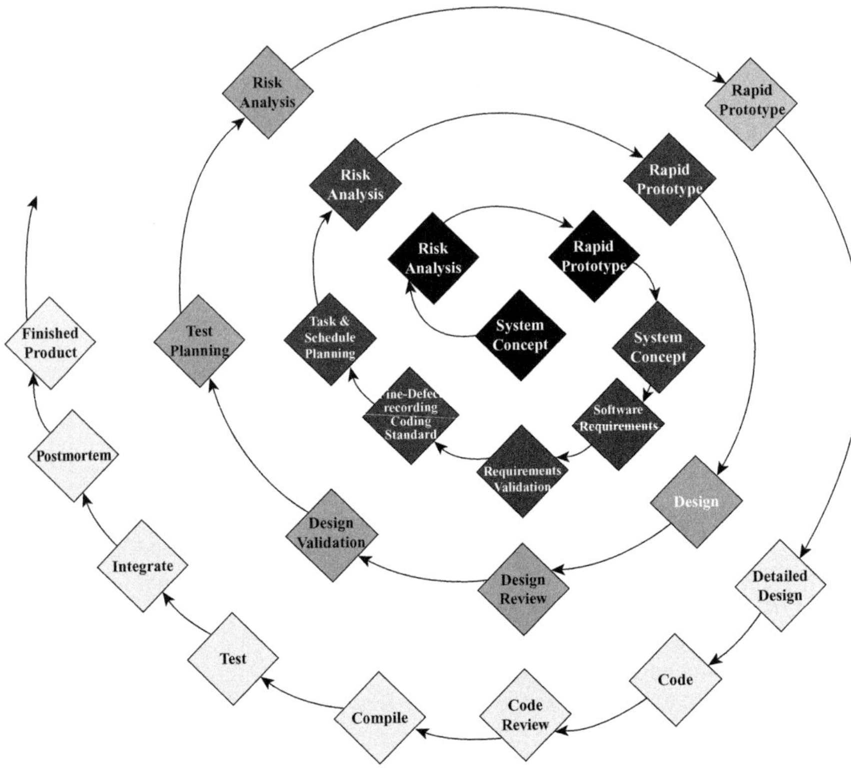

FIGURE 10.9 Practicing PSP & TSP using the Spiral Model.

team, transmission controls team, hybrid controls team, and OBDII compliance team to discuss high-level requirements. The discussion included type of hybrid vehicle, hybrid modes of the vehicle and power requirements, system requirements, hardware and software interfaces between subsystems, subsystem boundaries/overlaps, design guidelines, vehicle standards (SAE & ISO), communication protocols and safety standards, application implementation and integration environment, and team leaders/interfaces. Most requirements were finalized during the first few weeks and agreed to between various teams. Once the high-level requirements were finalized, each of the requirements was discussed thoroughly with internal and external interfaces.

Power-train vehicle architecture concepts were visited during this phase. As a part of this discussion, it was determined that the typical internal combustion controls tasks should be handled as is by the engine control unit, whereas a separate electronic control unit should carry out hybrid functionality with a core functionality to determine the torque arbitration. It also was identified and determined that a separate electronic control unit should be used to tackle alternative energy source controls. Only the hardware and software interfaces for power-train controls and motor controls were discussed and determined. The hybrid transmission controls, engine controls, and motor controls activities were carried out by different groups and were not the scope of this chapter.

The following were the requirements:

1. Engine Control Unit
   - Software interfaces with hybrid controls—part of scope
   - Typical engine controls software and hardware work—out of scope
   - Software interfaces with transmission controls—out of scope
2. Hybrid Control Unit for vehicle hybrid functionality (in scope)
   - Sensor(s) interfaces with hybrid control unit—This section details the requirement, related to interfacing of high-voltage sensors, temperature sensors, and current sensors with the electronic control unit.
     - High-Voltage Sensor(s)—type, operating range, resolution, supply voltages, number of sensors required, interface, placement, environment operating temperature range, and enclosure requirements
     - Current Sensor(s)—type, operating range, resolution, supply voltages, number of sensors required, interface, placement, environment operating temperature range, and enclosure requirements
     - Temperature Sensors (s)—type, operating range, resolution, supply voltages, number of sensors required, interface, placement, environment operating temperature range, and enclosure requirements
     - Redundant sensing—part of scope
   - The detailed software interface requirements document was prepared for software variables related to sensor(s) measurement, resolution, accuracy, error diagnostics, and local/global information handling. Also, a detailed algorithm and controls document was prepared for controls related to local and global software variables, error diagnostics, and software interfaces with other software modules.
     - Control interfaces with hybrid control unit—in scope
     - Software interfaces with engine control unit—in scope
     - Software interfaces with transmission control unit—in scope
     - Embedded controls for hybrid control unit (application layer)—in scope
     - Algorithm for arbitration of power between internal combustion engine and alternative energy source—in scope
     - Safety controls—part of scope
   - The following are the high-level software variables that were further detailed in either the sensor interface or algorithm and controls requirements document.
     - Minimum torque limit
     - Maximum torque limit
     - Torque demanded

  - Current energy-level status of alternative energy source
  - Torque available
  - Torque spilt between ICE and motor
  - Mode determination (ICE only or motor only or hybrid-ICE/motor)
  - Alternative energy—status of charge calculation
  - Redundant software/algorithm threads
  - Redundant software/algorithm processing
- Overview of hardware design (understand the limitations of hardware and if required provide input to the hardware team)—part of scope
- Diagnostics conditions
  - High-voltage interface diagnostics
  - High-voltage out-of-range diagnostics
  - Safety interlocks
  - Sensor failures
  - Digital input/output failures
  - Analog input/output failures
  - PWM input/output failures
  - Short to GND
  - Short to VSS
  - Loss of high-voltage isolation detection
  - Torque data integrity
  - Supply voltage fault
  - Micro-controller fault
  - Power-On RAM diagnostics
  - Power-On EEPROM diagnostics
  - Hardware watchdog timeout and reset
  - Software watchdog timeout and reset
- EMC requirements (h/w)
- Environmental requirements (h/w)
- Size and shape requirements (h/w)
- Placement requirements (h/w)
- Hardware—safety requirements
- Hardware—redundant control requirements
- Hardware—redundant processing requirements
- Hardware—default condition requirements
- Low-level software—safety requirements
- Low-level software—redundant thread requirements
- Low-level software—redundant processing requirements

- Low-level software—default condition requirements
- Communication protocols and diagnostics
- Module connector type and pins requirements (h/w)
- Control voltage wiring harness requirements—type, length, routing, protection, insulation, EMC—grounding and shielding (h/w & vehicle)
- Sensor interface wiring harness requirements—type, length, routing, protection, insulation, EMC—grounding and shielding (h/w & vehicle)

3. Alternative Energy Control Unit (in scope)
   - Sensor interfaces of alternative energy source—This section details requirement related to interfacing of low-voltage sensor, high-voltage sensor, alternative energy source temperature sensor, ambient air temperature sensor, cooling system temperature sensor, explosive gas detection sensor, local temperature sensor for alternative energy source, and current sensor with electronic control unit.
     - Low-Voltage Sensor—type, operating range, resolution, supply voltages, number of sensors required, interface, placement, environment operating temperature range, and enclosure requirements
     - High-Voltage Sensor—type, operating range, resolution, supply voltages, number of sensors required, interface, placement, environment operating temperature range, and enclosure requirements
     - Current Sensor—type, operating range, resolution, supply voltages, number of sensors required, interface, placement, environment operating temperature range, and enclosure requirements
     - Ambient Air Temperature Sensor—type, operating range, resolution, supply voltages, number of sensors required, interface, placement, environment operating temperature range, and enclosure requirements
     - Alternative Energy Source Temperature Sensor(s)—type, operating range, resolution, supply voltages, number of sensors required, interface, placement, environment operating temperature range, and enclosure requirements
     - Local Temperature Sensor(s) for Alternative Energy Source—type, operating range, resolution, supply voltages, number of sensors required, interface, placement, environment operating temperature range, and enclosure requirements
     - Cooling System Temperature Sensor—type, operating range, resolution, supply voltages, number of sensors required, interface, placement, environment operating temperature range, and enclosure requirements
     - Explosive Gas Detection Sensor—type, operating range, resolution, supply voltages, number of sensors required, interface, placement, environment operating temperature range, and enclosure requirements
     - Redundant sensing—part of scope

- The detailed software interface requirements document was prepared for software variables related to sensor(s) measurement, resolution, accuracy, error diagnostics, and local/global information handling. Also, a detailed algorithm and controls document was prepared for controls-related local and global software variables, error diagnostics, and software interfaces with other software modules.
  - Control interfaces of alternative energy source—in scope
  - Software interfaces of alternative energy source—in scope
  - Redundant controls software/algorithm—in scope
  - Redundant controls software threads processing—in scope
  - Measurement and calculation of energy source—in scope
  - Current energy-level status of energy source—in scope
  - Redundant measurement and calculation of energy source—in scope
  - Reliability checks for RAM, EEPROM, CPU, ALU, Register, Vehicle Data, and Communication Protocols—in scope
- Overview of hardware design (understand the limitations of hardware and if required provide input to the hardware team)—in scope
- Diagnostics conditions
  - Voltage interface diagnostics
  - Voltage out-of-range diagnostics
  - Current interface diagnostics
  - Current out-of-range diagnostics
  - Temperature interface diagnostics
  - Temperature out-of-range diagnostics
  - Explosive gas detection interface diagnostics
  - Explosive gas detection out-of-range diagnostics
  - Safety interlocks
  - Sensor failures
  - Input/output failures
  - Motor overvoltage
  - Motor overcurrent
  - Motor overtemperature
  - Module overtemperature
  - Short to GND
  - Short to VCC
  - Loss of high-voltage isolation detection
  - Torque limits
  - Supply voltage fault
  - Micro-controller fault

    - Power-On RAM diagnostics
    - Power-On EEPROM diagnostics
  - Hardware watchdog timeout and reset
  - Software watchdog timeout and reset
  - EMC requirements (h/w)
  - Environmental requirements (h/w)
  - Size and shape requirements (h/w)
  - Placement requirements (h/w)
  - Hardware—safety requirements
  - Hardware—redundant control requirements
  - Hardware—redundant processing requirements
  - Hardware—default condition requirements
  - Low-level software—safety requirements
  - Low-level software—redundant thread requirements
  - Low-level software—redundant processing requirements
  - Low-level software—default condition requirements
  - Communication protocols and diagnostics
  - Connector type and pins requirements (h/w)
  - High-voltage wiring harness requirements—type, length, routing, protection, insulation, EMC—grounding and shielding (h/w)
  - Control voltage wiring harness requirements—type, length, routing, protection, insulation, EMC—grounding and shielding (h/w)
  - Sensor interface wiring harness requirements—type, length, routing, protection, insulation, EMC—grounding and shielding (h/w)
4. Electronic Control Unit—Power—These are general ECU hardware requirements related to power, sleep current, wake-up, efficiency, hardware input/output, cold crank operation, and EMC.
  - Onboard—5-V supply (sensor and controls)
  - Onboard—3.3-V supply (internal)
  - Vehicle—12-V supply (sensor and controls)
  - PWM (control)
  - Low and high-side drivers
  - Efficiency of power supply
  - EMC compliance
  - Module operational temperature range
  - Cold-crank operation

DFSS Conceptualize Phase—Here requirements were first classified for software and hardware and then subclassified for redundancy and safety. Based on the engineering judgment, understanding the requirements in detail, the "Program Plan"

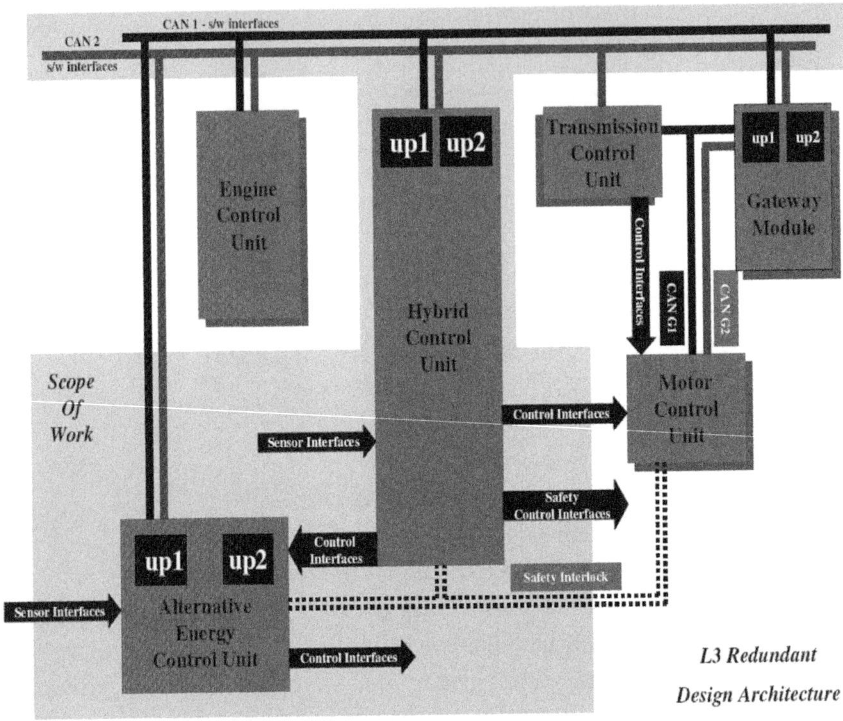

FIGURE 10.10 System design architecture.

consisting of the time line, deliverable(s) at each milestone, and final buy-off plan were prepared.

Eight to ten personnel at a time were working on average eight-man weeks. During each phase, for different tasks, different personnel with subject experts were involved to take advantage of acquired technical skills in order to improve quality and reliability. The bigger challenge was to apply PSP and TSP with personnel involved during various phases as well as personnel involved on the supplier sides.

DFSS Optimize Phase—As shown in Figure 10.10, *System Design Architecture,* the area with the light gray background was decided as part of the scope of this example project. During the design phase, various possible hybrid vehicle architectures were discussed with their trade-offs keeping in mind the difficulty to implement the above-mentioned architecture, cost, current technology, and future availability of various hardware components and sensor(s) among the cross-functional teams for a given organizational direction.

Concerns related to safety and reliability also were raised by various team leaders within the organization of this architecture as well as concerns regarding the maturity of the technology. And hence, safety and reliability requirements were discussed at length, while dealing with alternative energy sources and the hazard they posed in order to provide propulsion power to the vehicle.

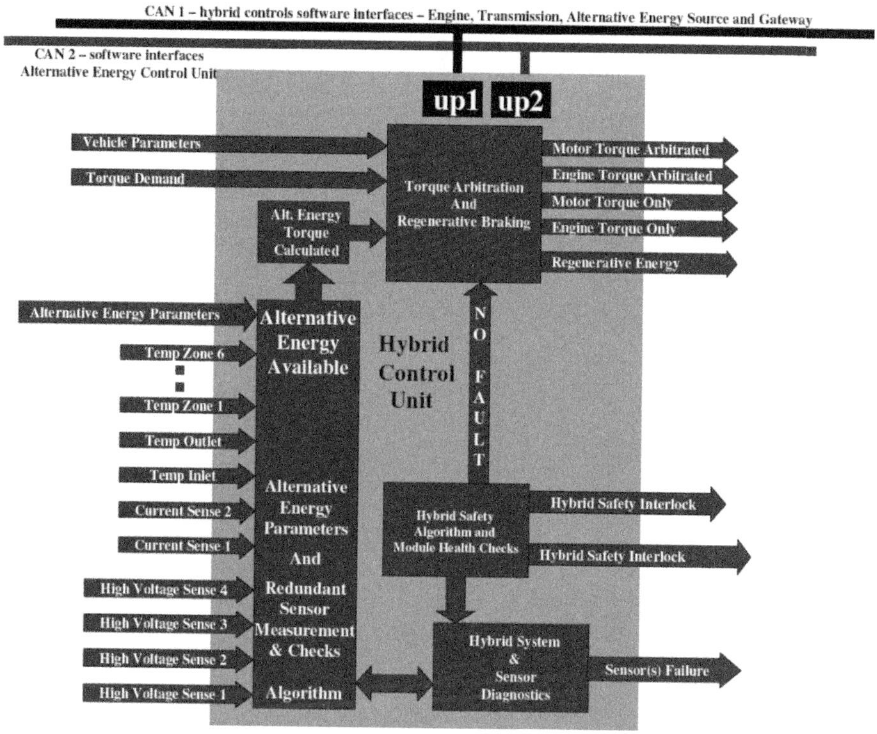

FIGURE 10.11 Hybrid control unit design architecture.

Figure 10.11 shows the details of the hybrid control unit design proposed architecture. In this design, four high-voltage sense lines for sensing high voltage, two current sensors for sensing current, six temperature sensors to sense six zones, an inlet temperature sensor, and an outlet temperature sensor were interfaced with the alternative energy redundant sensor measurement block. In addition, various alternative energy parameters were fed to this block for redundancy checks as well as for precise calculation of energy available from an alternative energy source. A sensor diagnostics block was designed to perform a power-on sensor health and a periodic sensor health check and to report sensor errors upon detection. If sensors were determined to be good, and no hybrid and motor safety interlock fault or ECU health check faults were set, then a "NO FAULT" flag was SET. Depending on the alternative energy available, available alternative energy torque was calculated and fed to the "torque arbitration and regenerative braking" algorithm block. In addition, vehicle parameters such as rpm, vehicle speed, acceleration, deceleration, emergency situation parameters, and vehicle torque demand also were fed to this block to calculate the arbitrated torque required from the motor and the engine. Three hybrid-operating modes were determined for which four required torques were calculated, which were Motor Torque Only, Engine Torque Only, Motor Torque Arbitrated, and Engine Torque Arbitrated.

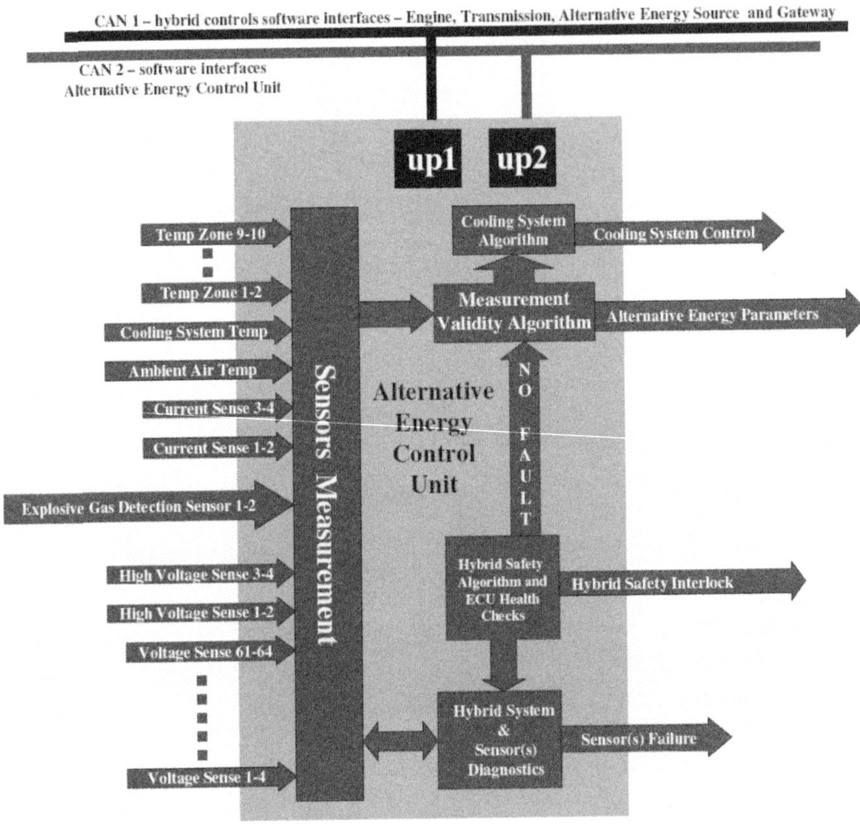

FIGURE 10.12 Alternative energy control unit design architecture.

This block also calculated the regenerative brake energy available during different vehicle operation scenarios.

As shown in Figure 10.12, the alternative energy control unit design architecture, 4 high-voltage sense lines for sensing high voltage, 64 low-voltage sense lines for sensing low voltage, 4 current sensors for sensing current, 10 temperature sensors to sense 10 zones, an ambient air temperature sensor, a cooling system temperature sensor, and an explosive gas detection sensor were interfaced with the sensor measurement block. The sensor diagnostics block was designed to perform a power-on sensor health and a periodic sensor health check and to report sensor errors upon detection. If sensors were determined to be good, and no hybrid and motor safety interlock fault or ECU health check faults were set, then a "NO FAULT" flag was SET. Depending on the alternative energy available, available alternative energy torque was calculated and fed to the torque arbitration and regenerative braking algorithm block. In addition, vehicle parameters such as rpm, vehicle speed, acceleration, deceleration, emergency situation parameters, and vehicle torque demand also were fed to this block to calculate the arbitrated torque required from the motor and the engine.

Three hybrid-operating modes were determined for which four required torques were calculated, which were Motor Torque Only, Engine Torque Only, Motor Torque Arbitrated, and Engine Torque Arbitrated. This block also calculated the regenerative brake energy available during different vehicle operation scenarios. The Measurement validity algorithm block was designed to determine the validity of the sensor measurement. Sensors measurements related to a cooling system were forwarded to the cooling system algorithm control block to keep the system within a specified temperature range.

During the Design phase, elaborate discussions were held while reviewing the current market demand and trend keeping in mind core requirements (i.e., fuel cost and its availability in the United States). Also, various energy storage solutions were discussed for day-to-day workability for a given vehicle platform and the hazard it posed to operator, passenger, the public, and the environment. Keeping all these things in mind, the final architecture was determined and designed. Next, a real-time operating system, application development environment, coding language, boundaries of various subsystems, partitions, and its overlaps were discussed and finalized.

DFSS Optimize Phase—Here details of the software implementation and the code itself are not at the center of the discussion because the intention is to evaluate the software process and its effectiveness on the software product quality and reliability and not on the coding and implementation details. Also, in this particular software development, operating systems as well as lower layer software development were used from previously designed, developed, and tried out concepts. It was decided to prototype most concepts by hand coding in C++. Proprietary compilation tools and to build environment were chosen to develop the software. Detail logs were maintained for the time consumed as well as for the type and number of errors injected and removed during the software code, compile, integration, and testing phases.

The system was divided into subsystem modules, and the best-suited knowledgeable team member was chosen to work on the given software (algorithm) modules. The coder on a bench primarily carried out unit testing while separate personnel were engaged to write test cases during the bottom-up integration testing, validation testing, and system testing. Scripts were prepared for reducing testing errors and to improve quality. Automatic bench testing was carried out on black box testing and white box testing method concepts while carrying out hardware-in-loop testing. Test logs were submitted to the coder for review. Final reviews were held with the cross-functional team.

The Time Recording Log, Defect Recording Log, and PSP Project Plan Summary were used to determine Planned, Actual, To Date, and To Date% PSP process parameters during this project. In this case, PSP processes results were planned for 20 persons for 20 weeks, whereas in actuality, 22 persons for 26 weeks were required to work on this project. Their combined efforts in terms of time, defects injected, and defects removed were logged. Also, defects were identified and removed related to code errors, compile errors, and testing errors. All these details were logged as shown in Table 10.6. For Table 10.6 and Table 10.7 calculations, please refer to Appendix 10.A1, 10.A2, and 10.A3.

TABLE 10.6 Complex and Large-Size PSP Project Plan Summary

Complex and Large-Size Project Sample PSP Project Plan Summary Farm				
Program Size (LOC):	Plan	Actual	To Date	
Base(B)	0 (Measured)	0 (Measured)		
Deleted (D)	0 (Estimated)	0 (Counted)		
Modified (M)	0 (Estimated)	0 (Counted)		
Added (A)	0 (N−M)	0 (T−B+D−R)		
Reused (R)	10000 (Estimated)	8600 (Counted)	0	
Total New & Changed (N)	90000 (Estimated)	95000 (A+M)	95000	
Total LOC (T)	90000 (N+B−M−D+R)	95000 (Measured)	95000	
Total New Reused	0	0	0	
Time in Phase (minute)	Plan	Actual	To Date	To Date %
Planning	4800	4800	4800	0.35
Design	48000	60000	60000	4.37
Design review	12000	14400	14400	1.05
Code	218400	312000	312000	22.73
Code review	96000	108000	108000	7.87
Compile	96000	144000	144000	10.49
Test	480000	720000	720000	52.45
Postmortem	4800	9600	9600	0.70
Total	960000	1372800	1372600	100.00
Defects Injected	Plan	Actual	To Date	To Date %
Planning	0	0	0	0.00
Design	10	8	8	1.04
Design review	0	0	0	0.00
Code	400	360	360	46.88
Code review	0	0	0	0.00
Compile	0	0	0	0.00
Test	300	400	400	52.08
Total Development	710	768	768	100.00
Defects Removed	Plan	Actual	To Date	To Date %
Planning	0	0	0	0.00
Design	0	0	0	0.00
Design review	0	0	0	0.00
Code	0	0	0	0.00
Code review	25	20	20	2.22
Compile	500	400	400	44.44
Test	120	480	480	53.33
Total Development	645	900	900	100.00
After Development	0	0	0	

TABLE 10.7 Complex and Large Size-Project Result

	Results using PSP and TSP Complex and Large Size Project	
Project	**Plan**	**Actual**
Size (LOC)	**90000**	**95000**
Effort (People)	**20**	**22**
Schedule (Weeks)	**20**	**26**
	Project Quality (Defect/KLOC removed in phase) Complex and Large-Size Project	
Integration	**0.005 Defects/KLOC**	**0.006 Defect/KLOC**
System Test	**0.0025 Defect/KLOC**	**0.002 Defect/KLOC**
Field Trial	**0 Defects/KLOC**	**0.001 Defect/KLOC**
Operation	**0 Defects/KLOC**	**0.001 Defect/KLOC**

Following PSP and TSP provided a very good initialization during the early stage of the project, whereas it also was realized that various important aspects of the software process method during the middle and later stages were not going to be fulfilled as observed during previous applications of PSP and TSP for moderate and medium-sized software projects. Since the project did not have a long life cycle, it was agreed to follow the concepts of other software process and methods. The shortcomings and possible improvisation to PSP and TSP are discussed in Chapter 2. In addition to the above views, while following PSP and TSP, it posed challenges to use the process methods while working with cross-functional teams and suppliers that were based globally. As shown in Table 10.7, the results were near to the plan but not encouraging compared with Six Sigma. The reliability was less than industry acceptable standards, which was proved during the series of vehicle-level testing. It was then determined to analyze current design, find out the flaws, and determine possible resolutions.

10.5 THE RELATION OF SIX SIGMA TO CMMI/PSP/TSP FOR SOFTWARE

Various researchers have experience with PSP/TSP, CMMI, and Six Sigma in the area of software systems in terms of complexity affecting reliability and safety, human errors, and changing regulatory and public views of safety. Although PSP/TSP covers the engineering and project management process areas generally well, they do not adequately cover all process management and support process areas of CMMI. Although a few elements of the Six Sigma for Software toolkit are invoked within the PSP/TSP framework (e.g., regression analysis for development of estimating models), there are many other tools available in the Six Sigma for Software toolkit that are not suggested or incorporated in PSP/TSP. Although PSP/TSP refers to and may employ some statistical techniques, specific training in statistical thinking and methods generally is not a part of PSP/TSP, whereas that is a central feature of software DFSS.

Whereas Six Sigma for Software incorporates the DFSS approach to improving the feature/function/cost trade-off in definition and design of the software product, this aspect is not addressed by CMMI/PSP/TSP. Tools such as KJ analysis, quality function deployment (QFD), conjoint analysis, design of experiments (DOE), and many others have high leverage applications in the world of software, but they are not specifically addressed by CMMI/PSP/TSP.

CMMI/PSP/TSP is among the several potential choices of software development process definition that can lead to improved software project performance. The full potential of the data produced by these processes cannot be fully leveraged without applying the more comprehensive Six Sigma for Software tool kit.

The relation of Six Sigma for Software to CMMI/PSP/TSP also might be characterized as a difference in goals, in which the goals of CMMI/PSP/TSP may be a subset of those associated with Six Sigma for Software. The primary goals of CMMI/PSP/TSP are continuous improvement in the performance of software development teams in terms of software product cost, cycle time, and delivered quality. The goals of Six Sigma for Software may include the goals of CMMI/PSP/TSP, but they do not specify any particular process definition to achieve those goals. In addition, Six Sigma for Software may be applied to achieve many other business objectives, such as improved customer service after delivery of the software, or improved customer satisfaction and value realization from the software product feature set delivered. Six Sigma for Software applies to the software process, the software product, and to balancing the "voice of the customer" and the "voice of the business" to maximize overall business value resulting from processes and products.

An additional distinction is that Six Sigma typically is applied to selected projects, whereas CMMI, PSP, and TSP are intended for all projects. Six Sigma may, for example, be used to plan and evaluate pilot implementation of CMMI/PSP/TSP, and CMMI/PSP/TSP can provide an orderly and defined vehicle to institutionalize the lessons learned from Six Sigma projects. The most fundamental tenet of Six Sigma is that it must be "managed by fact." This view is consistent with that of TSP/PSP, but it has not yet been established that PSP/TSP is the "best" alternative in every context, only that it is better than some alternatives.

APPENDIX 10.A

Software Support

Register at the SEI Web site to get the necessary software support package for student or instructor. After the necessary registering procedure, download the package from the SEI Web site "PSP-for-Engineers-Public-Student-V4.1.zip" (there could be a newer version now.

Version V4.1 contains three folders, namely, "Release Information," "Student Workbook," and "Support Materials." The release information folder has "Release information for V4.1" and "Configuration Document" where general information about various available documents and their locations within the package could be found.

The Student Workbook folder contains "PSP Student" and "Optional Excel Student" subfolders. PSP Student is the important folder, which contains Microsoft Access database, templates, forms, and scripts for various activities of PSP0, PSP1, PSP2, and PSP3 processes. Within this subfolder, PSP Course Materials is another important folder that is very useful for someone new to understand the PSP processes. This folder contains PowerPoint presentations Lecture 1 to Lecture 10 for the beginner to learn the PSP processes and to get a detailed understanding of it, although if learned from a qualified instructor, it could be much faster, but it does provide all the details one needs to begin with. In addition, this folder also contains ASGKIT1 to ASGKIT8 assignment program kits to practice PSP and then the ASGKIT Review Checklist. In addition, there are PowerPoint slides along with lectures on using PSP0, PSP0.1, PSP1, PSP1.1, PSP2, and PSP2.1. The Detail information is provided in Table 10.A1.

TABLE 10.A1 Content Details of Package V4.1

File/Folder	Type	Pages Slides	File Size (bytes)	Date and Time
PSP for Eng Student V4.1	**File Folder**			
Release information	**File Folder**		**\Release information**	
Release Notes for V4.1.doc	Word Document	1	43520	1/3/2007 8:38:39 AM
Student Workbook	**File Folder**		**\Student Workbook**	
Optional Excel Student Workbook - Interim Version	**File Folder**		**\Student Workbook\Optional Excel Student Workbook - Interim Version**	
Stuwbk.20040615.v5.xls	Excel Worksheet		1008640	10/16/2006 8:55:02 AM
PSP Student Workbook.2006.10.07	**File Folder**		**\Student Workbook\PSP Student Workbook.2006.10.07**	
PSP Student Workbook.20061007.Release Notes.doc	Word Document	1	45568	11/9/2006 1:16:22 PM
PSP Student Workbook.mde	Office Access MDE Database		13262848	11/9/2006 1:16:22 PM

(Continued)

TABLE 10.A1 Content Details of Package V4.1 ***(Continued)***

File/Folder	Type	Pages Slides	File Size (bytes)	Date and Time
STUn.XLS	Excel Worksheet		23552	11/9/2006 1:16:28 PM
PSP Assignments MDB	**File Folder**		**\Student Workbook\PSP Student Workbook.2006.10.07\PSP Assignments MDB**	
PSP Assignments_be.mdb	Office Access Application		1765376	11/9/2006 1:16:14 PM
PSP Course Materials	**File Folder**		**\Student Workbook\PSP Student Workbook.2006.10.07\PSP Course Materials**	
ASGKIT Coding Std.doc	.doc	9	90112	11/9/2006 1:16:15 PM
ASGKIT Counting Std.doc	.doc	11	195584	11/9/2006 1:16:15 PM
ASGKIT Final Report .doc	.doc	11	151040	11/9/2006 1:16:15 PM
ASGKIT Interim Report.doc	.doc	10	189952	11/9/2006 1:16:15 PM
ASGKIT PROG1.doc	.doc	12	180224	11/9/2006 1:16:15 PM
ASGKIT PROG2.doc	.doc	10	112640	11/9/2006 1:16:15 PM
ASGKIT PROG3.doc	.doc	18	383488	11/9/2006 1:16:16 PM
ASGKIT PROG4.doc	.doc	17	422400	11/9/2006 1:16:16 PM
ASGKIT PROG5.doc	.doc	20	368640	11/9/2006 1:16:16 PM
ASGKIT PROG6.doc	.doc	22	367616	11/9/2006 1:16:16 PM
ASGKIT PROG7.doc	.doc	25	493568	11/9/2006 1:16:16 PM
ASGKIT PROG8.doc	.doc	30	591872	11/9/2006 1:16:16 PM
ASGKIT Review Checklists.doc	.doc	15	172544	11/9/2006 1:16:16 PM
Course Overview I.ppt	PowerPoint	19	233984	11/9/2006 1:16:17 PM
Course Overview II.ppt	PowerPoint	18	202240	11/9/2006 1:16:17 PM
L1 Introduction to PSP.ppt	PowerPoint	27	168448	11/9/2006 1:16:17 PM
L10 Using the PSP.ppt	PowerPoint	60	340480	11/9/2006 1:16:17 PM
L2 Process Measurement.ppt	PowerPoint	37	246784	11/9/2006 1:16:17 PM
L3 PROBE I.ppt	PowerPoint	44	254464	11/9/2006 1:16:17 PM
L4 PROBE II.ppt	PowerPoint	37	249344	11/9/2006 1:16:18 PM
L5 Using PSP Data.ppt	PowerPoint	46	401992	11/9/2006 1:16:18 PM
L6 Software quality.ppt	PowerPoint	43	196096	11/9/2006 1:16:18 PM
L7 Software Design I.ppt	PowerPoint	47	388096	11/9/2006 1:16:18 PM
L8 Software Design II.ppt	PowerPoint	51	335360	11/9/2006 1:16:18 PM
L9 Design verification.ppt	PowerPoint	47	314880	11/9/2006 1:16:18 PM
Using PSP0.1.ppt	PowerPoint	16	319488	11/9/2006 1:16:18 PM
Using PSP0.ppt	PowerPoint	51	1309696	11/9/2006 1:16:19 PM
Using PSP1.1.ppt	PowerPoint	12	224768	11/9/2006 1:16:19 PM
Using PSP1.ppt	PowerPoint	24	600576	11/9/2006 1:16:19 PM
Using PSP2.1.ppt	PowerPoint	11	267776	11/9/2006 1:16:20 PM
Using PSP2.ppt	PowerPoint	21	528384	11/9/2006 1:16:20 PM

TABLE 10.A1 Content Details of Package V4.1 ***(Continued)***

File/Folder	Type	Pages Slides	File Size (bytes)	Date and Time
PSP Data MDB	**File Folder**		**\Student Workbook\PSP Student Workbook.2006.10.07\PSP Data MDB**	
PSP Student Workbook_be.mdb	Office Access Application		2428928	11/9/2006 1:16:20 PM
PSP Scripts and Forms	**File Folder**		**\Student Workbook\PSP Student Workbook.2006.10.07\PSP Scripts and Forms**	
PSP Materials.doc	Word Document	83	1786880	11/9/2006 1:16:21 PM
Support Materials	**File Folder**		**\Support Materials**	
Code Review Checklist Template.doc	Word Document	1	45568	8/28/2005 12:23:12 PM
Coding Standard Template.doc	Word Document	2	39424	3/2/2005 2:38:47 PM
Design Review Checklist Template.doc	Word Document	1	36352	8/28/2005 12:23:12 PM
Final Report Templates.doc	Word Document	3	117248	11/7/2006 11:27:35 AM
Interim Report Templates.doc	Word Document	4	117248	3/3/2005 6:47:58 PM
PSP BOK.pdf	Adobe Acrobat 7.0 Document		940948	2/28/2006 11:07:57 AM
PSP Materials.doc	Word Document	83	1797632	10/26/2006 10:17:41 AM
Size Counting Standard Template.doc	Word Document	1	54272	3/2/2005 2:38:48 PM
Total Word pages = 390				
Total PPT slides = 611				

Along with process forms and scripts for PSP processes, it also contained important information about C++ coding standards to follow as detailed in Table 10.A2.

TABLE 10.A2 C++ Coding Standards

Purpose	To guide implementation of C++ programs.
Program Headers	Begin all programs with a descriptive header.
Header Format	/**/ /* Program Assignment: the program number /* Name: your name */ /* Date: the date you started developing the program */ /* Description: a short description of the program and what it does */ /**/
Listing Contents	Provide a summary of the listing contents.
Contents Example	/**/ /* Listing Contents: */ /* Reuse instructions */ /* Modification instructions */ /* Compilation instructions */ /* Includes */ /* Class declarations: */ /* CData */ /* ASet */ /* Source code in c:/classes/CData.cpp: */ /* CData */ /* CData() */ /* Empty() */ /**/
Reuse Instructions	– Describe how the program is used: declaration format, parameter values, types, and formats. – Provide warnings of illegal values, overflow conditions, or other conditions that could potentially result in improper operation.

TABLE 10.A2 C++ Coding Standards *(Continued)*

Reuse Instruction Example	/***/ /* Reuse Instructions */ /* int PrintLine(Char *line_of_character) */ /* Purpose: to print string, 'line_of_character', on one print line */ /* Limitations: the line length must not exceed LINE_LENGTH */ /* Return 0 if printer not ready to print, else 1 */ /***/
Identifiers	Use descriptive names for all variable, function names, constants, and other identifiers. Avoid abbreviations or single-letter variables.
Identifier Example	Int number_of_students; /* This is GOOD */ Float: x4, j, ftave; /* This is BAD */
Comments	– Document the code so the reader can understand its operation. – Comments should explain both the purpose and behavior of the code. – Comments variable declarations to indicate their purpose.
Good Comment	If(record_count > limit) /* have all records been processed ? */
Bad Comment	If(record_count > limit) /* check if record count exceeds limit */
Major Sections	Precede major program sections by a block comment that describes the processing done in the next section.
Example	/***/ /* The program section examines the contents of the array 'grades' and calcu- */ /* lates the average class grade. */ /***/
Blank Spaces	– Write programs with sufficient spacing so they do not appear crowded. – Separate every program construct with at least one space.

APPENDIX 10.A1

PSP1 Plan Summary

Example PSP1 Project Plan Summary

Student	______________	Date	______________
Program	______________	Program #	______________
Instructor	______________	Language	______________

Summary	***Plan***	*Actual*	*To Date*
Size/Hour			

Program Size	***Plan***	*Actual*	*To Date*
Base (B)	***(Measured)***	(Measured)	
Deleted (D)	***(Estimated)***	(Counted)	
Modified (M)	***(Estimated)***	(Counted)	
Added (A)	***(A+M − M)***	(T − B + D − R)	
Reused (R)	***(Estimated)***	(Counted)	
Added and Modified (A+M)	***(Projected)***	(A + M)	
Total Size (T)	***(A+M + B − M − D + R)***	(Measured)	
Total New Reusable			
Estimated Proxy Size (E)			

Time in Phase (min.)	**Plan**	**Actual**	**To Date**	**To Date %**
Planning				
Design				
Code				
Compile				
Test				
Postmortem				
Total				

Defects Injected	**Actual**	**To Date**	**To Date %**
Planning			
Design			
Code			
Compile			
Test			
Total Development			

Defects Removed	**Actual**	**To Date**	**To Date %**
Planning			
Design			
Code			
Compile			
Test			
Total Development			
After Development			

PSP2 Plan Summary Instructions

Purpose	To hold the plan and actual data for programs or program parts.
General	– Use the most appropriate size measure, either LOC or element count. – "To Date" is the total actual to-date values for all products developed. – ***A part could be a module, component, product, or system.***
Header	– Enter your name and the date. – Enter the program name and number. – Enter the instructor's name and the programming language you are using.
Summary	– ***Enter the added and modified size per hour planned, actual, and to-date.***
Program Size	– ***Enter plan base, deleted, modified, reused, new reusable, and total size from the Size Estimating template.*** – Enter the plan added and modified size value (A+M) ***from projected added and modified size (P) on the Size Estimating template.*** – ***Calculate plan added size as A+M–M.*** – ***Enter estimated proxy size (E) from the Size Estimating template.*** – Enter actual base, deleted, modified, reused, total, and new reusable size Calculate actual added size as T-B+D-R and actual added and modified size as A+M. – Enter to-date reused, added and modified, total, and new reusable size.
Time in Phase	– Enter ***plan total time in phase from*** the estimated total development time ***on the Size Estimating template***. – Distribute the estimated total time across the development phases according to the To Date % for the most recently developed program. – Enter the actual time by phase and the total time. – To Date: Enter the sum of the actual times for this program plus the to-date times from the most recently developed program. – To Date %: Enter the percentage of to-date time in each phase.
Defects Injected	– Enter the actual defects by phase and the total actual defects. – To Date: Enter the sum of the actual defects injected by phase and the to-date values for the most recent previously developed program. – To Date %: Enter the percentage of the to-date defects injected by phase.
Defects Removed	– To Date: Enter the actual defects removed by phase plus the to-date values for the most recent previously developed program. – To Date %: Enter the percentage of the to-date defects removed by phase. – After development, record any defects subsequently found during program testing, use, reuse, or modification.

APPENDIX 10.A2

PROBE Estimating Script

Purpose	To guide the size and time-estimating process using the PROBE method.
Entry Criteria	– Requirements statement. – Size Estimating template and instructions. – Size per item data for part types. – Time Recording Log. – Historical size and time data.
General	– This script assumes that you are using added and modified size data as the size-accounting types for making size and time estimates. – If you choose some other size-accounting types, replace every "added and modified" in this script with the size-accounting types of your choice.

Step	Activities	Description
1	Conceptual Design	Review the requirements and produce a conceptual design.
2	Parts Additions	Follow the Size Estimating Template instructions to estimate the parts additions and the new reusable parts sizes.
3	Base Parts and Reused Parts	– For the base program, estimate the size of the base, deleted, modified, and added code. – Measure and/or estimate the side of the parts to be reused.
4	Size Estimating Procedure	– If you have sufficient estimated proxy size and actual added and modified size data (three or more points that correlate), use procedure 4A. – If you do not have sufficient estimated data but have sufficient plan added and modified and actual added and modified size data (three or more points that correlate), use procedure 4B. – If you have insufficient data or they do not correlate, use procedure 4C. – If you have no historical data, use procedure 4D.
4A	Size Estimating Procedure 4A	– Using the linear-regression method, calculate the β_0 and β_1 parameters from the estimated proxy size and actual added and modified size data. – If the absolute value of β_0 is not near 0 (less than about 25% of the expected size of the new program), or β_1 is not near 1.0 (between about 0.5 and 2.0), use procedure 4B

Step	Activities	Description
4B	Size Estimating Procedure 4B	– Using the linear-regression method, calculate the β_0 and β_1 parameters from the plan added and modified size and actual added and modified size data. – If the absolute value of β_0 is not near 0 (less than about 25% of the expected size of the new program), or β_1 is not near 1.0 (between about 0.5 and 2.0), use procedure 4C.
4C	Size Estimating Procedure 4C	If you have any data on plan added and modified size and actual added and modified size, set $\beta_0 = 0$ and $\beta_1 =$ (actual total added and modified size to date/plan total added and modified size to date).
4D	Size Estimating Procedure 4D	If you have no historical data, use your judgment to estimate added and modified size.
5	Time Estimating Procedure	– If you have sufficient estimated proxy size and actual development time data (three or more points that correlate), use procedure 5A. – If you do not have sufficient estimated size data but have sufficient plan added and modified size and actual development time data (three or more points that correlate), use procedure 5B. – If you have insufficient data or they do not correlate, use procedure 5C. – If you have no historical data, use procedure 5D.
5A	Time Estimating Procedure 5A	– Using the linear-regression method, calculate the β_0 and β_1 parameters from the estimated proxy size and actual total development time data. – If β_0 is not near 0 (substantially smaller than the expected development time for the new program), or β_1 is not within 50% of 1/(historical productivity), use procedure 5B.
5B	Time Estimating Procedure 5B	– Using the linear-regression method, calculate the β_0 and β_1 regression parameters from the plan added and modified size and actual total development time data. – If β_0 is not near 0 (substantially smaller than the expected development time for the new program), or β_1 is not within 50% of 1/(historical productivity), use procedure 5C.
5C	Time Estimating Procedure 5C	– If you have data on estimated–added and modified size and actual development time, set $\beta_0 = 0$ and $\beta_1 =$ (actual total development time to date/estimated–total added and modified size to date). – If you have data on plan–added and modified size and actual development time, set $\beta_0 = 0$ and $\beta_1 =$ (actual total development time to date/plan total added and modified size to date). – If you only have actual time and size data, set $\beta_0 = 0$ and $\beta_1 =$ (actual total development time to date/actual total added and modified size to date).

Step	Activities	Description
5D	Time Estimating Procedure 5D	If you have no historical data, use your judgment to estimate the development time from the estimated added and modified size.
6	Time and Size Prediction Intervals	– If you used regression method A or B, calculate the 70% prediction intervals for the time and size estimates. – If you did not use the regression method or do not know how to calculate the prediction interval, calculate the minimum and maximum development time estimate limits from your historical maximum and minimum productivity for the programs written to date.
Exit Criteria		– Completed estimated and actual entries for all pertinent size categories – Completed PROBE Calculation Worksheet with size and time entries – Plan and actual values entered on the Project Plan Summary

PROBE Calculation Worksheet (Added and Modified)

Student ____________________________ Program ____________

PROBE Calculation Worksheet (Added and Modified)		Size	Time
Added size (A):	A = BA+PA	______	
Estimated Proxy Size (E):	E = BA+PA+M	______	
PROBE estimating basis used:	(A, B, C, or D)	______	______
Correlation: (R^2)		______	______
Regression Parameters:	β_0 Size and Time	______	______
Regression Parameters:	β_1 Size and Time	______	______
Projected Added and Modified Size (P):	$P = \beta_{0size} + \beta_{1size}{}^{*}E$	______	
Estimated Total Size (T):	T = P + B − D − M + R	______	
Estimated Total New Reusable (NR):	sum of * items	______	
Estimated Total Development Time:	$\text{Time} = \beta_{0time} + \beta_{1time}{}^{*}E$	______	______
Prediction Range:	Range	______	______
Upper Prediction Interval:	UPI = P + Range	______	______
Lower Prediction Interval:	LPI = P − Range	______	______
Prediction Interval Percent:		______	______

Size Estimating Template Instructions

Purpose	Use this form with the PROBE method to make size estimates.
General	– A part could be a module, component, product, or system. – Where parts have a substructure of methods, procedures, functions, or similar elements, these lowest-level elements are called items. – Size values are assumed to be in the unit specified in size measure. – Avoid confusing base size with reuse size. – Reuse parts must be used without modification. – Use base size if additions, modifications, or deletions are planned. – If a part is estimated but not produced, enter its actual values as zero. – If a part is produced that was not estimated, enter it using zero for its planned values.
Header	– Enter your name and the date. – Enter the program name and number. – Enter the instructor's name and the programming language you are using. – Enter the size measure you are using.
Base Parts	– If this is a modification or enhancement of an existing product – measure and enter the base size (more than one product may be entered as base) – estimate and enter the size of the deleted, modified, and added size to the base program – After development, measure and enter the actual size of the base program and any deletions, modifications, or additions.
Parts Additions	– If you plan to add newly developed parts – enter the part name, type, number of items (or methods), and relative size – for each part, get the size per item from the appropriate relative size table, multiply this value by the number of items, and enter in estimated size – put an asterisk next to the estimated size of any new-reusable additions – After development, measure and enter – the actual size of each new part or new part items – the number of items for each new part
Reused Parts	– If you plan to include reused parts, enter the – name of each unmodified reused part – size of each unmodified reused part – After development, enter the actual size of each unmodified reused part.

PROBE Calculation Worksheet Instructions

Purpose	Use this form with the PROBE method to make size and resource estimate calculations.
General	– The PROBE method can be used for many kinds of estimates. Where development time correlates with added and modified size – use the Added and Modified Calculation Worksheet – enter the resulting estimates in the Project Plan Summary – enter the projected added and modified value (P) in the added and modified plan space in the Project Plan Summary – If development time correlates with some other combination of size-accounting types – define and use a new PROBE Calculation Worksheet – enter the resulting estimates in the Project Plan Summary – use the selected combination of size accounting types to calculated the projected size value (P) – enter this P value in the Project Plan Summary for the appropriate plan size for the size-accounting types being used
PROBE Calculations: Size (Added and Modified)	– Added Size (A): Total the added base code (BA) and Parts Additions (PA) to get Added Size (A). – Estimated Proxy Size (E): Total the added (A) and modified (M) sizes and enter as (E). – PROBE Estimating Basis Used: Analyze the available historical data and select the appropriate PROBE estimating basis (A, B, C, or D). – Correlation: If PROBE estimating basis A or B is selected, enter the correlation value (R^2) for both size and time. – Regression Parameters: Follow the procedure in the PROBE script to calculate the size and time regression parameters (β_0 and β_1), and enter them in the indicated fields. – Projected Added and Modified Size (P): Using the size regression parameters and estimated proxy size (E), calculate the projected added and modified size (P) as $P = \beta_{0Size} + \beta_{1Size} * E$. – Estimated Total Size (T): Calculate the estimated total size as $T = P+B-D-M+R$. – Estimated Total New Reusable (NR): Total and enter the new reusable items marked with *.
PROBE Calculations: Time (Added and Modified)	– PROBE Estimating Basis Used: Analyze the available historical data and select the appropriate PROBE estimating basis (A, B, C, or D). – Estimated Total Development Time: Using the time regression parameters and estimated proxy size (E), calculate the estimated development time as $Time = \beta_{0Time}\beta_{1Time} * E$.

PROBE Calculations: Prediction Range	– Calculate and enter the prediction range for both the size and time estimates. – Calculate the upper (UPI) and lower (LPI) prediction intervals for both the size and time estimates. – Prediction Interval Percent: List the probability percent used to calculate the prediction intervals (70% or 90%).
After Development (Added and Modified)	Enter the actual sizes for base (B), deleted (D), modified (M), and added base code (BA), parts additions (PA), and reused parts (R).

APPENDIX 10.A3

PSP Defect Recording

PSP Defect Recording Log Instructions

Purpose	– Use this form to hold data on the defects that you find and correct. – These data are used to complete the Project Plan Summary form.
General	– Record each defect separately and completely. – If you need additional space, use another copy of the form.
Header	– Enter your name and the date. – Enter the program name and number. – Enter the instructor's name and the programming language you are using.
Project	– Give each program a different name or number. – For example, record test program defects against the test program.
Date	Enter the date on which you found the defect.
Number	– Enter the defect number. – For each program or module, use a sequential number starting with 1 (or 001, etc.).
Type	– Enter the defect type from the defect type list summarized in the top left corner of the form. – Use your best judgment in selecting which type applies.
Inject	– Enter the phase when this defect was injected. – Use your best judgment.
Remove	Enter the phase during which you fixed the defect. (This will generally be the phase when you found the defect.)
Fix Time	– Enter the time that you took to find and fix the defect. – This time can be determined by stopwatch or by judgment.
Fix Ref.	– If you or someone else injected this defect while fixing another defect, record the number of the improperly fixed defect. – If you cannot identify the defect number, enter an X.
Description	Write a succinct description of the defect that is clear enough to later remind you about the error and help you to remember why you made it.

PSP Defect Type Standard

Type Number	Type Name	Description
10	Documentation	Comments, messages
20	Syntax	Spelling, punctuation, typos, instruction formats
30	Build, Package	Change management, library, version control
40	Assignment	Declaration, duplicate names, scope, limits
50	Interface	Procedure calls and references, I/O, user formats
60	Checking	Error messages, inadequate checks
70	Data	Structure, content
80	Function	Logic, pointers, loops, recursion, computation, function defects
90	System	Configuration, timing, memory
100	Environment	Design, compile, test, or other support system problems

Expanded Defect Type Standard

Purpose		**To facilitate causal analysis and defect prevention.**	
Note		– The types are grouped in ten general categories. – If the detailed category does not apply, use the general category. – The % column lists an example type distribution.	
No.	**Name**	**Description**	%
10	Documentation	Comments, messages, manuals	1.1
20	Syntax	General syntax problems	0.8
21	Typos	Spelling, punctuation	32.1
22	Instruction formats	General format problems	5.0
23	Begin-end	Did not properly delimit operation	0
30	Packaging	Change management, version control, system build	1.6
40	Assignment	General assignment problem	0
41	Naming	Declaration, duplicates	12.6
42	Scope		1.3
43	Initialize and close	Variables, objects, classes, and so on	4.0
44	Range	Variable limits, array range	0.3
50	Interface	General interface problems	1.3
51	Internal	Procedure calls and references	9.5
52	I/O	File, display, printer, communication	2.6
53	User	Formats, content	8.9
60	Checking	Error messages, inadequate checks	0
70	Data	Structure, content	0.5
80	Function	General logic	1.8
81	Pointers	Pointers, strings	8.7
82	Loops	Off-by-one, incrementing, recursion	5.5
83	Application	Computation, algorithmic	2.1
90	System	Timing, memory, and so on	0.3
100	Environment	Design, compile, test, other support system problems	0

PSP Defect Recording Log

Student		Date	
Program		Program #	
Instructor		Language	

Project	Date	Number	Type	Inject	Remove	Fix Time	Fix Ref.

Description:

Project	Date	Number	Type	Inject	Remove	Fix Time	Fix Ref.

Description:

Project	Date	Number	Type	Inject	Remove	Fix Time	Fix Ref.

Description:

Defect Types	
10 Documentation	60 Checking
20 Syntax	70 Data
30 Build, Package	80 Function
40 Assignment	90 System
50 Interface	100 Environment

APPENDIX 10.A4

PSP2

PSP2 Development Script

Purpose	To guide the development of small programs.
Entry Criteria	– Requirements statement. – Project Plan Summary form with estimated program size and development time. – For projects lasting several days or more, completed Task Planning and Schedule Planning templates. – Time and Defect Recording logs. – Defect Type standard and Coding standard.

Step	Activities	Description
1	Design	– Review the requirements and produce a design to meet them. – Record in the Defect Recording Log any requirements defects found. – Record time in the Time Recording Log.
2	***Design Review***	***– Follow the Design Review script and checklist to review the design.*** ***– Fix all defects found.*** ***– Record defects in the Defect Recording Log.*** ***– Record time in the Time Recording Log.***
3	Code	– Implement the design following the Coding standard. – Record in the Defect Recording Log any requirements or design defects found. – Record time in the Time Recording Log.
4	***Code Review***	***– Follow the Code Review script and checklist to review the code.*** ***– Fix all defects found.*** ***– Record defects in the Defect Recording Log.*** ***– Record time in the Time Recording Log.***
5	Compile	– Compile the program until there are no compile errors. – Fix all defects found. – Record defects in the Defect Recording Log. – Record time in the Time Recording Log.
6	Test	– Test until all tests run without error. – Fix all defects found. – Record defects in the Defect Recording Log. – Record time in the Time Recording Log. – Complete a Test Report template on the tests conducted and the results obtained.
Exit Criteria		– A thoroughly tested program that conforms to the Coding standard. ***– Completed Design Review and Code Review checklists.*** – Completed Test Report template. – Completed Time and Defect Recording logs.

PSP2 Design Review Script

Purpose	To guide you in reviewing detailed designs.
Entry Criteria	– Completed program design. – Design Review checklist. – Design standard. – Defect Type standard. – Time and Defect Recording logs.

General	Where the design was previously verified, check that the analyses – covered all of the design. – were updated for all design changes. – are correct. – are clear and complete.

Step	Activities	Description
1	Preparation	Examine the program and checklist and decide on a review strategy.
2	Review	– Follow the Design Review checklist. – Review the entire program for each checklist category; do not try to review for more than one category at a time! – Check off each item as you complete it. – Complete a separate checklist for each product or product segment reviewed.
3	Fix Check	– Check each defect fix for correctness. – Re-review all changes. – Record any fix defects as new defects and, where you know the defective defect number, enter it in the fix defect space.
Exit Criteria		– A fully reviewed detailed design. – One or more Design Review checklists for every design reviewed. – All identified defects fixed and all fixes checked. – Completed Time and Defect Recording logs.

Code Review Script

Purpose	To guide you in reviewing programs.
Entry Criteria	– A completed and reviewed program design. – Source program listing. – Code Review checklist. – Coding standard. – Defect Type standard. – Time and Defect Recording logs.
General	Do the code review with a source-code listing; do not review on the screen!

Step	Activities	Description
1	Review	– Follow the Code Review checklist. – Review the entire program for each checklist category; do not try to review for more than one category at a time! – Check off each item as it is completed. – For multiple procedures or programs, complete a separate checklist for each.
2	Correct	– Correct all defects. – If the correction cannot be completed, abort the review and return to the prior process phase. – To facilitate defect analysis, record all of the data specified in the Defect Recording Log instructions for every defect.
3	Check	– Check each defect fix for correctness. – Re-review all design changes. – Record any fix defects as new defects and, where you know the number of the defect with the incorrect fix, enter it in the fix defect space.
Exit Criteria		– A fully reviewed source program. – One or more Code Review checklists for every program reviewed – All identified defects fixed. – Completed Time and Defect Recording logs.

PSP2 Postmortem Script

Purpose	To guide the PSP postmortem process.
Entry Criteria	– Problem description and requirements statement. – Project Plan Summary form with program size, development time*, and defect* data. – For projects lasting several days or more, completed Task Planning and Schedule Planning templates. – Completed Test Report template. – ***Completed Design Review and Code Review checklists.*** – Completed Time and Defect Recording logs. – A tested and running program that conforms to the coding and size measurement standards.

Step	Activities	Description
1	Defect Recording	– Review the Project Plan Summary to verify that all of the defects found in each phase were recorded. – Using your best recollection, record any omitted defects.
2	Defect Data Consistency	– Check that the data on every defect in the Defect Recording log are accurate and complete. – Verify that the numbers of defects injected and removed per phase are reasonable and correct. ***– Determine the process yield and verify that the value is reasonable and correct.*** – Using your best recollection, correct any missing or incorrect defect data.
3	Size	– Count the size of the completed program. – Determine the size of the base, reused, deleted, modified, added, total, added and modified, and new reusable code. – Enter these data in the Project Plan Summary form.
4	Time	– Review the completed Time Recording log for errors or omissions. – Using your best recollection, correct any missing or incomplete time data.
Entry Criteria		– A thoroughly tested program that conforms to the coding and size measurement standards. ***– Completed Design Review and Code Review checklists.*** – Completed Test Report template. – Completed Project Plan Summary form. – Completed PIP forms describing process problems, improvement suggestions, and lessons learned. – Completed Time and Defect Recording logs.

PSP2 Project Plan Summary

Student ______________________ Date ______________________

Program ______________________ Program # ______________________

Instructor ______________________ Language ______________________

Summary	**Plan**	**Actual**	**To Date**
Size/Hour	________	________	________
Planned Time	________	________	________

Actual Time	______	______	______
CPI (Cost-Performance Index)	______	______	______
			(Planned/Actual)
% Reuse	______	______	______
% New Reusable	______	______	______
Test Defects/KLOC or equivalent	______	______	______
Total Defects/KLOC or equivalent	______	______	______
Yield %	______	______	______

Program Size	**Plan**	**Actual**	**To Date**
Base (B)	______	______	
	(Measured)	(Measured)	
Deleted (D)	______	______	
	(Estimated)	(Counted)	
Modified (M)	______	______	
	(Estimated)	(Counted)	
Added (A)	______	______	
	(A+M − M)	(T − B + D − R)	
Reused (R)	______	______	______
	(Estimated)	(Counted)	
Added and Modified (A+M)	______	______	______
	(Projected)	(A + M)	
Total Size (T)	______	______	______
	(A+M + B − M − D + R)	(Measured)	
Total New Reusable	______	______	______
Estimated Proxy Size (E)	______	______	______

Time in Phase (min.)	**Plan**	**Actual**	**To Date**	**To Date %**
Planning				
Design				
Design Review				
Code				
Code Review				
Compile				
Test				
Postmortem				
Total				

Defects Injected	**Plan**	**Actual**	**To Date**	**To Date %**
Planning				
Design				
Design Review				
Code				
Code Review				
Compile				
Test				
Total Development				

Defects Removed	**Plan**	**Actual**	**To Date**	**To Date %**
Planning				
Design				
Design Review				
Code				

Code Review				
Compile				
Test				
Total Development				
After Development				

Defect Removal Efficiency	**Plan**	**Actual**	**To Date**
Defects/Hour – Design Review			
Defects/Hour – Code Review			
Defects/Hour – Compile			
Defects/Hour – Test			
DRL (DLDR/UT)			
DRL (Code Review/UT)			
DRL (Compile/UT)			

PSP2 Plan Summary Instructions

Purpose	To hold the plan and actual data for programs or program parts.
General	– Use the most appropriate size measure, either LOC or element count. "To Date" is the total actual to-date values for all products developed. – A part could be a module, component, product, or system.
Header	– Enter your name and the date. – Enter the program name and number. – Enter the instructor's name and the programming language you are using.
Summary	– Enter the added and modified size per hour planned, actual, and to-date. – Enter the planned and actual times for this program and prior programs. – For planned time to date, use the sum of the current planned time and the ***to-date*** planned time for the ***most recent*** prior program. – CPI = (To Date Planned Time)/(To Date Actual Time).

	– Reused % is reused size as a percentage of total program size. – New Reusable % is new reusable size as a percentage of added and modified size. ***– Enter the test and total defects/KLOC or other appropriate measure.*** ***– Enter the planned, actual, and to-date yield before compile.***
Program Size	– Enter plan base, deleted, modified, reused, new reusable, and total size from the Size Estimating template. – Enter the plan added and modified size value (A+M) from projected added and modified size (P) on the Size Estimating template. – Calculate plan added size as A+M–M. – Enter estimated proxy size (E) from the Size Estimating template. – Enter actual base, deleted, modified, reused, total, and new reusable size from the Size Estimating template. – Calculate actual added size as T-B+D-R and actual added and modified size as A+M. – Enter to-date reused, added and modified, total, and new reusable size.
Time in Phase	– Enter plan total time in phase from the estimated total development time on the Size Estimating template. – Distribute the estimated total time across the development phases according to the To Date % for the most recently developed program. – Enter the actual time by phase and the total time. – To Date: Enter the sum of the actual times for this program plus the to-date times from the most recently developed program. – To Date %: Enter the percentage of to-date time in each phase.
Defects Injected	***– Enter the total estimated defects injected.*** ***– Distribute the estimated total defects across the development phases according to the To Date % for the most recently developed program.*** – Enter the actual defects by phase and the total actual defects. – To Date: Enter the sum of the actual defects injected by phase and the to-date values for the most recent previously developed program. – To Date %: Enter the percentage of the to-date defects injected by phase.
Defects Removed	***– Enter the estimated total defects removed.*** ***– Distribute the estimated total defects across the development phases according to the To Date % for the most recently developed program.*** – To Date: Enter the actual defects removed by phase plus the to-date values for the most recent previously developed program. – To Date %: Enter the percentage of the to-date defects removed by phase. – After development, record any defects subsequently found during program testing, use, reuse, or modification.
Defect-Removal Efficiency	***– Calculate and enter the defects removed per hour in design review, code review, compile, and test.*** ***– For DRL, take the ratio of the review and compile rates with test.*** ***– Where there were no test defects, use the to-date test defect/hour value.***

REFERENCES

Chhaya, Tejas (2008), "Modified Spiral Model Using PSP, TSP and Six Sigma (MSPTS) Process Model for Embedded Systems Control," MS Thesis, University of Michigan.

Humphrey, Watts S. (1995), *A Discipline for Software Engineering*. Addison Wesley, Upper Saddle River, NJ.

Humphrey, Watts S. (2005), *PSP: A Self-improvement Process for Software*, Addison Wesley, Upper Saddle River, NJ.

Humphrey, Watts S. (1997), *Introduction to the Personal Software Process*, Addison Wesley, Upper Saddle River, NJ.

Humphrey, Watts S. (1999), *Introduction to the Team Software Process*, Addison Wesley, Upper Saddle River, NJ.

Shaout, Adnan and Chhaya, Tejas (2008), "A New Process Model for Embedded Systems Control in Automotive Industry," Proceedings of the 2008 International Arab Conference on Information Technology (ACIT'2008), Tunis, Dec.

Shaout, Adnan and Chhaya, Tejas (2009), "A new process model for embedded systems control for automotive industry," *International Arab Journal of Information Technology*, Volume *6*, #5, pp. 472–479.

Thórisson, Kristinn R., Benko, Hrvoje, Abramov, Denis, Arnold, Andrew, Maskey, Sameer, and Vaseekaran, Aruchunan (2004), Volume *25*, 4 *"Constructionist design methodology for interactive intelligences." A.I. Magazine, Winter.*

CHAPTER 11

SOFTWARE DESIGN FOR SIX SIGMA (DFSS) PROJECT ROAD MAP

11.1 INTRODUCTION

This chapter is written primarily to present the software Design for Six Sigma (DFSS) project road map to support the software Black Belt and his or her team and the functional champion in the project execution mode of deployment. The design project is the core of the DFSS deployment and has to be executed consistently using a road map that lays out the DFSS principles, tools, and methods within an adopted gated design process (Chapter 8). From a high-level perspective, this road map provides the immediate details required for a smooth and successful DFSS deployment experience.

The chart presented in Figure 11.1 depicts the road map proposed. The road map objective is to develop Six Sigma software-solution entities with an unprecedented level of fulfillment of customer wants, needs, and delights throughout its life cycle (Section 7.4).

The software DFSS road map has four phases Identify, Conceptualize, Optimize, and Verify and Validate, denoted ICOV in seven developmental stages. Stages are separated by milestones called the *tollgates* (TGs). Coupled with design principles and tools, the objective of this chapter is to mold all that in a comprehensive implementable sequence in a manner that enables deployment companies to achieve systematically desired benefits from executing projects. In Figure 11.1, a *design stage* constitutes a collection of design activities and can be bounded by entrance and exit tollgates. A TG represents a milestone in the software design cycle and has

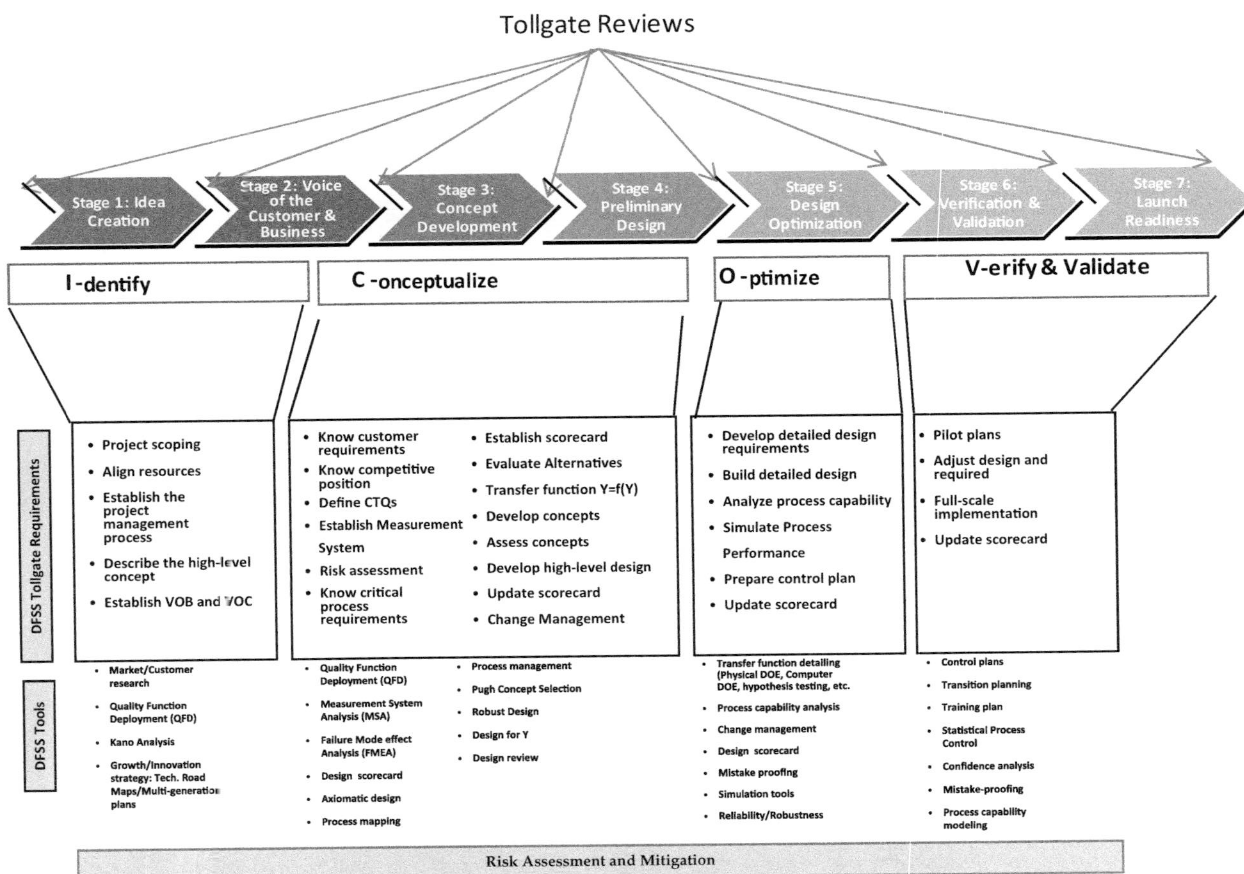

FIGURE 11.1 Software DFSS project road map.

some formal meaning defined by the company's own software development coupled with management recognition. The ICOV stages are an average of Dr. El-Haik studies of several deployments. It need not be adopted blindly but customized to reflect the deployment interest. For example, industry type, software production cycle, and volume are factors that can contribute to the shrinkage or elongation of some phases. Generally, the life cycle of a software or a process starts with some form of idea generation whether in free-invention format or using a more disciplined format such as multigeneration software planning and growth strategy.

Prior to starting on the DFSS road map, the Black Belt team needs to understand the rationale of the project. We advise that they ensure the feasibility of progressing the project by validating the project scope, the project charter, and the project resource plan (Section 8.3.2 Part d). A session with the champion is advised to take place once the matching between the Black Belt and project charter is done. The objective is to make sure that every one is aligned with the objectives and to discuss the next steps.

In software DFSS deployment, we will emphasize the synergistic software DFSS cross-functional team. A well-developed team has the potential to design winning Six Sigma level solutions. The growing synergy, which develops from ever-increasing numbers of successful teams, accelerates deployment throughout the company. The payback for up-front investments in team performance can be enormous. Continuous vigilance by Black Belt to improve and to measure team performance throughout the project life cycle will be rewarded with ever-increasing capability and commitment to deliver winning design solutions. Given time, there will be a transition from resistance to embracing the methodology, and the company culture will be transformed.

11.2 SOFTWARE DESIGN FOR SIX SIGMA TEAM[1]

It is well known that software intended to serve the same purpose and the same market may be designed and produced in radically different varieties. For example, compare your booking experience at different hotel websites or your mortgage experience shopping for a loan online. Why is it that two websites function and feel so differently? From the perspective of the design process, the obvious answer is that the website design derives a series of decisions and that different decisions made at the tollgates in the process result in such differentiation. This is common sense; however, it has significant consequences. It suggests that a design can be understood not only in terms of the adopted design process but also in terms of the decision-making process used to arrive at it. Measures to address both sources of design variation need to be institutionalized. We believe that the adoption of the ICOV DFSS process presented in this chapter will address at least one issue: the consistency of development activities and derived decisions. For software design teams, this means that the company structures used to facilitate coordination during the project execution

[1] In this section, we discuss the soft aspects of the DFSS team. The technical aspects are discussed using the Personal Software Process (PSP) and Team Software Process (TSP) frameworks in Chapter 10.

have an effect on the core of the development process. In addition to coordination, the primary intent of an organizing design structure is to control the decision-making process. It is logical then to conclude that we must consider the design implications of the types of organizing structures in which we deploy the ICOV process to manage design practice. When flat organizing structures are adopted with design teams, members must negotiate design decisions among themselves because a top-down approach to decision making may not be available. Members of a software design team negotiating decisions with one another during design projects is an obvious practice. A common assumption seems to be that these decision-making negotiations proceed in a reasonable manner—this being a basic premise of concurrent software design (what do you mean? Concurrent design means that more than one member of the design team is working on a different part of the design). Patterns and outcomes of decision making are best explained as a dynamic behavior of the teams. Even if two teams develop similar software using the same process, members of the otherwise comparable design teams may have varying levels of influence as decisions are made. The rank differences among members of a design team can play a substantial role in team dynamics from the perspective of day-to-day decisions. It is the responsibility of the Black Belt to balance such dynamics in his or her team. As team leaders, Black Belts and Master Black Belts need to understand that design teams must make decisions, and invariably, some set of values must drive those decisions.

Decision making and team structure in companies that use hierarchical structures follow known patterns. Although day-to-day decision making is subject to team dynamics, the milestone decisions are not. In the latter, decisions are made based upon the formal rank. That is, decisions made by higher ranking individuals override those made by lower ranking individuals. Such an authoritative decision-making pattern makes sense as long as the ranks cope with expertise and appreciation to company goals. This pattern also will ensure that those higher in rank can coordinate and align the actions of others with the goals of the company. We adopted this model for DFSS deployment in Chapter 9. Despite these clear benefits, several factors make this traditional form of hierarchical structure less attractive, particularly in the context of the design team. For example, risk caused by increased technological complexity of the software being designed, market volatility, and others make it difficult to create a decision-making structure for day-to-day design activities. To address this problem, we suggest a flatter, looser structure that empowers team members, Black Belts, and Master Black Belts to assert their own expertise when needed on day-to-day activities. In our view, an ideal design team should consist of team members who represent every phase of a software life cycle. This concurrent structure combined with the road map will assure company consistency (i.e., minimal design process variation and successful DFSS deployment). This approach allows information to flow freely across the bounds of time and distance, in particular, for geographically challenged companies. It also ensures that representatives of later stages of the life cycle have a similar influence in making design decisions as do those representatives of earlier stages (e.g., maintenance, vendors, aftermarket, etc.). Although obvious benefits such as these can result from a flattened structure, it does not need to be taken to the extreme. It is apparent that having no structure means the absence of a sound decision-making process. Current practice indicates that a design project is far from a rational process of simply identifying day-to-day activities and then assigning the expertise required to handle them. Rather, the truly important design decisions are more likely to be subjective decisions made based on judgments, incomplete information, or personally biased values even though we strive to minimize these gaps in voice of the customer (VOC) and technology road mapping. In milestones, the final say over decisions in a flat design team remains with the champions or TG approvers. It must not happen at random but rather in organized ways.

Our recommendation is twofold. First, a deployment company should adopt a common design process that is customized with their design needs with flexibility to adapt the DFSS process to obtain design consistency and to assure success. Second, it should choose flatter, looser design team structures that empower team members to assert their own expertise when needed. This practice is optimum in companies servicing advanced development work in high-technology domains.

A cross-functional synergistic design team is one of the ultimate objectives of any deployment effort. The Belt needs to be aware of the fact that full participation in design is not guaranteed simply because members are assigned into a team. The structural barriers and interests of others in the team are likely to be far too formidable as the team travels down the ICOV DFSS process.

The success of software development activities depends on the performance of this team that is fully integrated with representation from internal and external (suppliers and customers) members. Special efforts may be necessary to create a multifunctional DFSS team that collaborates to achieve a shared project vision. Roles, responsibilities, membership, and resources are best defined up front, collaboratively, by the teams. Once the team is established, however, it is just as important to maintain the team to improve continuously its performance. This first step, therefore, is an ongoing effort throughout the software DFSS ICOV cycle of planning, formulation, and production.

The primary challenge for a design organization is to learn and to improve faster than the competitor. Lagging competitors must go faster to catch up. Leading competitors must go faster to stay in front. A software DFSS team should learn rapidly not only about what needs to be done but about how to do it—how to implement pervasively the DFSS process.

Learning without application is really just gathering information, not learning. No company becomes premier by simply knowing what is required but rather by practicing, by training day in and day out, and by using the best contemporary DFSS methods. The team needs to monitor competitive performance using benchmarking software and processes to help guide directions of change and employ lessons learned to help identify areas for their improvement. In addition, they will benefit from deploying program and risk-management practices throughout the project life cycle (Figure 11.1). This activity is a key to achieving a winning rate of improvement by avoiding the elimination of risks. The team is advised to practice continuously design principles and systems thinking (i.e., thinking in terms of the total software profound knowledge).

11.3 SOFTWARE DESIGN FOR SIX SIGMA ROAD MAP

In Chapter 8, we learned about the ICOV process and the seven developmental stages spaced by bounding tollgates indicating a formal transition between entrance and exit. As depicted in Figure 11.2, tollgates or design milestones events include reviews to assess what has been accomplished in the current developmental stage and to prepare the next stage. The software design stakeholders including the project champion, design owner, and deployment champion conduct tollgate reviews. In a tollgate review, three options are available to the champion or his delegate of tollgate approver:

- Proceed to next stage of development
- Recycle back for further clarification on certain decisions
- Cancel the project

This is what I am talking about, which is to include this "Recycle back for further clarification on certain decisions" in Figures 11.1, 7.2, and 7.3.

In TG reviews, work proceeds when the exit criteria (required decisions) are made. Consistent exit criteria from each tollgate blend both software DFSS deliverables

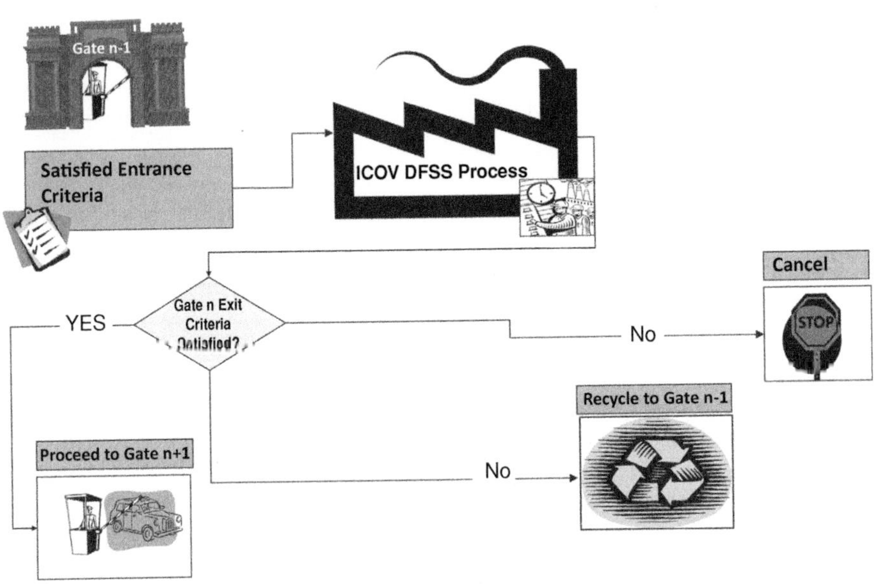

FIGURE 11.2 DFSS tollgate process.

caused by the application of the approach itself and the business unit, or function specific deliverables are needed.

In this section, we will first expand on the ICOV DFSS process activities by stage with comments on the applicable key DFSS tools and methods over what was baselined in Chapter 8. A subsection per phase is presented in the following sections.

11.3.1 Software DFSS Phase I: Identify Requirements

This phase includes two stages: idea creation (Stage 1) and voices of the customer and business (Stage 2).

- ***Stage 1: Idea Creation***
  Stage 1 Entrance Criteria
  Entrance criteria may be tailored by the deploying function for the particular program/project provided the modified entrance criteria, in the opinion of the function, are adequate to support the exit criteria for this stage. They may includes:
  - A target customer or market
  - A market vision with an assessment of marketplace advantages
  - An estimate of development cost
  - Risk assessment[2]

  TG "1"—Stage 1 Exit Criteria
  - Decision to collect the voice of the customer to define customer needs, wants, and delights
  - Verification of adequate funding is available to define customer needs
  - Identification of the tollgate keepers[3] leader and the appropriate staff
- **Stage 2: Customer and Business Requirements Study**
  Stage 2 Entrance Criteria
  - Closure of Tollgate 1: Approval of the gate keeper is obtained
  - A software DFSS project charter that includes project objectives, software design statement, Big Y and other business levers, metrics, resources and team members, and so on.

    This is almost the same criteria required with define, measure, analyze, improve, and control (DMAIC) Six Sigma type of projects. However, project duration is usually longer, and initial cost is probably higher. The DFSS team,

[2]See Chapter 15.

[3]A tollgate keeper is an individual or a group who will assess the quality of work done by the design team and initiate a decision to approve, reject or cancel, or recycle the project to an earlier gate. Usually, a project champion(s) is tasked with this mission.

relative to DMAIC, typically experiences longer project cycle time. The goal here is either designing or redesigning a different entity not just patching up the holes of an existing one. Higher initial cost is because the value chain is being energized from software development and not from production arenas. There may be new customer requirements to be satisfied, adding more cost to the developmental effort. For DMAIC projects, we may only work on improving a very limited subset of the critical-to-satisfaction (CTS) characteristics, also called the Big Ys.

- Completion of a market survey to determine customer needs CTS—VOC.
  In this step, customers are fully identified and their needs are collected and analyzed with the help of quality function deployment (QFD) and Kano analysis (Chapter 12). Then the most appropriate set of CTSs or Big Ys metrics are determined to measure and evaluate the design. Again, with the help of QFD and Kano analysis, the numerical limits and targets for each CTS are established. In summary, here is the list of tasks in this step. The detailed explanation is provided in later chapters:
- Determine methods of obtaining customer needs and wants
- Obtain customer needs and wants and transform them into a list of the VOC
- Finalize requirements
- Establish minimum requirement definitions
- Identify and fill gaps in customer-provided requirements
- Validate application and usage environments
- Translate the VOC to CTSs as critical-to-quality, critical-to-delivery, critical-to-cost, and so on.
- Quantify CTSs or Big Ys
- Establish metrics for CTSs
- Establish acceptable performance levels and operating windows
- Start flow-down of CTSs
- An assessment of required technologies
- A project development plan (through TG2)
- Risk assessment
- Alignment with business objectives—Voice of the Business (VOB) relative to growth and innovation strategy

TG "2"—Stage 2 Exit Criteria

- Assessment of market opportunity
- Command a reasonable price or be affordable
- Commitment to development of the conceptual designs
- Verification that adequate funding is available to develop the conceptual design
- Identification of the gate keepers leader (gate approver) and the appropriate staff
- Continue flow-down of CTSs to functional requirements

11.3.1.1 Identify Phase Road Map. DFSS tools used in this phase include (Figure 11.1):

- Market/customer research
- QFD: Phase I[4]
- Kano analysis[5]
- Growth/innovation strategy

11.3.1.2 Software Company Growth and Innovation Strategy: Multigeneration Planning (MGP)[6]. Even within best-in-class companies, there is a need and opportunity to strengthen and to accelerate progress. The first step is to establish a set of clear and unambiguous guiding growth principles as a means to characterize the company position and focus. For example, growth in emerging markets might be the focus abroad, whereas effectiveness and efficiency of resource usage within the context of enterprise productivity and sustainability may be the local position. Growth principles and vision at the high level are adequate to find agreement and to focus debate within the zone of interest and to exclude or diminish nonrealistic targets. The second key step is to assess the current knowledge and solutions of the software portfolio in the context of these growth principles. An inventory is developed of what the senior leadership team knows they have and how it integrates in the set of guiding growth principles. Third, a vision is established of the ultimate state for the company. Finally, a *multigeneration plan* is developed to focus the research, product development, and integration efforts in planned steps to move toward that vision. The multigeneration plan is key because it helps the deploying company stage progress in realistic developmental stages one DFSS project at a time but always with an eye on the ultimate vision.

In today's business climate, successful companies must be efficient and market-sensitive to supersede their competitors. By focusing on new software, companies can create custom solutions to meet customer needs, enabling customers to keep in step with new software trends and changes that affect them. As the design team engages the customers (surveys, interviews, focus groups, etc.) and processes the QFD, they gather competitive intelligence. This information helps increase the design teams awareness of competing software products or how they stack up competitively with a particular key customer. By doing this homework, the team identifies potential gaps in their development maturity. Several in-house tools to manage the life cycle of each software product from the cradle to the grave need to be developed to include the multigeneration plan and a customized version of the ICOV DFSS process. The multigeneration plan evaluates the market size and trends, software positioning, competition, and technology requirements. This tool provides a means to identify easily

[4]See Chapter 12.

[5]See Chapter 12.

[6]http://216.239.57.104/search?q=cache:WTPP0iD4WTAJ:cipm.ncsu.edu/symposium/docs/Hutchins_text.doc+product+multi-generation+plan&hl=en by *Scott H. Hutchins*.

any gaps in the portfolio while directing the DFSS project roadmap. The multigeneration plan needs to be supplemented with a decision-analysis tool to determine the financial and strategic value of potential new applications across a medium time horizon. If the project passes this decision-making step, it can be lined up with others in the Six Sigma project portfolio for a start schedule.

11.3.1.3 Research Customer Activities. This step is usually done by the software planning departments (software and process) or by the market research experts who should be on the DFSS team. The Belt and his team start by brainstorming all possible customer groups of the product, Using the affinity diagram method to group the brainstormed potential customer groups. Categories of markets, user types, or software and process applications types will emerge. From these categories, the DFSS team should work toward a list of clearly defined customer groups from which individuals can be selected.

External customers might be drawn from: customer centers, independent sales organizations, regulatory agencies, societies, and special interest groups. Merchants and, most importantly, the end user should be included. The selection of external customers should include existing and loyal customers, recently lost customers, and new conquest customers within the market segments. Internal customers might be drawn from: production, functional groups, facilities, finance, employee relations, design groups, distribution organizations, and so on. Internal research might assist in selecting internal customer groups that would be most instrumental in identifying wants and needs in operations and software operations.

The ideal software definition, in the eye of the customer, may be extracted from customer engagement activities. This will help turn the knowledge gained from continuous monitoring of consumer trends, competitive benchmarking, and customer likes and dislikes into a preliminary definition of ideal software. In addition, it will help identify areas for further research and dedicated efforts. The design should be described from a customer's viewpoint (external and internal) and should provide the first insight into what good software should look like. Concept models and design studies using an axiomatic design (Chapter 13) are good sources for evaluating consumer appeal and areas of likes or dislikes.

The array of customer attributes should include all customer and regulatory requirements as well as social and environmental expectations. It is necessary to understand the requirement and prioritization similarities and differences to understand what can be standardized and what needs to be tailored.

11.3.2 Software DFSS Phase 2: Conceptualize Design

This phase spans the following two stages: concept development (Stage 3) and preliminary design (Stage 4).

- **Stage 3: Concept Development**
  Stage 3 Entrance Criteria

- Closure of Tollgate 2: Approval of the gate keeper is obtained.
- Defined system technical and operational requirements.

  Translate customer requirements (CTSs or Big Ys) to software/process functional requirements: Customer requirements CTSs give us ideas about what will make the customer satisfied, but they usually cannot be used directly as the requirements for product or process design. We need to translate customer requirements to software and process functional requirements. Another phase of QFD can be used to develop this transformation. The axiomatic design principle also will be very helpful for this step.
- A software conceptual design (functional requirements, design parameters, flowcharts, etc.).
- Tradeoff of alternate conceptual designs with the following steps:
  - Generate design alternatives: After determining the functional requirements for the new design entity (software), we need to conceptualize (develop) products, which can deliver those functional requirements. In general, there are two possibilities. The first is that the existing technology or known design concept can deliver all the requirements satisfactorily, and then this step becomes almost a trivial exercise. The second possibility is that the existing technology or known design cannot deliver all requirements satisfactorily, and then a new design concept has to be developed. This new design should be "creative" or "incremental," reflecting the degree of deviation from the baseline design, if any. The axiomatic design (Chapter 13) will be helpful to generate many innovative design concepts in this step.
  - Evaluate design alternatives: Several design alternatives might be generated in the last step. We need to evaluate them and make a final determination of which concept will be used. Many methods can be used in design evaluation, which include the Pugh concept selection technique, design reviews, and failure mode and effects analysis (FMEA). After design evaluation, a winning concept will be selected. During the evaluation, many weaknesses of the initial set of design concepts will be exposed, and the concepts will be revised and improved. If we are designing a process, then process management techniques also will be used as an evaluation tool.
- Functional, performance, and operating requirements allocated to software design components (subprocesses).
- Develop cost estimate (Tollgate 2 through Tollgate 5).
- Target product/software unit production cost assessment.
- Market:
  - Profitability and growth rate.
  - Supply chain assessment.
  - Time-to-market assessment.
  - Share assessment.
- Overall risk assessment.

- A project management plan (Tollgate 2 through Tollgate 5) with a schedule and a test plan.
- A team member staffing plan.

TG "3"—Stage 3 Exit Criteria

- Assessment that the conceptual development plan and cost will satisfy the customer base.
- Decision that the software design represents an economic opportunity (if appropriate).
- Verification that adequate funding will be available to perform preliminary design.
- Identification of the tollgate keeper and the appropriate staff.
- An action plan to continue the flow-down of the design functional requirements.

- **Stage 4: Preliminary Design**

  Stage 4 Entrance Criteria

  - Closure of Tollgate 3: Approval of the gate keeper is obtained
  - Flow-down of system functional, performance, and operating requirements to subprocesses and steps (components)
  - Documented design data package with configuration management[7] at the lowest level of control
  - Development-to-production operations transition plan published and, in effect,
  - Subprocesses (steps) functionality, performance, and operating requirements are verified
  - Development testing objectives are completed under nominal operating conditions
  - Design parametric variations are tested under critical operating conditions
    - Tests might not use the intended operational production processes
  - Design, performance, and operating transfer functions
  - Reports documenting the design analyses as appropriate
  - A procurement strategy (if applicable)
  - Make/buy decision
  - Sourcing (if applicable)
  - Risk assessment

  TG "4"—Stage 4 Exit Criteria

  - Acceptance of the selected software solution/design
  - Agreement that the design is likely to satisfy all design requirements
  - Agreement to proceed with the next stage of the selected software solution/design

[7] A systematic approach to define design configurations and to manage the change process.

- An action plan to finish the flow-down of the design functional requirements to design parameters and process variables

  DFSS tools used in this phase:
  - QFD[8]
  - Axiomatic design[9]
  - Measurement system analysis (MSA)
  - (FMEA)
  - Design scorecard
  - Process mapping (flowcharting)
  - Process management
  - Pugh concept selection
  - Robust design[10]
  - Design for reusability[11]
  - Design reviews

  Software DFSS Phase 3: Optimize the design[12]

  This phase spans stage 5 only—the "design optimization" stage.

- **Stage 5: Design Optimization**

  Stage 5 Entrance Criteria
  - Closure of Tollgate 4: Approval of the gate keeper is obtained.
  - Design documentation defined: The design is complete and includes the information specific to the operations processes (in the opinion of the operating functions).
  - Design documents are under the highest level of control.
  - A formal change configuration is in effect.
  - Operations are validated by the operating function to preliminary documentations.
  - Demonstration test plan is put together that must demonstrate functionality and performance under operational environments. Full-scale testing and load testing.
  - Risk assessment.

  TG "5"—Stage 5 Exit Criteria
  - Agreement that functionality and performance meet the customers' and business's requirements under the intended operating conditions.
  - Decision to proceed with a verification test of a pilot built for preliminary operational process documentation.
  - Analyses to document the design optimization to meet or exceed functional, performance, and operating requirements.

[8] See Chapter 12.
[9] See Chapter 13.
[10] See Chapter 18.
[11] See Chapter 14.
[12] See Chapter 17.

- Optimized transfer functions: Design of Experiments (DOE) is the backbone of the process design and the redesign improvement. It represents the most common approach to quantify the transfer functions between the set of CTSs and/or requirements and the set of critical factors, the Xs, at different levels of the design hierarchy. DOE can be conducted by hardware or software (e.g., simulation). From the subset of a few vital X's, experiments are designed to manipulate the inputs actively to determine their effect on the outputs (Big Ys or small ys). This phase is characterized by a sequence of experiments, each based on the results of the previous study. "Critical" variables are identified during this process. Usually, a small number of Xs accounts for most of the variation in the outputs.

The result of this phase is an optimized software entity with all functional requirements released at Six Sigma performance level. As the concept design is finalized, there are still a lot of design parameters that can be adjusted and changed. With the help of computer simulation and/or hardware testing, DOE modeling, robust design methods, and response surface methodology, the optimal parameter settings will be determined. Usually this parameter optimization phase will be followed by a tolerance optimization step. The objective is to provide a logical and an objective basis for setting the requirements and process tolerances. If the design parameters are not controllable, we may need to repeat stages 1–3 of software DFSS.

DFSS tools used in this phase:

- Transfer function detailing (physical DOE, computer DOE, hypothesis testing, etc.)
- Process capability analysis
- Design scorecard
- Simulation tools
- Mistake-proofing plan
- Robustness assessment

Software DFSS Phase 4: Verify and Validate the Design[13]

This phase spans the following two stages: Verification (Stage 6) and Launch Readiness (Stage 7).

- **Stage 6: Verification**

Stage 6 Entrance Criteria

- Closure of Tollgate 5: Approval of the gate keeper is obtained
- Risk assessment

TG "6"—Stage 6 Exit Criteria

After the optimization is finished, we will move to the final verification and validation activities, including testing. The key actions are:

- The pilot tests are audited for conformance with design and operational documentation.

[13]See Chapter 19.

- Pilot test and refining: No software should go directly to market without first piloting and refining. Here we can use software failure mode effect analysis (SFMEA[14]) as well as pilot- and small-scale implementations to test and evaluate real-life performance.
- Validation and process control: In this step, we will validate the new entity to make sure that the software, as designed, meets the requirements and to establish process controls in operations to ensure that critical characteristics are always produced to specification of the optimize phase.

- **Stage 7: Launch Readiness**

  Stage 7 Entrance Criteria

  - Closure of Tollgate 6: Approval of the gate keeper is obtained.
  - The operational processes have been demonstrated.
  - Risk assessment.[15]
  - All control plans are in place.
  - Final design and operational process documentation has been published.
  - The process is achieving or exceeding all operating metrics.
  - Operations have demonstrated continuous operations without the support of the design development personnel.
  - Planned sustaining development personnel are transferred to operations.
    - Optimize, eliminate, automate, and/or control vital few inputs deemed in the previous phase.
    - Document and implement the control plan.
    - Sustain the gains identified.
    - Reestablish and monitor long-term delivery capability.
    - A transition plan is in place for the design development personnel.
    - Risk assessment.[16]

  TG "7" Exit Criteria

  - The decision is made to reassign the DFSS Black Belt.
  - Full commercial roll out and transfer to new design owner: As the design entity is validated and process control is established, we will launch a full-scale commercial roll out, and the newly designed software together with the supporting operations processes can be handed over to the design and design owners, complete with requirements settings and control and monitoring systems.
  - Closure of Tollgate 7: Approval of the gate keeper is obtained.

    DFSS tools used in this phase:
  - Process control plan
  - Control plans

[14] See Chapter 16.
[15] See Chapter 15.
[16] See Chapter 15.

- Transition planning
- Training plan
- Statistical process control
- Confidence analysis[17]
- Mistake-proofing
- Process capability modeling

11.4 SUMMARY

In this chapter, we presented the software design for the Six Sigma road map. The road map is depicted in Figure 11.1, which highlights at a high level, the identify, conceptualize, optimize, and verify and validate phases—the seven software development stages (idea creation, voices of the customer and business, concept development, preliminary design, design optimization, verification, and launch readiness). The road map also recognizes the tollgate design milestones in which DFSS teams update the stockholders on developments and ask for decisions to be made on whether to approve going into the next stage, to recycle back to an earlier stage, or to cancel the project altogether.

The road map also highlights the most appropriate DFSS tools with the ICOV phase. It indicates where the tool usage is most appropriate to start.

[17] See Chapter 6.

CHAPTER 12

SOFTWARE QUALITY FUNCTION DEPLOYMENT

12.1 INTRODUCTION

In this chapter, we will cover the history of quality function deployment (QFD), describe the methodology of applying QFD within the software Design for Six Sigma (DFSS) project road map (Chapter 11), and apply QFD to our software example. Within the context of DFSS, El-Haik and Roy (2005) and El-Haik and Mekki detailed the application of QFD for industrial products. The application of QFD to software design requires more than a copy and paste of an industrial model. Several key lessons have been learned through experience about the potentials and pitfalls of applying QFD to software development.

QFD in software applications focuses on improving the quality of the software development process by implementing quality improvement techniques during the Identify DFSS phase. These quality improvement techniques lead to increased productivity, fewer design changes, a reduction in the number of errors passed from one phase to the next, and quality software products that satisfy customer requirements. These new quality software systems require less maintenance and allow information system (IS) departments to shift budgeted dollars from maintenance to new project development, leading to a (long-term) reduction in the software development backlog. Organizations that have published material concerning the use of QFD application to software development include Hewlett-Packard (Palo Alto, CA) Rapid application development tool and project rapid integration & management application (PRIMA), a data integration network system (Betts, 1989; Shaikh,

Software Design for Six Sigma: A Roadmap for Excellence, By Basem El-Haik and Adnan Shaout

TABLE 12.1 Comparison of Results Achieved Between Traditional Approaches and QFD

Result Achieved	Mean Traditional Rating	Mean SQFD Rating
Communication satisfactory with technical personnel	3.7	4.09
Communication satisfactory with users	3.6	4.06
User requirements met	3.6	4.00
Communication satisfactory with management	3.4	3.88
Systems developed within budget	3.4	3.26
Systems easy to maintain	3.4	3.42
Systems developed on time	3.3	3.18
Systems relatively error-free	3.3	3.95
Systems easy to modify	3.3	3.58
Programming time reduced	3.2	3.70
Testing time reduced	3.0	3.29
Documentation consistent and complete	2.7	3.87

1989), IBM (Armonk, NY) automated teller machines (Sharkey 1991), and Texas Instruments' (Dallas, TX) products to support engineering process improvements (Moseley & Worley 1991). There are many cited benefits of QFD in software development. Chief among them are representing data to facilitate the use of metrics, creating better communication among departments, fostering better attention to customers' perspectives, providing decision justification, quantifying qualitative customer requirements, facilitating cross-checking, avoiding the loss of information, reaching consensus of features faster, reducing the product definition interval, and so on. These findings are evident by the results in Table 12.1 (Hagg et al., 1996). The table provides a comparison of the results achieved using traditional approaches and using QFD (given on a 5-point Likert scale, with 1 being the result was not achieved and 5 being the result was achieved very well). QFD achieves significantly higher results in the areas of communications satisfaction with technical personnel, communications satisfaction with users, user requirements being met, communications satisfaction with management, systems being relatively error-free, programming time being reduced, and documentation being consistent and complete. The remaining areas yielded only minor differences. Despite the fact that these two studies were undertaken 5 years apart, these new data indicate that the use of QFD improves the results achieved in most areas associated with the system development process (Hagg et al., 1996).

QFD is a planning tool that allows the flow-down of high-level customer needs and wants to design parameters and then to process variables that are critical to fulfilling the high-level needs. By following the QFD methodology, relationships are explored between the quality characteristics expressed by customers and the substitute quality requirements expressed in engineering terms (Cohen, 1988, 1995). In the context of DFSS, we call these requirements "critical-to" characteristics. These critical-to characteristics can be expanded along the dimensions of speed (critical-to-delivery, CTD), quality (critical to quality [CTQ]), cost (critical to cost [CTC]), as well as the other dimensions introduced in Figure 1.1. In the QFD methodology, customers

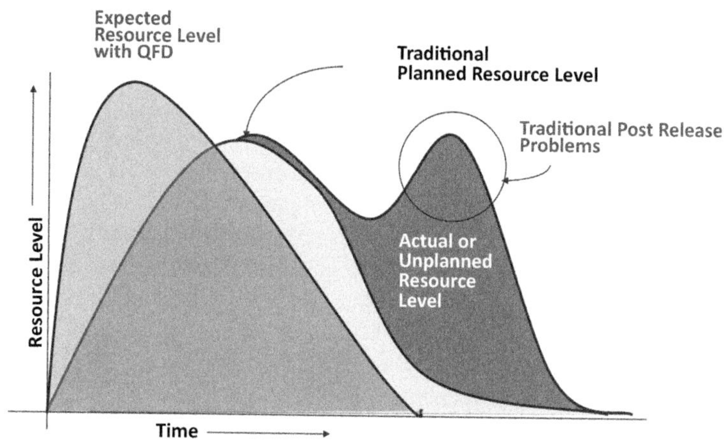

FIGURE 12.1 The time phased effort for DFSS vs traditional design.

define their wants and needs using their own expressions, which rarely carry any actionable technical terminology. The voice of the customer can be affinitized into a list of needs and wants that can be used as the input in a relationship matrix, which is called QFD's house of quality (HOQ).

Knowledge of customer's needs and wants is paramount in designing effective software with innovative and rapid means. Using the QFD methodology allows the developer to attain the shortest development cycle while ensuring the fulfillment of the customers' needs and wants.

Figure 12.1 shows that teams who use QFD place more emphasis on responding to problems early in the design cycle. Intuitively, it incurs more effort, time, resources, and energy to implement a design change at the production launch than at the concept phase because more resources are required to resolve problems than to prevent their occurrence in the first place. QFD is a front-end requirements solicitation technique, adaptable to any software engineering methodology that quantifiably solicits and defines critical customer requirements.

With QFD, quality is defined by the customer. Customers want products and services that, throughout their lives, meet their needs and expectations at a value that exceeds cost. QFD methodology links the customer needs through design and into process control. QFD's ability to link and prioritize at the same time provides laser focus to show the design team where to focus energy and resources.

In this chapter, we will provide the detailed methodology to create the four QFD houses and evaluate them for completeness and goodness, introduce the Kano model for voice of the customer (VOC), and relate the QFD with the DFSS road map introduced in Chapter 11.

12.2 HISTORY OF QFD

QFD was developed in Japan by Dr. Yoji Akao and Shigeru Mizuno in 1966 but was not westernized until the 1980s. Their purpose was to develop a quality assurance method that would design customer satisfaction into a product before it was manufactured. For six years, the methodology was developed from the initial concept of Kiyotaka Oshiumi of Bridgestone Tire Corporation (Nashuille, TN). After the first publication of "Hinshitsu Tenkai," quality deployment by Dr. Yoji Akao (1972), the pivotal development work was conducted at Kobe Shipyards for Mitsubishi Heavy Industry (Tokyo, Japan). The stringent government regulations for military vessels coupled with the large capital outlay forced the management at the shipyard to seek a method of ensuring upstream quality that cascaded down throughout all activities. The team developed a matrix that related all the government regulations, critical design requirements, and customer requirements to company technical-controlled characteristics of how to achieve these standards. Within the matrix, the team depicted the importance of each requirement that allowed for prioritization. After the successful deployment within the shipyard, Japanese automotive companies adopted the methodology to resolve the problem with rust on cars. Next it was applied to car features, and the rest, as we say, is history. In 1978, the detailed methodology was published (Mizuno & Akao, 1978, 1994) in Japanese and was translated to English in 1994.

12.3 QFD OVERVIEW

The benefits of using QFD methodology are, mainly, ensuring that high-level customer needs are met, that the development cycle is efficient in terms of time and effort, and that the control of specific process variables is linked to customer wants and needs for continuing satisfaction.

To complete a QFD, three key conditions are required to ensure success. Condition 1 is that a multidisciplinary software DFSS team is required to provide a broad perspective. Condition 2 is that more time is expended upfront in the collecting and processing of customer needs and expectations. Condition 3 is that the functional requirements defined in HOQ2 will be solution-free.

All of this theory sounds logical and achievable; however, there are three realities that must be overcome to achieve success. Reality 1 is that the interdisciplinary DFSS team will not work well together in the beginning. Reality 2 is the prevalent culture of heroic problem solving in lieu of drab problem prevention. People get visibly rewarded and recognized for fire fighting and receive no recognition for problem prevention, which drives a culture focused on correction rather than prevention. The final reality is that the software DFSS team members and even customers will jump right to solutions early and frequently instead of following the details of the methodology and remaining solution-free until design requirements are specified.

12.4 QFD METHODOLOGY

Quality function deployment is accomplished by multidisciplinary software DFSS teams using a series of matrixes, called *houses of quality*, to deploy critical customer

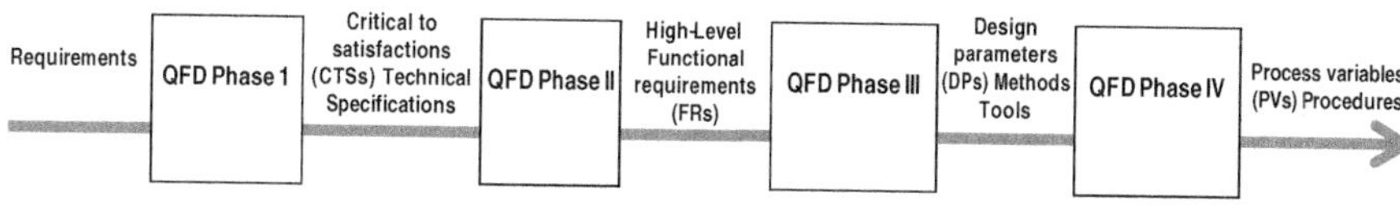

FIGURE 12.2 QFD 4 phases I/O relationship.

needs throughout the phases of the design development. The QFD methodology is deployed through a four-phase sequence shown in Figure 12.3 The four planning phases are:

- Phase I—Critical to satisfaction planning—House 1
- Phase II—functional requirements planning—House 2
- Phase III—design parameters planning—House 3
- Phase IV—process variable planning—House 4

These phases are aligned with axiomatic design mapping in Chapter 13. Each of these phases will be covered in detail within this chapter. The input/output(I/O) relationship among the phases is depicted in Figure 12.2.

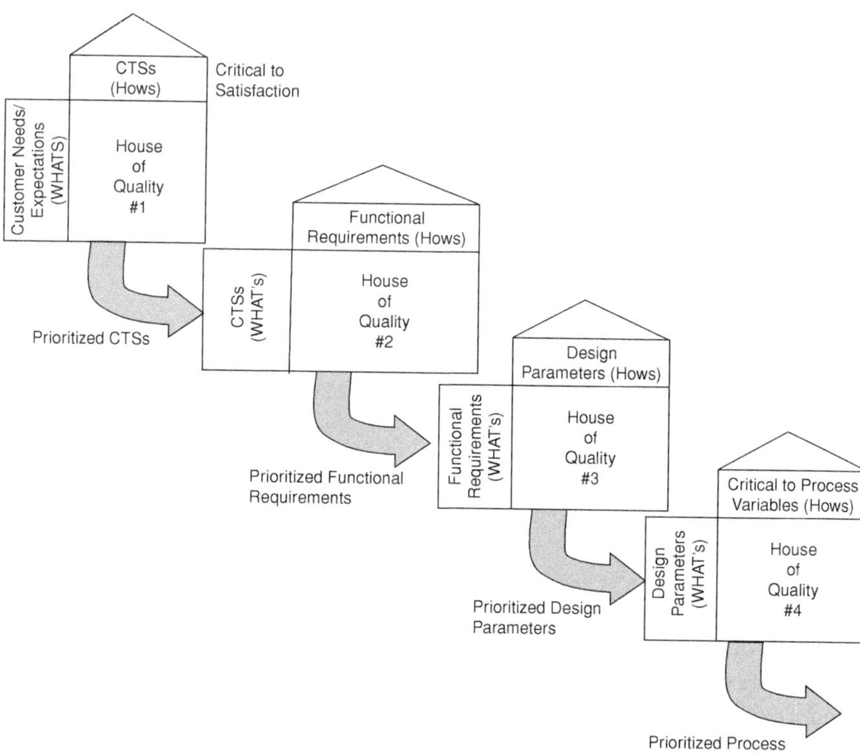

FIGURE 12.3 The four phases of QFD.

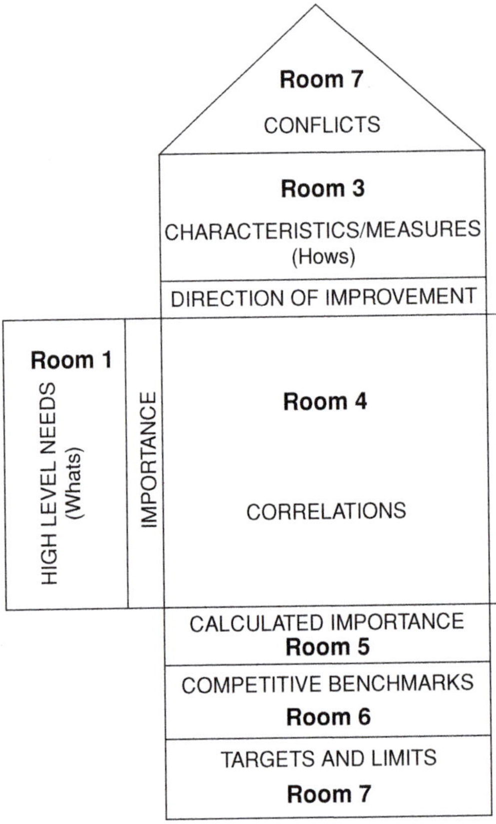

FIGURE 12.4 House of quality.

It is interesting to note that the QFD is linked to VOC tools at the front end as well as to design scorecards and customer satisfaction measures throughout the design effort. These linkages along with adequate analysis provide the feed forward (requirements flow-down) and feed backward (capability flow-up) signals that allow for the synthesis of software design concepts (Suh, 1990).

Each of these four phases deploys the HOQ with the only content variation occurring in Room #1 and Room #3. Figure 12.4 depicts the generic HOQ. Going room by room, we see that the input is into Room #1 where we answer the question "What?" These "Whats" are either the results of VOC synthesis for HOQ 1 or a rotation of the "Hows" from Room #3 into the following HOQs. These "Whats" are rated in terms of their overall importance and placed in the Importance column. Based on customer survey data, the VOC priorities for the stated customer needs, wants, and delights are developed. Additional information may be gathered at this point from the customers concerning assessments of competitors' software products. Data also may be gathered from the development team concerning sales and improvement indices.

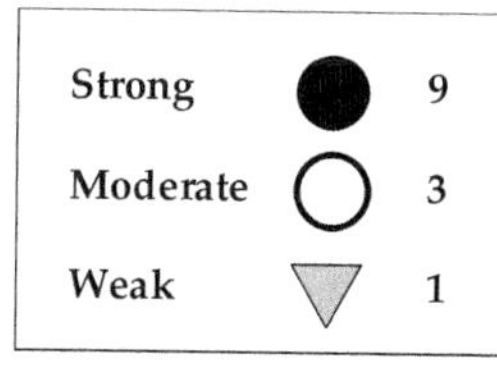

FIGURE 12.5 Rating values for affinities.

Next we move to Room #2 and compare our performance and the competitions' performance against these "Whats" in the eyes of the customer. This is usually a subjective measure and is generally scaled from 1 to 5. A different symbol is assigned to the different providers so that a graphical representation is depicted in Room #2. Next we must populate Room #3 with the "Hows," For each "What" in Room #1, we ask "How can we fulfill this?" We also indicate which direction the improvement is required to satisfy the "What"—maximize, minimize, or on target. This classification is in alignment with robustness methodology (Chapter 18) and indicates an optimization direction.

In HOQ1, these become "How does the customer measure the What?" In HOQ1, we call these CTS measures. In HOQ2, the "Hows" are measurable and are solution-free functions required to fulfill the "Whats" of CTSs. In HOQ3, the "Hows" become DPs and in HOQ4 the Hows become PVs. A word of caution: Teams involved in designing new softwares or processes often jump to specific solutions in HOQ1. It is a challenge to stay solution-free until HOQ3. There are some rare circumstances in which the VOC is a specific function that flows straight through each house unchanged.

Within Room #4, we assign the weight of the relationship between each "What" and each "How," using 9 for strong, 3 for moderate, and 1 for weak. In the actual HOQ, these weightings will be depicted with graphical symbols, the most common being the solid circle for strong, an open circle for moderate and a triangle for weak (Figure 12.5).

Once the relationship assignment is completed, by evaluating the relationship of every "What" to every "How," then the calculated importance can be derived by multiplying the weight of the relationship and the importance of the "What" and summing for each "How." This is the number in Room #5. For each of the "Hows," a company also can derive quantifiable benchmark measures of the competition and itself in the eyes of industry experts; this is what goes in Room #6. In Room #7, we can state the targets and limits of each of the "Hows." Finally, in Room #8, often called the roof, we assess the interrelationship of the "Hows" to each other. If we were to maximize one of the "Hows," then what happens to the other "Hows"? If it also were to improve in measure, then we classify it as a synergy, whereas if it were to move away from the direction of improvement then it would be classified as a compromise. In another example, "easy to learn" is highly correlated to "time to complete tutorial" (a high correlation may receive a score of 9 in the correlation matrix) but not "does landscape printing" (which would receive a score of 0 in the correlation matrix). Because there are many customers involved in this process, it is important to gain "consensus" concerning the strength of relationships.

Wherever a relationship does not exist, it is left blank. For example, if we wanted to improve search time by adding or removing interfaces among databases, then the data integrity error rate may increase. This is clearly a compromise. Although it would be ideal to have correlation and regression values for these relationships, often they are based on common sense, tribal knowledge, or business laws. This completes each of the eight rooms in the HOQ. The next steps are to sort based on the importance in Room #1 and Room #5 and then evaluate the HOQ for completeness and balance.

12.5 HOQ EVALUATION

Completing the HOQ is the first important step; however, the design team should take the time to review their effort for quality and checks and balances as well as design resource priorities. The following diagnostics can be used on the sorted HOQ:

1. Is there a diagonal pattern of strong correlations in Room #4? This will indicate good alignment of the "Hows" (Room #3) with the "Whats" (Room #1).
2. Do all "Hows" (Room #3) have at least one correlation with "Whats" (Room #1)
3. Are there empty or weak rows in Room #4? This indicates unaddressed "Whats" and could be a major issue. In HOQ1, this would be unaddressed customer wants or needs.
4. Evaluate the highest score in Room #2. What should our design target be?
5. Evaluate the customer rankings in Room #2 versus the technical benchmarks in Room #6. If Room #2 values are lower than Room #6 values, then the design team may need to work on changing the customer's perception, or the correlation between the Want/Need and CTS is not correct.
6. Review Room #8 tradeoffs for conflicting correlations. For strong conflicts/synergies, changes to one characteristic (Room #3) could affect other characteristics.

12.6 HOQ 1: THE CUSTOMER'S HOUSE

Quality function deployment begins with the VOC, and this is the first step required for HOQ 1. The customers would include end users, managers, system development personnel, and anyone who would benefit from the use of the proposed software product. VOC can be collected by many methods and from many sources. Some common methods are historical research methods, focus groups, interviews, councils, field trials, surveys, and observations. Sources range from passive historical records of complaints, testimonials, customers' records, lost customers, or target customers. The requirements are usually short statements recorded specifically in the

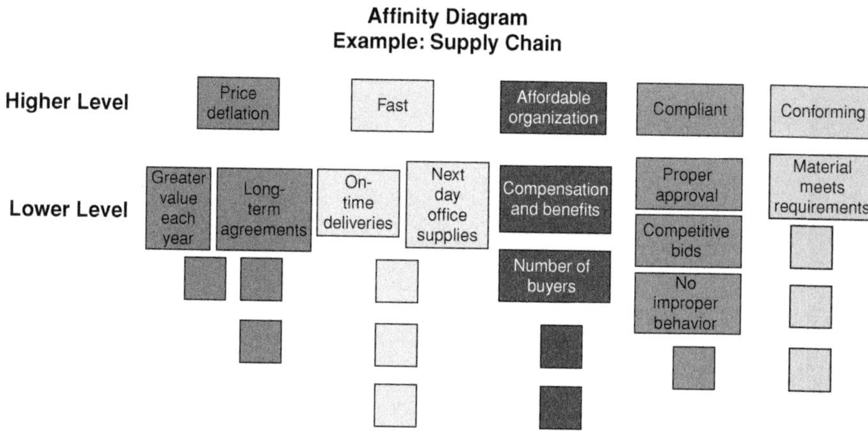

FIGURE 12.6 Affinity diagram.

customers' terminology (e.g., "easy to learn") and are accompanied by a detailed definition—the QFD version of a data dictionary. Stick with the language of the customer and think about how they speak when angered or satisfied; this is generally their natural language. These voices need to be prioritized and synthesized into rank order of importance. The two most common methods are the affinity diagram (see Figure 12.6) and Kano analysis. We will cover the Kano model (see Figure 12.6) before taking the prioritized CTSs into Room #1 of HOQ 2.

When collecting the VOC, make sure that it is not the voice of code or voice of boss. Although the QFD is a robust methodology, if you start with a poor foundation, then it will be exacerbated throughout the process.

12.7 KANO MODEL

In the context of DFSS, customer attributes are potential benefits that the customer could receive from the design and are characterized by qualitative and quantitative data. Each attribute is ranked according to its relative importance to the customer. This ranking is based on the customer's satisfaction with similar design entities featuring that attribute.

The understanding of customer expectations (wants and needs) and delights (wow factors) by the design team is a prerequisite to further development and is, therefore, the most important action prior to starting the other conceptual representation (Chapters: 4 and 13). The fulfillment of these expectations and the provision of differentiating delighters (unspoken wants) will lead to satisfaction. This satisfaction ultimately will determine what software functionality and features the customer is going to endorse and buy. In doing so, the software DFSS team needs to identify constraints that limit the delivery of such satisfaction. Constraints present opportunities to exceed expectations and create delighters. The identification of customer

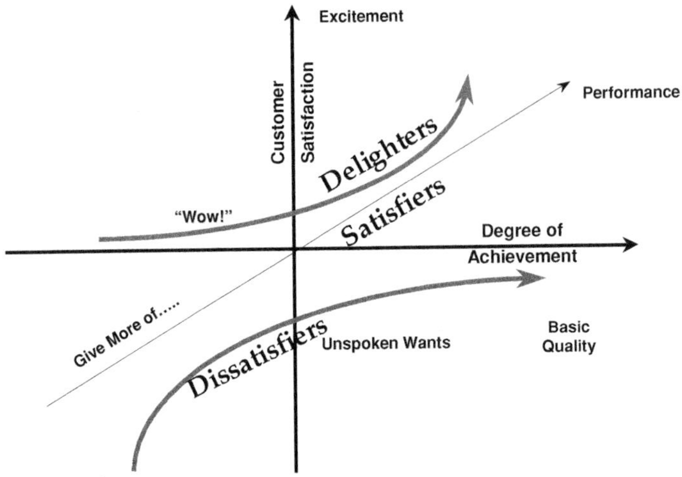

FIGURE 12.7 Kano model.

expectations is a vital step for the development of Six Sigma level software the customer will buy in preference to those of the competitors. Noriaki Kano, a Japanese consultant, has developed a model relating design characteristics to customer satisfaction (Cohen, 1995). This model (see Figure 12.7) divides characteristics into categories, each of which affects customers differently—dissatifiers, satisfiers, and delighters.

"Dissatisfiers" also are known as basic, "must-be," or expected attributes and can be defined as a characteristic that a customer takes for granted and causes dissatisfaction when it is missing. "Satisfiers" are known as performance, one-dimensional, or straight-line characteristics and are defined as something the customer wants and expects; the more, the better. "Delighters" are features that exceed competitive offerings in creating unexpected, pleasant surprises. Not all customer satisfaction attributes are equal from an importance standpoint. Some are more important to customers than others in subtly different ways. For example, dissatisfiers may not matter when they are met but may subtract from overall design satisfaction when they are not delivered.

When customers interact with the DFSS team, delighters are often surfaced that would not have been independently conceived. Another source of delighters may emerge from team creativity, as some features have the unintended result of becoming delighters in the eyes of customers. Any software design feature that fills a latent or hidden need is a delighter and, with time, becomes a want. A good example of this is the remote controls first introduced with televisions. Early on, these were differentiating delighters; today they are common features with televisions, radios, and even automobile ignitions and door locks. Today, if you received a package without installation instructions, then it would be a dissatisfier. Delighters can be sought in areas of weakness and competitor benchmarking as well as technical, social, and strategic innovation.

The DFSS team should conduct a customer evaluation study. This is hard to do in new design situations. Customer evaluation is conducted to assess how well the current or proposed design delivers on the needs and desires of the end user. The most frequently used method for this evaluation is to ask the customer (e.g., focus group or a survey) how well the software design project is meeting each customer's expectations. To leap ahead of the competition, the DFSS team must also understand the evaluation and performance of their toughest competition. In the HOQ 1, the team has the opportunity to grasp and compare, side by side, how well the current, proposed, or competitive design solutions are delivering on customer needs.

The objective of the HOQ 1 Room 2 evaluation is to broaden the team's strategic choices for setting targets for the customer performance goals. For example, armed with meaningful customer desires, the team could aim their efforts at either the strengths or the weaknesses of best-in-class competitors, if any. In another choice, the team might explore other innovative avenues to gain competitive advantages.

The list of customer wants and needs should include all types of the customer as well as the regulatory requirements and the social and environmental expectations. It is necessary to understand the requirements and prioritization similarities and differences to understand what can be standardized and what needs to be tailored.

Customer wants and needs, in HOQ1, social, and other company wants can be refined in a matrix format for each identified market segment. The "customer importance rating" in Room #1 is the main driver for assigning priorities from both the customer's and the corporate perspectives, as obtained through direct or indirect engagement forms with the customer.

The traditional method of conducting the Kano model is to ask functional and dysfunctional questions around known wants/needs or CTSs. Functional questions take the form of "How do you feel if the 'CTS' is present in the software?" Dysfunctional questions take the form of "How do you feel if the 'CTS' is NOT present in the software?" Collection of this information is the first step, and then detailed analysis is required beyond the scope of this book. For a good reference on processing the voice of the customer, see Burchill et al. (1997).

In the Kano analysis plot, the y-axis consists of the Kano model dimensions of must be, one-dimensional, and delighters. The top item, indifferent, is where the customer chooses opposite items in the functional and dysfunctional questions. The x-axis is based on the importance of the CTSs to the customer. This type of plot can be completed from the Kano model or can be arranged qualitatively by the design team, but it must be validated by the customer, or we will fall into the trap of voice of the engineer again.

12.8 QFD HOQ 2: TRANSLATION HOUSE

The customer requirements are then converted to a technical and measurable set of metrics, the CTSs, of the software product. For example, "easy to learn" may be converted to "time to complete the tutorial," "number of icons," and "number of online help facilities." It is important to note here that some customer requirements

Requirements / Use-Cases \ Engineering Measures			Kano classification	Importance	Applications: Data integrity error rate	Applications: Database interface extensibility	Applications: Route optimization effectiveness	Applications: Path exception error rate
					↓	↑	↑	↓
Applications	1	Translating I/O to and from database protocols	S	4	3	9		
	2	Adding and removing interfaces	S	5		9		
	3	Verifying data content and integrity	S	6	9			
	4	Optimizing routing	D	2.5			9	9
	5	Managing exceptions en route	S	3.5			3	9
	6	Logging performance data	D	3.5		3		

FIGURE 12.8 HOQ 2 VOCs translation to CTSs.[1]

may be converted to multiple technical product specifications, making it crucial to have extensive user involvement. Additionally, the technical product specifications must be measurable in some form. The metrics used are usually numerically based but also may be Boolean. For example, the customer requirement "provides multiple print formats" may be converted to "number of print formats" (using a numerically based metric) and "does landscape printing" (measured using "Yes" or "No").

The CTSs list is the set of metrics derived by the design team from the customer to answer the customer attributes list. The CTSs list rotates into HOQ 2 Room #1 in this QFD phase. The objective is to determine a set of functional requirements (FRs), with which the CTSs requirements can be materialized. The answering activity translates customer expectations into requirements such as waiting time, number of mouse clicks for online purchasing service, and so on. For each CTS, there should be one or more FRs, that describe a means of attaining customer satisfaction. A QFD translation example is given in Figure 12.8. A complete QFD example is depicted in Figure 12.9.

At this stage, only overall CTSs that can be measured and controlled need to be used. We will call these CTSs, technical CTSs. As explained earlier, CTSs are traditionally known as substitute quality characteristics. Relationships between technical CTSs and FRs often are used to prioritize CTSs filling the relationship matrix of HOQ2 rooms. For each CTS, the design team has to assign a value that reflects the extent to which the defined FRs contribute to meeting it. This value, along with the

[1] Hallowell, D. on http://software.isixsigma.com/library/content/c040707b.asp.

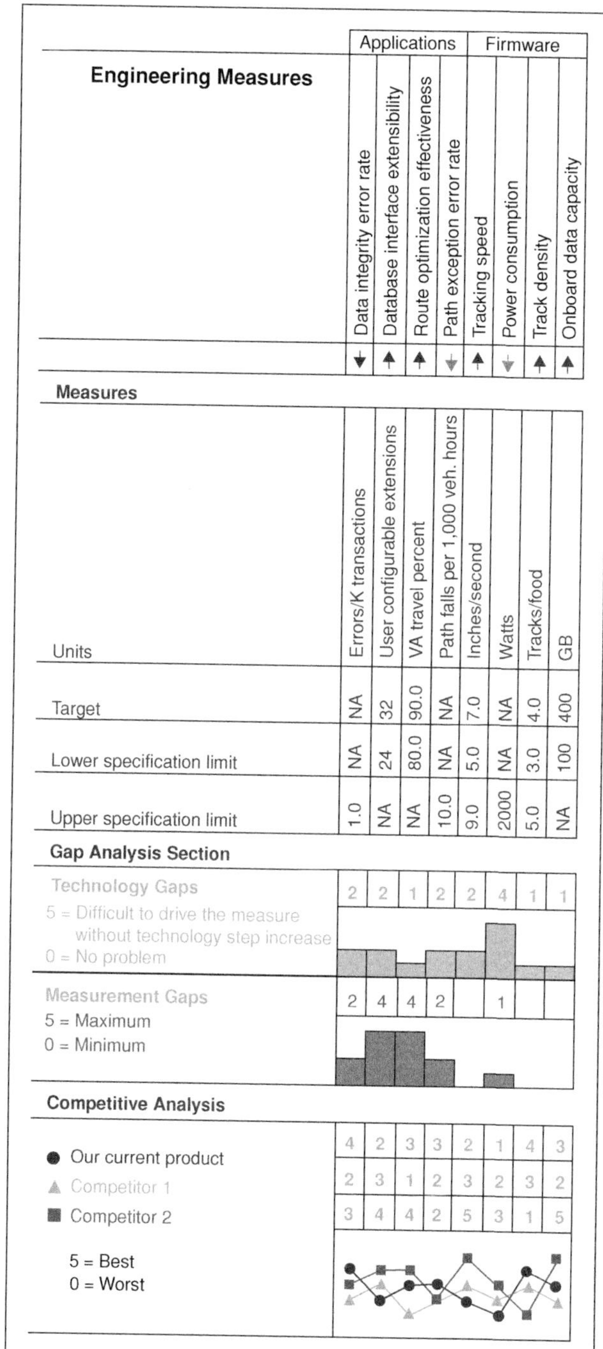

FIGURE 12.9 QFD exxample.[2]

[2]Hallowell, D. on http://software.isixsigma.com/library/content/c040707b.asp.

calculated importance index of the CTS, establishes the contribution of the FRs to the overall satisfaction and can be used for prioritization.

The analysis of the relationships of FRs and CTSs allows a comparison with other indirect information, which needs to be understood before prioritization can be finalized. The new information from the Room #2 in the QFD HOQ needs to be contrasted with the available design information (if any) to ensure that the reasons for modification are understood.

The purpose of the QFD HOQ2 activity is to define the design functions in terms of customer expectations, benchmark projections, institutional knowledge, and interface management with other systems as well as to translate this information into software technical functional requirement targets and specifications. This will facilitate the design mappings (Chapter 13). Because the FRs are solution-free; their targets and specifications for them are flowed down from the CTs. For example, if a CTS is for "*Speed of Order*" and the measure is hours to process and we want order processing to occur within four hours, then the functional requirements for this CTS, the "Hows," could include *Process Design* in which the *number of automated process steps* (via software) and the *speed of each step* would be the flow-down requirements to achieve "*Speed of Order*." Obviously, the greater the number of process steps, the shorter each step will need to be. Because at this stage we do not know what the process will be and how many steps will be required, we can allocate the sum of all process steps multiplied by their process time not to exceed four hours. A major reason for customer dissatisfaction is that the software design specifications do not adequately link to customer use of the software.

Often, the specification is written after the design is completed. It also may be a copy of outdated specifications. This reality may be attributed to the current planned design practices that do not allocate activities and resources in areas of importance to customers and waste resources by spending too much time in activities that provide marginal value—a gap that is filled nicely by the QFD activities. The targets and tolerance setting activity in QFD Phase 2 also should be stressed.

12.9 QFD HOQ3—DESIGN HOUSE

The FRs are the list of solution-free requirements derived by the design team to answer the CTS array. The FRs list is rotated into HOQ3 Room #1 in this QFD phase. The objective is to determine a set of design parameters that will fulfill the FRs. Again, the FRs are the "Whats," and we decompose this into the "Hows." This is the phase that most design teams want to jump right into, so hopefully, they have completed the prior phases of HOQ 1 and HOQ 2 before arriving here. The design requirements must be tangible solutions.

12.10 QFD HOQ4—PROCESS HOUSE

The DPs are a list of tangible functions derived by the design team to answer the FRs array. The DPs list is rotated into HOQ4 Room #1 in this QFD phase. The objective is to determine a set of process variables that, when controlled, ensure the DRs. Again, the DRs are the "Whats," and we decompose this into the "Hows."

12.11 SUMMARY

QFD is a planning tool used to translate customer needs and wants into focused design actions. This tool is best accomplished with cross-functional teams and is key in preventing problems from occurring once the design is operationalized. The structured linkage allows for rapid design cycle and effective use of resources while achieving Six Sigma levels of performance.

To be successful with the QFD, the team needs to avoid "jumping" right to solutions and needs to process HOQ1 and HOQ2 thoroughly and properly before performing detailed design. The team also will be challenged to keep the functional requirements solution neutral in HOQ2.

It is important to have the correct voice of the customer and the appropriate benchmark information. Also, a strong cross-functional team willing to think out of the box is required to obtain truly Six Sigma capable products or processes. From this point, the QFD is process driven, but it is not the charts that we are trying to complete, it is the total concept of linking voice of the customer throughout the design effort.

REFERENCES

Akao, Yoji (1972), "New product development and quality assurance–quality deployment system." *Standardization and Quality Control*, Volume *25*, #4, pp. 7–14.

Betts, M. (1989), "QFD Integrated with Software Engineering," Proceedings of the Second Symposium on Quality Function Deployment, June, pp. 442–459.

Brodie, C.H. and Burchill, G. (1997), *Voices into Choices: Acting on the Voice of the Customer*, Joiner Associates Inc., Madison, WI.

Cohen, L. (1988), "Quality function deployment and application perspective from digital equipment corporation." *National Productivity Review*, Volume *7*, #3, pp. 197–208.

Cohen, L. (1995), "*Quality Function Deployment: How to Make QFD Work for You*," Addison-Wesley Publishing Co., Reading, MA.

El-Haik, Basem and Mekki, K. (2008), "*Medical Device Design for Six Sigma: A Road Map for Safety and Effectiveness*," 1st Ed., Wiley-Interscience, New York.

El-Haik, Basem and Roy, D. (2005), "*Service Design for Six Sigma: A Roadmap for Excellence*," Wiley-Interscience, New York.

Hagg, S, Raja, M.K., and Schkade, L.L. (1996), "QFD usage in software development." *Communications of the ACM*, Volume *39*, #1, pp. 41–49.

Mizuno, Shigeru and Yoji Akao (eds.) (1978), *Quality Function Deployment: A Company Wide Quality Approach* (in Japanese). Juse Press, Toyko, Japan.

Mizuno, Shigeru and Yoji Akao (eds.) (1994), *QFD: The Customer-Driven Approach to Quality Planning and Deployment* (Translated by Glenn H. Mazur). Asian Productivity Organization, Tokyo, Japan.

Moseley, J. and Worley, J. (1991), "Quality Function Deployment to Gather Customer Requirements for Products that Support Software Engineering Improvement," Third Symposium on Quality Function Deployment, June, pp. 243–251.

Shaikh, K.I. (1989), "Thrill Your Customer, Be a Winner," Symposium on Quality Function Deployment, June, pp. 289–301.

Sharkey, A.I. (1991), "Generalized Approach to Adapting QFD for Software," Third Symposium on Quality Function Deployment, June, pp. 379–416.

Suh N. P. (1990), "*The Principles of Design* (Oxford Series on Advanced Manufacturing)," Oxford University Press, USA.

CHAPTER 13

AXIOMATIC DESIGN IN SOFTWARE DESIGN FOR SIX SIGMA (DFSS)

13.1 INTRODUCTION

Software permeates in every corner of our daily life. Software and computers are playing central roles in all industries and modern life technologies. In manufacturing, software controls manufacturing equipment, manufacturing systems, and the operation of the manufacturing enterprise. At the same time, the development of software can be the bottleneck in the development of machines and systems because current industrial software development is full of uncertainties, especially when new products are designed. Software is designed and implemented by making prototypes based on the experience of software engineers. Consequently, they require extensive "debugging"—a process of correcting mistakes made during the software development process. It costs unnecessary time and money beyond the original estimate (Pressman, 1997). The current situation is caused by the lack of fundamental principles and methodologies for software design, although various methodologies have been proposed.

In current software development practices, both the importance and the high cost of software are well recognized. The high cost is associated with the long software development and debugging time, the need for maintenance, and uncertain reliability. It is a labor-intensive business that is in need of a systematic software development approach that ensures high quality, productivity, and reliability of software systems a priori. The goals of software Design for Six Sigma (DFSS) is twofold: first, enhance algorithmic efficiency to reduce execution time and, second, enhance productivity

Software Design for Six Sigma: A Roadmap for Excellence, By Basem El-Haik and Adnan Shaout

to reduce the coding, extension, and maintenance effort. As computer hardware rapidly evolves and the need for large-scale software systems grows, productivity is increasingly more important in software engineering. The so-called "software crisis" is closely tied to the productivity of software development (Pressman, 1997).

Software development requires the translation of good abstract ideas into clear design specifications. Subsequent delivery of the software product in moderate-to-large-scale projects requires effective definition, requirements of translation into useful codes, and assignments for a team of software belts and engineers to meet deadlines in the presence of resource constraints. This section explores how axiomatic design may be integrated into the software DFSS process (Chapter 11). An approach to mapping the functional requirements and design parameters into code is described.

The application of axiomatic design to software development was first presented at the 1991 CIRP General Assembly (Kim et al., 1991), and the system design concepts were presented in the 1997 CIRP General Assembly (Suh, 1997).

This section presents a new software design methodology based on axiomatic design theory that incorporates object-oriented programming. This methodology overcomes the shortcomings of various software design strategies discussed in Chapter 2—extensive software development and debugging times and the need for extensive maintenance. It is not heuristic in nature and provides basic principles for good software systems. The axiomatic design framework for software overcomes many shortcomings of current software design techniques: high maintenance costs, limited reusability, low reliability, the need for extensive debugging and testing, poor documentation, and limited extensibility of the software, in addition to the high development cost of software. The methodology presented in this section has helped software engineers to improve productivity and reliability.

In this section, we will start by reviewing the basic principles of axiomatic design as applied to hardware product development. It explains why software DFSS belts should apply this methodology, and then we proceed to discuss how it applies to software DFSS. In the context of DFSS, the topic of axiomatic design was discussed extensively by El-Haik (2005), El-Haik and Roy (2005), and El-Haik and Mekki (2008).

13.2 AXIOMATIC DESIGN IN PRODUCT DFSS: AN INTRODUCTION

Axiomatic design is a prescriptive engineering[1] design method. Systematic research in engineering design began in Germany during the 1850s. The recent contributions in the field of engineering design include axiomatic design (Suh, 1984, 1990, 1995, 1996, 1997, 2001), product design and development (Ulrich & Eppinger, 1995), the mechanical design process (Ulman, 1992), Pugh's total design (Pugh, 1991, 1996), and TRIZ (Altshuller, 1988, 1990), Rantanen, 1988, and Arciszewsky, 1988. These contributions demonstrate that research in engineering design is an active field that

[1] Prescriptive design describes how a design should be processed. Axiomatic design is an example of prescriptive design methodologies. Descriptive design methods like design for assembly are descriptive of the best practices and are algorithmic in nature.

has spread from Germany to most industrialized nations around the world. To date, most research in engineering design theory has focused on design methods. As a result, Several design methods now are being taught and practiced in both industry and academia. However, most of these methods overlook the need to integrate quality methods in the concept stage. Therefore, the assurance that only healthy concepts are conceived, optimized, and validated with no (or minimal) vulnerabilities cannot be guaranteed.

Axiomatic design is a design theory that constitutes basic and fundamental design elements knowledge. In this context, a scientific theory is defined as a theory comprising fundamental knowledge areas in the form of perceptions and understandings of different entities and the relationship between these fundamental areas. These perceptions and relations are combined by the theorist to produce consequences that can be, but are not necessarily, predictions of observations. Fundamental knowledge areas include mathematical expressions, categorizations of phenomena or objects, models, and so on. and are more abstract than observations of real-world data. Such knowledge and relations between knowledge elements constitute a theoretical system. A theoretical system may be one of two types—axioms or hypotheses—depending on how the fundamental knowledge areas are treated. Fundamental knowledge that are generally accepted as true, yet cannot be tested, is treated as an axiom. If the fundamental knowledge areas are being tested, then they are treated as hypotheses (Nordlund et al., 1996). In this regard, axiomatic design is a scientific design method, however, with the premise of a theoretic system based on two axioms.

Motivated by the absence of scientific design principles, Suh (1984, 1990, 1995, 1996, 1997, 2001) proposed the use of axioms as the pursued scientific foundations of design. The following are two axioms that a design needs to satisfy:

Axiom 1: The Independence Axiom

Maintain the independence of the functional requirements

Axiom 2: The Information Axiom

Minimize the information content in a design

In the context of this book, the independence axiom will be used to address the conceptual vulnerabilities, whereas the information axiom will be tasked with the operational type of design vulnerabilities. Operational vulnerability is usually minimized and cannot be totally eliminated. Reducing the variability of the design functional requirements and adjusting their mean performance to desired targets are two steps to achieve such minimization. Such activities also results in reducing design information content, a measure of design complexity per axiom 2. Information content is related to the probability of successfully manufacturing the design as intended by the customer. The design process involves three mappings among four domains (Figure 13.1). The first mapping involves the mapping between customer attributes (CAs) and the functional requirements (FRs). This mapping is very important as it yields the definition of the high-level minimum set of functional requirements needed to accomplish the design intent. This definition can be accomplished by the

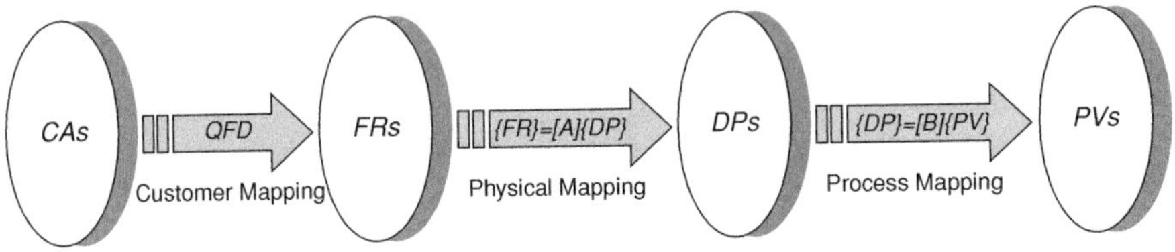

FIGURE 13.1 The design mapping process.

application of quality function deployment (QFD). Once the minimum set of FRs is defined, the *physical mapping* may be started. This mapping involves the FRs domain and the design parameter codomain (DPs). It represents the product development activities and can be depicted by design matrices; hence, the term "mapping" is used. This mapping is conducted over design hierarchy as the high-level set of FRs, defined earlier, is cascaded down to the lowest hierarchical level. Design matrices reveal coupling, a conceptual vulnerability (El-Haik, 2005: Chapter 2), and provides a means to track the chain of effects of design changes as they propagate across the design structure.

The *process mapping* is the last mapping of axiomatic design and involves the DPs domain and the process variables (PVs) codomain. This mapping can be represented formally by matrices as well and provides the process elements needed to translate the DPs to PVs in manufacturing and production domains. A conceptual design structure called the physical structure usually is used as a graphical representation of the design mappings.

Before proceeding further, we would like to define the following terminology relative to axiom 1 and to ground the readers about terminology and concepts that already vaguely are grasped from the previous sections. They are:

- Functional requirements (FRs) are a minimum set of independent requirements that completely characterize the functional needs of the design solution in the functional domain within the constraints of safety, economy, reliability, and quality.
  - How to define the functional requirements?

    In the context of the Figure 13.1 first mapping, customers define the product using some features or attributes that are saturated by some or all kinds of linguistic uncertainty. For example, in an automotive product design, customers use the term quiet, stylish, comfortable, and easy to drive in describing the features of their dream car. The challenge is how to translate these features into functional requirements and then into solution entities. QFD is the tool adopted here to accomplish an actionable set of the FRs.

    In defining their wants and needs, customers use some vague and fuzzy terms that are hard to interpret or attribute to specific engineering terminology, in particular, the FRs. In general, functional requirements are technical terms extracted from the voice of the customer. Customer expressions are not dichotomous or crisp in nature. It is something in between. As a result, uncertainty may lead to an inaccurate interpretation and, therefore, to vulnerable or unwanted design. There are many classifications for a customer's linguistic inexactness. In general, two major sources of imprecision in human knowledge—linguistic inexactness and stochastic uncertainty (Zimmerman, 1985)—usually are encountered. Stochastic uncertainty is well handled by the probability theory. Imprecision can develop arise from a variety of sources: incomplete knowledge, ambiguous definitions, inherent stochastic characteristics, measurement problems, and so on.

This brief introduction to linguistic inexactness is warranted to enable design teams to appreciate the task on hand, assess their understanding of the voice of the customer, and seek clarification where needed. The ignorance of such facts may cause several failures to the design project and their efforts altogether. The most severe failure among them is the possibility of propagating inexactness into the design activities, including analysis and synthesis of wrong requirements.

- Design parameters (DPs) are the elements of the design solution in the physical domain that are chosen to satisfy the specified FRs. In general terms, standard and reusable DPs (grouped into design modules within the physical structure) often are used and usually have a higher probability of success, thus improving the quality and reliability of thc design.
- Constraints (Cs) are bounds on acceptable solutions.
- Process variables (PVs) are the elements of the process domain that characterize the process that satisfies the specified DPs.

The design team will conceive a detailed description of what functional requirements the design entity needs to perform to satisfy customer needs, a description of the physical entity that will realize those functions (the DPs), and a description of how this object will be produced (the PVs).

The mapping equation $\boldsymbol{FR} = \boldsymbol{f(DP)}$ or, in matrix notation $\{\boldsymbol{FR}\}_{mx1} = [\boldsymbol{A}]_{mxp}$ $\{\boldsymbol{DP}\}_{px1}$, is used to reflect the relationship between the domain, array $\{\boldsymbol{FR}\}$, and the codomain array $\{\boldsymbol{DP}\}$ in the physical mapping, where the array $\{\boldsymbol{FR}\}_{mx1}$ is a vector with m requirements, $\{\boldsymbol{DP}\}_{px1}$ is the vector of design parameters with p characteristics, and $\boldsymbol{A}$ is the design matrix. Per axiom 1, the ideal case is to have a one-to-one mapping so that a specific DP can be adjusted to satisfy its corresponding FR without affecting the other requirements. However, perfect deployment of the design axioms may be infeasible because of technological and cost limitations. Under these circumstances, different degrees of conceptual vulnerabilities are established in the measures (criteria) related to the unsatisfied axiom. For example, a degree of *coupling* may be created because of axiom 1 violation, and this design may function adequately for some time in the use environment; however, a conceptually weak system may have limited opportunity for continuous success even with the aggressive implementation of an operational vulnerability improvement phase.

When matrix $\boldsymbol{A}$ is a square diagonal matrix, the design is called *uncoupled* (i.e., each FR can be adjusted or changed independent of the other FRs). An uncoupled

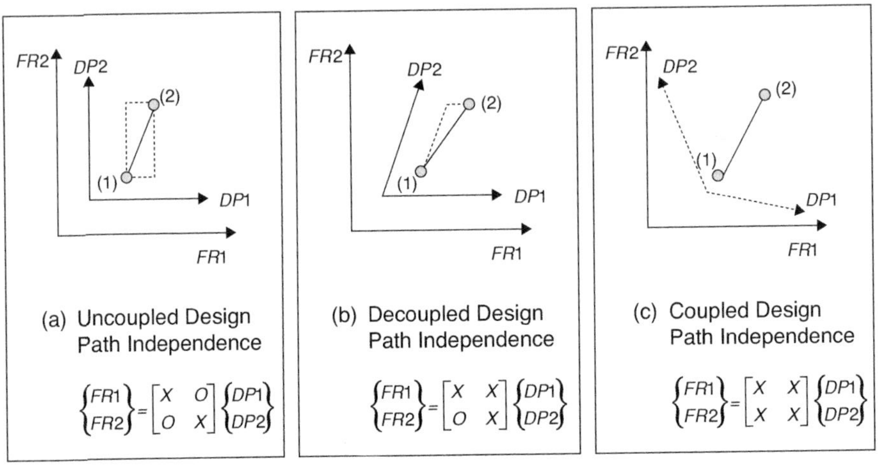

FIGURE 13.2 Design categories according to axiom 1.

design is a one-to-one mapping. Another design that obeys axiom 1, though with a known design sequence, is called *decoupled*. In a decoupled design, matrix $\boldsymbol{A}$ is a *lower* or an *upper triangular* matrix. The decoupled design may be treated as an uncoupled design when the DPs are adjusted in some sequence conveyed by the matrix. Uncoupled and decoupled design entities possess conceptual robustness (i.e., the DPs can be changed to affect specific requirements without affecting other FRs unintentionally). A coupled design definitely results in a design matrix with several requirements, m, greater than the number of DPs, p. Square design matrices ($m = p$) may be classified as a coupled design when the off-diagonal matrix elements are nonzeros. Graphically, the three design classifications are depicted in Figure 13.2 for 2×2 design matrix case. Notice that we denote the nonzero mapping relationship in the respective design matrices by "*X*." On the other hand, "0" denotes the absence of such a relationship.

Consider the uncoupled design in Figure 13.2(a). The uncoupled design possesses the path independence property, that is, the design team could set the design to level (1) as a starting point and move to setting (2) by changing DP1 first (moving east to the right of the page or parallel to DP1) and then changing DP2 (moving toward the top of the page or parallel to DP2). Because of the path independence property of the uncoupled design, the team could move start from setting (1) to setting (2) by changing DP2 first (moving toward the top of the page or parallel to DP2) and then changing DP1 second (moving east or parallel to DP1). Both paths are equivalent, that is, they accomplish the same result. Notice also that the FR's independence is depicted as orthogonal coordinates as well as perpendicular DP axes that parallel its respective FR in the diagonal matrix.

Path independence is characterized mathematically by a diagonal design matrix (uncoupled design). Path independence is a very desirable property of an uncoupled design and implies full control of the design team and ultimately the customer (user) over the design. It also implies a high level of design quality and reliability because the interaction effects between the FRs are minimized. In addition, a failure in one (FR, DP) combination of the uncoupled design matrix is not reflected in the other mappings within the same design hierarchical level of interest.

For the decoupled design, the path independence property is somehow fractured. As depicted in Figure 13.2(b), decoupled design matrices have a design settings sequence that needs to be followed for the functional requirements to maintain their independence. This sequence is revealed by the matrix as follows: First, we need to set FR2 using DP2 and fix DP2, and second set FR1 by leveraging DP1. Starting from setting (1), we need to set FR2 at setting (2) by changing DP2 and then change DP1 to the desired level of FR1.

The previous discussion is a testimony to the fact that uncoupled and decoupled designs have a conceptual robustness, that is, coupling can be resolved with the proper selection of DPs, path sequence application, and employment of design theorems (El-Haik, 2005).

The coupled design matrix in Figure 13.2(c) indicates the loss of the path independence resulting from the off-diagonal design matrix entries (on both sides), and the design team has no easy way to improve the controllability, reliability, and quality of their design. The design team is left with compromise practices (e.g., optimization) among the FRs as the only option because a component of the individual DPs can be projected on all orthogonal directions of the FRs. The uncoupling or decoupling step of a coupled design is a conceptual activity that follows the design mapping and will be explored later on.

An example of design coupling is presented in Figure 13.3 in which two possible arrangements of the generic water faucet[2] (Swenson & Nordlund, 1996) are displayed. There are two functional requirements: water flow and water temperature. The Figure 13.3(a) faucet has two design parameters: the water valves (knobs) (i.e., one for each water line). When the hot water valve is turned, both flow and temperature are affected. The same would happen if the cold water valve is turned. That is, the functional requirements are not independent, and a coupled design matrix below the schematic reflects such a fact. From the consumer perspective, optimization of the temperature will require reoptimization of the flow rate until a satisfactory compromise amongst the FRs, as a function of the DPs settings, is obtained over several iterations.

Figure 13.3(b) exhibits an alternative design with a one-handle system delivering the *FRs*, however, with a new set of design parameters. In this design, flow is adjusted by lifting the handle while moving the handle sideways to adjust the temperature. In this alternative, adjusting the flow does not affect temperature and vice versa. This design is better because the functional requirements maintain their independence per axiom 1. The uncoupled design will give the customer path independence to set either requirement without affecting the other. Note also that in the uncoupled design case, design changes to improve an FR can be done independently as well, a valuable design attribute.

[2]See El-Haik, 2005: Section 3.4 for more details.

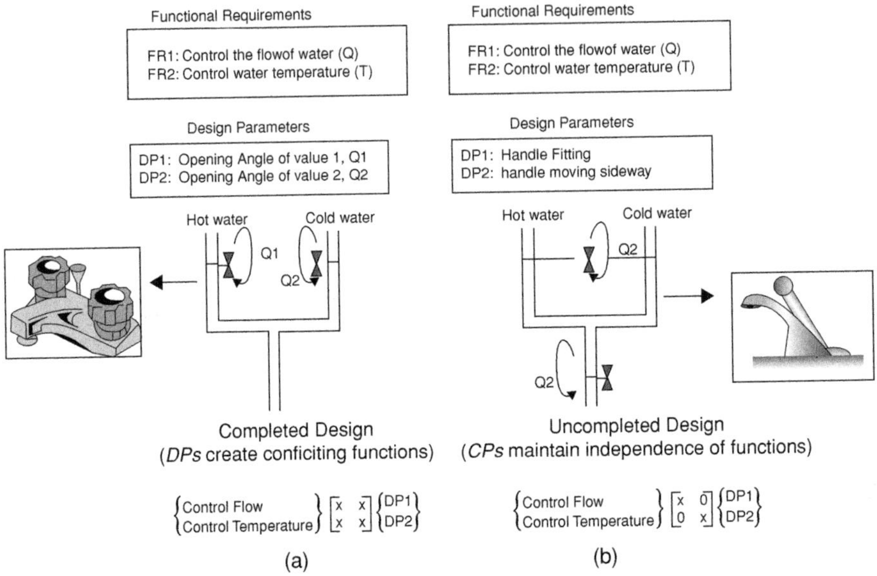

FIGURE 13.3 Faucet coupling example.

In general, the mapping process can be written mathematically as the following matrix equations:

$$\begin{Bmatrix} FR_1 \\ \cdot \\ \cdot \\ FR_m \end{Bmatrix} = \begin{bmatrix} X & 0 & \cdot & 0 \\ 0 & X & & \cdot \\ \cdot & & \cdot & 0 \\ 0 & \cdot & 0 & X \end{bmatrix}_{mxp} \begin{Bmatrix} DP_1 \\ \cdot \\ \cdot \\ DP_m \end{Bmatrix} \tag{13.1}$$

(Uncoupled design)

$$\begin{Bmatrix} FR_1 \\ \cdot \\ \cdot \\ FR_m \end{Bmatrix} = \begin{bmatrix} X & 0 & \cdot & 0 \\ X & X & 0 & \cdot \\ \cdot & \cdot & \cdot & 0 \\ X & X & \cdot & X \end{bmatrix}_{mxp} \begin{Bmatrix} DP_1 \\ \cdot \\ \cdot \\ DP_m \end{Bmatrix} \tag{13.2}$$

(Decoupled design)

$$\begin{Bmatrix} FR_1 \\ \cdot \\ \cdot \\ FR_m \end{Bmatrix} = \begin{bmatrix} X & X & \cdot & X \\ X & X & & \cdot \\ \cdot & & \cdot & X \\ X & \cdot & X & X \end{bmatrix}_{\mathrm{mxp}} \begin{Bmatrix} DP_1 \\ \cdot \\ \cdot \\ DP_p \end{Bmatrix} \tag{13.3}$$

(Coupled design)

TABLE 13.1 Software Functional Requirements (FRs) Examples

Functional Requirement Category	Example
Operational requirement	Outline of what the product will do for the user.
Performance requirement	Speed or duration of product use.
Security requirements	Steps taken to prevent improper or unauthorized use.
Maintainability requirements	Ability for product to be changed.
Reliability requirements	The statement of how this product prevents failure attributed to system defects.
Availability requirements	Ability for product to be used in its intended manner.
Database requirements	Requirements for managing, storing, retrieving, and securing data from use.
Documentation requirements	Supporting portions of products to enable user references.
Additional requirements	Can include many categories not covered in other sections.

where $\{\boldsymbol{FR}\}_{mx1}$ is the vector of independent functional requirements with m elements, $\{\boldsymbol{DP}\}_{px1}$ is the vector of design parameters with p elements. Examples of FRs and DPs are listed in Tables[3] 13.1 & 13.2.

The shape and dimension of matrix $\boldsymbol{A}$ is used to classify the design into one of the following categories: *uncoupled*, *decoupled*, *coupled*, and *redundant*. For the first two categories, the number of functional requirements, m, equals the number of design parameters, p. In a redundant design, we have $m < p$. A design that completely complies with the independence axiom is called an *uncoupled (independent)* design. The resultant design matrix in this case, $\boldsymbol{A}$, is a square diagonal matrix, where $m = p$ and $A_{ij} = X \neq 0$ when $i = j$ and 0 elsewhere as in (13.1). An uncoupled design is an ideal (i.e., square matrix) design with many attractive attributes. First, it enjoys the path independence property that enables the traditional quality methods objectives of reducing functional variability and mean adjustment to target through only one parameter per functional requirement, its respective DP. Second, the complexity of the design is additive (assuming statistical independence) and can be reduced through axiomatic treatments of the individual DPs that ought to be conducted separately. This additivity property is assured because complexity may be measured by design information content, which in turn is a probabilistic function. Third, cost and other constraints are more manageable (i.e., less binding) and are met with significant ease, including high degrees of freedom for controllability and adjustability.

A violation of the independence axiom occurs when an FR is mapped to a DP, that is, coupled with another FR. A design that satisfies axiom 1, however, with path dependence[4] (or sequence), is called a *decoupled* design as in (13.2). In a decoupled design, matrix $\boldsymbol{A}$ is a square triangular (lower or upper, sparse or otherwise). In an

[3]Zrymiak, D. @ http://www.isixsigma.com/library/content/c030709a.asp

[4]See Theorem 7 in Section 2.5 as well as Section 1.3.

TABLE 13.2 Software Design Parameters (DPs) Examples

Design Parameter Considerations	DP Example
User	Product design for user profile.
Subject-matter expert	Product design for consistency with expert opinion.
Designer	Reflection of product.
Customer	Reflection of customer preferences beyond product.
Functionality	Individual independent tasks performed by the product.
Integrated functions	Combined tasks necessary to complete a transaction or other function.
Menu	User interface display permitting access to features.
Domain	Coverage of specific information.
Equivalence classes	Determination of inputs generating a consistent product behavior.
Boundaries	Parameters where product behavior is altered.
Logic	Sequence of actions following a consistent pattern.
State-based	Use conditions indicating different function availability or product behavior.
Configuration	Ability for product to work in different intended operating environments.
Input constraints	Determine how user or system can enter data.
Output constraints	Determine how data or information is displayed.
Computation constraints	Determine how data is computed.
Storage or data constraints	Determine limitations to data.
Regression	Impact of incremental design changes on the product.
Scenario	Complex fulfillment of a particular set of tasks.
Business cycle	Scenario intended to replicate product use for an entire business cycle.
Installation	Initial application of product in its intended operating environment.
Load	Ability to handle excessive activity.
Long sequence	Sustained product use over an extended period.
Performance	Speed and duration of product use.
Comparison with results	Determination of variations to external references.
Consistency	Determination of variances to internal product references.
Oracle	Comparison to common acceptance indicators.

extreme situation, $\boldsymbol{A}$ could be a complete, that is, nonsparse full lower or upper, triangular matrix. For example, in a *full lower triangle matrix,* the maximum number of nonzero entries, $p(p-1)/2$ where $A_{ij} = X \neq 0$ for $j = 1, i$ and $i = 1, \ldots, p$. A lower (upper) triangular decoupled design matrix is characterized by $A_{ij} = 0$ for $i < j$ (for $i > j$). A rectangular design matrix with $(m > p)$ is classified as a *coupled* design, as in (13.3).

A case study is presented in Hintersteiner and Nain, (2000) and is reproduced here. In this study, axiom 1 was applied for hardware and software systems to the design of a photolithography tool manufactured by SVG Lithography Systems, Inc(Wilton, CT). The system uses one 6 degrees of freedom (DOF) robot to move wafers between different wafer processing areas in a work cell as well as moving the wafers into and out of the system. A second robot also is used in a similar fashion for transporting reticles (i.e., wafer field masks). The example outlines the design of the robot calibration routine for these robots. This routine is responsible for initializing and calibrating the robot with respect to the discrete locations in each work cell.

Constraints imposed on the design of the robot calibration routine include the use of a standard robot accessory (a teaching pendant with display, known as the metacarpal-phalangeal [MCP] joint control pad) for the user interface, speed and trajectory limitations, restrictions on robot motions at each discrete location in the work cell, and implied constraints for minimizing the necessary time required to calibrate the locations. Efforts were made early on in the design process to establish and reconcile the functional requirements dictated by various departments, including engineering, assembly, field servicing, and so on. For example, requirements from engineering emerged from the design of the work cell itself, whereas field service requirements focused more on ease of use and maintaining a short learning curve.

The top-level decomposition is shown in (13.4). The programs are the blocks of code that perform the value-added functions of selecting the locations (DP1), moving the robot between locations (DP2), calibrating the locations (DP3), and recording the locations (DP4). The only interface defined here is the user interface (DP5), which displays information gathered by and given to the user during different phases of the calibration. The control logic is DP6. The support programs (DP7) constitute the elements required to maintain the continuity thread between the various programs and the control logic. These include global variables, continuous error recovery logic, library functions, and so forth.

The corresponding design matrix, shown in (13.4) indicates that the robot calibration routine is a decoupled design. The off-diagonal "X" terms indicate that, for example, the locations to be calibrated must be established before the motion to the locations and the calibration and recording routines for those locations are designed. This has ramifications not only for how the programs interact, but also for the user interface.

$$\begin{Bmatrix} \text{FR1: Select locations} \\ \text{FR2: Move robot} \\ \text{FR3: Calibrate location} \\ \text{FR4: Record location} \\ \text{FR5: Provide user interface} \\ \text{FR6: Control processes} \\ \text{FR7: Integrate and support} \end{Bmatrix} = \begin{bmatrix} X & 0 & 0 & 0 & 0 & 0 & 0 \\ X & X & 0 & 0 & 0 & 0 & 0 \\ X & 0 & X & 0 & 0 & 0 & 0 \\ X & X & X & X & 0 & 0 & 0 \\ X & X & X & X & X & 0 & 0 \\ X & X & X & X & X & X & 0 \\ X & X & X & X & X & X & X \end{bmatrix} \begin{Bmatrix} \text{DP1: Location selection list} \\ \text{DP2: Robot motion algorithm} \\ \text{DP3: Calibration algorithm} \\ \text{DP4: Record algorithm} \\ \text{DP5: MCP interface} \\ \text{DP6: Control logic diagram} \\ \text{DP7: Support programs} \end{Bmatrix} \quad (13.4)$$

Similarities between the information exchanged with the user for each program give rise to the creation of basic building blocks for developing the interface. Although not shown here, the decomposition has been performed to the low-level design for

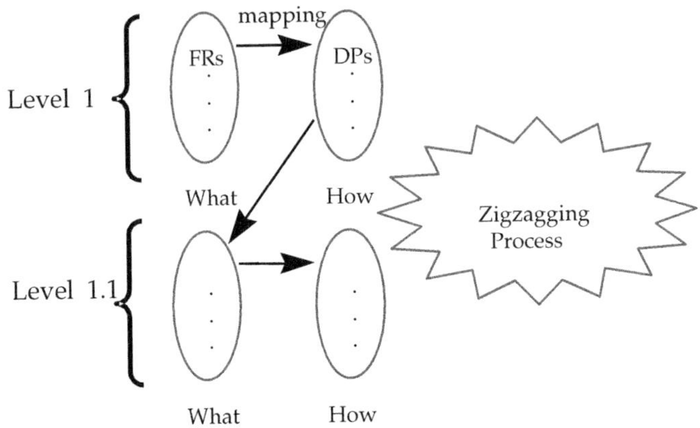

FIGURE 13.4 The zigzagging process.

this software, and the system representation for software holds at every hierarchical level.

The importance of the design mapping has many perspectives. Chief among them is the identification of coupling among the functional requirements, which result from the physical mapping process with the design parameters, in the codomain. Knowledge of coupling is important because it provides the design team clues with which to find solutions, make adjustments or design changes in proper sequence, and maintain their effects over the long term with minimal negative consequences.

The design matrices are obtained in a hierarchy and result from employment of the *zigzagging* method of mapping, as depicted in Figure 13.4 (Suh, 1990). The zigzagging process requires a solution-neutral environment, where the DPs are chosen after the FRs are defined and not vice versa. When the FRs are defined, we have to *zig* to the physical domain, and after proper DPs selection, we have to *zag* back to the functional domain for further decomposition or cascading, though at a lower hierarchical level. This process is in contrast with the traditional cascading processes that use only one domain at a time, treating the design as the sum of functions or the sum of parts.

At lower levels of hierarchy, entries of design matrices can be obtained mathematically from basic physical and engineering quantities enabling the definition and detailing of transfer functions, an operational vulnerability treatment vehicle. In some cases, these relationships are not readily available, and some effort needs to be paid to obtain them empirically or via modeling.

13.3 AXIOM 1 IN SOFTWARE DFSS[5]

Several design methodologies for software systems have been proposed in the past. Two decades ago, structured methods, such as structured design and structured

[5]See and 2000.

analysis, were the most popular idea (DeMarco, 1979). As the requirement for productive software systems has increased, the object-oriented method has become the basic programming tool (Cox, 1986). It emphasizes the need to design software right during the early stages of software development and the importance of modularity. However, even with object-oriented methods, there are many problems that intelligent software programmers face in developing and maintaining software during its life cycle. Although there are several reasons for these difficulties, the main reason is that the current software design methodology has difficulty explaining the logical criterions of good software design.

Modularity alone does not ensure good software because even a set of independent modules can couple software functions. The concept of the axiomatic design framework has been applied successfully to software design (Kim et al., 1991; Do & Park, 1996; Do, 1997). The basic idea used for the design and development of software systems is exactly the same as that used for hardware systems and components, and thus, the integration of software and hardware design becomes a straightforward exercise.

The methodology presented in this section for software design and development uses both the axiomatic design framework and the object-oriented method. It consists of three steps. First, it designs the software system based on axiomatic design (i.e., the decomposition of FRs and DPs) the design matrix, and the modules as defined by axiomatic design (Suh, 1990, 2001). Second, it represents the software design using a full-design matrix table and a flow diagram, which provide a well-organized structure for software development. Third is the direct building of the software code based on a flow diagram using the object-oriented concept. This axiomatic approach enhances software productivity because it provides the road map for designers and developers of the software system and eliminates functional coupling.

A software design based on axiomatic design is self-consistent, provides uncoupled or decoupled interrelationships and arrangements among "modules," and is easy to change, modify, and extend. This is a result of having made correct decisions at each stage of the design process (i.e., mapping and decomposition [Suh, 1990; El-Haik, 2005]).

Based on axiomatic design and the object-oriented method, Do and Suh (2000) have developed a generic approach to software design. The software system is called "axiomatic design of object-oriented software systems (ADo-oSS)" that can be used by any software designers. It combines the power of axiomatic design with the popular software programming methodology called the object-oriented programming technique (OOT) (Rumbaugh et al., 1991) (Booch, 1994). The goal of ADo-oSS is to make the software development a subject of science rather than an art and, thus, reduce or eliminate the need for debugging and extensive changes.

ADo-oSS uses the systematic nature of axiomatic design, which can be generalized and applied to all different design task, and the infrastructure created for object-oriented programming. It overcomes many of the shortcomings of the current software design techniques, which result in a high maintenance cost, limited reusability, an extensive need to debug and test, poor documentation, and limited extensionality of the software. ADo-oSS overcomes these shortcomings.

One of the final outputs of ADo-oSS is the system architecture, which is represented by the flow diagram. The flow diagram can be used in many different applications for a variety of different purposes such as:

- Improvement of the proposed design through identification of coupled designs.
- Diagnosis of the impending failure of a complex system.
- Reduction of the service cost of maintaining machines and systems.
- Engineering change orders.
- Job assignment and management of design tasks.
- Management of distributed and collaborative design tasks.
- Reusability and extensionality of software.

In axiomatic design, a "module" is defined as the row of design matrix that yields the FR of the row when it is multiplied by the corresponding DP (i.e., data). The axiomatic design framework ensures that the modules are correctly defined and located in the right place in the right order. A "V" model for software, shown in Figure 13.5 (El-Haik, 1999), will be used here to explain the concept of ADo-oSS. The first step is to design the software following the top-down approach of axiomatic design, build the software hierarchy, and then generate the full-design matrix (i.e., the design matrix that shows the entire design hierarchy) to define modules.

The final step is to build the object-oriented model with a bottom-up approach, following the axiomatic design flow diagram for the designed system. Axiomatic design of software can be implemented using any software language. However, in the 1990s most software is written using an object-oriented programming language such as C++ or Java. Therefore, axiomatic design of software is implemented using object-oriented methodology.

To understand ADo-oSS, it is necessary to review the definitions of the words used in OOT and their equivalent words in axiomatic design. The fundamental construct for the object-oriented method is object 2, which is equivalent to FRs. An object-oriented design decomposes a system into objects. Objects "encapsulate" both data

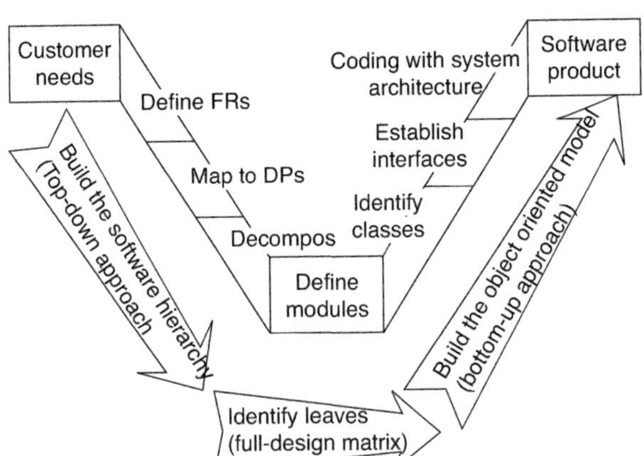

FIGURE 13.5 Axiomatic design process for object-oriented software system (the V model).

(equivalent to DPs), and method (equivalent to the relationship between FRi and DPi, that is module) in a single entity. Object retains certain information on how to perform certain operations, using the input provided by the data and the method imbedded in the object. (In terms of axiomatic design, this is equivalent to saying that an object is [FRi = Aij DPj].)

An object-oriented design generally uses four definitions to describe its operations: identity, classification, polymorphism, and relationship. Identity means that data—equivalent to DPs—are incorporated into specific objects. Objects are equivalent to an FR—with a specified [$FR_i = A_{ij}\ DP_j$] relationship—of axiomatic design, where DPs are data or input and A_{ij} is a method or a relationship. In an axiomatic design, the design equation explicitly identifies the relationship between FRs and DPs. Classification means that objects with the same data structure (attributes) and behavior (operations or methods) are grouped into a class. The object is represented as an instance of specific class in programming languages. Therefore, all objects are instances of some classes. A class represents a template for several objects and describes how these objects are structured internally. Objects of the same class have the same definition both for their operations and for their information structure.

Sometimes an "object" also is called a tangible entity that exhibits some well-defined "behavior." "Behavior" is a special case of FR. The relationship between "objects" and "behavior" may be compared with the decomposition of FRs in the FR hierarchy of axiomatic design. "Object" is the "parent FR" relative to "Behavior," which is the "child FR." That is, the highest FR between the two layers of decomposed FRs is "object," and the children FRs of the 'object FR' are "behavior."

The distinction between "super class," "class," "object" and "behavior" is necessary in OOT to deal with FRs at successive layers of a system design. In OOT, class represents an abstraction of objects and, thus, is at the same level as an object in the FR hierarchy. However, object is one level higher than behavior in the FR hierarchy. The use of these key words, although necessary in OOT, adds unnecessary complexity when the results of axiomatic design are to be combined with OOT. Therefore, we will modify the use of these key words in OOT.

In ADo-oSS, the definitions used in OOT are slightly modified. We will use one key word "object," to represent all levels of FRs (i.e., class, object, and behavior). "Objects with indices" will be used in place of these three key words. For example, class or object may be called "object i," which is equivalent to FR_i, Behavior will be denoted as "Object ij" to represent the next level FRs, FR_{ij}.

Conversely, the third level FRs will be denoted as "Object $_{ijk}$." Thus, "Object $_{i}$," "Object $_{ij}$," and "Object $_{ijk}$" are equivalent to FR_i, FR_{ij}, and FR_{ijk}, which are FRs at three successive levels of the FR hierarchy.

To summarize, the equivalence between the terminology of axiomatic design and those of OOT may be stated as:

- An FR can represent an object.
- DP can be data or input for the object, (i.e., FR).
- The product of a module of the design matrix and DP can be a method (i.e., FR = A × DP).
- Different levels of FRs are represented as objects with indices.

The ADo-oSS shown in Figure 13.5 involves the following steps:

a. Define FRs of the software system: The first step in designing a software system is to determine the customer attributes in the customer domain that the software system must satisfy. Then, the (FR) of the software in the functional domain and constraints (Cs) are established to satisfy the customer needs.
b. Mapping between the domains and the independence of software functions: The next step in axiomatic design is to map these FRs of the functional domain into the physical domain by identifying the DPs. DPs are the "hows" of the design that satisfy specific FRs. DPs must be chosen to be consistent with the constraints.
c. Decomposition of {FRs}, {DPs}, and {PVs}: The FRs, DPs, and PVs must be decomposed until the design can be implemented without further decomposition. These hierarchies of {FRs}, {DPs}, {PVs}, and the corresponding matrices represent the system architecture. The decomposition of these vectors cannot be done by remaining in a single domain but can only be done through zigzagging between domains.
d. Definition of modules—full-design matrix: One of the most important features for the axiomatic design framework is the design matrix, which provides the relationships between the FRs and the DPs. In the case of software, the design matrix provides two important bases in creating software. One important basis is that each element in the design matrix can be a method (or operation) in terms of the object-oriented method. The other basis is that each row in the design matrix represents a module to satisfy a specific FR when a given DP is provided. The off-diagonal terms in the design matrix are important because the sources of coupling are these off-diagonal terms. It is important to construct the full-design matrix based on the leaf-level FR-DP-Aij to check for consistency of decisions made during decomposition.
e. Identify objects, attributes, and operations: Because all DPs in the design hierarchy are selected to satisfy FRs, it is relatively easy to identify the objects. The leaf is the lowest level object in a given decomposition branch, but all leaf-level objects may not be at the same level if they belong to different decomposition branches. Once the objects are defined, the attributes (or data)—DPs—and operations (or methods)—products of module times DPs—for the object should be defined to construct the object model. This activity should use the full-design matrix table. The full-design matrix with FRs and DPs can be translated into the OOT structure, as shown in Figure 13.6.
f. Establish interfaces by showing the relationships between objects and operations: Most efforts are focused on this step in the object-oriented method because the relationship is the key feature. The axiomatic design methodology presented in this case study uses the off-diagonal element in the design matrix as well as the diagonal elements at all levels. A design matrix element represents a link or association relationship between different FR branches that have totally different behavior.

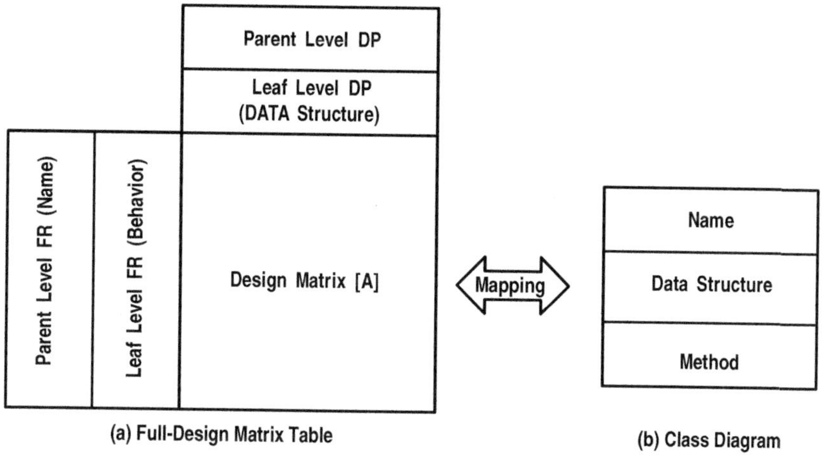

FIGURE 13.6 The correspondence between the full design matrix and the OOT diagram.

The sequence of software development begins at the lowest level, which is defined as the leaves. To achieve the highest level FRs, which are the final outputs of the software, the development of the system must begin from the inner most modules shown in the flow diagram that represent the lowest level leaves then move to the next higher level modules (i.e., next inner most box), following the sequence indicated by the system architecture (i.e., go from the inner most boxes to the outer most boxes). In short, the software system can be developed in the following sequence:

1. Construct the core functions using all diagonal elements of the design matrix.
2. Make a module for each leaf FR, following the sequence given in the flow diagram that represents the system architecture.
3. Combine the modules to generate the software system, following the module junction diagram.

When this procedure is followed, the software developer can reduce the coding time because the logical process reduces the software construction into a routine operation.

13.3.1 Example: Simple Drawing Program

In the preceding section, the basic concept for designing software based on ADo-oSS was presented. In this section, a case study involving the simple drawing software design based on ADo-oSS will be presented.

a. Define FRs of the software system: Let us assume the customer attributes as follows:

   CA1= We need software to draw a line or a rectangle or a circle at a time.

CA2 = The software should work with the mouse using push, drag, and release action

Then, the desired first-level functional requirements of the software can be described as follow:

FR1 = Define element.

FR2 = Specify drawing environment.

b. Mapping between the domains and the independence of software functions: The mapping for the first level can be derived as shown in (13.5). The upper character in the design matrix area represents a diagonal relationship and the lower in table character represents an off-diagonal relationship.

$$\begin{Bmatrix} \text{FR1: Define element} \\ \text{R2: Specify drawing environment} \end{Bmatrix} = \begin{bmatrix} A & 0 \\ a & B \end{bmatrix} \begin{Bmatrix} \text{DP1: Element characteristics} \\ \text{DP2: GUI with window} \end{Bmatrix} \tag{13.5}$$

c. Decomposition of {FRs}, {DPs}, and {PVs}: The entire decomposition information can be summarized in (13.6)–(13.12), with the entire design hierarchy depicted in Figure 13.7.

$$\begin{Bmatrix} \text{FR11: Define line element} \\ \text{FR12: Define rectangle element} \\ \text{FR13: Define circle element} \end{Bmatrix} = \begin{bmatrix} C & 0 & 0 \\ 0 & D & 0 \\ 0 & 0 & E \end{bmatrix} \begin{Bmatrix} \text{DP11: Line chracteristic} \\ \text{DP12: Rectangle chracteristic} \\ \text{DP13: Circle characteristic} \end{Bmatrix} \tag{13.6}$$

$$\begin{Bmatrix} \text{FR21: Identify the drawing type} \\ \text{FR22: Detect drawing location} \\ \text{FR23: Draw an element} \end{Bmatrix} = \begin{bmatrix} F & 0 & 0 \\ b & G & 0 \\ c & 0 & H \end{bmatrix} \begin{Bmatrix} \text{DP21: Ratio buttons} \\ \text{DP22: Mouse click information} \\ \text{DP23: Drawing area (i.e. canvas)} \end{Bmatrix} \tag{13.7}$$

$$\begin{Bmatrix} \text{FR111: Define start} \\ \text{FR112: Define end} \end{Bmatrix} = \begin{bmatrix} I & 0 \\ 0 & J \end{bmatrix} \begin{Bmatrix} \text{DP111: Start point} \\ \text{DP112: End point} \end{Bmatrix} \tag{13.8}$$

$$\begin{Bmatrix} \text{FR121: Define upper left corner} \\ \text{FR122: Define lower left corner} \end{Bmatrix} = \begin{bmatrix} K & 0 \\ 0 & L \end{bmatrix} \begin{Bmatrix} \text{DP121: Upper left point} \\ \text{DP122: Lower right point} \end{Bmatrix} \tag{13.9}$$

$$\begin{Bmatrix} \text{FR131: Define center} \\ \text{FR132: Define radius} \end{Bmatrix} = \begin{bmatrix} M & 0 \\ 0 & N \end{bmatrix} \begin{Bmatrix} \text{DP131: Center point} \\ \text{DP132: } \textit{Radius} \end{Bmatrix} \tag{13.10}$$

$$\begin{Bmatrix} \text{FR211: Identify line} \\ \text{FR212: Identify rectangle} \\ \text{FR213: Identify circle} \end{Bmatrix} = \begin{bmatrix} O & 0 & 0 \\ 0 & P & 0 \\ 0 & 0 & Q \end{bmatrix} \begin{Bmatrix} \text{DP211: Line button} \\ \text{DP212: Rectangle button} \\ \text{DP213: Circle button} \end{Bmatrix} \tag{13.11}$$

$$\begin{Bmatrix} \text{FR221: Detect mouse push} \\ \text{FR222: Detect mouse release} \end{Bmatrix} = \begin{bmatrix} R & 0 \\ 0 & S \end{bmatrix} \begin{Bmatrix} \text{DP221: Event for push} \\ \text{DP222: Event for release} \end{Bmatrix} \tag{13.12}$$

d. Definition of modules—Full-design matrix: When the decomposition process finishes, an inconsistency check should be done to confirm the decomposition.

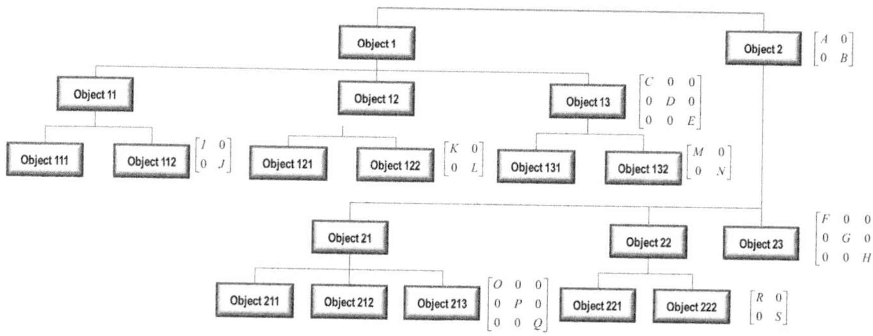

FIGURE 13.7 The design hierarchy.

The fulldesign matrix shown in Figure 13.8 indicates that the design has no conflicts between hierarchy levels. By definition, each row in the full-design matrix represents a module to fulfill corresponding FRs. For example, FR_{23} (draw an element) only can be satisfied if all DPs, except DP_{221} and DP_{222}, are present.

e. Identify objects, attributes, and operations: Figure 13.9 shows how each design matrix elements was transformed into programming terminology. Unlike the other design cases, the mapping between the physical domain and the process

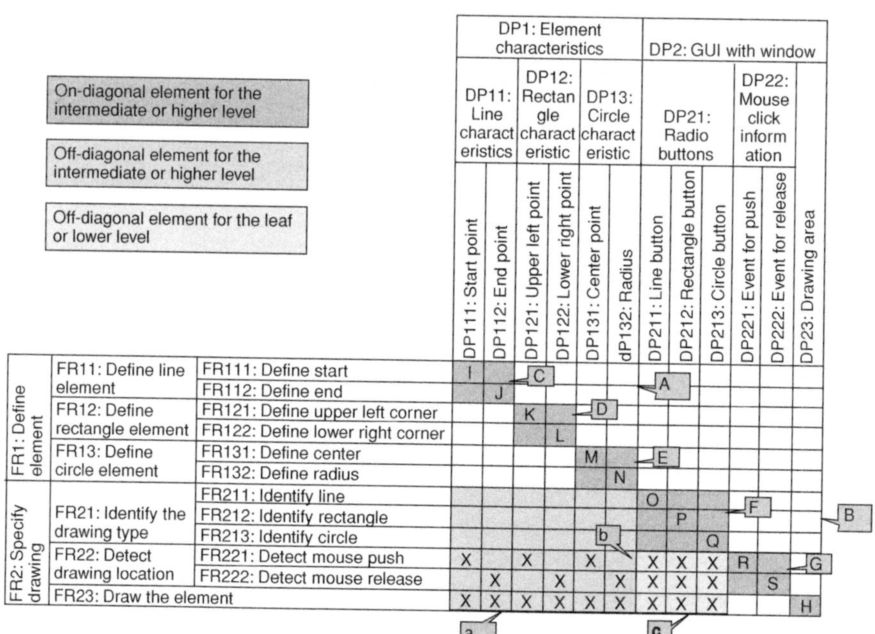

FIGURE 13.8 The full-design matrix.

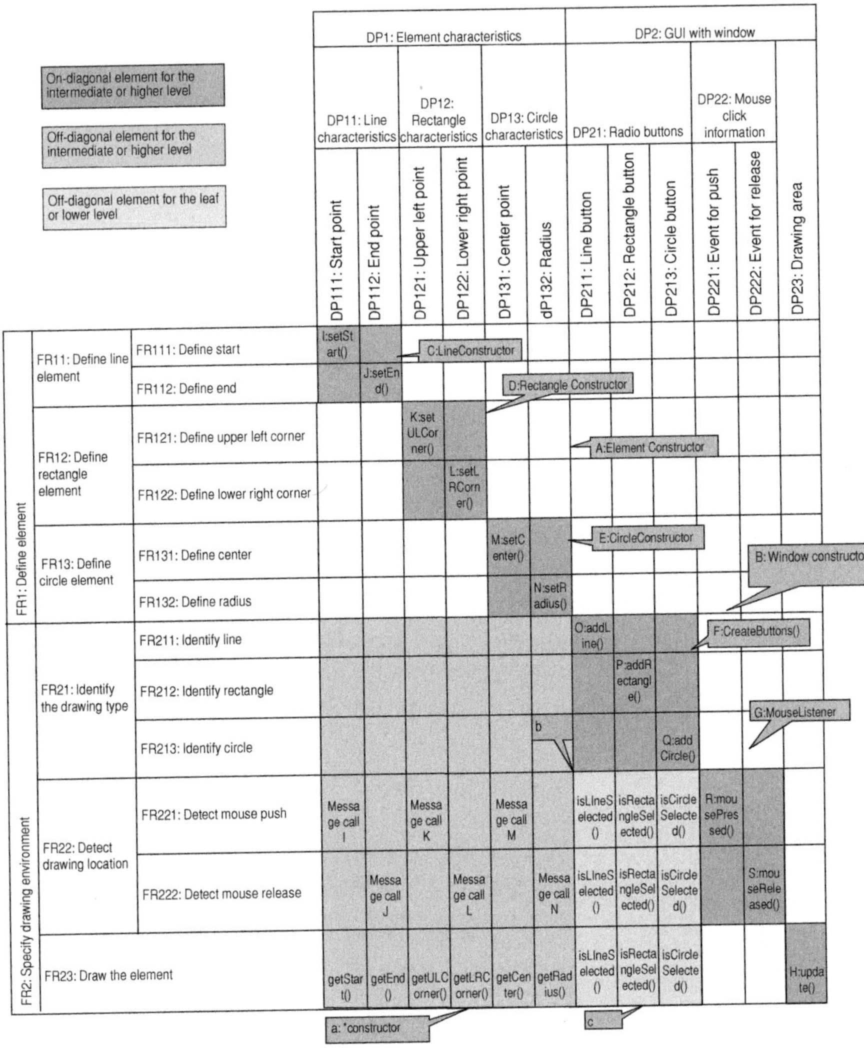

FIGURE 13.9 The method representation.

domain is pretty straightforward in a software design case because the process variables for software are the real source codes. These source codes represent each class in an object-oriented programming package. Whenever the software designer categorizes module groups as classes using the full-design matrix, they define the process variables for corresponded design hierarchy levels. Designers can assume that the design matrixes for DP/PV mapping are identical to those for FR/DP.

f. Establish interfaces by showing the relationships between objects and operations: Figure 13.9 represents the additional information for FR/DP mapping.

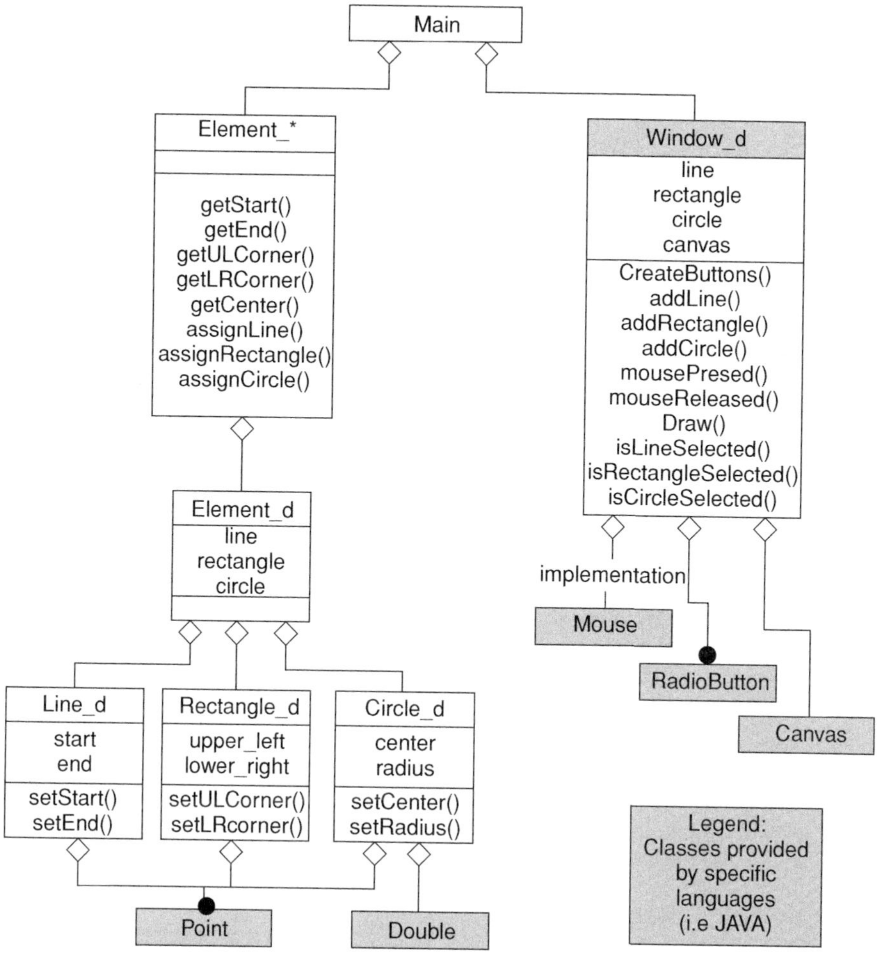

FIGURE 13.10 Object-oriented model generation.

The same rule can be introduced to represent the interface information such as aggregation, generalization, and so forth in the design matrix for DP/PV mapping. Figure 13.10 shows a class diagram for this example based on the matrix for DP/PV mapping. The flow diagram in Figure 13.11 shows the developing process depicting how the software can be programmed sequentially.

g. Table 13.3 categorizes the classes, attributes, and operations from Figure 13.9 using this mapping process. The first row in Table 13.3 represents the PV. The sequences in Table 13.3 (i.e., left to right) also show the programming sequences based on the flow diagram. Figure 13.11 shows classes diagram for this example based on the matrix for DP/PV mapping.

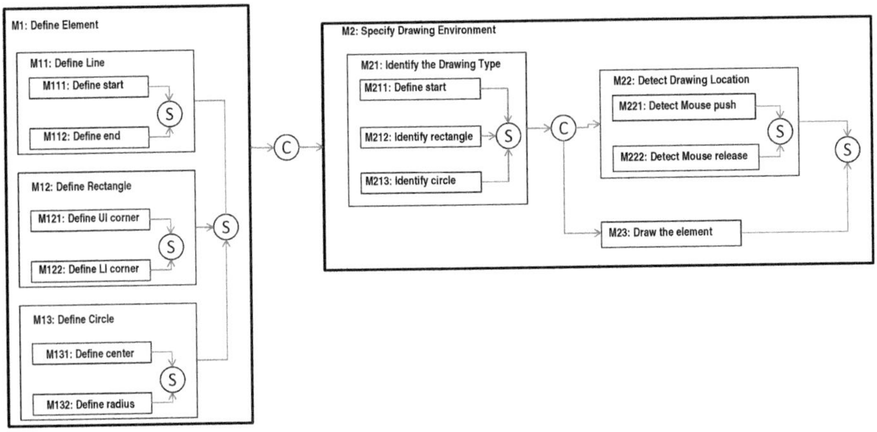

FIGURE 13.11 Flow diagram for the simple drawing example.

In this case study, the axiomatic design framework has been applied to the design and development of an object-oriented software system. The current software development methodologies demand that each individual module be independent. However, modularity does not mean functional independence, and therefore, the existing methodologies do not provide a means to achieve the independence of functional

TABLE 13.3 Class Identification

Object	Object 111/112/ 121/122/ 131	Object 132	Object 11		Object 12		Object 13	
Name	Point	Double	Lin_d		Rectangle_d		Circle_d	
Attribute			DP111	Point start	DP121	Point upper left	DP131	Point center
			DP112	Point end	DP122	Point lower right	DP132	Double radius
Method			C	Line()	D	Rectangle()	E	Center()
			I	setStart()	K	setULCorner()	M	setCenter()
			J	setEnd()	L	SetLRCorner()	N	setRadius()

requirements. To have good software, the relationship between the independent modules must be designed to make them work effectively and explicitly. The axiomatic design framework supplies a method to overcome these difficulties systematically and ensures that the modules are in the right place in the right order when the modules are established as the row of design matrix. The axiomatic design methodology for software development can help software engineers and programmers to develop effective and reliable software systems quickly.

13.4 COUPLING MEASURES

Coupling is a measure of how interconnected modules are. Two modules are coupled if a change to a DP in one module may require changes in the other module. The lowest coupling is desirable.

In hardware, coupling is defined on a continuous scale. Rinderle (1982) and Suh and Rinderle (1982) proposed the use of reangularity "*R*" and semangularity "*S*" as coupling measures. Both *R* and *S* are defined in (13.13) and (13.14), respectively. *R* is a measure of the orthogonality between the DPs in terms of the absolute value of the product of the geometric sines of all the angles between the different DP pair combinations of the design matrix. As the degree of coupling increases, "*R*" decreases. Semangularity, *S*, however, is an angular measure of parallelism of the pair

TABLE 13.3 (*Continued*)

Object 1		Object 2		Object 211/ 212/ 213	Object 22	Object 23	Object 1*
Element_d		Window_d		Radio B	Mouse	Canvas	Element_*
DP11	Line1	DP211	Radiobutton line				
DP12	Rectangle r	DP212	Radiobutton rectangle				
DP13	Circle c	DP213	Radiobutton circle				
		DP22	Mouse m				
		DP23	Canvas c				
A	Element	B	Window()				a Element*()
		F	CreateButtons()				getStart()
		O	addLine()				getEnd()
		P	addRectangle				getULCorner()
		Q	addCircle()				getLRCorner()
		G	implement MouseLisner				getCenter()
		R	mousePressed()				getRadius()
		S	mouseReleased()				assignLine()
		H	draw()				assignRectangle()
		b/c	isLineSelected()				assignCircle()
		b/c	inRectangleSelected()				
		b/c	inCircleSelected()				

DP and FR (see Figure 1.2). When $R = S = 1$, the design is completely uncoupled. The design is decoupled when $R = S$ (Suh, 1991).

$$R = \prod_{\substack{j=1,p-1 \\ k=1+i,p}} \sqrt{\left[1 - \left(\sum_{k=1}^{p} A_{kj} A_{kj}\right)^2 / \left(\sum_{k=1}^{p} A_{kj}^2\right)\left(\sum_{k=1}^{p} A_{kj}^2\right)\right]} \quad (13.13)$$

$$S = \prod_{j=1}^{p} \left(|A_{jj}| / \sqrt{\sum_{k=1}^{p} A_{kj}^2}\right) \quad (13.14)$$

Axiom 1 is best satisfied if A is a diagonal matrix depicting an uncoupled design. For a decoupled design, axiom 1 can be satisfied if the DPs can be set (adjusted) in a specific order conveyed by the matrix to maintain independence. A design that violates axiom 1 as it distances itself from uncoupled and decoupled categories is, by definition, a coupled design. The vulnerability of coupling is assured whenever the number of DPs, p, is less than the number of FRs, m (El-Haik, 2005: See Theorem 1 and Theorem 2, Section 2.5). In other words, the desired bi-jection one-to-one mapping property between two design domains cannot be achieved without an axiomatic treatment. An axiomatic treatment can be produced by the application of design theories and corollaries deduced from the axioms (El-Haik, 2005).

For a unifunctional design entity ($m = 1$), the independence axiom is always satisfied. Optimization, regardless of whether being deterministic or probabilistic, of a multifunctional module is complicated by the presence of coupling (lack of independence). Uncoupled design matrices may be treated as independent modules for optimization (where DPs are the variables) and extreme local or global DPs settings in the direction of goodness can be found. In a decoupled design, the optimization of a modular element cannot be carried out in one routine. Many optimization algorithms, in fact m routines, need to be invoked sequentially starting from the DP at the head of the triangular matrix and proceeding to the base. The coupling that we need to guard against in software design is the *content* type. The *content* coupling is bad as in hardware and should be avoided. It occurs when one module (a DP) directly affects the workings of another (another DP) or when a module (a DP) changes another module's data.

In addition to the content type, several types of software coupling are listed as follows:

- *Common:* Two modules have shared data (e.g., global variables).
- *External:* Modules communicate through an external medium, like a file.
- *Control:* One module directs the execution of another by passing control information (e.g., via flags).
- *Stamp:* Complete data structures or objects are passed from one module to another.
- *Data:* Only simple data is passed between modules.

In software, several measures of coupling were proposed. For example, in the OOT case, such as the study in Section 13.3, we propose the following coupling measure (C_F) between the software classes (Figure 13.9)

$$C_F = \frac{\sum_{i=1}^{p} \sum_{J=1}^{p} is\_rel\left(c_i, c_j\right)}{p^2 - p} \tag{13.15}$$

where p is the total number of objects (DPs) in the concerned software, and

$$is\_rel = \begin{cases} 1 & \text{If class } i \text{ has a relation with class } j \\ 0 & \text{Otherwise} \end{cases}$$

The relation might be that class i calls a method in class j or has a reference to class j or to an attribute in class j. In this case, C_F measures the strength of intermodule connections with the understanding that a high coupling indicates a strong dependence between classes, which implies that we should study modules as pairs. In general, a low coupling indicates independent modules, and generally, we desire less coupling because it is easier to design, comprehend, and adapt.

Dharma (1995) proposed the following coupling metric:

$$m_c = \frac{k}{M} \tag{13.16}$$

$$M = d_i + 2 \times c_i + d_o + 2 \times c_o + g_d + 2 \times g_c + w + r \tag{13.17}$$

with the following arguments:

- *Data and control flow coupling*
  - d_i = number of input data parameters
  - c_i = number of input control parameters
  - d_o = number of output data parameters
  - c_o = number of output control parameters
- *Global coupling*
  - g_d = number of global variables used as data
  - g_c = number of global variables used as control
- *Environmental coupling*
  - w = number of modules called (fan-out)
  - r = number of modules calling the module under consideration (fan-in)

The more situations encountered, the greater the coupling and the smaller m_c. One problem is parameters and calling counts do not guarantee the module is linked to the FRs of other modules.

13.5 AXIOM 2 IN SOFTWARE DFSS

13.5.1 Axiom 2: The Information Axiom

13.5.1.1 Minimize the Information Content in a Design. The second axiom of axiomatic design stated previously provides a selection metric based on design information content. Information content is defined as a measure of complexity, and it is related to the probability of certain events occurring when information is supplied. Per axiom 2, the independent design that minimizes the information content is the best. However, the exact deployment of design axioms might not be feasible because of technological and/or cost limitations. Under these circumstances, different degrees of conceptual vulnerabilities are established in the measures (criteria) related to the unsatisfied axioms. For example, a degree of design complexity may exist as a result of an axiom 2 violation. Such a vulnerable design entity may have questionable quality and reliability performance even after thorough operational optimization. Quality and reliability improvements of weak conceptual software entities usually produce marginal results. Before these efforts, conceptual vulnerability should be reduced, if not eliminated. Indeed, the presence of *content* functional coupling and complexity vulnerabilities aggravates the symptomatic behavior of the software entities.

13.5.2 Axiom 2 in Hardware DFSS: Measures of Complexity

In hardware design, the selection problem between alternative design solution entities (concepts) of the same design variable (project) will occur in many situations. Even in the ideal case, a pool of uncoupled design alternatives, the design team still needs to select the best solution. The selection process is criteria based, hence axiom 2. The information axiom states that the design that results in the highest probability of FRs success (*Prob(FR1), Prob(FR2), . . ., Prob(FRm)*) is the best design. Information and probability are tied together via entropy, H . Entropy H may be defined as

$$H = -\log_{\nu}(\mathrm{Pr}\,ob) \tag{13.18}$$

Note that probability "Prob" in (13.18) takes the Shannon (1948) entropy form of a discrete random variable supplying the information, the source. Note also that the logarithm is to the base ν, a real nonnegative number. If $\nu = 2(e)$, [6] , then H is measured in *bits (nats)*.

The expression of information and, hence, design *complexity* in terms of probability hints to the fact that FRs are random variables themselves, and they have to be met with some tolerance accepted by the customer. The array $\{\boldsymbol{FR}\}$ are also functions of (the physical mapping) random variables, and the array $\{\boldsymbol{DP}\}$, which in turn, are functions (the process mapping) of another vector of random variables, the array $\{\boldsymbol{PV}\}$. The PVs downstream variation can be induced by several sources

[6] *e* is the natural logarithm base

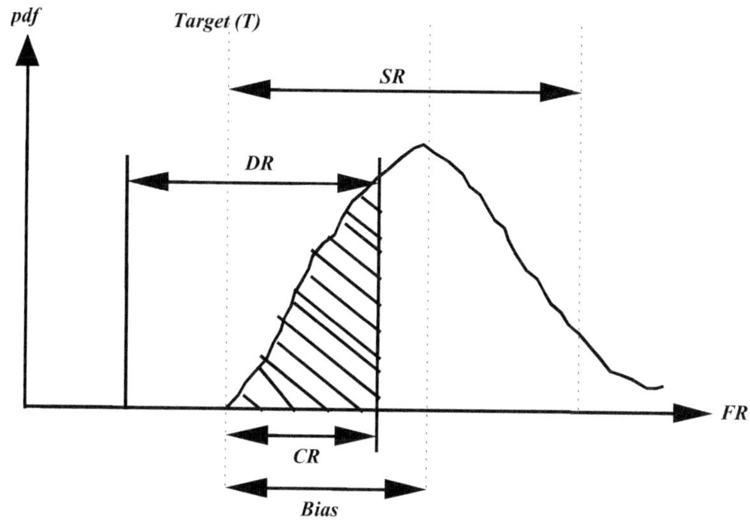

FIGURE 13.12 The probability of success definition.

such as manufacturing process variation, including tool degradation and environmental factors—the noise factors. Assuming statistical independence, the overall (total) design information content of a given design hierarchical level is additive because its probability of success is the multiplication of the individual FRs probability of success belonging to that level. That is, to reduce complexity, we need to address the largest contributors to the total (the sum). When the statistical independence assumption is not valid, the system probability of success is not multiplicative; rather, it is conditional.

A solution entity is characterized as complex when the probability of success of the total design (all hierarchical levels) is low. Complex design solution entities require more information to manufacture them. That is, complexity is a design vulnerability that is created in the design entity caused by the violation of axiom 2. Note that complexity here has two arguments: the number of FRs as well as their probability of success.

Information content is related to tolerances and process capabilities because probabilities are arguments of process capabilities indices. The probability of success may be defined as the probability of meeting design specifications, the area of intersection between the *design range* (voice of the customer) and the *system range* (voice of the process). System range is denoted "SR" and the design range is denoted "*DR*" (see Figure 13.12). The overlap between the design range and the system range is called the *common range* "CR." The probability of success is defined as the area ratio of the common range to system range, $\frac{CR}{SR}$. Substituting this definition in (13.19), we have:

$$H = \log_{\nu} \frac{SR}{CR} \tag{13.19}$$

McCabe's cyclomatic number, Henry-Kafura Information Flow, and Halstead's Software Science are different complexity measures that can be used in axiom 2 applications. These were discussed in Chapter 5.

REFERENCES

Altshuller G.S. (1988), *Creativity as Exact Science*, Gordon & Breach, New York, NY.

Altshuller, G.S. (1990), "On the Theory of Solving Inventive Problems," *Design Methods and Theories*, Volume *24*, #2, pp. 1216–1222.

Arciszewsky, T. (1988), "ARIZ 77: An Innovative Design Method," *Design Methods and Theories*, Volume 22, #2, pp. 796–820.

Booch, G. (1994), *Object-Oriented Analysis and Design with Applications*, 2nd Ed., The Benjamin/Cummings Publishing Company, San Francisco, CA.

Cox, B.J. (1986), *Object-Oriented Programming*, Addison Wesley, Reading, MA.

DeMarco, T. (1979), *Structural Analysis and System Specification*, Prentice Hall, Upper Saddle River, NJ.

Do, S.H. (1997), "Application of Design Axioms to the Design for Manufacturability for the Television Glass Bulb," Ph.D Dissertation, Hanyang University, Seoul, Korea.

Do, S.H. and Park (1996),

Do, S.H. and Suh, N.P. (2000), "Object Oriented Software Design with Axiomatic Design," Proceedings of the ICAD, p. 27.

El-Haik, Basem S. (1999), "The Integration of Axiomatic Design in the Engineering Design Process," 11th Annual RMSL Workshop, May.

El-Haik, Basem S.. (2005), *Axiomatic Quality & Reliability: Integrating Principles of Design. Six Sigma. Reliability, and Quality Engineering*, John Wiley & Sons, New York.

El-Haik, Basem S. and Mekki, K.S. (2008), *Medical Device Design For Six Sigma. A Road Map for Safety and Effectiveness*, Wiley-Interscience, New York.

El-Haik, Basem S. and Roy, D. (2005), *Service Design for Six Sigma*, John Wiley & Sons, New York.

Hintersteiner, J. and Nain, A. (2000), "Integrating Software into Systems: An Axiomatic Design Approach," Proceeding of the ICAD, Apr.

Kim, S.J., Suh, N.P., and Kim, S.-K. (1991), "Design of software systems based on axiomatic design." *Annals of the CIRP*, Volume *40*, #1 [also *Robotics & Computer-Integrated Manufacturing*, Volume *3*, pp. 149–162, 1992].

Nordlund, M., Tate, D. and Suh, N.P. (1996), "*Growth of Axiomatic Design Through Industrial Practice*," 3rd CIRP Workshop on Design and Implementation of Intelligent Manufacturing Systems, June 19–21, Tokyo, Japan, pp. 77–84.

Pressman, R.S. (1997), *Software Engineering. A Practioner's Approach*, 4th Ed., McGraw Hill, New York.

Pugh, S. (1991), *Total Design: Integrated Methods for successful Product Engineering*, Addison-Wesley, Reading, MA.

Pugh, S. (1996), *Creating Innovative Products Using Total Design*, edited by Clausing, D. and Andrade, R., Addison-Wesley, Reading, MA.

Rantanen, K. (1988), "*Altshuler's Methodology in Solving Inventive Problems*," ICED-88, Budapest.

Reinderle, J.R. (1982), "*Measures of Functional Coupling in Design*," Ph.D. dissertation, Massachusetts Institute of Technology, June.

Rumbaugh, J., Blaha, M., Premerlani, W., Eddy. F., and Lorensen, W. (1991), *Object Oriented Modeling and Design*, Prentice Hall, Upper Saddle River, NJ.

Suh, N.P. (1984), "Development of the science base for the manufacturing field through the axiomatic approach," *Robotics & Computer Integrated Manufacturing*, Volume *1*, # 3/4, pp. 397–415.

Suh, N.P. (1990), *The Principles of Design*, 1st Ed., Oxford University Press, New York.

Suh, N.P. (1995), "Design and operation of large systems," *Journal of Manufacturing Systems*, Volume *14*, #3, pp. 203–213.

Suh, N.P. (1996). "Impact of Axiomatic Design", 3rd CIRP Workshop on Design and the Implementation of Intelligent Manufacturing Systems, June 19–22, Tokyo, Japan, pp. 8–17.

Suh, N.P. (1997), "Design of systems," *Annals of CIRP*, Volume *46*, #1, pp. 75–80.

Suh, N.P. (2001), "*Axiomatic Design: Advances and Applications*," 1st Ed., Oxford University Press, New York.

Suh, N.P. (1997), "Design of systems." *Annals of CIRP*, Volume *46*, #1.

Suh, N.P. (1990), *The Principles of Design*, Oxford University Press, New York.

Suh, N.P. (2001), *Axiomatic Design: Advances and Applications*, Oxford University Press, New York.

Rinderle, J.R. and Suh, N.P. (1982), "Measures of Functional Coupling in Design," *ASME Journal of Engineering for Industry*, Volume *104*, pp. 383–388.

Swenson, A. and Nordlund, M. (1996), "*Axiomatic Design of Water Faucet*," unpublished report, Linkoping, Sweden.

Ulrich, K.T. and Eppinger, S.D. (1995), "*Product Design and Development*," McGraw-Hill, Inc., New York, NY.

Ullman, D.G. (1992), "*The Mechanical design Process*," 1st Ed., McGraw-Hill, Inc., New York, NY.

Zimmerman H.-J. (1985). Fuzzy Set Theory and its Application, 1st Ed., Springer, New York.

BIBLIOGRAPHY

Ulrich, K. T. and Tung, K., (1994), "*Fundamentals of Product Modularity*," ASME Winter Annual Meeting, DE Volume *39*, Atlanta, pp. 73–80.

Ulrich, K. T. and Seering, (1987), W. P., "Conceptual Design: Synthesis of systems Components," *Intelligent and Integrated Manufacturing Analysis and Synthesis*," American Society of Mechanical Engineers, New York, pp. 57–66.

Ulrich, K. T. and Seering, W. P., (1988), "*Function Sharing in Mechanical Design*," 7th National Conference on Artificial Intelligence, AAAI-88, Minneapolis, MN.

Ulrich, K. T. and Seering, W. P., (1989), "*Synthesis of SchematicDescription in Mechanical Design*," Research in Engineering Design, Volume *1*, #1.

CHAPTER 14

SOFTWARE DESIGN FOR X

14.1 INTRODUCTION

We will focus on vital few members of the DFX family. The letter "X" in software Design for X-ability (DFX) is made up of two parts: software processes (x) and performance measure (ability) (i.e., X = x + abilty such as test – ability, reliability, etc.). They parallel design for manufacturability, design for inspectability, design for environmentablity, design for recycl-ability, and so on in hardware Design for Six Sigma (DFSS) (Yang & El-Haik, 2003). Many software DFSS teams find that the concepts, tools, and approaches discussed in hardware are useful analogies in many ways serving as eye openers by stimulating out-of-the-box thinking.

The Black Belt continually should revise the DFSS team membership to reflect the concurrent design, which means team members are key, equal team members. DFX techniques are part of detail design and are ideal approaches to improve life-cycle cost[1] and quality, increase design flexibility, and increase efficiency and productivity. Benefits usually are pinned as competitiveness measures, improved decision making, and enhanced software development and operational efficiency. Software DFX focuses on vital business elements of software engineering maximizing the use of limited resources available to the DFSS team.

[1]Life-cycle cost is the real cost of the design. It includes not only the original cost of development and production but the associated costs of defects, litigations, buy backs, distributions support, warranty, and the implementation cost of all employed DFX methods.

Software Design for Six Sigma: A Roadmap for Excellence, By Basem El-Haik and Adnan Shaout

The DFX family of tools collect and present facts about both the design entity and its production processes, analyze all relationships between them, measure the critical-to-quality (CTQs) of performance as depicted by the software architecture, generate alternatives by combining strengths and avoiding vulnerabilities, provide a redesign recommendation for improvement, provide if–then scenarios, and do all that with many iterations.

The objective of this chapter is to introduce the vital few of the software DFX family. The software DFSS team should take advantage of, and strive to design into, the existing capabilities of suppliers, internal plants, and assembly lines. It is cost-effective, at least for the near-term. The idea is to create software sufficiently robust to achieve Six Sigma performance from current capability.

The key “design for” activities to be tackled by the team are:

1. Use DFX as early as possible in the software DFSS process.
2. Start with software design for reliability (DFR).
3. Based on the findings of (2), determine what DFX to use next. This is a function of DFSS team competence. Time and resources need to be provided to carry out the “design for” activities. The major challenge is implementation.

A danger lurks in the DFX methodologies that can curtail or limit the pursuit of excellence. Time and resource constraints can tempt software DFSS teams to accept the unacceptable on the premise that the shortfall can be corrected in one of the subsequent steps—the second chance syndrome. Just as wrong concepts cannot be recovered by brilliant detail design, bad first-instance detail designs cannot be recovered through failure mode analysis, optimization, or fault tolerancing.

14.2 SOFTWARE RELIABILITY AND DESIGN FOR RELIABILITY

Software reliability is a key part in software quality. Software quality measures how well software is designed (quality of design), and how well the software conforms to that design (quality of conformance), although there are several different definitions. Whereas quality of conformance is concerned with implementation, quality of design measures how valid the design and requirements are in creating a worthwhile product. ISO 9126 is an international standard for the evaluation of software quality. The fundamental objective of this standard is to address some of the well-known human biases that adversely can affect the delivery and perception of a software development project. These biases include changing priorities after the start of a project or not having any clear definitions of “success.” By clarifying then agreeing on the project priorities and subsequently converting abstract priorities (compliance) to measurable values (output data can be validated against schema X with zero intervention), ISO 9126 tries to develop a common understanding of the projects objectives and goals.

The standard is divided into four parts: quality model, external metrics, internal metrics, and quality in use metrics. Each quality subcharacteristic (e.g., adaptability) is divided further into attributes. An attribute is an entity that can be verified or measured in the software product. Attributes are not defined in the standard, as they vary between different software products.

A software product is defined in a broad sense; it encompasses executables, source code, architecture descriptions, and so on. As a result, the notion of user extends to operators as well as to programmers, which are users of components as software libraries.

The standard provides a framework for organizations to define a quality model for a software product. On doing so, however, it leaves up to each organization the task of precisely specifying its own model. This may be done, for example, by specifying target values for quality metrics, which evaluates the degree of presence of quality attributes.

The quality model established in the first part of the standard (ISO 9126-1) classifies software quality in a structured set of characteristics and subcharacteristics as follows:

- *Functionality*—A set of attributes that bear on the existence of a set of functions and their specified properties. The functions are those that satisfy stated or implied needs.
  - Suitability
  - Accuracy
  - Interoperability
  - Compliance
  - Security
- *Usability*—A set of attributes that bear on the effort needed for use and on the individual assessment of such use, by a stated or implied set of users.
  - Learnability
  - Understandability
  - Operability
- *Efficiency*—A set of attributes that bear on the relationship between the level of performance of the software and the amount of resources used under stated conditions.
  - Time behavior
  - Resource behavior
- *Maintainability*—A set of attributes that bear on the effort needed to make specified modifications.
  - Stability
  - Analyzability
  - Changeability
  - Testability
- *Portability*—A set of attributes that bear on the ability of software to be transferred from one environment to another.

- Installability
- Replaceability
- Adaptability

- Conformance (similar to compliance, but here related specifically to portability, e.g., conformance to a particular database standard)
- *Reliability*—A set of attributes that bear on the capability of software to maintain its level of performance under stated conditions for a stated period of time.
  - Maturity
  - Recoverability
  - Fault tolerance

Much of what developers call software reliability has been borrowed or adapted from the more mature field of hardware reliability. The influence of hardware is evident in the current practitioner community where hardware-intensive systems and typical hardware-related concerns predominate.

Two issues dominate discussions about hardware reliability: time and operating conditions. Software reliability—the probability that a software system will operate without failure for a specified time under specified operating conditions—shares these concerns (Musa et al., 1987). Because of the fundamental differences between hardware and software, it is legitimate to question these two pillars of software reliability.

The study of software reliability can be categorized into three parts: modeling, measurement, and improvement. Software reliability modeling has matured to the point that meaningful results can be obtained by applying suitable models to the problem. Many models exist, but no single model can capture the necessary amount of software characteristics. Assumptions and abstractions must be made to simplify the problem. There is no single model that is universal to all situations. Software reliability measurement is immature. Measurement is far from commonplace in software, as in other engineering fields. Software reliability cannot be directly measured, so other related factors are measured to estimate software reliability and compare it with products. Development process, faults, and failures found are all factors related to software reliability.[2]

Because more and more software is creeping into embedded systems, we must make sure they do not embed disasters. If not considered carefully, then software reliability can be the reliability bottleneck of the whole system. Ensuring software reliability is no easy task. As hard as the problem is, promising progresses still are being made toward more reliable software. More standard components and better processes are introduced in the software engineering field.

Many belts draw analogies between hardware reliability and software reliability. Although it is tempting to draw an analogy between both, software and hardware have basic differences that make them different in failure mechanisms and, hence, in

[2] See Jiantao Pan, http://www.ece.cmu.edu/~koopman/des_s99/sw_reliability/presentation.pdf.

TABLE 14.1 Software Distinct Characteristics as Compared with Hardware[3]

Characteristic	Differentiation from Hardware
Wear Out	Software does not have energy-related wear.
Reliability prediction	Software reliability cannot be predicted from any physical basis because it depends completely on human factors in design.
Redundancy	We simply cannot improve software reliability if identical software components are used.
Failure cause	Software defects are mainly design defects.
Repairable system concept	Periodic restarts can help fix software problems.
Time dependency and life cycle	Software reliability is not a function of operational time.
Environmental factors	Do not affect software reliability, except it might affect program inputs.
Interfaces	Software interfaces are purely conceptual other than visual.
Failure rate motivators	Usually not predictable from analyses of separate statements.
Built with standard components	Well-understood and extensively tested standard parts will help improve maintainability and reliability. But in software industry, we have not observed this trend. Code reuse has been around for some time, but to a very limited extent. Strictly speaking, there are no standard parts for software, except some standardized logic structures.

reliability estimation, analysis, and usage. Hardware faults are mostly physical faults, whereas software faults are design faults, which are harder to visualize, classify, detect, and correct (Dugan and Lyu, 1995). In software, we can hardly find a strict corresponding counterpart for "manufacturing" as a hardware manufacturing process if the simple action of uploading software modules into place does not count. Therefore, the quality of software will not change once it is uploaded into the storage and start running. Trying to achieve higher reliability by simple redundancy (duplicating the same software modules) will not enhance reliability; it may actually make it worse. Table 14.1 presents a partial list of the distinct characteristics of software compared with hardware is presented in (Keene, 1994) and in Figure 14.1:

All software faults are from design, not manufacturing or wear. Software is not built as an assembly of preexisting components. Off-the-shelf software components do not provide reliability characteristics. Most "reused" software components are modified and are not recertified before reuse. Extending software designs after product deployment is commonplace. Software updates are the preferred avenue for product extensions and customizations. Software updates provide fast development turnaround and have little or no manufacturing or distribution costs.

[3]See Jiantao Pan, at http://www.ece.cmu.edu/~koopman/des_s99/sw_reliability/presentation.pdf.

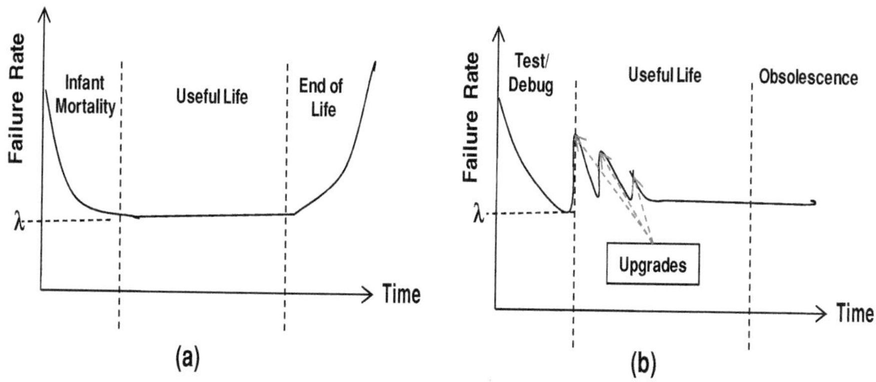

FIGURE 14.1 Bath tub curve for (a) hardware (b) software.[4,5]

As software permeates to every corner of our daily life, software-related problems and the quality of software products can cause serious problems. The defects in software are significantly different than those in hardware and other components of the system; they are usually design defects, and a lot of them are related to problems in specification. The unfeasibility of completely testing a software module complicates the problem because bug-free software cannot be guaranteed for a moderately complex piece of software. No matter how hard we try, a defect-free software product cannot be achieved. Losses caused by software defects cause more and more social and legal concerns. Guaranteeing no known bugs is certainly not an adequate approach to the problem.

Although software reliability is defined as a probabilistic function and comes with the notion of time, we must note that it is different from traditional hardware reliability. Software reliability is not a direct function of time such as electronic and mechanical parts that may age and wear out with time and usage. Software will not rust or wear out during its life cycle and will not change across time unless intentionally changed or upgraded. *Software reliability* can be defined as the probability of failure-free software operation for a specified period of time in a specified environment (Dugan and Lyu, 1995). Software reliability is also an important factor affecting system reliability. It differs from hardware reliability in that it reflects the design perfection, rather than the manufacturing perfection. The high complexity[6] of software is the major contributing factor of software-reliability problems. Because computers and computer systems have become a significant part of our modern society, it is virtually impossible to conduct many day-to-day activities without the aid of computer systems controlled by software. As more reliance is placed on these software systems, it is essential that they operate in a reliable manner. Failure to do so can result in high monetary, property, or human loss.

[4] See Jiantao Pan at http://www.ece.cmu.edu/~koopman/des_s99/sw_reliability/presentation.pdf.
[5] See Jiantao Pan at http://www.ece.cmu.edu/~koopman/des_s99/sw_reliability/.
[6] See software metrics (Chapter 5).

Software reliability as a discipline of software assurance has many attributes: 1) it defines the requirements for software-controlled system fault/failure detection, isolation, and recovery; 2) it reviews the software development processes and products for software error prevention and/or reduced functionality states; and, 3) it defines the process for measuring and analyzing defects and defines/derives the reliability and maintainability factors.

The modeling technique for software reliability is reaching its prosperity, but before using the technique, we must carefully select the appropriate model that can best suit our case. Measurement in software is still in its infancy. No good quantitative methods have been developed to represent software reliability without excessive limitations. Various approaches can be used to improve the reliability of software; however, it is hard to balance development time and budget with software reliability.

This section will provide software DFSS belts with a basic overview of software reliability, tools, and resources on software reliability as a prerequisite for covering DFR.

14.2.1 Basic Software Reliability Concepts

Software reliability is a measure of the software nonconformances that are visible to a customer and prevent a system from delivering essential functionality. Nonconformances can be categorized as:

- *Defects*: A flaw in software requirements, design, or source code that produces unintended or incomplete run-time behavior. This includes defects of commission and defects of omission. Defects of commission are one of the following: Incorrect requirements are specified, requirements are incorrectly translated into a design model, the design is incorrectly translated into source code, and the source code logic is flawed. Defects of omission are one of the following: Not all requirements were used in creating a design model, the source code did not implement all the design, or the source code has missing or incomplete logic. Defects are static and can be detected and removed without executing the source code. Defects that cannot trigger software failures are not tracked or measured for reliability purposes. These are quality defects that affect other aspects of software quality such as soft maintenance defects and defects in test cases or documentation.
- *Faults*: A fault is the result of triggering a software defect by executing the associated source code. Faults are NOT customer-visible. An example is a memory leak or a packet corruption that requires retransmission by the higher layer stack. A fault may be the transitional state that results in a failure. Trivial, simple defects (e.g., display spelling errors) do not have intermediate fault states.
- *Failures*: A failure is a customer (or operational system) observation or detection that is perceived as an unacceptable departure of operation from the designed software behavior. Failures are the visible, run-time symptoms of faults. Failures MUST be observable by the customer or another operational system. Not all

failures result in system outages. Note that for the remainder of this chapter, the term "failure" will refer only to the failure of essential functionality, unless otherwise stated.

There are three types of run-time defects/failures:

1. Defects/failures that are never executed (so they do not trigger faults)
2. Defects/failures that are executed and trigger faults that do NOT result in failures
3. Defects/failures that are executed and trigger faults that result in failures

Typically, we focus solely on defects that have the potential to cause failures by detecting and removing defects that result in failures during development and by implementing fault-tolerance techniques to prevent faults from producing failures or mitigating the effects of the resulting failures. Software fault tolerance is the ability of software to detect and recover from a fault that is happening or already has happened in either the software or hardware in the system where the software is running to provide service in accordance with the specification. Software fault tolerance is a necessary component to construct the next generation of highly available and reliable computing systems from embedded systems to data warehouse systems. Software fault tolerance is not a solution unto itself, however, and it is important to realize that software fault tolerance is just one piece in the design for reliability.

Software reliability is an important attribute of software quality as well as all other abilities such as functionality, usability, performance, serviceability, capability, maintainability, and so on. Software reliability is hard to achieve as complexity increases. It will be hard to reach a certain level of reliability with any system of high complexity. The trend is that system developers tend to push complexity into the software layer with the rapid growth of system size and ease of doing so by upgrading the software. Although the complexity of software is inversely related to software reliability, it is directly related to other important factors in software quality, especially functionality, capability, and so on. Emphasizing these features will tend to add more complexity to software (Rook, 1990).

Across time, hardware exhibits the failure characteristics shown in Figure 14.1(a), known as the bathtub curve.[7] The three phases in a bathtub curve are: infant mortality phase, useful life phase, and end-of-life phase. A detailed discussion about the curve can be found in (Kapur & Lamberson, 1977). Software reliability, however, does not show the same characteristics. A possible curve is shown in Figure 14.1(b) if we depict software reliability on the same axes. There are two major differences between hardware and software bath tub curves: 1) In the last phase, software does not have an increasing failure rate as hardware does because software is approaching obsolescence, and usually there are no motivations for any upgrades or changes to the software. As a result, the failure rate will not change; 2) In the useful-life phase,

[7]The name is derived from the cross-sectional shape of the eponymous device. It does not hold water!

software will experience a drastic increase in failure rate each time an upgrade is made. The failure rate levels off gradually, partly because of the defects found and fixed after the upgrades.[8]

The upgrades in Figure 14.1(b) imply that software reliability increases are a result of feature or functionality upgrades. With functionality upgrading, the complexity of software is likely to increase. Functionality enhancement and bug fixes may be a reason for additional software failures when they develop failure modes of their own. It is possible to incur a drop in software failure rate if the goal of the upgrade is enhancing software reliability, such as a redesign or reimplementation of some modules using better engineering approaches, such as the clean-room method.

More time gives the DFSS team more opportunity to test variations of input and data, but the length of time is not the defining characteristic of complete testing. Consider the software module that controls some machinery. You would want to know whether the hardware would survive long enough. But you also would want to know whether the software has been tested for every usage scenario that seems reasonable and for as many scenarios as possible that are unreasonable but conceivable. The real issue is whether testing demonstrates that the software is fit for its duty and whether testing can make it fail under realizable conditions.

What criteria could better serve software reliability assessment? The answer is that it depends on (Whittaker & Voas, 2000):

- *Software Complexity*[9]: If you are considering a simple text editor, for example, without fancy features like table editing, figure drawing, and macros, then 4,000 hours might be a lot of testing. For modern, feature-rich word processors, 4,000 hours is not a match.
- *Testing Coverage*: If during those 4,000 hours the software sat idle or the same features were tested repeatedly, then more testing is required. If testers ran a nonstop series of intensive, minimally overlapping tests, then release might be justified.
- *Operating Environment*: Reliability models assume (but do not enforce) testing based on an operational profile. Certified reliability is good only for usage that fits that profile. Changing the environment and usage within the profile can cause failure. The operational profile simply is not adequate to guarantee reliability. We propose studying a broader definition of usage to cover all aspects of an application's operating environment, including configuring the hardware and other software systems with which the application interacts.

The contemporary definition of software reliability based on time-in-test assumes that the testers fully understand the application and its complexity. The definition also assumes that teams applied a wide variety of tests in a wide variety of operating conditions and omitted nothing important from the test plan. As Table 14.2 shows,

[8] See Jiantao Pan at http://www.ece.cmu.edu/~koopman/des_s99/sw_reliability/.

[9] See Chapter 5.

TABLE 14.2 Software Reliability Growth Models

Model Name	Formula for Hazard Function	Data and/or Estimation Required	Limitations and Constraints
Musa Basic	$\lambda_0[1 - \mu/\nu_0]$	• Number of detected faults at some time x (μ). • Estimate of λ_0	• Software must be operational. • Assumes no new faults are introduced in correction. • Assumes number of residual faults decreases linearly across time.
Musa Logarithmic	$\lambda_0 \exp(-\phi\mu)$	• Number of detected faults at some time x (μ). • Estimate of λ_0 • Relative change of failure rate over time (ϕ)	• Software must be operational. • Assumes no new faults are introduced in correction. • Assumes number of residual faults decreases exponentially across time.
General Exponential (General form of the Shooman, Jelinski-Moranda, and Keene-Cole exponential models)	$K(E_0 - E_c(x))$	• Number of corrected faults at some time x. • Estimate of E_0	• Software must be operational. • Assumes no new faults are introduced in correction. • Assumes number of residual faults decreases linearly across time.
Littlewood/Verrall	$\dfrac{\alpha}{(t + \Psi(i))}$	• Estimate of α (number of failures) • Estimate of Ψ (reliability growth) • Time between failures detected or the time of the failure occurrence.	• Software must be operational. • Assumes uncertainty in correction process.
Schneidewind model	$\alpha \exp(-\beta\iota)$	• Faults detected in equal interval i	• Software must be operational.

(Continued)

TABLE 14.2 ***(Continued)***

Model Name	Formula for Hazard Function	Data and/or Estimation Required	Limitations and Constraints
		• Estimation of α (failure rate at start of first interval) • Estimation of β(proportionality constant of failure rate over time)	• Assumes no new faults are introduced in correction. • Rate of fault detection decreases exponentially across time.
Duane's model	$\dfrac{\lambda t^b}{t}$	• Time of each failure occurrence • b estimated by $n/\Sigma \ln(t_n+t_i)$from i = 1 to number of detected failures n.	• Software must be operational.
Brook's and Motley's IBM model	Binomial Model Expected number of failures = $\binom{R_i}{n_i} q_i^{n_i}(1-q_i)^{R_i-n_i}$ Poisson model Expected number failures = $\dfrac{(R_i\phi_i)^{n_i}\exp^{-R_i\phi_i}}{n_i!}$	• Number faults remaining at start of ith test (R_i) • Test effort of each test (K_i) • Total number of faults found in each test (n_i) • Probability of fault detection in ith test • Probability of correcting faults without introducing new ones	• Software developed incrementally. • Rate of fault detection assumed constant across time. • Some software modules may have different test effort then others.
Yamada, Ohba, and Osaki's S-Shaped model	$ab_2t\exp^{-bt}$	• Time of each failure detection • Simultaneous solving of a and b	• Software is operational. • Fault detection rate is S shaped across time.
Weibull model	$\text{MTTF} = \dfrac{b}{a}\Gamma\left(\dfrac{1}{a}\right)$	• Total number faults found during each testing interval • The length of each testing interval • Parameter estimation of a and b	• Failure rate can be increasing, decreasing, or constant.

TABLE 14.2 (*Continued*)

Model Name	Formula for Hazard Function	Data and/or Estimation Required	Limitations and Constraints
Geometric model	$\Delta\phi^{i-1}$	• Either time between failure occurrences X_i or the time of the failure occurrence • Estimation of constant Δ, which decreases in geometric progression ($0<\phi<1$) as failures are detected.	• Software is operational. • Inherent number of faults assumed to be infinite. • Faults are independent and unequal in probability of occurrence and severity.
Thompson and Chelson's Bayesian Model	$(f_i + f_0 + 1)/(T_i + T_0)$	• Number of failures detected in each interval (f_i) • Length of testing time for each interval i (T_i)	• Software is corrected at end of testing interval. • Software is operational. • Software is relatively fault free.

most reliability growth equations assume that as time increases, reliability increases, and the failure intensity of the software decreases. Instead of having a reliability theory that makes these assumptions, it would be better to have a reliability measure that actually had these considerations built into it. The notion of time is only peripherally related to testing quality. Software reliability models typically ignore application complexity and test coverage.

Software failures may be a result of errors, ambiguities, oversights, misinterpretations of the specification that the software is supposed to satisfy, carelessness or incompetence in writing code, inadequate testing, incorrect or unexpected usage of the software, or other unforeseen problems (Keiller & Miller, 1991). Reliable software has the following three characteristics:

A. Operates within the reliability specification that satisfies customer expectations. This is measured in terms of failure rate and availability level. The goal is rarely "defect free" or "ultrahigh reliability."
B. "Gracefully" handles erroneous inputs from users, other systems, and transient hardware faults and attempts to prevent state or output data corruption from "erroneous" inputs.
C. Quickly detects, reports, and recovers from software and transient hardware faults. Software provides system behavior as continuously monitoring, self-diagnosing," and "self-healing." It prevents as many run-time faults as possible from becoming system-level failures.

TABLE 14.3 Difference between Software Reliability (A) Trending models and (B) Predictive models[10]

Factor	Trending Models	Predictive Models
Data Source	Uses data from the current software development effort	Uses historical data
Development Cycle Usage	Usually made later in life cycle(after some data have been collected); not typically used in concept or development phases	Usually made prior to development or test phases; can be used as early as concept phase
Time Frame	Estimate reliability at either present or some future time	Predict reliability at some future time

14.2.2 Software Reliability Modeling Techniques

Because software reliability is one of the most important aspects of software quality, reliability engineering approaches are practiced in the software field as well, and software reliability engineering is the quantitative study of the operational behavior of software-based systems with respect to user requirements concerning reliability.

A proliferation of software reliability models have emerged as people try to understand the characteristics of how and why software fails and try to quantify software reliability. Hundreds of models have been developed since the early 1970s, but how to quantify software reliability still remains largely unsolved. Interested readers may refer to Dugan and Lyu (1995). Although there are many models, and many more are emerging, none of the models can capture a satisfying amount of the complexity of software; constraints and assumptions have to be made for the quantifying process. Therefore, there is no single model that can be used in all situations. No model is complete or even representative. One model may work well for a set of certain software but may be completely off track for other kinds of problems. Most software models contain the following parts: assumptions, factors, and a mathematical function that relates the reliability with the factors. The mathematical function is usually higher order exponential or logarithmic.

There are two major categories of reliability modeling techniques: 1) trending techniques and 2) predictive techniques. In practice, reliability trending is more appropriate for software, whereas predictive reliability is more suitable for hardware. Both kinds of modeling techniques are based on observing and accumulating failure data and analyzing with statistical inference. The major differences of the two models are shown in Table 14.3.

A. **Trending reliability models** track the failure data produced by the software system to develop a reliability operational profile of the system during a

[10] See Jiantao Pan at http://www.ece.cmu.edu/~koopman/des_s99/sw_reliability/presentation.pdf.

specified time. Representative estimation models include exponential distribution models, the Weibull distribution model, Thompson and Chelson's model, and so on. Exponential models and the Weibull distribution model usually are named as classical fault count/fault rate estimation models, whereas Thompson and Chelson's model belong to Bayesian fault rate estimation models. Trending reliability can be further classified into four categories:

- *Error Seeding*: Estimates the number of errors in a program by using multistage sampling. Errors are divided into indigenous and induced (seeded) errors. The unknown number of indigenous errors is estimated from the number of induced errors, and the ratio of errors obtained from debugging data.

  This technique simulates a wide variety of anomalies, including programmer faults, human operator errors, and failures of other subsystems (software and hardware) with which the software being tested interacts. For example, seeding programmer faults can be accomplished by testing the stoppage criteria based on test effectiveness. One of the earliest applications of software fault seeding was mutation testing (DeMillo et al., 1978). Mutation testing builds a test suite that can detect all seeded, syntactic program faults. Because there are multiple definitions of what it means to detect all simulated syntactic programmer faults, there are multiple types of mutation testing. Once mutation testing builds the test suite, the suite is used during testing. Seeded programmer errors are nothing more than semantic changes to the code itself. For example, changing $x = x - 1$ to $x = x + 1$ is a seeded fault. By making such modifications, the DFSS team can develop a set of test cases that distinguish these mutant programs from the original. The hypothesis is that test cases that are good at detecting hypothetical (seeded) errors are more likely to be good at detecting real errors. Using error seeding to measure test effectiveness, the team needs to:

  1. Build test suites based on the effectiveness of test cases to reveal the seeded errors.
  2. Use the test cases to test for real faults.

  Just as all test cases are not equally effective for fault detection, not all seeded faults are of equal value. This brings us to the notion of fault size. The size of a real fault (or seeded fault) is simply the number of test cases needed to detect the fault. When we inject a large number of errors, most test cases can catch it. Therefore, it is more beneficial to inject a smaller number of errors and create a test suite that reveals them. Small errors are harder to detect, and 10 test cases that detect tiny faults are more valuable than a 20-member test suite that catches only huge errors. A test that detects small errors almost certainly will detect huge errors. The reverse is not necessarily true.

- *Failure Rate*: Is used to study the program failure rate per fault at the failure intervals. As the number of remaining faults change, the failure rate of the program changes accordingly.

- *Curve Fitting*: Uses statistical regression analysis to study the relationship between software complexity and the number of faults in a program as well as the number of changes, or failure rate.
- *Reliability Growth*: Measures and predicts the improvement of reliability programs through the testing process. Reliability growth also represents the reliability or failure rate of a system as a function of time or the number of test cases. Reliability growth for software is the positive improvement of software reliability across time and is accomplished through the systematic removal of software faults. The rate at which the reliability grows depends on how fast faults can be uncovered and removed. A software reliability growth model allows project management to track the progress of the software's reliability through statistical inference and to make projections of future milestones.

  If the assessed growth falls short of the planned growth, then management will have sufficient notice to develop new strategies, such as the reassignment of resources to attack identified problem areas, adjustment of the project time frame, and reexamination of the feasibility or validity of requirements.

  Measuring and projecting software reliability growth requires the use of an appropriate software reliability model that describes the variation of software reliability with time. The parameters of the model can be obtained either from prediction performed during the period preceding the system test or from estimation performed during the system test. Parameter estimation is based on the times at which failures occur.

  The use of a software reliability growth testing procedure to improve the reliability of a software system to a defined reliability goal implies that a systematic methodology will be followed for a significant duration. To perform software reliability estimation, a large sample of data must be generated to determine statistically, with a reasonable degree of confidence, that a trend has been established and is meaningful. Commonly used reliability growth models are listed in Table 14.2. It is recommended that the leader familiarize himself (herself) with basic reliability modeling mathematics in Appendix 14.A. The mathematics of a hazard function can be explained best using a bathtub curve.

B. **Predictive reliability** models assign probabilities to the operational profile of a software system; for example, the system has a 10% chance of failure during the next 120 operational hours. Representative prediction models include Musa's execution time model (Musa, 1975), Putnam's model (Putnam & Ware, 2003), and Rome laboratory models TR-92-51 and TR-92-15, and so on. Using prediction models, software reliability can be predicted early in the development phase, and enhancements can be initiated to improve the reliability.

The software reliability field has matured to the point that software models can be applied in practical situations and give meaningful results and that there is no one model that is best in all situations. Because of the complexity of software, any model has to have extra assumptions. Only limited factors can be put into consideration. Most software reliability models ignore the software development process and focus on the results—the observed faults and/or failures. By doing so, complexity is reduced and abstraction is achieved; however, the models tend to specialize to be applied to only a portion of the situations and to a certain class of the problems. We have to choose carefully the right model that suits our specific case. Furthermore, the modeling results cannot be blindly adopted.

Design for Six Sigma methods (such as Axiomatic Design, Fault Tree Analysis, FMEA, etc.) largely can improve software reliability. Before the deployment of software products, testing, verification, and validation are necessary steps. Software testing is used heavily to trigger, locate, and remove software defects. Software testing is still in its infant stage; testing is crafted to suit specific needs in various software development projects in an ad hoc manner. Various analysis tools such as trend analysis, fault-tree analysis, Orthogonal defect classification and formal methods, and so on, also can be used to minimize the possibility of defect occurrence after release and, therefore, improve software reliability.

After deployment of the software product, field data can be gathered and analyzed to study the behavior of software defects. Fault tolerance or fault/failure forecasting techniques will be helpful techniques and will guide rules to minimize fault occurrence or impact of the fault on the system.

14.2.3 Software Reliability Measurement and Metrics[11,12]

Measurement is a process to collect data for tracking or to calculate meta-data (metrics) such as defect counts. Metrics are variables with information derived from measurements (metadata) such as failure rate, defect removal efficiency, and defect density. Reliability measurements and metrics accomplish several goals:

- Provide estimates of software reliability prior to customer deployment.
- Track reliability growth throughout the life cycle of a release.
- Identify defect clusters based on code sections with frequent fixes.
- Determine where to focus improvements based on analysis of failure data.

Tools for software configuration management and defect tracking should be updated to facilitate the automatic tracking of this information. They should allow for data entry in all phases, including development. Also, they should distinguish code-based updates for critical defect repair versus any other changes, (e.g., enhancements, minor defect repairs, coding standards updates, etc.).

Measurement is commonplace in other engineering field but not in software, though the quest of quantifying software reliability has never ceased. Measuring

[11] See Chapter 5 for software metrics.

[12] See Rosenberg et al. (1998).

software reliability remains a difficult problem because we do not have a good understanding of the nature of software. There is no clear definition to what aspects are related to software reliability. We cannot find a suitable way to measure software reliability and most of the aspects related to software reliability. Even the most obvious product metrics such as software size have no uniform definition. The next good thing is to measure something related to reliability to reflect the characteristics if we cannot measure reliability directly.

Software reliability metrics can be categorized as static code and dynamic metrics as follows:

A. *Static Code Metrics*: Software size is thought to be reflective of complexity, development effort, and reliability. Lines of code (LOC), or LOC in thousands (KLOC), is an intuitive initial approach to measure software size. But there is no standard way of counting. Typically, source code is used (SLOC, KSLOC), and comments and other nonexecutable statements are not counted. This method cannot faithfully compare software not written in the same language. The advent of new technologies of code reuses and code generation techniques also cast doubt on this simple method. Test coverage metrics are estimate fault and reliability by performing tests on software products based on the assumption that software reliability is a function of the portion of software that successfully has been verified or tested.

   The static code metric is divided into three categories with measurements under each: Line count, complexity and structure, and object-oriented metrics.

   - Line count:
     - Lines of code
     - Source Lines of code
   - Complexity and structure: Complexity is related directly to software reliability, so representing complexity is important. Complexity-oriented metrics is a method of determining the complexity of a program's control structure by simplifying the code into a graphical representation.
     - Cyclomatic complexity
     - Number of modules
     - Number of go-to statements
   - Object-oriented: Object-oriented functional point metrics is a method of measuring the functionality of a proposed software development based on a count of inputs, outputs, master files, inquires, and interfaces. The method can be used to estimate the size of a software system as soon as these functions can be identified. It is a measure of the functional complexity of the program. It measures the functionality delivered to the user and is independent of the programming language. It is used primarily for business systems; it is not proven in scientific or real-time applications.
     - Number of classes
     - Weighted methods per class

- Coupling between objects
- Response for a class
- Number of child classes
- Depth of inheritance tree

B. *Dynamic Metrics*: The dynamic metric has two major measurements: failure rate data and problem reports. The goal of collecting fault and failure metrics is to be able to determine when the software is approaching failure-free execution. Minimally, both the number of faults found during testing (i.e., before delivery) and the failures (or other problems) reported by users after delivery are collected, summarized, and analyzed to achieve this goal. Test strategy is highly relative to the effectiveness of fault metrics because if the testing scenario does not cover the full functionality of the software, then the software may pass all tests and yet be prone to failure once delivered. Usually, failure metrics are based on customer information regarding failures found after release of the software. The failure data collected, therefore, is used to calculate failure density, mean time between failures (MTBF), or other parameters to measure or predict software reliability.

14.2.4 DFR in Software DFSS

In the context of DFR, we found the following practices are dominating the software industry, in particular for software products containing embedded code:

- *Quality through testing*: Quality through software testing is the most prevalent approach for implementing software reliability within small or unstructured development companies. This approach assumes that reliability can be increased by expanding the types of system tests (e.g., integration, performance, and loading) and increasing the duration of testing. Software reliability is measured by various methods of defect counting and classification. Generally, these approaches fail to achieve their software reliability targets.
- *Traditional reliability programs*: Traditional software reliability programs treat the development process as a software-generating black box. Predictive models are generated, usually by a separate team of reliability engineers, to provide estimates of the number of faults in the resulting software; greater consistency in reliability leads to increased accuracy in the output modeling. Within the black box, a combination of reliability techniques like failure analysis, (e.g., Failure Mode and Effects Analysis [FMEAs], Fault Tree Analysis [FTAs], defect tracking, and operational profile testing) are used to identify defects and produce software reliability metrics.
- *Process control*: Process control assumes a correlation between process maturity and latent defect density in the final software. Companies implementing Capability Maturity Model (CMM) Level 3 processes generate software

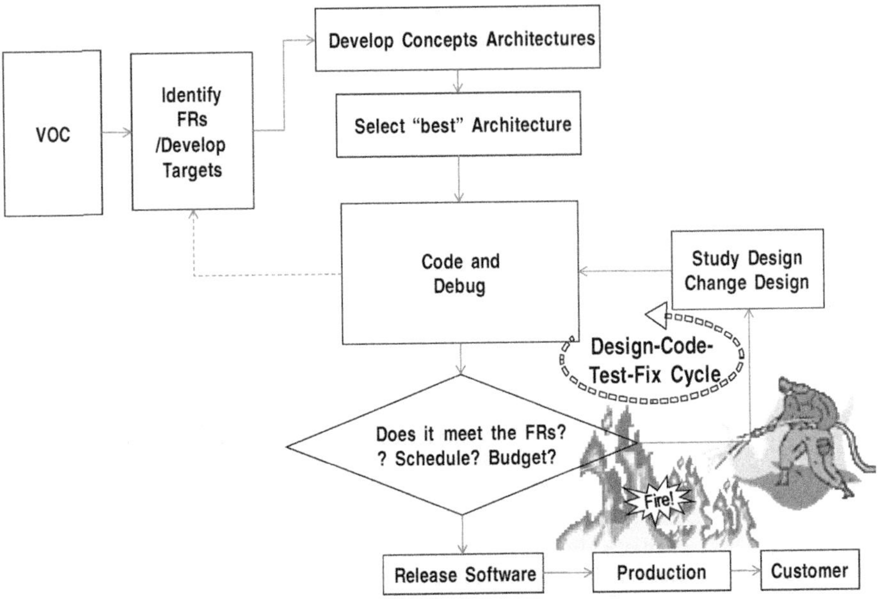

FIGURE 14.2 Software design-code-test-fix cycle.

containing 2.0–3.5 faults per KSLOC.[13] If the current process level does not yield the desired software reliability, then audits and stricter process controls are implemented.

None of these industry "best practices" are before the fact, leading the team and, hence, their home organization to spend their time, effort, and valuable resources fixing what they already designed as depicted in Figure 14.2. The team will assume the role of fire-fighters switching from their prescribed design team. In these practices, software design teams find that their software engineers spend more time debugging than designing or coding, and accurate software reliability measurements are not available at deployment to share with customers. We recommend that the DFSS team assess their internal development practices against industry best practices to ensure they have a solid foundation upon which to integrate DFR. To do so in a DFSS environment, it will be helpful for a software DFSS team to fill in gaps by identifying existing internal best practices and tools to yield the desired results and integrating them with the DFSS road map presented in Chapter 11. A set of reliability practices to move defect prevention and detection as far upstream of the development cycle as possible is always the target.

Reliability is a broad term that focuses on the ability of software to perform its intended function. Mathematically speaking, assuming that software is performing

[13]KSLOC = 1,000 source lines of code.

TABLE 14.4 Defect Removal Techniques Efficiency

Defect Removal Technique	Efficiency Range
Design Inspection	45%–60%
Code Inspection	45%–60%
Unit Testing	15%–45%
Regression test	15%–30%
Integration Test	25%–40%
Performance Test	20%–40%
System Testing	25%–55%
Acceptance Test	25%–35%

its intended function at time equals zero, reliability can be defined as the probability that the software will continue to perform its intended function without failure for a specified period of time under stated conditions.

Though the software has a reliable design, it is effectively unreliable when fielded, which is actually the result of a substandard development process. Evaluating and finding ways to attain high reliability are all aspects of software reliability.

The best option for software design for reliability is to optimize the returns from software development "best practices." Table 14.4[14] shows the difference in defect removal efficiency between inspections and testing.

Most commercial companies do not measure defect removal in pretesting phases. This leads to inspections that provide very few benefits. Unstructured inspections result in weak results. Software belts simply do not know how to apply effectively their efforts as reviewers to find defects that will lead to run-time failures. Inspection results are increased by incorporating prevalent defect checklists based on historical data and assigning reviewer perspectives to focus on vulnerable sections of designs and code. By performing analysis techniques, such as failure analysis, static code analysis, and maintenance reviews for coding standards compliance and complexity assessments, code inspections become smaller in scope and uncover more defects. Once inspection results are optimized, the combined defect removal results with format testing and software quality assurance processes have the potential to remove up to 99% of all inherent defects.

By redirecting their efforts upstream, most development organizations will see greater improvements in software reliability with investments in design and code inspections than further investments in testing (Table 14.5[15]).

Software DFR practices increase confidence even before the software is executed. The view of defect detection changes from relying solely on the test phases and customer usage to one of phase containment across all development phases. By

[14]See Silverman and De La Fuente, http://www.opsalacarte.com/pdfs/Tech_Papers/Software_Design_for_Reliability_-_Paper.pdf.

[15]See Silverman and De La Fuente, http://www.opsalacarte.com/pdfs/Tech_Papers/Software_Design_for_Reliability_-_Paper.pdf.

TABLE 14.5 Best-In-Class Defect Removal Efficiency

Application Type	"Best in Class" Defect Removal Efficiency
Outsourced software	92%
IT software	73%
Commercial software	90%
System software	94%
Military software	96%
Web software	72%
Embedded software	95%

measuring phase containment of defects, measurements can be collected to show the separation between defect insertion and discovery phases.[16]

14.2.4.1 DFSS Identify Phase DFR Practices. The requirements for software reliability are to: identify important software functionality, including essential, critical, and nonessential; explicitly define acceptable software failure rates; and specify any behavior that impacts software availability (see Section 14.3). We must define acceptable durations for software upgrades, reboots, and restarts, and we must define any operating cycles that apply to the system and the software to define opportunity for software restarts or rejuvenation such as maintenance or diagnostic periods, off-line periods, and shutdown periods.

In this identity, conceptualize, optimize, and verify/validate (ICOV) DFSS phase, the software team should define system-level reliability and availability software goals, which are different from hardware goals. These goals become part of the project reliability and integration plan and are applied to the conceptualize and optimize phases. The two major activities in this phase are:

- Software reliability goal setting
- Software reliability program and integration plan

14.2.4.2 DFSS Conceptualize Phase DFR Practices. The reliability engineering activity should be an ongoing process starting at the conceptualize phase of a DFSS-design project and continuing throughout all phases of a device life cycle. The goal always needs to be to identify potential reliability problems as early as possible in the device life cycle. Although it may never be too late to improve the reliability of software, changes to a design are orders of magnitude less expensive in the early part of a design phase rather than once the product is released.

A reliability prediction can be performed in the conceptualize DFSS phase to "ballpark" the expected reliability of the software. A reliability prediction is simply

[16] See Silverman and De La Fuente, http://www.opsalacarte.com/pdfs/Tech_Papers/Software_Design_for_Reliability_-_Paper.pdf.

the analysis of parts and components (e.g., objects and classes) in an effort to predict and calculate the rate at which an item will fail. A reliability prediction is one of the most common forms of reliability analyses for calculating failure rate and MTBF. If a critical failure is identified, then a reliability block diagram analysis can be used to see whether redundancy should be considered to mitigate the effect of a single-point failure. A reliable design should anticipate all that can go wrong. We view DFR as a means to maintain and sustain the Six Sigma capability across time.

The software designs should evolve using a multitiered approach such as[17]:

- System architecture[18]: Identify all essential system-level functionality that requires software and identify the role of software in detecting and handling hardware failure modes by performing system-level failure mode analysis. These can be obtained from quality function deployment (QFD) and axiomatic design deployments.
- High-level design: Identify modules based on their functional importance and vulnerability to failures. Essential functionality is executed most frequently. Critical functionality is executed infrequently but implements key system operations (e.g., boot or restart, shutdown, backup, etc.). Vulnerability points are points that might flag defect clusters (e.g., synchronization points, hardware and module interfaces, initialization and restart code, etc.). Identify the visibility and access major data objects outside of each module.
- Low-level design: Define the availability behavior of the modules (e.g., restarts, retries, reboots, redundancy, etc.). Identify vulnerable sections of functionality in detail.

Functionality is targeted for fault-tolerance techniques. Focus on simple implementations and recovery actions. For software DFSS belts, the highest return on investment (ROI) for defect and failure detection and removal is low-level design. It defines sufficient module logic and flow-control details to allow analysis based on common failure categories and vulnerable portions of the design. Failure handling behavior can be examined in sufficient detail. Low-level design bridges the gap between traditional design specs and source code. Most design defects that were caught previously during code reviews now will be caught in the low-level Design review. We are more likely to fix correctly design defects because the defect is caught in the conceptualize phase. Most design defects found after this phase are not fixed properly because the scheduling costs are too high. Design defects require returning to the design phase to correct and review the design and then correcting, rereviewing, and unit testing the code! Low-level design also can be reviewed for testability. The goal

[17] See Silverman and De La Fuente, http://www.opsalacarte.com/pdfs/Tech_Papers/Software_Design_for_Reliability_-_Paper.pdf.

[18] See Chapters 4 and 13.

of defining and validating all system test cases as part of a low-level design review is achievable.[19]

In this ICOV DFSS phase, team predesign review meetings provide members with forums to expand their knowledge base of DFSS design techniques by exchanging design templates. Design review results will be greatly improved if they are preceded by brief, informal reviews that are highly interactive at multiple points throughout the progression from system architecture through low-level design. Prior to the final stage of this phase, software failure analysis is used to identify core and vulnerable sections of the software that may benefit from additional runtime protection by incorporating software fault-tolerance techniques. The major activities in this phase are:

- Inclusion of DFR in team design reviews
- Software failure analysis
- Software fault tolerance

14.2.4.3 DFSS Optimize Phase DFR Practices. Code reviews should be carried out in stages to remove the most defects. Properly focused design reviews coupled with techniques to detect simple coding defects will result in shorter code reviews. Code reviews should focus on implementation issues and not design issues. Language defects can be detected with static and with dynamic analysis tools.

Maintenance defects that are caught with coding standards prereviews in which authors review their own code significantly reduces simple code defects and possible areas of code complexity. The inspection portion of a review tries to identify missing exception handling points. Software failure analysis will focus on the robustness of the exception handling behavior. Software failure analysis should be performed as a separate code inspection once the code has undergone initial testing.

In this ICOV DFSS phase, reliability reviews target only the core and vulnerable sections of code to allow the owner of the source code to develop sufficient synergy with a small team of developers in finding defects. Unit testing efforts focus on efficient detection of software faults using robustness and coverage testing techniques for thorough module-level testing. The major activities in this phase are:

- Code reliability reviews
- Software robustness (Chapter 18)
- Coverage testing techniques

14.2.4.4 DFSS Verify and Validate Phase DFR Practices. Unit testing can be driven effectively using code coverage techniques. It allows software belts to define and execute unit testing adequacy requirements in a manner that is meaningful and easily measured. Coverage requirements can vary based on the critical nature of

[19] See Silverman and De La Fuente, http://www.opsalacarte.com/pdfs/Tech_Papers/Software_Design_for_Reliability_-_Paper.pdf.

a module. System-level testing should measure reliability and validate as many customer operational profiles as possible. It requires that most of the failure detection be performed prior to system testing. System integration becomes the general functional validation phase.[20]

In this ICOV DFSS phase, reliability measurements and metrics are used to track the number of remaining software defects, the software mean time to failure (MTTF), and to anticipate when the software is ready for deployment. The test engineers in the DFSS team can apply usage profiling mechanisms to emphasize test cases based on their anticipated frequency of execution in the field. The major activities in this phase are:

- Software reliability measurements (after metrics definition)
- Usage of profile-based testing
- Software reliability estimation techniques
- Software reliability demonstration tests

14.3 SOFTWARE AVAILABILITY

Reliability analysis concerns itself with quantifying and improving the reliability of a system. There are many aspects to reliability, and the reliability profile of one system may be quite different from that of another. Two major aspects of reliability that are common to all systems are *availability* (i.e., the proportion of time that all critical functions are available) and *reliability* (i.e., no transaction or critical data be lost or corrupted). These two characteristics are independent. A system may have a very high availability, but transactions may be lost or corrupted during the unlikely occurrence of a failure. However, a system may never lose a transaction but might be down often.

Let us make the following definitions:

MTBF = Mean time between failures (MTBF), or uptime.
MTTR = mean time to repair the system (MTR), or downtime.
A = software system availability.

System availability is the proportion of time that the system is up. Because the system only can be either up or down, the following is true:

$$A = \frac{MTBF}{MTBF + MTTR} = \frac{1}{1 + \dfrac{MTTR}{MTBF}} \qquad (14.1)$$

[20]See Silverman and De La Fuente, http://www.opsalacarte.com/pdfs/Tech_Papers/Software_Design_for_Reliability_-_Paper.pdf.

If n systems with availability A_1, A, $A_3 \ldots A_n$ must operate for the system to be up, then the combined system availability is

$$A = \prod_{i=1}^{n} A_i \tag{14.2}$$

For a software of two components with availabilities A_1 and A_2, the probability that either system has failed is $(1 - A_1)$ or $(1 - A_2)$. If either system must operate, but not both, then the combined system has failed only if both systems have failed. This will occur with probability $(1 - A_1)(1 - A_2)$, and the combined system availability is, therefore,

$$A = 1 - (1 - A_1)(1 - A_2) \tag{14.3}$$

Equation (14.3) is a generalization for a system of n components

$$A = 1 - \prod_{i=1}^{n} (1 - A_i) \tag{14.4}$$

14.4 SOFTWARE DESIGN FOR TESTABILITY

Design for test is a name for design techniques that add certain testability features to a hardware product design (Design for Test, 2009). Testability means having reliable and convenient interfaces to drive the execution and verification of tests (Pettichord, 2002). IEEE defines software testability as "the degree to which a system or component facilitates the establishment of test criteria and the performance of tests to determine whether those criteria have been met."(ANSI/IEEE Std 610.12-1990).

Pettichord (2000) identified three keys to system test automation:

1. A constructive relationship between testers and developers,
2. A commitment by the whole team for successful automation, and
3. A chance for testers to engage early in the product development cycle.

Testability should be integrated in the DFSS design process instead of dealing with it after the product has been designed. Pettichord (2002) stated that "it is more efficient to build testability support into a product than to construct the external scaffolding that automation would otherwise require." To incorporate testability in a design, the following recommendations need to be followed:

1. Cooperation from developers and customers to add testability features. This means that testable features should be expected in the design requirements,
2. A realization that testability issues blocking automation warrant attention from the whole team, and

3. A chance to uncover these challenges early when the product is still open for design changes, which means that testability must be included in every phase of the software design cycle.

Design for testability for a system design has many advantages including the following:

1. Makes the design easier to develop,
2. Allows the application of manufacturing tests for the design, which are used to validate that the product hardware contains no defects,
3. Facilitate for usability, so that a testable component of a testable design may be reused in another system design,
4. Possible cost reduction of the product,
5. Allows the manufacturer to use efficiently its design engineers, and
6. Reduces time-to-market for the product

Tests are applied at several steps in the hardware manufacturing flow and, for certain products, also may be used for hardware maintenance in the customer's environment (Design for Test, 2009). A software product is testable if it supports acceptable criteria and evaluation of performance. For a software product to have this software quality, the design must not be complex.

$\mu_{\text{Testability}}$ can be used to measure the testability of a system design as was presented in Chapter 1. $\mu_{\text{Testability}}$ can be interpreted as follows:

$\mu_{\text{Testability}} = 0$ and that the system design is not testable at all,

$\mu_{\text{Testability}} = 1$ and that the system design is fully testable; otherwise the system is partially testable with a membership (confident value) equal to $\mu_{\text{Testability}}$.

14.5 DESIGN FOR REUSABILITY

In any system design, reusability can be defined as the likelihood of using a segment of source code or a hardware module again to a new system design with slight or no modification. Reusable modules and classes or hardware units reduce implementation time (Reusability, 2010), increase the likelihood that prior testing and use has eliminated bugs, and localizes code modifications when a change in implementation is required. Hardware description languages (HDLs) commonly are used to build complex designs using simple designs. The HDLs allow the creation of reusable models, but the reusability of a design does not come with language features alone. It requires design disciplines to reach an efficient reusable design (Chang & Agun, 2000).

For software systems, subroutines or functions are the simplest form of reuse. A chunk of code is organized regularly using modules or namespaces layers. Proponents claim that objects and software components offer a more advanced form of reusability, although it has been tough to measure objectively and to define levels or scores of reusability (Reusability, 2010). The ability to reuse software modules or hardware

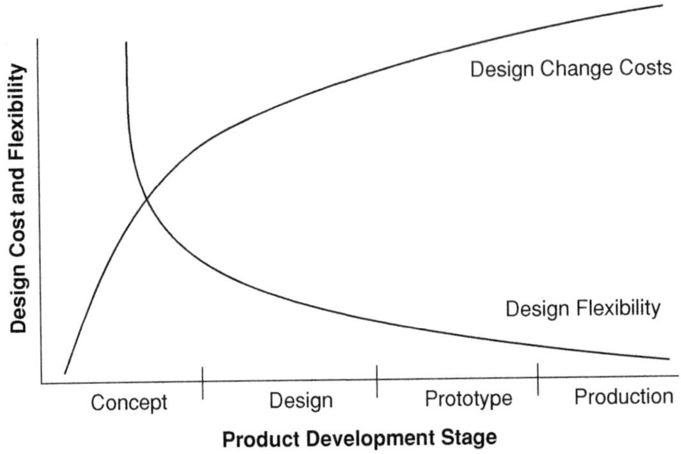

FIGURE 14.3 Product phase versus product costs/flexibility.

components depends mainly on the ability to build larger things from smaller parts and the ability to identify commonalities among those parts.

There are many attributes to good system design, even if we only concentrate on issues involving implementation. Reusability often involves a longer term because it concerns productivity (Biddle & Temper, 1998). The reuse of hardware units can improve the productivity in system design. However, without careful planning, units rarely are designed for reuse (Chang & Agun, 2000).

Reusability is a required characteristic for a successful manufacturing product and often should be included in the DFSS design process. Reusability brings several aspects to software development that does not need to be considered when reusability is not required (Reusability, 2010).

14.6 DESIGN FOR MAINTAINABILITY

Maintainability is to provide updates to satisfy new requirements. A maintainable software product should be well documented, and it should not be complex. A maintainable software product should have spare capacity of memory storage and processor utilization and other resources.

Maintainability is the degree to which the system design can be maintained or repaired easily, economically, and efficiently. Some maintainability objects can be as follows[21]:

- Identify and prioritize maintenance requirements.
- Increase product availability and decrease maintenance time.

[21] See http://www.theriac.org/DeskReference/viewDocument.php?id=222.

TABLE 14.6 Design for Maintainability Features/Benefits Matrix

Design for Maintainability Benefits	Design for Maintainability Features
Easy access to serviceable items	• Maintenance time and costs reduced • Product availability increases • Technician fatigue/injury reduced
No or minimal adjustment	• Maintenance time and costs reduced • Product availability increases • Maintenance training curve reduced
Components/modules quick and easy to replace	• Technician fatigue/injury reduced • Product availability increases • Problem identification improves
Mistake proofing, part/module installs one way only	• Probability of damage to the part or product reduced • Reliability improves • Maintenance training curve reduced
Self-diagnostics or built in test or indicators to find problems quickly	• Maintenance time and costs reduced • Product availability increases • Customer satisfaction improves
No or few special hand tools	• Maintenance investment reduced • Customer satisfaction improves • Tool crib inventory reduced
Standard fasteners and components	• Number of spare parts in inventory reduced • Product cost reduced • Maintenance time and costs reduced
Reduce number of components in final assembly	• Product cost reduced • Reliability improves • Spare parts inventory reduced

- Increase customer satisfaction.
- Decrease logistics burden and life cycle costs.

The effectiveness of a design for maintainability strategy can be measured using maintenance metrics and industry benchmarks. Fuzzy logic can be used to measure system design maintainability. The membership function for measuring the software quality with respect to maintainability is presented in Chapter 1.

The design for maintainability should be considered early when flexibility is high and design change costs are low. The design flexibility is the greatest in the conceptual stage of the product, and design change costs are the lowest as show in Figure 14.3.[22]

Maintainability features should be considered as early as possible in the DFSS design process. Maintainability may increase the cost during the design phase, but it should reduce the end users maintenance costs throughout the product's life. Table 14.6[23] lists typical design for maintainability features used in the product development stage and the benefits these features provide to the designer and the customer.

A system design that has the maintainability feature can reduce or eliminate maintenance costs, reduce downtime, and improve safety.

APPENDIX 14.A

Reliability engineering is an engineering field that deals with the study of reliability, of the ability of a system or component to perform its required functions under stated conditions for a specified period of time. It often is reported in terms of a probability. Mathematically reliability *R(t)* is the probability that a system will be successful in the interval from time 0 to time t:

$$R(t) = P(t > T) = \int_{T}^{\infty} f(t)dt, t \geq 0 \qquad (14.A.1)$$

where T is a random variable denoting the time-to-failure or failure time, *f(t)* is the failure probability density function, and t is the length of time (which is assumed to start from time zero).

The unreliability *F(t)*, a measure of failure, is defined as the probability that the system will fail by time t.

$$F(t) = P(t \leq T), \quad t \geq 0 \qquad (14.A.2)$$

In other words, *F(t)* is the failure distribution function. The following relationship applies to reliability in general. Reliability *R(t)*, is related to failure probability *F(t)* by:

$$R(t) = 1 - F(t) \qquad (14.A.3)$$

We note the following four key elements of reliability (Reliability Engineering, 2010):

1. Reliability is a probability. This means that failure is regarded as a random phenomenon; it is a recurring event, and we do not express any information

[22] See http://www.theriac.org/DeskReference/viewDocument.php?id=222.
[23] See http://www.theriac.org/DeskReference/viewDocument.php?id=222.

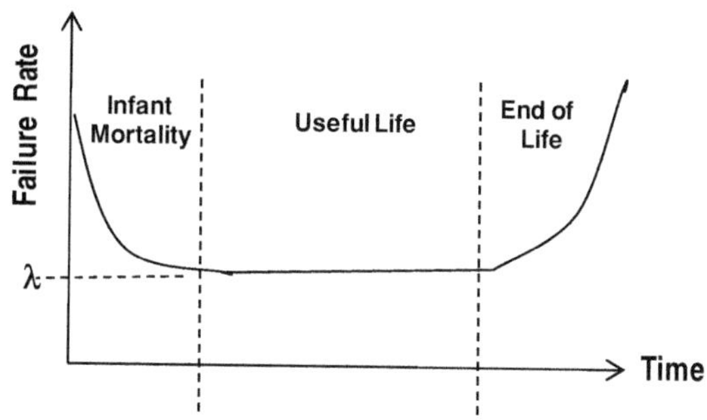

FIGURE 14.A.1 Bath tub curve.

on individual failures, the causes of failures, or relationships between failures, except that the likelihood for failures to occur varies across time according to the given probability function. Reliability engineering is concerned with meeting the specified probability of success at a specified statistical confidence level.

2. Reliability is predicated on "intended function;" generally, this is taken to mean operation without failure. However, even if no individual part of the system fails, but the system as a whole does not do what was intended, then it is still charged against the system reliability. The system requirements specification is the criterion against which reliability is measured.
3. Reliability applies to a specified period of time. In practical terms, this means that a system has a specified chance that it will operate without failure before time. Reliability engineering ensures that components and materials will meet the requirements during the specified time. Units other than time sometimes may be used. The automotive industry might specify reliability in terms of miles; the military might specify reliability of a gun for a certain number of rounds fired. A piece of mechanical equipment may have a reliability rating value in terms of cycles of use.
4. Reliability is restricted to operation under stated conditions. This constraint is necessary because it is impossible to design a system for unlimited conditions. A Mars Rover will have different specified conditions than the family car. The operating environment must be addressed during design and testing. Also, that same rover may be required to operate in varying conditions requiring additional scrutiny.

The bath tub curve (Figure 14.A.1) is used widely in reliability engineering. It describes a particular form of the hazard function that comprises three phases:

1. The first phase is a decreasing failure rate, known as early failures.
2. The second phase is a constant failure rate, known as random failures.
3. The third phase is an increasing failure rate, known as wear-out failures.

The bath tub curve is generated by mapping the rate of early "infant mortality" failures when first introduced, the rate of random failures with a constant failure rate during its "useful life," and finally the rate of "wear out" failures as the product exceeds its design lifetime.

In less technical terms, in the early life of a product adhering to the bath tub curve, the failure rate is high but rapidly decreasing as defective products are identified and discarded, and early sources of potential failure such as handling and installation error are surmounted. In the mid-life of a product—generally, once it reaches consumers—the failure rate is low and constant. In the late life of the product, the failure rate increases as age and wear take their toll on the product. Many consumer products strongly reflect the bath tub curve, such as computer processors.

For hardware, the bath tub curve often is modeled by a piecewise set of three hazard functions:

$$h(t) = \begin{cases} c_o - c_1 t + \lambda & 0 \leq t \leq c_o/c_1 \\ \lambda & c_o/c_1 < t \leq t_o \\ c_2(t - t_o) + \lambda & t_o < t \end{cases} \tag{14.A.4}$$

For software, you can replace the piecewise approximation by the applicable hazard function from Table 14.2: Software Reliability Growth Models and in light of Figure 14.1.

REFERENCES

Biddle, Robert and Temperd, Ewan (1998), "Evaluating Design by Reusability," Victoria University of Wellington, New Zealand. http://www.mcs.vuw.ac.nz/research/design1/1998/submissions/biddle/.

DeMillo, R.A., Lipton, R.J., and Sayward, F.G. (1978), "Hints on test data selection: Help for the practicing programmer." *Computer*, Volume *11*, #4, pp. 34–41.

Design for Test (2009), *Wikipedia, the Free Encyclopedia*. http://en.wikipedia.org/wiki/Design_for_Test.

Dugan J.B. and Lyu, M.R. (1995), "*Dependability Modeling for Fault-Tolerant Software and Systems*," in Software Fault Tolerance, Wiley & Sons, pp. 109–137.

ANSI/IEEE Std. 610.12-1990 (1990), "Standard Glossary of Software Engineering Terminology," IEEE, Washington, DC.

Kapur, K.C. and Lamberson, L.R. (1977), "*Reliability In Engineering Design*," John Wiley & Sons, Inc., New York

Keene, S. and Cole, G.F. (1994), Reliability growth of fielded software, *Reliability Review*, Volume *14*, pp. 5–26.

Keiller, P. and Miller, D. (1991), "On the use and the performance of software reliability growth models." *Software Reliability and Safety*, pp. 95–117.

Morris, Chang and Agun, Kagan (2000), "On Design-for-Reusability in Hardware Description Languages," VLSI Proceedings, IEEE Computer Society Workshop, Apr.

Musa, J.D. (1975), "A theory of software reliability and its application." *IEEE Transactions on Software Engineering*, Volume *1*, #3, pp. 312–327.

Musa, J.D. et al. (1987), *Software Reliability Measurement Prediction Application*, McGraw Hill, New York.

Petticord, Bret (2000), Three Keys to Test Automation, *Stickyminds.com.* http://www.stickyminds.com/sitewide.asp?ObjectID=2084&ObjectType=COL&Function=edetail.

Petticord, Bret (2002), "Design for Testability," Pacific Northwest Software Quality Conference, Portland, OR, Oct.

Putnam, L. and Ware, M. (2003), *Five Core Metrics: The Intelligence Behind Successful Software Management*, Dorset House Publishing, Dorset, VT.

Reusability (2010), *Wikipedia, the Free Encyclopedia.* http://en.wikipedia.org/wiki/Reusability.

Rook, P. (1990), *Software Reliability Handbook*, Centre for Software Reliability, City University, London, UK.

Rosenberg, L., Hammer, T., and Shaw, J. (1998), "Software metrics and reliability," 9th International Symposium, November, Germany.

Software Quality (2010), *Wikipedia, the Free Encyclopedia.* http://en.wikipedia.org/wiki/Software_quality.

Whittaker, J.A. and Voas, J. (2000), "Toward a more reliable theory of software reliability." *IEEE Computer*, Volume *33*, #12, pp. 36–42.

Yang, K. and El-Haik, Basem (2008), "*Design for Six Sigma: A Roadmap for Product Development*," 2nd Ed., McGraw-Hill Professional, New York.

APPENDIX REFERENCE

Reliability Engineering (2010), *Wikipedia, the Free Encyclopedia.* http://en.wikipedia.org/wiki/Reliability_engineering.

BIBLIOGRAPHY

Fitzgerald, Andy (2001), "Design for Maintainability." http://www.theriac.org/DeskReference/viewDocument.php?id=222.

CHAPTER 15

SOFTWARE DESIGN FOR SIX SIGMA (DFSS) RISK MANAGEMENT PROCESS

15.1 INTRODUCTION

Risk management is an activity that spans all identify, conceptualize, optimize, and verify/validate Design for Six Sigma (ICOV DFSS) phases. Computers and, therefore, software are introduced into applications for the many advantages that they provide. It is what lets us get cash from an automated teller machine (ATM), make a phone call, and drive our cars. A typical cell phone now contains 2 million lines of software code; by 2010 it likely will have 10 times as many. General Motors Corporation (Detroit, MI) estimates that by then its cars will each have 100 million lines of code. But these advantages do not come without a price. The price is the risk that the computer system brings with it. In addition to providing several advantages, the increased risk has the potential for decreasing the reliability and, therefore, the quality of the overall system. This can be dangerous in safety-critical systems where incorrect computer operation can be catastrophic.

The average company spends about 4%–5% revenue on information technology (IT), with those that are highly IT dependent—such as financial and telecommunications companies—spending more than 10% on it. In other words, IT is now one of the largest corporate expenses outside labor costs. What are the risks involved, and how they can be mitigated?

Governments, too, are big consumers of software. The U.S. government cataloged 1,200 civilian IT projects costing more than $60 billion, plus another $16 billion for military software. What are the risks involved, and how they can be mitigated?

Any one of these projects can cost more than $1 billion. To take two current examples, the computer modernization effort at the U.S. Department of Veterans Affairs is projected to run $3.5 billion. Such megasoftware projects, once rare, are now much more common, as smaller IT operations are joined in "systems of systems." Air traffic control is a prime example because it relies on connections among dozens of networks that provide communications, weather, navigation, and other data. What are the risks involved, and how they can be mitigated?

In general, software quality, reliability, safety, and effectiveness only can be considered in relative terms. Safety by definition is the freedom from unacceptable risk where risk is the combination of likelihood of harm and severity of that harm. Subsequently, *hazard* is the potential for an adverse event, a source of harm. All designed software, carry a certain degree of risk and could cause problems in a definite situation. Many software problems cannot be detected until extensive market experience is gained. For example, on June 4, 1996 an unmanned Ariane 5 rocket launched by the European Space Agency exploded just 40 seconds after lift off from Kourou, French Guiana. The rocket was on its first voyage after a decade of development, whic cost $7 billion. The destroyed rocket and its cargo were valued at $500 million. A board of inquiry investigated the causes of the explosion and in two weeks issued a report. It turned out that the cause of the failure was a software error in the inertial reference system. Specifically a 64-bit floating point number relating to the horizontal velocity of the rocket with respect to the platform was converted to a 16-bit signed integer. The number was larger than 32,767, the largest integer storable in a 16-bit signed integer, and thus, the conversion failed.

Attention starts shifting from improving the performance during the later phases of the software life cycle to the front-end phases where development takes place at a higher level of abstraction. It is the argument of "pay now" or "pay later" or prevention versus problem solving. This shift also is motivated by the fact that the software design decisions made during the early stages of the design life cycle have the largest impact on the total cost and the quality of the system. For industrial and manufactured products, it often is claimed that up to 80% of the total cost is committed in the concept development phase (Fredrikson, 1994). For software, it is the experience of the authors that at least 70%–80% of the design quality also is committed in the early phases, as depicted in Figure 15.1 for generic systems. The potential is defined as the difference between the commitment and the ease of change for metrics such as performance, cost, technology, schedule, and so on. The potential is positive but decreasing as development progresses, implying reduced design freedom across time. As financial resources are committed (e.g., buying production machines and facilities, hiring staff, etc.) the potential starts changing signs, going from positive to negative. In the consumer hand, the potential is negative, and the cost overcomes the impact tremendously. At this phase, design changes for corrective actions only can be achieved at high cost including customer dissatisfaction, warranty, marketing promotions, and in many cases, under the scrutiny of the government (e.g., recall costs).

The software equivalent of Figure 15.1 is depicted Figure 15.2. The picture is as blurry as for general systems. However, the research area of software development

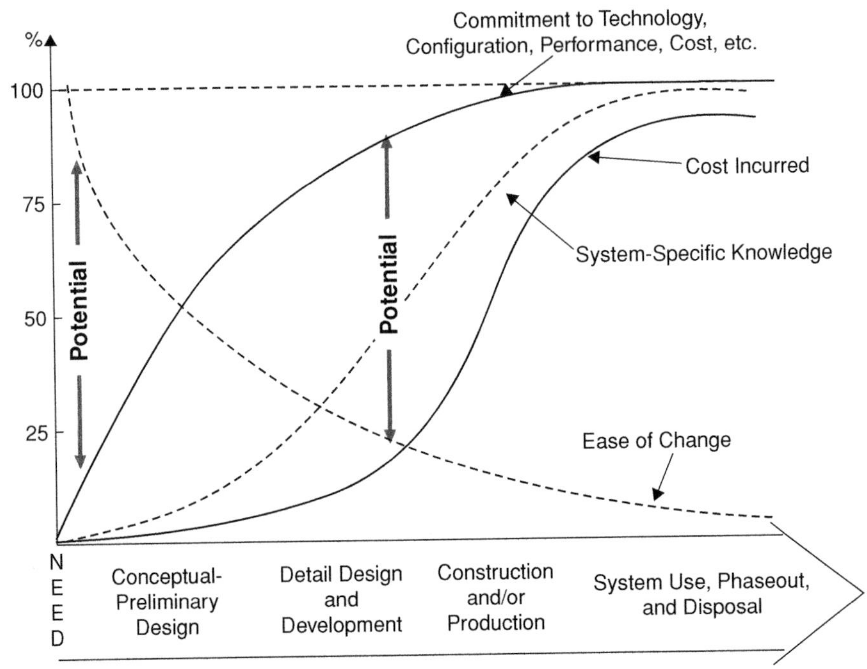

FIGURE 15.1 Risk of delaying risk management for systems (Blanchard & Fabrycky, 1981).

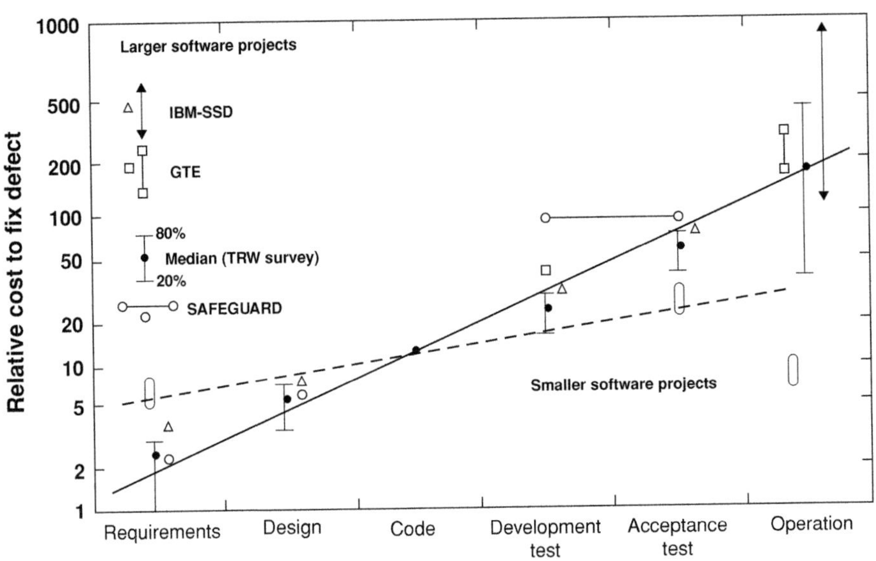

FIGURE 15.2 Risk of delaying risk management for software (Blanchard & Fabrycky, 1981).

FIGURE 15.3 Software risk management elements.

currently is receiving increasing focus to address industry efforts to shorten lead times, cut development and production costs, lower total life-cycle cost, and improve the quality of the software entities.

The current approach to software risk mitigation is to manage all potential risks becoming a hazard that could result in safety problems and harm. This approach of *risk management* plasters broad categories of risk such as project risks, technical risks, and environmental risks and domain specific software such as medical device risks and many others. In this chapter, we elected to combine all risks pertaining to the environment and to humans into a category called safety risks and all other risks into a category called business risks and then used the Design for Six Sigma methodology to manage both types of risks.

Software development houses generally are required to have a quality management system as well as processes for addressing software-related risks. Figure 15.3, illustrates the integration of the risk management process into a quality management system.

A software risk management begins with ***planning*** for the software based on the quality system objectives to include the risk acceptability criteria defined by management then followed by ***risk analysis*** to identify all potential hazards associated with the software, followed by ***risk evaluation*** to estimate the risk for each hazard. Risk evaluation is based on experience, evidence, testing, calculation, or even subjective

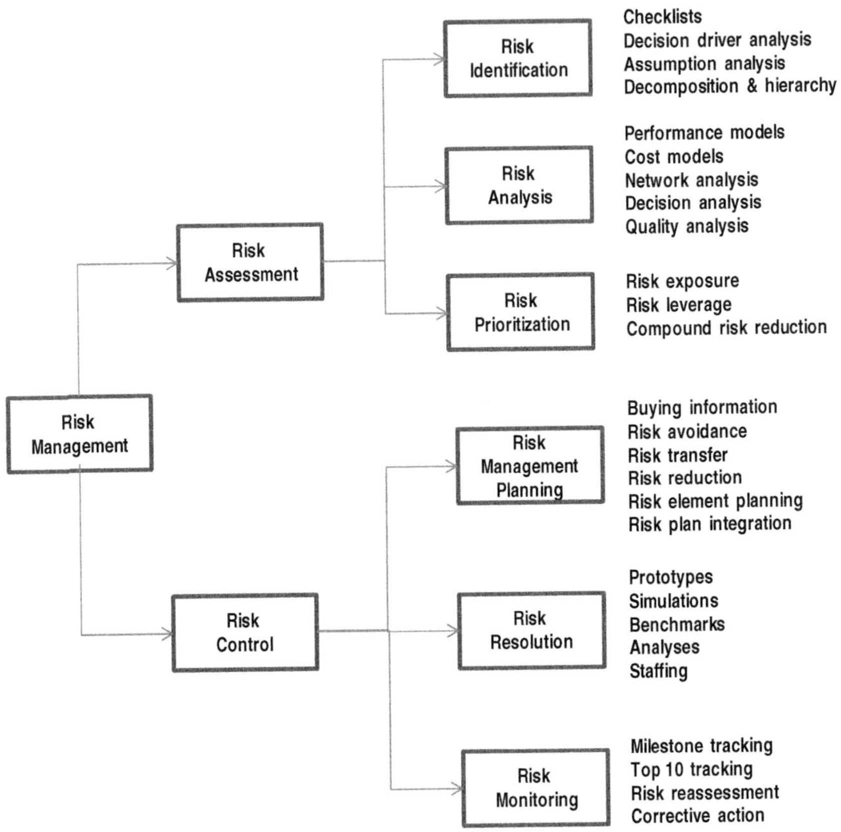

FIGURE 15.4 Software risk management.

judgment. Risk assessment is complex, as it can be influenced by personal perception and other factors such as political climates, economic conditions, and cultural background. It is highly recommended to base risk assessment of software on an expert's knowledge and safety-centric engineering. The causal relationship among the harm, the hazard, and the cause of the hazard plays a great role in risk management in which causes may occur in the absence of failures or as a result of one or more failure modes. Naturally, hazards are inherent in products, and many unplanned attempts to overcorrect a hazardous event tend to increase the potential risk of creating new hazards. Therefore, the focus should be on the cause of the hazard, not the actual harm itself. Figure 15.4 depicts the risk management elements, some of which are to be discussed in this chapter.

As is the case with software reliability[1], this chapter is concerned with the software nonconformances that are visible to a customer and prevent a system from delivering essential functionality causing risk. In Chapter 14, we classified nonconformance into

[1] See Chapter 14.

three categories. Although the definitions imply reliability treatment, nevertheless, they are risks, and from a severity and safety standpoint, and for some applications, defects can be more hazardous than faults and failures. These are repeated below:

- *Defects*: A flaw in software requirements, design, or source code that produces unintended or incomplete run-time behavior. This includes defects of commission and defects of omission. Defects of commission are one of the following: Incorrect requirements are specified, requirements are incorrectly translated into a design model, the design is incorrectly translated into source code, and the source code logic is flawed. Defects of omission are one of the following: Not all requirements were used in creating a design model, the source code did not implement all the design, or the source code has missing or incomplete logic.
- *Faults*: A fault is the result of triggering a software defect by executing the associated source code. Faults are NOT customer-visible. An example is a memory leak or a packet corruption that requires retransmission by the higher layer stack.
- *Failures*: A failure is a customer (or operational system) observation or detection that is perceived as an unacceptable departure of operation from the designed software behavior. Failures are the visible, run-time symptoms of faults. Failures MUST be observable by the customer or another operational system.

Other definitions and terminology that may be useful in reading the rest of this chapter are listed in Appendix 15.A.

15.2 PLANNING FOR RISK MANAGEMENT ACTIVITIES IN DESIGN AND DEVELOPMENT

A ***business risk*** can be defined as a potential threat to achieving business objectives for the software under development, these risks are related to but not limited to technology maturity, software complexity, software reliability and availability, performance and robustness, and finally, project timelines. ***Safety risk*** is defined as the potential threat to software-produced health and environment failures for the product under development; these risks are related to but not limited to failure, defect, fault nonconformities, customer misuse and abuse, systems integration, and so on. When considering safety risks, it is apparent that it can be classified as business risk resulting from the criteria mentioned. It is isolated in a category by itself for its profound effect on the end user. The emphasis on decoupling safety risk from business risk is to manage the complexity of the rigor applied to reduce and eliminate safety risks as a result of the software regulatory expectations for risk management in industries such as aerospace. A structured approach for risk management, as described throughout this chapter, is required by a software DFSS teams when safety risk and regulatory compliance are impacted. Contrasting rigor is required when dealing with only business risks.

The risk management process starts early on during the voice of the customer (VOC) stage (see Chapter 11 for DFSS project road map) by identifying potential hazards and establishing risk assessment criteria. A risk management plan defines the process for ensuring that hazards resulting from errors in the customer usage environment, foreseeable software misuses, and the development and production of nonconformities are addressed. A risk management plan should include the following:

1. The scope of the plan in the context of the software development life cycle as applicable
2. A verification plan, allocation of responsibilities and requirements for activities review
3. Criteria for risk acceptability

Risk management plans are performed on software platforms where activities are reviewed for effectiveness either as part of a standard design review process or as independent stand-alone reviews. Sometimes the nature of hazards and their causes are unknown, so the plan may change as knowledge of the software is accumulated. Eventually, hazards and their controls should be linked to verification and validation plans.

At the DFSS identify phase, risk estimation establishes a link between requirements and hazards and ensures the safety requirements are complete. Then risk assessment is performed on the software as a design activity. Subsequently, risk mitigation, including risk elimination and/or reduction, ensures that effective traceability between hazards and requirements are established during verification and validation. Risk acceptability and residual risks are reviewed at applicable milestones (see Chapter 11 for DFSS tollgate in ICOV process). It is very important for management to determine responsibilities, establish competent resources, and review risk management activities and results to ensure that an effective management process is in place. This should be an on-going process in which design reviews and DFSS gate reviews are decision-making milestones.

A risk management report summarizes all results from risk management activities such as a summary of the risk-assessment-techniques, risk-versus-benefit analysis, and the overall residual risk assessment. The results of all risk management activities should be recorded and maintained in a software risk management file. See Section 15.7 for more details on the roles and responsibilities that can be assumed by the software DFSS team members in developing a risk management plan.

15.3 SOFTWARE RISK ASSESSMENT TECHNIQUES

Risk assessment starts with a definition of the intended use of the software and their potential risks or hazards, followed by a detailed analysis of the software functionality or characteristics that cause each of the potential hazards, and then finally, a well-defined rating scale to evaluate the potential risk. The risk in both normal and fault conditions then is estimated. In risk evaluation, the DFSS team decides whether risk reduction is needed. Risk assessment includes risk identification, analysis, and evaluation. Brainstorming is a useful tool for identifying hazards. Requirement documents are another source for hazard identification because many hazards are associated with the nonfulfillment or partial fulfillment of each requirement. For example, in infusion medicine instruments, there may be software requirements for medication delivery and hazards associated with overdelivery or underdelivery. Estimating the risks associated with each hazard usually concludes the risk analysis part of the process. The next step is risk evaluation and assessment.

As defined earlier in this chapter, risk is the combination of the likelihood of harm and the severity of that harm. Risk evaluation can be *qualitative* or *quantitative* depending on when in the software life cycle the risk estimation is occurring and what information is available at that point of time. If the risk cannot be established or predicted using objective (quantitative) data, then expert judgment may be applied. Many risk analysis tools can be used for risk assessment; in this chapter we will discuss some common tools used in the software industry such as preliminary hazard analysis (PHA), hazard and operability (HAZOP) analysis, failure mode and effects analysis (FMEA), and fault tree analysis (FTA). We then will touch base on other risk analysis tools used by other industries as a gateway to the software industry.

15.3.1 Preliminary Hazard Analysis (PHA)

PHA is a qualitative risk assessment method for identifying hazards and estimating risk based on the intended use of the software. In this approach, risk is estimated by assigning severity ratings to the consequences of hazards and likelihood of occurrence ratings to the causes. PHA helps to identify risk reduction/elimination measures early in the design life cycle to help establish safety requirements and test plans.

15.3.2 Hazard and Operability Study (HAZOP)

The HAZOP technique (Center for Chemical Process Safety, 1992) can be defined as the application of a systematic examination of complex software designs to find actual or potentially hazardous procedures and operations so that they may be eliminated or mitigated. The methodology may be applied to any process or project although most practitioners and experts originate in chemical and offshore industries. This technique usually is performed using a set of key words (e.g., "more," "less," and "as well as"). From these key words, a scenario that may result in a hazard or an operational problem is identified. Consider the possible flow problems in a process control software controlling the flow of chemicals; the guide word "more" will correspond to a high flow rate, whereas that for "less" will correspond to a low flow rate. The consequences of the hazard and the measures to reduce the frequency with which the hazard will occur are evaluated.

15.3.3 Software Failure Mode and Effects Analysis (SFMEA)[2]

For a long time, the measured failure rate has been the standard for software quality and, hence, reliability. In current computing environments and in the DFSS era, it can be lagging, inapplicable, and even misleading. Consider server software with an expanding number of clients. More users are likely to cause an increase in the failure rate, though the software is not changed. Another example is software controlling a machine tool. The machine tool is aging, causing more exception conditions to be encountered by the program and, hence, more failures. The machine shop supervisor sees a higher failure rate, even though the software remains the same.

Because there are problems with using failure rate as an indicator of quality in existing software, we looked for alternatives for predicting software quality during development that would continue to be valid in operation. The severity of failure effects needed to be taken into account so that preventive DFSS actions could focus on avoidance of the most severe failures. This latter requirement suggested a look at software risk management, including tools such as FMEA. But although FMEA for hardware is used widely (Yang & El-Haik, 2008), it rarely is encountered for software. An obvious reason is that hardware generally is made up of parts with well-known failure modes; there is no equivalent of this in software. Instead, software is analyzed by "functions." But these are subjective partitions, and there is usually no certainty that all functions that can contribute to failure have been included.

FMEA[3] is a systematic method used to analyze products and processes by qualitatively determining their failure modes, causes of failure, and potential effects then quantitatively classifying their risk estimate to prioritize better corrective and preventive actions and risk reduction measures required by the analysis.

Software FMEA (SFMEA) determines software effects of each failure mode of each code component, one by one, identifies failures leading to specific end events, has rules that differ from hardware analysis rules and is complex, its effects dependent on time and state.

When SFMEA is extended further by a criticality analysis, the resulting technique then is called failure mode and effects criticality analysis (FMECA). Failure mode and effects analysis has gained wide acceptance by most industries. In fact, the technique has adapted itself in many other forms such as concept FMEA, robust design FMEA,[4] process (manufacturing and service) FMEA, and use FMEA.

What makes SFMEA different from other applications?

- Extended effects: Variables can be read and set in multiple places
- Failure mode applicability: There can be different failure modes in different places
- Time dependency: Validity can depend on what is happening around it

[2] See Chapter 16 for more details.

[3] FMEAs were formally introduced in the late 1940s with the introduction of the United States Military Procedure MIL-P-1629.

[4] See Mekki (2006). See also Yang and El-Haik (2008) and El-Haik and Roy (2005).

- Unpredictable results: cannot always determine effects
- Purchased software: How to assess failure effects?

FMEAs have gone through a metamorphosis of sorts in the last decade, as a focus on severity and occurrence has replaced risk priority number (RPN)-driven activities. In large part, this is a result of measurement risk outcomes, resulting from associated RPNs being misinterpreted, as so many practitioners of FMEAs believe that the RPN is the most important outcome. However, the FMEA methodology must consider taking action as soon as it is practical.

An FMEA can be described as complementary to the process of defining what software must do to satisfy the customer. In our case, the process of "defining what software must do to satisfy the customer" is what we entertain in the software DFSS project road map discussed in Chapter 11. The DFSS team may visit existing datum FMEA, if applicable, for further enhancement and updating. In all cases, the FMEA should be handled as a living document.

15.3.4 Fault Tree Analysis (FTA)

FTA is a technique for performing the safety evaluation of a system. It is a process that uses logical diagrams for identifying the potential causes of a risk or a hazard or an undesired event based on a method of breaking down chains of failures. FTA identifies a combination of faults based on two main types. First, several functional elements must fail together to cause other functional elements to fail (called an "and" combination), and second, only one of several possible faults needs to happen to cause another functional element to fail (called an "or" combination). Fault tree analysis is used when the effect of a failure/fault is known, and the software DFSS team needs to find how the effect can be caused by a combination of other failures. The probability of the top event can be predicted using estimates of failure rates for individual failure. It helps in identifying single-point failures and failure path sets to facilitate improvement actions and other measures of making the software under analysis more *robust*.

Fault tree analysis can be used as a qualitative or a quantitative risk analysis tool. The difference is that the earlier is less structured and does not require the use of the same rigorous logic as the later analysis. The FTA diagram shows faults as a hierarchy, controlled by gates because they prevent the failure event above them from occurring unless their specific conditions are met. The symbols that may be used in FTA diagrams are shown in Table 15.1.

FTA is an important and widely used safety analysis technique and is also the subject of active research. Using the architecture obtained from Chapter 13 and the failure probabilities of its components (modules), a system fault tree model is constructed and used to estimate the probability of occurrence of the various hazards that are of interest to the DFSS team.

The failure probabilities of modules (components) are either measured or estimated. The probability is estimated when it cannot be measured easily. An estimate

TABLE 15.1 FTA Symbols

Symbol	Name	Meaning
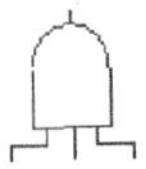	And gate	Event above happens only if all events below happen.
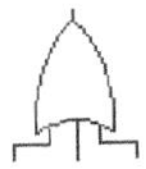	Or gate	Event above happens if one or more of events below are met.
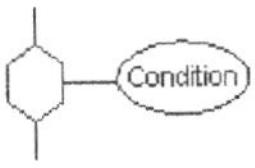	Inhibit gate	Event above happens if event below happens and conditions described in oval happen.
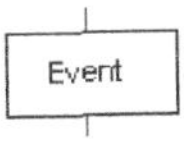	Combination gate	Event that results from combination of events passing through gate below it.
	Basic event	Event that does not have any contributory events.
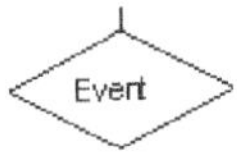	Undeveloped basic event	Event that does have contributory events, but which are not shown.
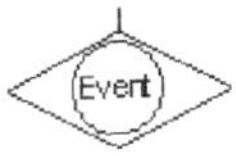	Remote basic event	Event that does have contributory events, but are shown in another diagram.
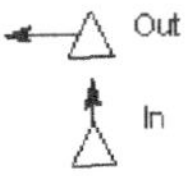	Transferred event	A link to another diagram or to another part of the same diagram.
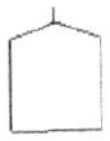	Switch	Used to include or exclude other parts of the diagram that may or may not apply in specific situations.

is developed by treating the component itself as a system made up of simpler refined components (modules) whose failure probabilities can be measured by further testing. These probabilities then are used in a model of the component of interest to produce the required estimate (Knight & Nakano, 1997).

FTA of systems that include computers can treat the computer hardware much like other components. Computer systems can fail, however, as a result of software defects as well as hardware defects, and this raises the question about how the software "components" of a system can be included in a fault-tree model. In practice, this has proved difficult.

To obtain the probabilistic data needed for fault tree analysis, it is tempting to analyze software in the same way that hardware is analyzed—either as a black-box component whose failure probability can be measured by sampling the input space (i.e., life testing) or as a component whose structure permits modeling of its failure probability from its architecture. Unfortunately, the quantification of software failure probability by life testing has been shown to be infeasible because of the *very large number of tests*[5] required to establish a useful bound on the probability of failure. The large number of tests derives from the number of combinations of input values that can occur. Also unfortunate is the fact that no general models predict software dependability from the software's design; the type of Markov models used in hardware analysis do not apply in most software cases. The reason for this is that the basic assumptions underlying Markov analysis of hardware systems do not apply to software systems (e.g., the assumption that independent components in a system fail independently does not apply) (Knight & Nakano, 1997).

It is possible to obtain the parameters needed for fault tree analysis by some means other than testing or modeling. Many techniques exist, usually within the field of formal methods (Diller, 1994) that can show that a particular software system possesses useful properties without testing the software. If these properties could be used to establish the parameters necessary for FTA, then the requirement of using testing and Markov models would be avoided.

From a testing estimation perspective, a major part of the problem derives from the size of modern software systems. Knight and Nakano (1997) suggested dealing with testing estimation complexity using a combination of the following concepts:

Protection-Shell Architecture: A protection she can be used to limit severely the amount of software on which a system depends for correct operation. As a result, the amount of critical software that has to be tested can be reduced significantly. The detailed description and analysis of this architecture is to restrict most of the implementation of safety and, thereby, the dependability analysis of a system to a conceptual shell that surrounds the application. Provided the shell is not starved of processor resources (by the operating system defect, for example), the shell ensures that safety policies are enforced no matter what action is taken by the rest of the software. In other words, provided the shell itself is dependable and can execute properly, safety will not be compromised by defects in the remainder of the software including the operating system and the application. With a protection shell in place, the testing of a system can be focused on the shell. It is no longer necessary to undertake testing to demonstrate ultradependability of the entire system. For many systems, this alone might bring the number of test cases required down to a feasible value.

Specification limitation: This technique deliberately limits the range of values that to a system input can take to the smallest possible set that is consistent with safe operation. In many cases, the range of values that an input can take is determined by an external physical device, such as a sensor, and the range might be unnecessarily wide. It is the combination of the ranges of input values

[5]It is literally the case that for most realistic systems, the number of tests required would take thousands of years to complete, even under the most optimistic circumstances.

that leads to the unrealistic number of test cases in the ultradependable range. Specification limitation reduces the number of inputs to the least possible.

Exhaustive testing: There are many circumstances in which it is possible to test all possible inputs that a piece of software could ever receive (i.e., to test exhaustively). Despite the relative simplicity of the idea, it is entirely equivalent to a proof of correct operation. If a piece of software can be tested exhaustively and that testing can be trusted (and that is not always the case), then the quantification needed in fault-tree analysis of the system, including that software, is complete—the probability of failure of the software is zero.

Life testing: Although initially we had to reject life testing as infeasible, with the application of the elements of restricted testing already mentioned, for many software components it is likely that life testing becomes feasible. What is required is that the sample space presented by the software's inputs be "small enough" that adequate samples can be taken to estimate the required probability with sufficient confidence (i.e., sufficient tests are executed to estimate the software's probability of failure).

There are many other tree analysis techniques used in risk assessment such as event tree analysis (ETA). ETA is a method for illustrating through graphical representation of the sequence of outcomes that may develop in a software code after the occurrence of a selected initial event. This technique provides an inductive approach to risk assessment as they are constructed using forward logic. Event tree analysis and fault tree analysis are closely linked. Fault trees often are used to quantify system events that are part of event tree sequences. The logical processes employed to evaluate an event tree sequence and to quantify the consequences are the same as those used in fault tree analysis.

Cause-consequence analysis (CCA) is a mix of fault tree and event tree analyses. This technique combines cause analysis, described by fault trees, and consequence analysis, described by event trees. The purpose of CCA is to identify chains of events that can result in undesirable consequences. With the probabilities of the various events in the CCA diagram, the probabilities of the various consequences can be calculated, thus establishing the risk level of the software or any subset of it.

Management oversight risk tree (MORT) is an analytical risk analysis technique for determining causes and contributing factors for safety analysis purposes in which it would be compatible with complex, goal-oriented management systems. MORT arranges safety program elements in an orderly and logical fashion, and its analysis is carried out similar to software fault tree analysis.

15.4 RISK EVALUATION

The risk evaluation analysis is a quantitative extension of the FMEA based on the severity of failure effects and the likelihood of failure occurrence, possibly augmented with the probability of the failure detection. For an automation system application, the severity is determined by the effects of automation function failures on the safety of the controlled process. Even if it is difficult, the severity of a single low-level component failure mode, in principle, can be concluded backward from the top level in a straight forward manner.

The likelihood of occurrence is much harder to define for a software-based system. The manifestation of an inherent software fault as a failure depends not only on the software itself but also on the operational profile of the system (i.e., on the frequency of the triggering that causes the fault to lead to failure). This frequency is usually not known. Luke (1995) proposed that a proxy such as McCabe's complexity (Chapter 5) value or Halstead's complexity measure (Chapter 5) be substituted for occurrence. Luke argued that there is really no way to know a software failure rate at any given point in time because the defects have not yet been discovered. He stated that design complexity is positively and linearly correlated to defect rate. Therefore, Luke suggested using McCabe's complexity value or Halstead's complexity measure to estimate the occurrence of software defects. Also, the probability of detection is hard to define because only a part of software failures can be detected with self-diagnostic methods.

Software risk evaluation starts once the components of risk for each hazard/harm have been identified then uses risk acceptability criteria, defined by the risk management plan, to rank order the risk to complete the risk evaluation. Given the different risk analysis techniques discussed, the evaluation of risk is totally dependent on the software company's culture and internal procedures, as regulations and standards cannot dictate one's approach for risk evaluation because of the difference in software applications within the software industry. Few of the most used standards for software risk management are ISO 9000-3, ISO/IEC 19770-1, IEC 60812, SAE J-1739, MIL-STD-1629A, and ISO 12207. In this chapter, we will discuss risk evaluation criteria based on our own *hybrid* approach.

To quantify risk consistently, we need to estimate the severity for each hazard and the likelihood of occurrence associated with their causes against a criteria set forth by the risk management plan defined at the product level.

Severity rating is the rank associated with the possible consequences of a hazard or harm. Table 15.2 lists a generic software severity ratings based on two commonly used scales: 1 to 5 and 1 to 10. Risk management teams can develop a severity rating that best suits their application.

Likelihood of occurrence rating is the rank associated with probability (or frequency) that a specific cause will occur and cause the potential hazard during a predetermined time period (typically the software life cycle). Table 15.3 lists a generic likelihood of occurrence ratings based on two commonly used scales: 1 to 5 and 1 to 10. Risk management teams can develop a likelihood of occurrence rating that best suits their application.

Risk classification is a process of categorizing risk in different criteria as defined by the risk management plan. Risk classification criteria define the foundation for risk acceptance or highlight the need for risk reduction. Table 15.4 lists the different risk classification criteria by event (intersection cell).

Risk acceptance is a relative term that the product is deemed acceptable if it is risk free or if the risks are *as low as reasonably practicable* (ALARP), and the benefits

TABLE 15.2 Software Severity Rating

Severity of Hazard/Harm			
Criteria	Description	Rating 1–5	Rating 1–10
Catastrophic	***Product Halts/Process Taken Down/Reboot Required*** : The product is completely hung up, all functionality has been lost and system reboot is required.	5	9–10
Serious	***Functional Impairment/Loss***: The problem will not resolve itself and no "work around" can bypass the problem. Functionality either has been impaired or lost, but the product can still be used to some extent.	4	7–8
Critical	***Functional Impairment/Loss***: The problem will not resolve itself, but a "work around" temporarily can bypass the problem area until it is fixed without losing operation.	3	5–6
Marginal	***Product Performance Reduction***: Temporary through time-out or system load–the problem will "go away" after a period of time.	2	3–4
Negligible	***Cosmetic Error***: No loss in product functionality. Includes incorrect documentation.	1	1–2

TABLE 15.3 Software Likelihood of Occurrence Rating

Likelihood of Occurrence Rating			
Criteria	Description	Rating 1–5	Rating 1–10
Frequent	***Hazard/harm likely to occur frequently***: 1 per 10 min (1/10) to 1+ per min (1/1)	5	9–10
Probable	***Hazard/harm will occur several times during the life of the software***: 1 per shift (1/480) to 1 per hour (1/60)	4	7–8
Occasional	***Hazard/harm likely to occur sometime during the life of the software***: 1 per week (1/10k) to 1 per day (1/1440)	3	5–6
Remote	***Hazard/harm unlikely but possible to occur during the life of the software***: 1 per 1 unit-year (1/525k) to 1 per 1 unit-month (1/43k)	2	3–4
Improbable	***Hazard/harm unlikely to occur during the life of the software***: 1 per 100 units- years (1/50m) to 1 per 10 unit-years (1/5m)	1	1–2

TABLE 15.4 Software Risk Classification Criteria

		Severity of Hazard/harm				
		1	2	3	4	5
	Likelihood of Occurrence	**Negligible**	**Marginal**	**Critical**	**Serious**	**Catastrophic**
5	**Frequent**	R_3	R_4	R_4	R_4	R_4
4	**Probable**	R_2	R_3	R_4	R_4	R_4
3	**Occasional**	R_1	R_2	R_3	R_3	R_4
2	**Remote**	R_1	R_1	R_2	R_2	R_4
1	**Improbable**	R_1	R_1	R_1	R_1	R_3
	R_4 Event	Intolerable- risk is unacceptable and must be reduced.				
	R_3 Event	Risk should be reduced as low as reasonably practicable - benefits must rationalize any residual risks even at a considerable cost.				
	R_2 Event	Risk is unacceptable and should be reduced as low as reasonably practicable - benefits must rationalize any residual risks at a cost that represents value.				
	R_1 Event	Broadly acceptable- No need for further risk reduction.				

associated with the software outweigh the residual risk. However, intolerable risks are not acceptable and must be reduced at least to the level of ALARP risks. If this is not feasible, then the software must be redesigned with fault prevention standpoint.

The concept of practicability in ALARP involves both technical and economic consideration, a part of what we defined as business risk earlier in this chapter in which technical refers to the availability and feasibility of solutions that mitigate or reduce risk, and economic refers to the ability to reduce risks at a cost that represent value.

Risk-versus-benefit determination must satisfy at least one of the following: 1) all practicable measures to reduce the risk have been applied, 2) risk acceptance has been met, and finally, 3) the benefit that the software provides outweighs the residual risk.

15.5 RISK CONTROL

Once the decision is made to reduce risk, control activities begin. Risk reduction should focus on reducing the hazard severity, the likelihood of occurrence, or both. Only a design revision or technology change can bring a reduction in the severity ranking. The likelihood of occurrence reduction can be achieved by removing or controlling the cause (mechanism) of the hazard. Increasing the design verification actions can reduce detection ranking.

Risk control should consist of an integrated approach in which software companies will use one or more of the following in the priority order listed: 1) inherent safety by design (a design in safety leads to a more robust software design), 2) protective design measures in which the product will fail safe and/or alarms when risk presents, 3) protective measures (e.g., input/output mistake proofing) and/or inherent correction test capabilities, 4) information for safety such as instructions for use and training.

15.6 POSTRELEASE CONTROL

Information gained about the software or similar software in the postrelease phase (see beyond stage 8 in the software life cycle shown in Chapter 8) performance should be reviewed and evaluated for possible relevance to safety for the following: 1) if new or previously unrecognized hazards or causes are present, 2) if the estimated risk resulting from a hazard is no longer acceptable, and 3) if the original assessment of risk is invalidated. If further action is necessary, then a Six Sigma project should be initiated to investigate the problem.

15.7 SOFTWARE RISK MANAGEMENT ROLES AND RESPONSIBILITIES

Table 15.5 outlines the responsibility for the deliverables created by the risk management process within the DFSS road map. RASCI stands for R = Responsible, A = Approver; S = can be Supportive, C = has to be Consulted, and I = has to be Informed.

15.8 CONCLUSION

The most significant aspects of building risk management into the flow of the software development process are to imbed the tradeoff concept of the risk-versus-benefit analysis as part of the design and development process.

The DFSS methodology helps in making data decision based and allows for logical tradeoffs and quantifiable risk-versus-benefits analysis. DFSS methodology provides traceability in which relationships among hazards, requirements, and verification and validation activities are identified and linked.

Risk management itself is a process centered on understanding risks and evaluating their acceptability, reducing any risks to as low as possible, and then evaluating residual risk and overall software safety against the benefits derived. Integrating risk management into the design and development process requires keeping risk issues at the forefront of the entire process from design planning to verification and validation testing. In this way, risk management becomes part of the software development process, evolves with the design, and provides a framework for decision making.

The software Design For Six Sigma process—the subject of this book—is used as a risk management toolkit in which it drives the data-driven approach behind decision making. It is well known that if we make decisions based on factual data, then the chances of negative consequences are reduced.

TABLE 15.5 Software Risk Management Roles and Responsibilities

Software Risk Management Deliverable	Quality	Regulatory	R&D	Service Process	Reliability	Marketing	Corrective Action Teams	Process Owners	Project Management
DFSS I-dentify Phase									
Hazard analysis	A	S	R	S	S	S	S	S	
Risk management file (for regulated industries)	S		R						S
DFSS C-onceptualize Phase									
Risk management plan	R	A	A	S	S	S	S	S	S
Hazard analysis	A	S	A	S	S	S	S	S	
Risk analysis documents	A	S	R		S	S	S	S	
Risk management report	R	A	A		S	S	C		
Risk management file (for regulated industries)	S		R						S
DFSS O-ptimize & Verify Phases									
Risk management plan	R	A	A		S	S	S		
Hazard analysis	A	S	A	S	S	S	S	S	
Risk analysis documents	A	S	R		S	S	S	S	
Post-market monitoring requirements	A	S	R		S	S	C	S	
Software failure modes and effect analysis (SFMEA)	A	A	A	R	S	S		R	
Process control plan	A		A	R	S			R	
Risk management report	R	A	A		S	S	C		
Risk management file (for regulated industries)	S		R						S
Release stage									
Risk management reviews	A		R	S	S		C	S	
Risk management file (for regulated industries)	S		R						S
On-Going Support									
Risk management reviews	A		R	S	S		C	S	
Risk management file (for regulated industries)	S		R						S

Finally, and most importantly, risk management reduces the potential for systematic errors in the development process and increases the likelihood that the DFSS team will get it right the first time.

APPENDIX 15.A

Risk Management Terminology

Harm: Physical injury or damage to the health of people or damage to property or to the environment caused by software failure, defect, or fault.

Hazard: The potential source of harm.

Hazard Analysis: A risk analysis activity that analyzes the software and the usage of the associated hardware, including any reasonably foreseeable misuse throughout the life cycle. The analysis is performed in-house or at usage level and results in mitigations that are at a functional or system requirements level. The primary emphasis is to identify the list of harms, the causes (hazards) of the harms, the users affected by the harm, the risk, and to ensure that the system's safety functions and requirements have been identified for further implementation.

Risk Management Plan: Includes the scope, defining identification, and a description of the system and the applicability of the plan, a link to the verification plan, allocation of responsibilities, risk management activities and the review(s), and the criteria for risk accessibility.

Mitigations: see risk controls.

Occurrence: The probability of occurrence of harm. The occurrence should include the probability that the cause creates the hazardous condition that result in the harm.

Postmarket: The time or activities after the release of a new software or software change (e.g., upgrade) into the market place.

Postmarket Monitoring Requirements: A document that identifies the safety and effectiveness parameters to be monitored during launch and support stages, the criteria for monitoring, and the actions to be taken if the acceptance criteria have not been met.

Postmarket Risk Analysis: Any risk analysis conducted based on post-market risk data. The postmarket risk analysis initiates the review and/or update of the appropriate risk management documents.

Postmarket Risk Data: Any data collected after the product has left the development stages, including production process data, supplier and supplied data, service data, complaint data, new customer requirements (actual or regulatory), advisories, warning and recalls, corrective and preventative action trends, field corrective action trends, customer's requests for information, and other similar types of data.

Software Requirements: The requirements are inputs from the Identify DFSS phase and include marketing or customer requirements, architecture documents, system requirements, subsystem or component requirements, formulations, production or servicing requirements, specifications, and so on.

Software Life cycle: All phases in the software life cycle from the initial development through pre- and postapproval until the product's discontinuation (see Chapter 8).

Residual Risk: The risk remaining after risk controls have been implemented.

Risk: The combination of the probability of occurrence of harm and the severity of that harm.

Risk Acceptance Criteria: A process describing how the severity, occurrence, risk, and risk acceptance decisions are determined. The risk acceptance criteria should be defined in the risk management plan.

Risk Management Process: This process applies to software risks. It is the process of identifying hazards associated with software, estimating and evaluating the associated risks, controlling these risks, and monitoring the effectiveness of the control throughout the life cycle of the software, including postmarket analysis.

Risk Analysis: The systematic use of information to identify sources and to estimate the risk. The risk analysis activity may include a hazard analysis to evaluate the clinical risks and the use of risk analysis tools to support the software product, production process, and/or postmarket analysis.

Risk Analysis Documents: Any outputs generated from the risk analysis activities.

Risk Analysis Tools: Risk analysis may use tools (risk analysis tools) such as FMEA, HAZOP, FTA, or other similar analysis methods.

Risk Evaluation: This activity involves the evaluation of estimated risks by using risk acceptability criteria to decide whether risk mitigation needs to be pursued. The risk evaluation may include the initial risk, the residual risk acceptance, and/or the overall product acceptance.

Risk Control: This involves risk reduction, implementation of risk control measure(s), residual risk evaluation, risk/benefit analysis, and completeness of risk evaluation. If a hazard cannot be mitigated completely, then the potential harms must be communicated to the user. Risk control should consist of an integrated approach in which one or more of the following, in the priority order, are used: inherent safety by design, protective measures in software itself or the associated processes, and information for safety.

Risk Management File: The software's design history file should document the location of the risk management file or provide traceability to the documentation and supporting data. The risk management file should include the appropriate record retention.

Safety: The freedom from unacceptable risk.

Severity: The measure of the possible consequences of a hazard.

User: A user includes the user and service personnel, internal personnel, bystanders, and environmental impact. The user is any person that interfaces with the software during the life cycle.

REFERENCES

Blanchard, B.S. and Fabrycky, W.J. (1981), *Systems Engineering and Analysis*, Prentice Hall, Upper Saddle River, NJ.

Center for Chemical Process Safety (1992), *Guidelines for Hazard Evaluation Procedures with Worked Examples*, 2nd Ed., John Wiley & Sons, New York.

Diller, A.Z. (1994), *An Introduction to Formal Methods*, 2nd Ed., John Wiley & Sons, New York.

El-Haik, Basem S. and Roy, D. (2005), *Service Design for Six Sigma: A Roadmap for Excellence*, Wiley-Interscience, New York.

Fredrikson, B. (1994), Holostic Systems Engineering in Product Development, The Saab-Scania Griffin, Saab-Scania, AB, Linkoping, Sweden.

Knight, J.C. and Nakano, L.G. (1997), "Software Test Techniques for System Fault-Tree Analysis," The 16th International Conference on Computer Safety, Reliability, and Security (SAFECOMP), Sept.

Luke, S.R. (1995), "Failure Mode, Effects and Criticality Analysis (FMECA) for Software." 5th Fleet Maintenance Symposium, Virginia Beach, VA, October 24–25, pp. 731–735.

Mekki, K.S. (2006), "Robust design failure mode and effects analysis in design for Six Sigma." International Journal Product Development, Volume *3*, #3&4, pp. 292–304.

Yang and El-Haik, Basem S. (2008). *Design for Six Sigma: A Roadmap for Product Development*, 2nd Ed., McGraw-Hill Professional, New york.

CHAPTER 16

SOFTWARE FAILURE MODE AND EFFECT ANALYSIS (SFMEA)

16.1 INTRODUCTION

Failure mode and effect analysis (FMEA) is a disciplined procedure that recognizes and evaluates the potential failure of a product, including software, or a process and the effects of a failure and identifies actions that reduce the chance of a potential failure from occurring. The FMEA helps the Design for Six Sigma (DFSS) team members improve their design and its delivery processes by asking "what can go wrong?" and "where can variation come from?" Software design and production, delivery, and other processes then are revised to prevent the occurrence of failure modes and to reduce variation. Input to an FMEA application includes past warranty or process experience, if any; customer wants, needs, and delights; performance requirements; specifications; and functional mappings.

In the hardware-(product) oriented DFSS applications (Yang & El-Haik, 2008), various FMEA types will be experienced by the DFSS team. They are depicted in Figure 16.1. The FMEA concept is used to analyze systems and subsystems in the early concept and design stages. It focuses on potential failure modes associated with the functions of a system caused by the design. The concept FMEA helps the DFSS team to review targets for the functional requirements (FRs), to select optimum physical architecture with minimum vulnerabilities, to identify preliminary testing requirements, and to determine whether hardware system redundancy is required for reliability target settings. Design FMEA (DFMEA) is used to analyze designs before they are released to production. In the DFSS algorithm, a DFMEA always should

Software Design for Six Sigma: A Roadmap for Excellence, By Basem El-Haik and Adnan Shaout

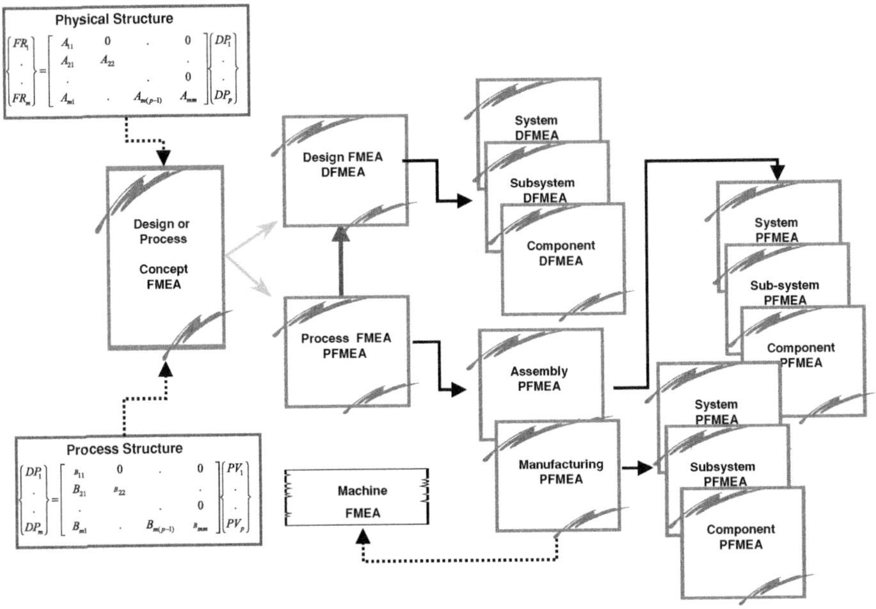

FIGURE 16.1 Product FMEA types (Yang & El-Haik, 2008).

be completed well in advance of a prototype build. The input to DFMEA is the array of FRs[1].

FMEA is well understood at the systems and hardware levels where the potential failure modes usually are known, and the task is to analyze their effects on system behavior. Today, more and more system functions are realized on the software level, which has aroused the urge to apply the FMEA methodology also to software-based systems. Software failure modes generally are unknown—"software modules do not fail, they only display incorrect behavior"—and depend on dynamic behavior of the application. These facts set special requirements on the FMEA of software-based systems and make it difficult to realize.

Performing FMEA for a mechanical or electrical system (Yang & El-Haik, 2008) or for a service (El-Haik & Roy, 2005) in a DFSS project environment is usually a more straightforward operation than what it is for a software-based system. The failure mode of components such as relays and resistors are generally well understood. Physics for the component failures are known, and their consequences may be studied. Mechanical and electrical components are supposed to fail as a result of some *noise factors*[2] such as wearing, aging, or unanticipated stress. The analysis may

[1] See Chapter 13.

[2] A term used in Taguchi methods. Taguchi calls common cause variation the "noise." Noise factors are classified into three categories: outer noise, inner noise, and between product noise. Taguchi's approach is not to eliminate or ignore the noise factors; Taguchi techniques aim to reduce the effect or impact of the noise on the product quality.

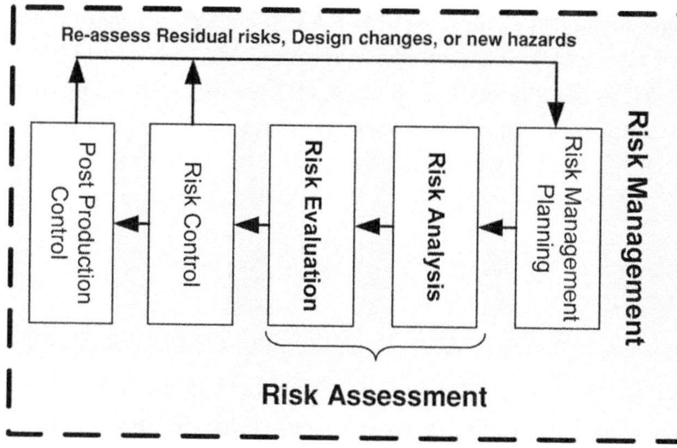

FIGURE 16.2 Risk management elements.

not always be easy, but at least the DFSS can rely on data provided by the component manufacturers, results of tests, and feedback of available operational experience. For software, the situation is different. The failure modes of software are generally unknown. The software modules *do not fail*; they only display incorrect behavior. To discover this incorrect behavior, the risk management process (Chapter 15) needs to be applied to mitigate risks and to set up an appropriate SFMEA approach.

For each software functional requirement or object (see Chapter 13), the team needs to ask "What can go wrong?" Possible design failure modes and sources of potential nonconformities must be determined in all software codes under consideration. The software DFSS team should modify software design to prevent errors from happening and should develop strategies to deal with different situations using risk management (Chapter 15) and mistake proofing (poka-yoke) of software and associated processes.

The main phases of SFMEA are similar to the phases shown in Figure 16.2. The SFMEA performer has to find the appropriate starting point for the analyses, set up a list of relevant failure modes, and understand what makes those failure modes possible and what their consequences are. The failure modes in SFMEA should be seen in a wide perspective that reflects the failure modes of incorrect behavior of the software as mentioned and not, for example, just as typos in the software code.

In this chapter, the failure mode and effects analysis is studied for use in the DFSS road map (Chapter 11) of software-based systems. Efforts to anticipate failure modes and sources of nonconformities are iterative. This action continues as the team strives to improve their design further and its developmental processes making SFMEA a living document.

We use SFMEA to analyze software in the concept and design stages (Figure 13.1). The SFMEA helps the DFSS team to review targets for the FRs[3], to select optimum architectures with minimum vulnerabilities, to identify preliminary testing requirements, and to determine whether risk mitigation is required for

[3]See Chapter 13.

reliability target settings. The input to SFMEA is the array of functional requirements that is obtained from quality function deployment (QFD) and axiomatic design analyses. Software FMEA documents and addresses failure modes associated with software functions. The outputs of SFMEA are 1) a list of actions to prevent causes or to detect failure modes and 2) a history of actions taken and future activity. The SFMEA helps the software DFSS team in:

1. Estimating the effects on users,
2. Assessing and selecting software design alternatives,
3. Developing an efficient validation phase within the DFSS algorithm,
4. Inputting the needed information for Design For X (e.g., design for reliability)
5. Prioritizing the list of corrective actions strategies that include mitigation, transferring, ignoring, or preventing the failure modes altogether,
6. Identifying the potential special design parameters from a failure standpoint and documenting the findings for future reference.

SFMEA is a team activity with representation from quality and reliability, operations, suppliers, and customers if possible. A Six Sigma operative, typically belts, leads the team. The software DFSS belt should own documentation.

16.2 FMEA: A HISTORICAL SKETCH[4]

A FMEA can be described as a systematic way to identify failure modes of a system, item, or function and to evaluate the effects of the failure modes on a higher level. The objective is to determine the causes for the failure modes and what could be done to eliminate or reduce the chance of failure. A bottom-up technique, such as FMEA, is an effective way to identify component failures or system malfunctions and to document the system under consideration.

The FMEA discipline originally was developed in the U.S. military (Military procedure MIL-P-1629)[5] . The method was used as a reliability evaluation technique to determine the effect of system and equipment failures. Failures were classified according to their impact on the military mission success and personnel/equipment safety. The military procedure MIL-P-1629 has functioned as a model for latter military standards MIL-STD-1629 and MIL-STD-1629A, which illustrate the most widely used FMEA procedures.

Outside the military, the formal application of FMEA first was adopted to the aerospace industry where FMEA was already used during the Apollo missions in the 1960s. In the early 1980s, U.S. automotive companies began to incorporate FMEA formally into their product development process. A task force representing Chrysler Corporation (Auborn Hills, MI), Ford Motor Company, (Dearborn, MI) and

[4] See Haapanen, P. and Helminen (2002).

[5] It was entitled procedures for performing a failure mode, effects, and criticality analysis and was issued on November 9, 1949.

TABLE 16.1 Hardware vs. Software FMEA Characteristics

Hardware FMEA	Software FMEA
• States the criticality and the measures taken to prevent or mitigate the consequences. • May be performed at functional level or part level. • Applies to a system considered as free from failed components. • Postulates failures of hardware components according to failure modes caused by ageing, wearing, or stress. • Analyzes the consequences of these failures at system level.	• States the criticality and describes the measures[6] taken to prevent or mitigate the consequences. • Is only practice at functional level. • Applies to a system considered as containing software faults that may lead to failure under triggering conditions. • Postulates failures of software components according to functional failure modes caused by potential software faults. • Analyzes the consequences of these failures at system level.

General Motors Corporation (Detroit, MI) developed the QS 9000 standard in an effort to standardize supplier quality systems. QS 9000 is the automotive analogy to the better known standard ISO 9000. QS 9000-compliant automotive suppliers must use FMEA in the advanced quality planning process and in the development of their quality control plans. The effort made by the task force led to an industry-wide FMEA standard SAE J-1739 issued by the Society of Automotive Engineers (SAE) in 1994.

Academic discussion on FMEA originates from the 1960s when studies of component failures were broadened to include the effects of component failures on the system of which they were a part. One of the earliest descriptions of a formal approach for performing an FMEA was given at the New York Academy of Sciences (Coutinho, 1964). In the late 1960s and early 1970s, several professional societies published formal procedures for performing the analysis. The generic nature of the method assisted the rapid broadening of FMEA to different application areas, and various practices fundamentally using the same analysis method were created. Along with the digital revolution, the FMEA was applied in the analysis of software-based systems, and one of the first articles regarding SFMEA was given in 1979 (Reifer, 1979). Even thought there is no explicit standard for SFMEA, the standard IEC 60812, published in 1985, often is referred to when carrying out FMEA for software-based systems.

The failure mode and effects analysis for hardware or software has certain distinguishing characteristics. Ristord and Esmenjaud (2001) discussed these characteristics, and they are listed in Table 16.1.

FMEA methodology was started on software-based systems. A historic progression of major contributions is listed in Table 16.2.

[6]Measures, for example, can show that a fault leading to the failure mode necessarily will be detected by the tests performed on the component or will demonstrate that there is no credible cause leading to this failure mode because of the software design and coding rules applied.

TABLE 16.2 Major Software Failure Mode and Effect Analysis Research Contributions

Year	Reference	Contribution
1993	• Goddard, P.L, Validating the safety of embedded real-time control systems using FMEA, Proceedings Annual Reliability and Maintainability Symposium, pp. 227–230, 1993.	• Goddard: • Described the use of software FMEA at Hughes Aircraft. Goddard noted that performing the software FMEA as early as possible allows early identification of potential failure modes. • Pointed out that a static technique like FMEA cannot fully assess the dynamics of control loops.
	• Fenelon, P. & McDermid, J.A., An Integrated Tool set for software safety analysis, The Journal of Systems and Software, 21, pp. 279–290, 1993.	• Fenelon and McDermid: • Pointed out that FMEA is highly labor intensive and relies on the experience of the analysts.
1995	• Banerjee, N., Utilization of FMEA concept in software lifecycle management. Proceedings of Conference on Software Quality Management, pp. 219–230, 1995.	• Banerjee: • Provided an insightful look at how teams should use FMEA in software development. FMEA requires teamwork and the pooled knowledge of all team members. Many potential failure modes are common to a class of software projects. • Pointed out that the corresponding recommended actions are also common. Good learning mechanisms in a project team or in an organization greatly increase the effectiveness of FMEA. FMEA can improve software quality by identifying potential failure modes. • Stated that FMEA can improve productivity through its prioritization of recommended actions.
	• Luke, S.R., Failure mode, effects and criticality analysis (FMECA) for software. 5th Fleet Maintenance Symposium, Virginia Beach, VA (USA), 24–25 Oct 1995, pp. 731–735, 1995.	• Luke: • Discussed the use of FMEA for software. He pointed out that early identification of potential failure modes is an excellent practice in software development because it helps in the design of tests to check for the presence of failure modes. In FMEA, a software failure may have effects on the current module, on higher level modules, and on the system as a whole. • Suggested that a proxy such as historical failure rate be substituted for occurrence.

TABLE 16.2 ***(Continued)***

Year	Reference	Contribution
	• Stamatis, D. H., Failure Mode and Effect Analysis: FMEA from Theory to Execution, Milwaukee, ASQC Quality Press, 1995.	• Stamatis: • Presented the use of FMEA with information systems. • Noted that computer industry failures may result from software development process problems, coding, systems analysis, systems integration, software errors, and typing errors. • Pointed out that failures may develop from the work of testers, developers, and managers. • Noted that a detailed FMEA analysis may examine the source code for errors in logic and loops, parameters and linkage, declarations and initializations, and syntax.
1996	• Becker, J.C. & Flick, G, A Practical Approach to Failure Mode, Effects and Criticality Analysis (FMECA) for computing systems. High–Assurance Systems Engineering Workshop, pp. 228–236, 1996.	• Becker and Flick: • Applied FMEA in Lockheed Martin's development of a distributed system for air-traffic control. • Described the failure modes and detection methods used in their FMEA. The classes of failure modes for their application included hardware or software stop, hardware or software crash, hardware or software hang, slow response, startup failure, faulty message, check-point file failure, internal capacity exceeded, and loss of service. • Listed several detection methods. A task heartbeat monitor is coordination software that detects a missed function task heartbeat. A message sequence manager checks message sequence numbers to flag messages that are not in order. A roll call method takes attendance to ensure that all members of a group are present. A duplicate message check looks for the receipt of duplicate messages.
	• Lutz, R.R & Woodhouse, R.M., Experience Report: Contributions of SFMEA to Requirements Analysis, Proceedings of ICRE '96, April 15–18,1996, Colorado Springs, CO, pp. 44–51, 1996.	• Lutz and Woodhouse: • Described their use of software FMEA in requirements analysis at the Jet Propulsion Laboratory. Software FMEA helped them with the early understanding of requirements, communication, and error removal. • Noted that software FMEA is a time-consuming, tedious, manual task.

(Continued)

TABLE 16.2 ***(Continued)***

Year	Reference	Contribution
		Software FMEA depends on the domain knowledge of the analyst. • Stated that a complete list of software failure modes cannot be developed.
	• Goddard, P.L., A Combined Analysis Approach to Assessing Requirements for Safety Critical Real-time Control Systems, Proceedings Annual Reliability and Maintainability Symposium, pp.110–115, 1996.	• Goddard: • (1996) reported that a combination of Petri nets and FMEA improved the software requirements analysis process at Hughes Aircraft.
1997	• Moriguti, S., Software Excellence: A Total Quality management guide. Portland, Productivity Press, 1997.	• Moriguti: • Provided a thorough examination of total quality management for software development. • Presented an overview of FMEA. The book pointed out that FMEA is a bottom-up analysis technique for discovering imperfections and hidden design defects. • Suggested performing the FMEA on subsystem-level functional blocks. • Noted that when FMEA is performed on an entire product, the effort often quite large. • Pointed out that using FMEA before the fundamental design is completed can prevent extensive rework. • Explained that when prioritization is emphasized in the FMEA, the model sometimes is referred to as failure modes, effects and criticality analysis (FMECA).
	• Ammar, H.H., Nikzadeh, T. & Dugan, J.B., A Methodology for Risk Assessment of functional Specification of software systems using Colored Petri Nets, Proceedings of Fourth International Soft-ware Metrics Symposium, pp. 108–117, 1997.	• Ammar, Nikzadeh, and Dugan: • Used severity measures with FMEA for a risk assessment of a large-scale spacecraft software system. • Noted that severity considers the worst potential consequence of a failure whether degree of injuries or system damages. • Used four severity classifications. Catastrophic failures are those that may cause death

TABLE 16.2 ***(Continued)***

Year	Reference	Contribution
		or system loss. Critical failures are failures that may cause severe injury or major system damage that result in mission loss. Marginal failures are failures that may cause minor injury or minor system damage that results in delay or loss of availability or mission degradation. Minor failures are not serious enough to cause injuries or system damage but result in unscheduled maintenance or repair.
	• Maier, T., FMEA and FTA to support Safe Design of Embedded software in Safety–Critical Systems. Safety and Reliability of Software Based Systems, Twelfth Annual CSR Workshop, pp. 351–367, 1997.	• Maier: • Described the use of FMEA during the development of robot control system software for a fusion reactor. • Used FMEA to examine each software requirement for all possible failure modes. Failure modes included an unsent message, a message sent too early, a message sent too late, a wrong message, and a faulty message. FMEA causes included software failures, design errors, and unforeseen external events. • Noted that for software failures, additional protective functions to be integrated in the software may need to be defined. For design errors, the errors may need to be removed, or the design may need to be modified. • Stated that unforeseen external events may be eliminated by protective measures or by changing the design. • Recommended that the methodology he presented be applied at an early stage of the software development process to focus development and testing efforts.
1998	• Pries, K.H., Failure Mode and Effects Analysis in Software Development, SAE Technical Paper Series No. 982816, Warrendale, PA, Society of Automotive Engineers, 1998.	• Pries: • Outlined a procedure for using software design FMEA. • Stated that software design FMEA should start with system or subsystem outputs listed in the item and function (left-most) columns of the FMEA. The next steps are to list potential failure modes, effects of failures, and potential causes.

(Continued)

TABLE 16.2 ***(Continued)***

Year	Reference	Contribution
		• Noted that current design controls can include design reviews, walk throughs, inspections, complexity analysis, and coding standards. • Argued that because reliable empirical numbers for occurrence values are difficult or impossible to establish, FMEA teams can set all occurrences to a value of 5 or 10. • Noted that detection numbers are highly subjective and heavily dependent on the experience of the FMEA team.
	• Bouti, A., Kadi, D.A. & Lefrançois, P., An Integrative Functional Approach for Automated Manufacturing systems modeling. Integrated Computer-Aided Engineering, 5(4), pp. 333–348, 1998.	• Bouti, Kadi, and Lefrançois: • Described the use of FMEA in an automated manufacturing cell. • Noted that a good functional description of the system is necessary for FMEA. • Recommended the use of an overall model that clearly specifies the system functions. • Suggested the use of system modeling techniques that facilitate communication and teamwork. • Argued that it is impossible to perform a failure analysis when functions are not well defined and understood. • Pointed out that failure analysis is possible during the design phase because the functions are well established by then. • Noted that when several functions are performed by the same component, possible failures for all functions should be considered.
	• Stålhane, T. & Wedde, K.J., Modification of Safety Critical Systems: An Assessment of three Approached, Microprocessors and Microsystems, 21(10), pp. 611–619, 1998.	• Stålhane and Wedde: • Used FMEA with a traffic control system in Norway. • Used FMEA to analyze changes to the system. • Noted that potentially any change involving an assignment or a procedure call can change system parameters in a way that could compromise the system's safety. The FMEA pointed out code segments or procedures requiring further investigation.

TABLE 16.2 ***(Continued)***

Year	Reference	Contribution
		• Stated that for an FMEA of code modifications, implementation, and programming language knowledge is very important.
	• Pfleeger, S.L., Software engineering: Theory and practice. Upper Saddle River, Prentice Hall, New Jersey, 1998.	• Pfleeger: • Pointed out that FMEA is highly labor intensive and relies on the experience of the analysts. Lutz and Woodhouse stated that a complete list of software failure modes cannot be developed.
2000	• Goddard, P.L., Software FMEA Techniques, Proceedings Annual Reliability and Maintainability Symposium, pp. 118–123, 2000.	• Goddard: • Stated that there are two types of software FMEA for embedded control systems: system software FMEA and detailed software FMEA. System software FMEA can be used to evaluate the effectiveness of the software architecture without all the work required for detailed software FMEA. • Noted that system software FMEA analysis should be performed as early as possible in the software design process. This FMEA analysis is based on the top-level software design. • Stated that the system software FMEA should be documented in the tabular format used for hardware FMEA. • Stated that detailed software FMEA validates that the software has been constructed to achieve the specified safety requirements. Detailed software FMEA is similar to component-level hardware FMEA. • Noted that the analysis is lengthy and labor intensive. • Pointed out that the results are not available until late in the development process. • Argued that detailed software FMEA is often cost effective only for systems with limited hardware integrity.

(Continued)

TABLE 16.2 ***(Continued)***

Year	Reference	Contribution
	• Ristord, L. & Esmenjaud, C., FMEA Per-oredon the SPINLINE3 Operational System Software as Part of the TIHANGE 1 NIS Refurbishment Safety Case. CNRA/CNSI Workshop 2001–Licens-ing and Operating Experience of Computer Based I&C Systems. Ceské Budejovice–September 25–27, 2001.	• Ristord & Esmenjaud • Stated that the software FMEA is practicable only at the (application) function level. They consider the SPINLINE 3 application software to consist of units called blocks of instructions (BIs) executed sequentially. The BIs are defined by having the following properties: • BIs are either "intermediate"—they are a sequence of smaller BIs—or "terminal"—they cannot be decomposed in smaller BIs. • They have only one "exit" point. They produce output results from inputs and possibly memorized values. Some BIs have direct access to hardware registers. • They have a bounded execution time (i.e., the execution time is always smaller than a fixed value). • They exchange data through memory variables. A memory variable most often is written by only one BI and may be read by one or several BIs. • List of five general purpose failure modes at processing unit level: • The operating system stops • The program stops with a clear message • The program stops without clear message • The program runs, producing obviously wrong results • The program runs, producing apparently correct but in fact wrong results.

16.3 SFMEA FUNDAMENTALS

The failure mode and effects analysis procedures originally were developed in the post-World War II era for mechanical and electrical systems and their production processes, before the emergence of software-based systems in the market. Common standards and guidelines, even today, only briefly consider the handling of the malfunctions caused by software faults and their effects in FMEA and often state that this is possible only to a limited extent (IEC 60812). The standards procedures

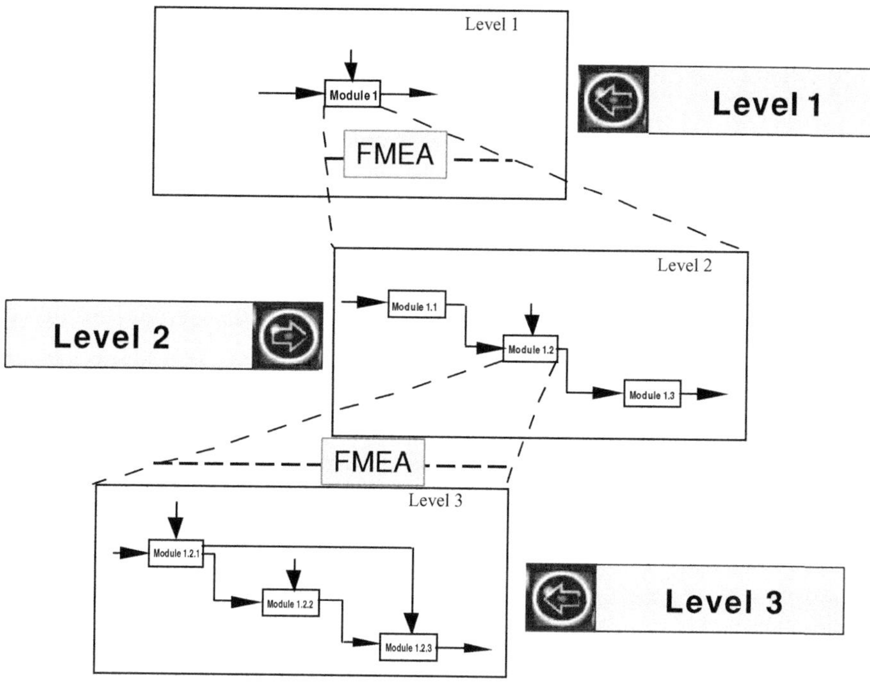

FIGURE 16.3 SFMEA hierarchy.

constitute a good starting point also for the FMEA for software-based systems. Depending on the objectives, level, and so on. of the specific FMEA this procedure easily can be adapted to the actual needs case by case (Haapanen et al., 2000).

In this section, we focus on the software failure modes, effects in the failure mode, effects analysis of a software-based control, and automation system application. A complete FMEA for a software-based automation system should include both the hardware and software failure modes and their effects on the final system function. In this section, however, we limit ourselves only to the software part of the analysis; the hardware part being discussed more in Yang and El-Haik (2008) and El-Haik and Mekki (2008) with the DFSS framework.

FMEA is documented on a tabular worksheet; an example of a typical FMEA worksheet is presented in Figure 16.3; this readily can be adapted to the specific needs of each actual FMEA, application.

Risk analysis is a quantitative extension of the (qualitative) FMEA, as described in Chapter 15. Using the failure effects identified by the FMEA, each effect is classified according to the severity of damage it causes to people, property, or the environment. The frequency of the effect to come about, together with its severity, defines the criticality. A set of severity and frequency classes are defined, and the results of the analysis is presented in the criticality matrix. The SAE J-1739 standard adds a third aspect to the criticality assessment by introducing the concept of a risk priority number (RPN), defined as the product of three entities, severity, occurrence (i.e., frequency), and detection (Haapanen et al., 2000).

A SFMEA can be described[7] as complementary to the process of defining what software must do to satisfy the user—the customer. In our case the process of "defining what software must do to satisfy the user—the customer" is what we entertain in the software DFSS project road map discussed in Chapter 11. The DFSS team may visit existing datum FMEA, if applicable, for further enhancement and updating. In all cases, the FMEA should be handled as a living document.

16.3.1 SFMEA Hierarchy

The FMEA is a bottom-up method in which the system under analysis first is divided hierarchically into components as in Figure 16.3. The division should be done in such a way that the failure modes of the components (modules) at the bottom level can be identified. We suggest the method of axiomatic design, as discussed in Chapter 13. The failure effects of the lower level components constitute the failure modes of the upper level components.

The basic factors influencing the selection of the proper lowest level of system decomposition are the purpose of the analysis and the availability of system design information.

When considering the SFMEA, the utmost purpose of the analysis usually is to find out whether there are some software faults that, in some situation, could jeopardize the proper functioning of the system. The lowest level components from which the analysis is started are then units of software executed sequentially in a single processor or concurrently in a parallel processor of the system (Haapanen et al., 2000).

For control-based software, a well-established way to realize software-based safety-critical automation applications is to implement the desired functions on an automation system platform (e.g., on a programmable logic system or on a more general automation system). The software in this kind of realization, is divided into system software and application software. The system software (a simple operating system) can be divided further into the system kernel and system services. Examples of the kernel functions include the system boot, initialization, self-tests and so on, whereas the system services, for example, take care of different data handling operations. The platform also includes a library of standardized software components with the function blocks ("modules"), of which the application is constructed by connecting ("configuring") adequate function blocks to form the desired application functions, which rest on the system service support (Haapanen et al., 2000).

A natural way of thinking then would suggest that the FMEA of a software-based application could be started from the function block diagrams by taking the individual function blocks as the lowest level components in the analysis. In practice, this procedure, however, seems unfeasible. First, this approach, in most cases, leads

[7] See AIAG FMEA Handbook, 2002.

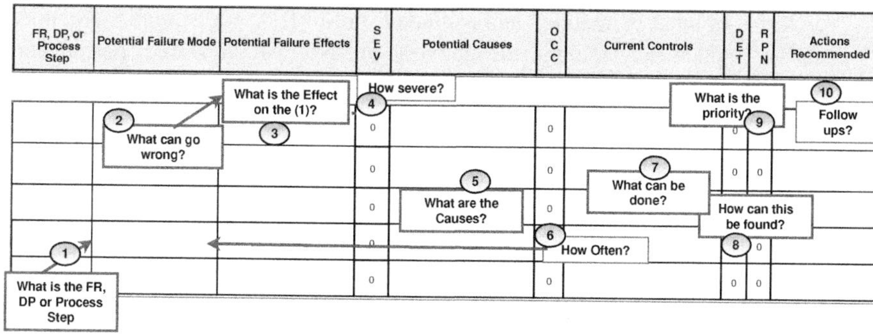

FIGURE 16.4 SFMEA worksheet.

to rather extensive and complicated analyses, and second, the failure modes of the function blocks are not known.

16.3.2 SFMEA Input

The IEC 60812[8] standard defines rather comprehensively the information needed for the general FMEA procedure. It emphasizes the free availability of all relevant information and the active cooperation of the designer. The main areas of information in this standard are: system structure, system initiation, operation, control and maintenance, system environment, modeling, system boundary, definition of the system's functional structure, representation of system structure, block diagrams, and failure significance and compensating provisions (Haapanen et al., 2000).

A well-documented software-based system design mostly covers these items, so it is more the question of the maturity of the design process than the specialties of software-based system.

16.3.3 SFMEA Steps

The fundamentals of an FMEA inputs, regardless of its type, are depicted in Figure 16.4 and in the list below:

1. Define scope, the software functional requirements and design parameters and process steps: For the DFSS team, this input column easily can be extracted from the functions and mappings discussed in Chapter 13. However, we suggest doing the FMEA exercise for the revealed design hierarchy resulting from the employment of mapping techniques of their choice. At this point, it may be useful to revisit the project scope boundary as input to the FMEA of interest

[8]IEC 60812 gives guidance on the definition of failure modes and contains two tables of examples of typical failure modes. They are, however, largely rather general and/or concern mainly mechanical system thus not giving much support for software FMEA.

in terms of what is included and excluded. In SFMEA, for example, potential failure modes may include the delivery of "No" FR delivered, partial and degraded FR delivery over time, intermittent FR delivery, and unintended FR (not intended in the mapping).

2. Identify potential failure modes: Failure modes indicate the loss of at least one software FR. The DFSS team should identify all potential failure modes by asking "in what way does the software fail to deliver its FRs?" as identified in the mapping. A potential failure mode can be a cause or an effect in a higher level subsystem, causing failure in its FRs. A failure mode may occur, but it must not necessarily occur. Potential failure modes may be studied from the baseline of past and current data, tests, and current baseline FMEAs.

   For the software components, such information does not exist, and failure modes are unknown (if a failure mode would be known, then it would be corrected). Therefore, the definition of failure modes is one of the hardest parts of the FMEA of a software-based system (Haapanen et al., 2000). The analysts have to apply their own knowledge about the software and postulate the relevant failure modes. Reifer (1979) suggested failure modes in major categories such as computational, logic, data I/O, data handling, interface, data definition, and database. Ristord and Esmenjaud (2001) proposed five general purpose failure modes at a processing unit level: 1) the operating system stops, 2) the program stops with a clear message, 3) the program stops without a clear message, 4) the program runs, producing obviously wrong results, and 5) the program runs, producing apparently correct but, in fact, wrong results. Lutz and Woodhouse (1999) divide the failure modes concerning either the data or the processing of data. For each input and each output of the software component, they considered four major failure modes classification: 1) missing data (e.g., lost message or data loss resulting from hardware failure), 2) incorrect data (e.g., inaccurate or spurious data), 3) timing of data (e.g., obsolete data or data arrives too soon for processing), and 4) extra data (e.g., data redundancy or overflow). For step in processing, they consider of the following four failure modes: 1) halt/abnormal termination (e.g., hung or deadlocked, at this point), 2) omitted event (e.g., event does not take place, but execution continues), 3) incorrect logic (e.g., preconditions are inaccurate; event does not implement intent), and 4) timing/order (e.g., event occurs in wrong order; event occurs too early or too late). Becker and Flick (1996) give the following classes of failure modes: 1) hardware or software stop, 2) hardware or software crash, 3) hardware or software hang, 4) slow response, 5) startup failure, 6) faulty message, 7) checkpoint file failure, 8) internal capacity exceeded, and 9) loss of service. They also listed a detection method based on Haapanen et al. (2002):

   - A task heartbeat monitor is coordination software that detects a missed function task heartbeat
   - A message sequence manager checks the sequence numbers for messages to flag messages that are not in order

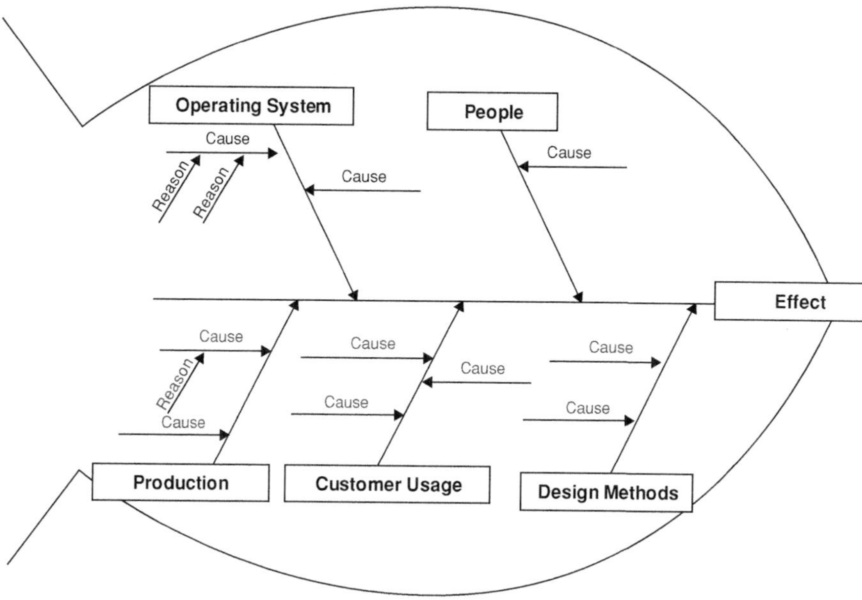

FIGURE 16.5 Cause and effect diagram.

- A roll call method takes attendance to ensure that all members of a group are present
- A duplicate message looks for the receipt of duplicate messages

  The FMEA also includes the identification and description of possible causes for each possible failure mode. Software failure modes are caused by inherent design faults in the software; therefore, when searching the causes of postulated failure modes, the design process should be looked at. IEC 60812 gives a table of possible failure causes, which largely are also applicable for software.

3. Potential failure effects(s): A potential effect is the consequence of the failure on other entities, as experienced by the user. The relation between effects and their causes usually is documented in a cause-and–effect (fishbone diagram/ishikawa diagram) diagram similar to the one depicted in Figure 16.5.
4. Severity: Severity is a subjective measure of "how bad" or "how serious" is the effect of the failure mode. Usually severity is rated on a discrete scale from 1 (no effect) to 10 (hazardous effect). Severity ratings of 9 or higher (4 or higher on 5 scale) indicate a potential special effect that needs more attention, and this typically is a safety or government regulation issue (Table 15.3 is reproduced as Table 16.2). Severe effects usually are classified as "catastrophic," "serious," "critical," "marginal," and "negligible." "Catastrophic" effects are usually a safety issue and require deeper study for all causes to the lowest level, possibly

using fault tree analysis[9] (FTA). "Serious" elements are important for the design itself. "Critical" elements are regulated by the government for any public concern.

The failure effects are propagated to the system level, such as the flight management system (FMS) in which severity designations are associated with each failure mode. A FMS crash probably will cause the mission to be abandoned, which conventionally is considered "Serious." Crash of a flight control system may jeopardize the safety of the aircraft and would be considered "Catastrophic." Failures that impair mission effectiveness (short of abandonment) are designated "Critical," and all others are considered "Marginal."

Depending on application, the reliability assessment can deal exhaustively with all failure modes that lead to severity "Catastrophic" and "Serious" failures (Table 16.3) and summarize the protection against other types of failure severity. For the highest severity failure modes, it is essential that detection (step 8 of this section) is direct and close to the source and that compensation is immediate and effective, preferably by access to an alternate routine or stand-by processor. For the lower severity failure modes, detection by effect (removed from the source) can be acceptable, and compensation by default value or retry can be used. Where gaps are found, the required corrective action in most cases is obvious. This severity treatment, tied to system effects, is appropriate for management review and may be preferred to one using failure rates.

The reliability assessment has an important legacy to test; once a failure mode is covered by detection and compensation provisions, the emphasis in test can shift to testing these provisions with fewer resources allocated to testing the functional code. Because detection and compensation provisions take a limited number of forms, test case generation is simplified, and the cost of test is reduced.

A control plan is needed to mitigate the risks for the catastrophic and the serious elements. The team needs to develop proactive design recommendations.

Potential causes: Generally, these are the set of noise factors and the deficiencies designed in resulting from the violation of design principles, axioms, and best practices (e.g., inadequate assumptions). The study of the effect of noise factors helps the software DFSS team identify the mechanism of failure. The analysis conducted by the team with the help of the functional decomposition (Chapter 13) allows for the identification of the interactions and coupling of their scoped project with the surrounding environment. For each potential failure mode identified in Column 2, the DFSS team needs to enter a cause in this column.

5. Occurrence: Occurrence is the assessed cumulative subjective rating of the software failures that could occur throughout the intended life. In other words, the likelihood of the event "the cause occurs." SFMEA usually assumes that if the cause happens then so does the failure mode. Based on this assumption,

[9] See Chapter 15.

TABLE 16.3 SFMEA Severity Rating

Severity of Hazard/Harm			
Criteria	Description	Rating 1–5	Rating 1–10
Catastrophic	***Product Halts/Process Taken Down/Reboot Required***: The product is completely hung up, all functionality has been lost, and system reboot is required	5	9–10
Serious	***Functional Impairment/Loss***: The problem will not resolve itself, and no "work around" can bypass the problem. Functionality either has been impaired or lost, but the product can still be used to some extent	4	7–8
Critical	***Functional Impairment/Loss***: The problem will not resolve itself, but a "work around" temporarily can bypass the problem area until fixed it is without losing operation	3	5–6
Marginal	***Product Performance Reduction***: Temporary through time-out or system load; the problem will "go away" after a period of time	2	3–4
Negligible	***Cosmetic Error***: No loss in product functionality. Includes incorrect documentation	1	1–2

occurrence also is the likelihood of the failure mode. Occurrence is rated on a scale of 1 (almost never) to 10 (almost certain) based on failure likelihood or probability, usually given in some probability metric as shown in Table 16.4[10]. In addition to this subjective rating, a regression correlation model can be used.

The occurrence rating is a ranking scale and does not reflect the actual likelihood. The actual likelihood or probability is based on the failure rate extracted from historical software or warranty data with the equivalent legacy software.

In SFMEA, design controls help in preventing or reducing the causes of failure modes, and the occurrence column will be revised accordingly.

6. Current controls: The objective of software design controls is to identify and detect the software nonconformities, deficiencies, and vulnerabilities as early as possible. Design controls usually are applied for *first-level failures in the respective hierarchy* (Figure 16.3). For hardware, a wide spectrum of controls is available like lab tests, project and design reviews, and modeling (e.g., simulation). In the case of a redesign software DFSS project, the team should review relevant (similar failure modes and detection methods experienced on surrogate software designs), historical information from the corporate memory. In the case of a white-sheet design, the DFSS team needs to brainstorm new

[10]Reproduced from Table 15.3.

TABLE 16.4 SFMEA Likelihood of Occurrence

Likelihood of Occurrence Rating			
Criteria	Description	Rating 1–5	Rating 1–10
Frequent	***Hazard/Harm likely to occur frequently***: 1 per 10 min (1/10) to 1+ per min (1/1)	5	9–10
Probable	***Hazard/Harm will occur several times during the life of the software***: 1 per shift (1/480) to 1 per hour (1/60)	4	7–8
Occasional	***Hazard/Harm likely to occur sometime during the life of the software***: 1 per week (1/10k) to 1 per day (1/1440)	3	5–6
Remote	***Hazard/Harm unlikely, but possible to occur during the life of the software***: 1 per 1 unit-year (1/525k) to 1 per 1 unit-month (1/43k)	2	3–4
Improbable	***Hazard/Harm unlikely to occur during the life of the software***: 1 per 100 unit-years (1/50m) to 1 per 10 unit-years (1/5m)	1	1–2

techniques for failure detection by asking: "In what means they can recognize the failure mode?" In addition, "how they can discover its occurrence?"

Design controls span a spectrum of different actions that include changes and upgrades (without creating vulnerabilities), special controls, design guidelines, DOEs, design verification plans, and modifications of standards, procedures, and best-practiced guidelines.

7. Detection: Detection is a subjective rating corresponding to the likelihood that the detection method will detect the first-level failure of a potential failure mode. This rating is based on the effectiveness of the control system through related events in the design algorithm; hence, FMEA is a living document. The DFSS team should:
8. Assess the capability of each detection method and how early in the DFSS endeavor each method will be used
9. Review all detection methods in column 8 and achieve a consensus on a detection rating
10. Rate the methods. Select the lowest detection rating in case of a tie.

    Examples of detection methods are assertions, code checks on incoming and outgoing data, and sequence checks on operations. See Table 16.5 for recommended ratings.
11. The ***product*** of severity (column 4), Occurrence (Column 6) and detection (Column 8) ratings. The range is between 1 and 1,000 (on a 1–10 scale) or between 1 and 125 (on a 1–5 scale).

TABLE 16.5 Software Detection Rating

Detection Rating Criteria	Description	Rating 1–5	Rating 1–10
Very remote detection	Detectable only once "online"	5	9–10
Remote detection	Installation and start-up	4	7–8
Moderate detection	System integration and test	3	5–6
High detection	Code walkthroughs/unit testing	2	3–4
Very high detection	Requirements/design reviews	1	1–2

RPN numbers are used to prioritize the potential failures. The severity, occurrence, and detection ratings are industry specific, and the belt should use his/her own company adopted rating system. A summary of the software ratings is provided in Table 16.6.

After the potential failure modes are identified, they are analyzed further by potential causes and potential effects of the failure mode (causes and effects analysis). For each failure mode, the RPN is assigned based on Tables 20.3–20.5. For all potential failures identified with an RPN score greater than a threshold (to be set by the DFSS team or accepted as a tribal knowledge), the FMEA team will propose recommended actions to be completed within the phase the failure was found (Step 10 below). A resulting RPN score must be recomputed after each recommended action to show that the risk has been mitigated significantly.

12. Actions recommended: The software DFSS team should select and manage recommended subsequent actions. That is, where the risk of potential failures is high, an immediate control plan should be crafted to control the situation.

    Here is a list of recommended actions:

- Transferring the risk of failure to other systems outside the project scope
- Preventing failure all together (e.g., software poka-yoke, such as protection shells)
- Mitigating risk of failure by:
    a. Reducing "severity" (most difficult)
    b. Reducing "occurrence" (redundancy and mistake-proofing)
    c. Increasing the "detection" capability (e.g., brainstorming sessions, concurrently or use top-down failure analysis like FTA[11])

    Throughout the course of the DFSS project, the team should observe, learn, and update the SFMEA as a dynamic living document. SFMEA is not retrospective but a rich source of information for corporate

[11] See Chapter 15.

TABLE 16.6 The Software FMEA Ratings

Rating	Severity of Effect	Likelihood of Occurrence	Detection
1	Cosmetic Error: No loss in product functionality. Includes incorrect documentation	1 per 100 unit-years (1/50m)	Requirements/ design reviews
2	Cosmetic Error: No loss in product functionality. Includes incorrect documentation	1 per 10 unit-years (1/5m)	Requirements/ design reviews
3	Product Performance Reduction: Temporary through time-out or system load the problem will "go away" after a period of time	1 perl unit-year (1/525k)	Code walk-throughs/unit testing
4	Product Performance Reduction: Temporary Through time-out or system load the problem will "go away" after a period of time	1 per 1 unit-month (1/43k)	Code walk-throughs/unit testing
5	Functional Impairment/Loss: The problem will not resolve itself, but a "work around" temporarily can bypass the problem area until fixed without losing operation	1 perweek (1/10k)	System integration and test
6	Functional Impairment/Loss: The problem will not resolve itself, but a "work around" temporarily can bypass the problem area until fixed without losing operation	1 per day (1/1440)	System integration and test
7	Functional Impairment/Loss: The problem will not resolve itself, and no "work around" can bypass the problem. Functionality either has been impaired or lost, but the product still can be used to some extent	1 per shift (1/480)	Installation and start-up
8	Functional Impairment/Loss: The problem will not resolve itself, and no "work around" can bypass the problem. Functionality either has been impaired or lost, but the product still can be used to some extent	1 per hour (1/60)	Installation and start-up
9	Product Halts/Process Taken Down/Reboot Required: The product is completely hung up, all functionality has been lost, and system reboot is required	1 per 10 min (1/10)	Detectable only once "online"
10	Product Halts/Process Taken Down/Reboot Required: The product is completely hung up, all functionality has been lost, and system reboot is required	1 + permin (1/1)	Detectable only once "online"

memory[12]. The DFSS team should document the SFMEA and store it in a widely acceptable format in the company in both electronic and physical media.

Software FMEA return on investment (ROI) is calculated in terms of a cost avoidance factor—the amount of cost avoided by identifying issues early in the software life cycle. This is calculated by multiplying the number of issues found by the software cost value of addressing these issues during a specific DFSS phase. The main purpose of doing a SFMEA is to catch defects in the associated DFSS phases (i.e., catching requirements defects in the identify phase, design defects in the conceptialize phase, and so on).

The ROI of SFMEA is many folds: more robust and reliable software, better quality of software, focus on defect prevention by identifying and eliminating defects in the software early developmental phases to help drive quality upstream, reduced cost of testing when measured in terms of *cost of poor quality* (COPQ). The proactive identification and elimination of software defects saves time and money. If a defect cannot occur, then there will be no need to fix it. In addition, dividends can be gained with enhanced productivity by way of developing a higher quality software in less time—a competitive edge. Prioritization of potential failures based on risk helps support the most effective allocation of people and resources to prevent them.

Because the SFMEA technique requires detailed analysis of expected failures, it results in a complete view of potential issues, leading to more informed and clearer understanding of risks in the software. Engineering knowledge persists in future software development projects and iterations. This helps an organization avoid relearning what is already known, guide design and development decisions, and gear testing to focus on areas where more testing is needed.

Practically, the potential time commitment required can discourage satellite DFSS team members' participation. Focus area documentation does not exist prior to the SFMEA session and needs to be created, adding to the time needed. Generally, the more knowledgeable and experienced the session participants are, the better the SFMEA results. The risk is that key individuals are often busy and, therefore, are unable or unwilling to participate and commit their time to the process.

16.4 SOFTWARE QUALITY CONTROL AND QUALITY ASSURANCE

Control plans are the means to sustain any software DFSS project findings. However, these plans are not effective if not implemented within a comprehensive software

[12]Companies should build a "Corporate Memory" that will record the design best practices, lesson learned, transfer functions and retain what corrective actions were attempted and what did and did not work and why. This memory should include pre-and postremedy costs and conditions including examples. This is a vital tool to apply when sustaining good growth and innovation strategies and avoiding attempted solutions that did not work. An online "Corporate Memory" has many benefits. It offers instant access to knowledge at every level of management and design staff.

quality operating system. A solid quality system can provide the means through which the DFSS project will sustain its long-term gains. Quality system certifications are becoming a customer requirement and a trend in many industries. The verify and validate phase of identify, conceptualize, optimize, and verify/validate (ICOV) DFSS algorithm requires that a solid quality system be employed in the DFSS project area.

The quality system objective is to achieve customer satisfaction by preventing nonconformity at all developmental stages. A quality system is the company's agreed upon method of doing business. It is not to be confused with a set of documents that is meant to satisfy an outside auditing organization (i.e., ISO 9000). That is, a quality system represents the actions not the written words of a company. The elements of an effective quality system include a quality mission statement, management reviews, company structure, planning, design control, data control, purchasing quality-related functions (e.g., supplier evaluation and incoming inspection), structure for traceability, process control, process monitoring and operator training, capability studies, measurement system analysis (MSA), audit functions, inspection and testing, software, statistical analysis, standards, and so on.

Two functions are needed: "assurance" and "control." Both can be assumed by different members of the team or outsourced to the respective concerned departments. In software, the "control" function is different from the "assurance" function.

Software quality assurance is the function of software quality that assures that the standards, processes, and procedures are appropriate for the project and are implemented correctly. Software quality assurance consists of a means of monitoring the software development processes and methods used to ensure quality. The methods by which this is accomplished are many and varied and may include ensuring conformance to one or more standards, such as ISO 9000 or capability maturity model integration (CMMI). However, software quality control is the function of software quality that checks that the project follows its standards processes and procedures and that the software DFSS project produces the required internal and external (deliverable) products. These terms seem similar, but a simple example highlights the fundamental difference. Consider a software project that includes requirements, user interface design, and an structured query language (SQL) database implementation. The DFSS team would produce a quality plan that would specify any standards, processes, and procedures that apply to the example project. These might include, for example, IEEE X specification layout (for the requirements), Motif style guide A (for the user interface design), and Open SQL standards (for the SQL implementation). All standards processes and procedures that should be followed are identified and documented in the quality plan; this is done by the assurance function.

When the requirements are produced, the team would ensure that the requirements, did infact, follow the documented standard (in this case, IEEE X). The same task, by team quality control function, would be undertaken for the user interface design and the SQL implementation; that is, they both followed the standard identified by the assurance function. Later, this function of the team could make audits to verify that IEEE X and not IEEE A indeed was used as the requirements standard. In this way, a difference between correctly implemented by the assurance function followed by a control function clearly can be drawn.

In addition, the software quality control definition implies software testing, as this is part of the project that produces the required internal and external (deliverable) products definition for software quality control. The term required refers not only to the functional requirements but also to the nonfunctional aspects of supportability, performance and usability, and so on. All requirements are verified or validated (V phase of ICOV DFSS road map, Chapter 11) by the control function. For the most part, however, it is the distinction around correctly implemented and followed for standards, processes, and procedures that gives the most confusion for the assurance and control function definitions. Testing normally is identified clearly with control, although it usually only is associated with functional requirement testing. We will discuss verification and validation matters in Chapter 19. The independent verification and validation (IV&V) and requirements verification matrix are used as verification and validation methods.

16.4.1 Software Quality Control Methods

Automated or manual control methods are used. The most used software control methods include:

- Rome laboratory software framework
- Goal question metric paradigm
- Risk management model
- The plan-do-check-action model of quality control
- Total software quality control
- Spiral model of software development

Control methods include fault tolerancing, mistake proofing (poka-yoke), statistical process control charting (SPC)[13]—with or without warning and trend signals applied to control the significant parameters/variables—standard operating procedures (SOP) for detection purposes, and short-term inspection actions. In applying these methods, the software DFSS team should revisit training to ensure proper control functions and to extract historical long-term and short-term information.

Control plans are the living documents in the production environment, which are used to document all control methods, as suggested by the SFMEA or yielded by other DFSS algorithm steps like optimization. The control plan is a written description of the systems for controlling software modules. The control plan should be updated to reflect changes of controls based on experience gained throughtout time. A customized form can be devised from Figure 16.6 (El-Haik and Mekki, 2008).

[13]SPC like X-bar and R or X and MR charts (manual or automatic), p and np charts (manual or automatic), c and u charts (manual or automatic), and so on.

Control Plan Worksheet

DF SS Team:								Date (Orig):			
								Date (Rev):			
							Current Control Plan				
Process Step	Input	Output	Process Spec (LSL, USL, Target)	Cpk / Date (Sample Size)	Measure ment System	%R&R or P/T	Current Control Method (from PEMEA)	Who?	Where?	When?	Reaction Plan?

FIGURE 16.6 Control plan worksheet.

16.5 SUMMARY

SFMEA is a proactive approach to defect prevention. SFMEA involves analyzing failure modes—potential or actual—rating and ranking the risk to the software, and taking appropriate actions to mitigate the risk. SFMEA is used to improve the quality of the work software during the DFSS road map and help reduce defects.

Failure modes are the ways or modes in which failures occur. Failures are potential or actual errors or defects. Effect analysis is the study of the consequences of these failures. Failures are prioritized according to how serious their consequences are, how frequently they occur, and how easily they can be detected. This technique helps software DFSS teams anticipate failure modes and assess their associated risks. Prioritized by potential risk, the riskiest failure modes then can be targeted to design them out of the software, or at least mitigate their effects. SFMEA also documents current knowledge and actions about the risks of failures for use in development and later on continuous improvements. Potential failure modes can be identified from many different sources. Some include brainstorming, bugs data, defect taxonomy, root cause analysis, security vulnerabilities and threat models, customer feedback, support issues, and corrective action fixes.

REFERENCES

AIAG FMEA Handbook, 2002.

Becker, J.C. and Flick, G. (1996), "A Practical Approach to Failure Mode, Effects and Criticality Analysis (FMEACA) for Computing Systems," High-Assurance Systems Engineering Workshop, Oct., pp. 228–236.

Coutinho, J.S. (1964), "Failure effect analysis." *Transactions of the New York Academy of Sciences*, pp. 564–585.

El-Haik, Basem S. and Mekki, K. (2008), *Medical Device Design for Six Sigma: A Road Map for Safety and Effectiveness*, 1st Ed., Wiley-Interscience, New York.

El-Haik, Basem S. and Roy, D. (2005), *Service Design for Six Sigma: A Roadmap for Excellence*, Wiley-Interscience, New York.

Haapanen, P. and Helminen, A. (2002), "*Failure Mode and Effects Analysis of Software-based Automation Systems*," STUK-YTO-TR 190, Helsinki, p. 35.

Haapanen, P., Helminen, A., and Pulkkinen, U. (2004), "Quantitative Reliability Assessment in the Safetycase of Computer-Based Automation Systems," VTT Industrial Systems STUK Report series, STUK-YTO-TR 202/May 2004, http://www.stuk.fi/julkaisut/tr/stuk-yto-tr202.pdf

Haapanen, P., Korhonen, J., and Pulkkinen, U. (2000), "*Licensing Process for Safety-critical Software-based Systems*," STUK-YTO-TR 171, Helsinki, p. 84.

IEC 60812 (2006), International Electrotechnical Commission (IEC), Second edition, 2006-01. Online: http://webstore.iec.ch/preview/info_iec60812%7Bed2.0%7Den_d.pdf

Lutz, R.R. and Woodhouse, R.M. (1999), "Bi-Directional Analysis for Certification of Safety-Criticial Software," Proceedings, ISACC'99, International Software Assurance Certification Conference, Feb.

Reifer, D.J. (1979), "Software failure modes and effects analysis." *IEEE Transactions on Reliability*, Volume R-*28*, #3, pp. 247–249.

Ristord, L. and Esmenjaud, C. (2001), "FMEA Per-oredon the SPINLINE3 Operational System Software as part of the TIHANGE 1 NIS Refurbishment Safety Case," CNRA/CNSI Workshop 2001-Licensing and Operating Experience of Computer Based I&C Systems, Ceske Budejovice, Czech Repubic, Sept.

Yang and El-Haik, Basem S. (2008). *Design for Six Sigma: A Roadmap for Product Development*, 2nd Ed., McGraw-Hill Professional, New York.

CHAPTER 17

SOFTWARE OPTIMIZATION TECHNIQUES

17.1 INTRODUCTION

Optimization is the third phase of the software identify, conceptualize, optimize, and verify/validate (ICOV) process (Chapter 11). Optimization is linked directly to software metrics (Chapter 5). In hardware, optimization has a very specific objective: minimizing variation and adjusting performance mean to the target, which may be static or dynamic in nature (El-Haik & Mekki, 2008). The DFSS methodology to achieve such an objective is called robust design. Application of robust design to software is presented in Chapter 18. However, software optimization is the process of modifying a software system in an effort to improve its efficiency.[1] One way software can be optimized is by identifying and removing wasteful computation in code, thereby reducing code execution time (LaPlante, 2005). However, there are several ways that software can be optimized, especially for real-time systems. Moreover, it also should be noted that software optimization can be executed on several different levels. That is, software can be optimized on a design level, a source code level, or even on a run-time level.

It is important to note that there may be tradeoffs when optimizing a system. For example, a very high level of memory may be compromised by another important factor, such as speed. More specifically, if a system's cache is increased, it improves the run-time performance but also will increase the memory consumption.

[1] http://www.wordiq.com/definition/Software_optimization

Software Design for Six Sigma: A Roadmap for Excellence, By Basem El-Haik and Adnan Shaout

This chapter discusses the following techniques in software optimization. The first topic, software optimization, discusses several popular metrics used to analyze how effective software actually is. This chapter also introduces several topics that are especially essential in a real-time system, including interrupt latency, time loading, memory requirements, performance analysis, and deadlock handling. In addition, this chapter discusses and gives several examples of performance and compiler optimization tools. Although all these topics are relevant to all types of computing systems, the effect each of these topics has on real-time systems will be highlighted.

17.2 OPTIMIZATION METRICS

Specifically, in software development, a metric is the measurement of a particular characteristic of a program's performance or efficiency.[2] It should be noted that there are several caveats to using software optimization metrics. First, as with just about any powerful tool, software metric must be used carefully. Sloppy metrics can lead to bad decision making and can be misused in an effort to "prove a point" (LaPlante, 2005). As LaPlante points out, a manager easily could point out that one of his or her team members is incompetent, based on some arbitrary metric, such as the number of lines of code written. Another caveat is the danger of measuring the correlation effects of a metric without a clear understanding of the causality. Metrics can be helpful and harmful at the same time; therefore, it is important to use them carefully and with a full understanding of how they work.

Some of the most common optimization metrics that will be discussed are:

1. Lines of source code
2. Function points
3. Conditional complexity
4. Halstead's Metrics
5. Cohesion
6. Coupling

Identifying performance is an important step before optimizing system. Some common parameters used to describe a system's performance include CPU utilization, turnaround time, waiting time, throughput, and response time.

17.2.1 Lines of Source Code[3]

One of the oldest metrics that has been used is the lines of source code (LOC) used. LOC, first was introduced in the 1960s and was used for measuring economics, productivity, and quality (Capers Jones & Associates, 2008). The economics of

[2]http://whatis.techtarget.com/definition/0,,sid9_gci212560,00.html

[3]See Chapter 5.

software applications were measured using "dollars per LOC," productivity was measured in terms of "lines of code per time unit," and quality was measured in terms of "defects per KLOC" where "K" was the symbol for 1,000 lines of code. However, as higher level programming languages were created, the LOC metric was not as effective. For example, LOC could not measure non coding activities such as requirements and design.

As time progressed from the 1960s until today, hundreds of programming languages developed, applications started to use multiple programming languages, and applications grew from less than 1,000 lines of code to millions of lines of code. As a result, the LOC metric could not keep pace with the evolution of software.

The lines of code metric does not work well when there is ambiguity in counting code, which always occurs with high-level languages and multiple languages in the same application. LOC also does not work well for large systems where coding is only a small fraction of the total effort. In fact, the LOC metric became less and less useful until about the mid-1980s, when the metric actually started to become harmful. In fact, in some types of situations, using the LOC metric could be viewed as a professional malpractice if more than one programming language is part of the study or the study seeks to measure real economic productivity. Today, a better metric to measure economic productivity for software is probably function point metrics, which is discussed in the next section.

17.2.2 Function Point Metrics

The function point metric generally is used to measure productivity and quality. Function points were introduced in the late 1970s as an alternative to the lines of code metric, and the basis of the function point metric is the idea that as a programming language becomes more powerful, fewer lines of code are necessary to perform a function (LaPlante, 2005). A function typically is defined as a collection of executable statements that perform a certain task.[4] The measure of software productivity is the number of functions a development team can produce given a certain amount of resources without regard to the number of lines of code. If the defect per unit of functions is low, then the software should have a better quality even though the defects per KLOC value could be higher.

The following five software characteristics for each module represent its function points:

Number of inputs to the application (I)
Number of outputs (O)
Number of under inquires (Q)
Number of files used (F)
Number of external interfaces (X)

[4] http://www.informit.com/articles/article.aspx?p=30306&rll=1

Each of these factors can be used to calculate the function point, where the calculation will depend on the weight of each factor. For example, one set of weighing factors might yield a function point value calculated as:

$$FP = 4I + 4O + 5Q + 10F + 7X$$

The complexity of a system can be adjusted accordingly and can be adapted to adjust for other types of applications, such as real-time systems. The function point metric mostly has been used in business processing; however, there is an increasing interest in using the function point metric in embedded systems. In particular, systems such as large-scale real-time databases, multimedia, and Internet support are data driven and behave like the large-scale transaction-based systems for which function points initially were developed.

Function point metrics have become the dominant metric for serious economic and quality studies (Capers Jones & Associates, 2008). However, several issues have kept function point metrics from becoming the industry standard for both economic and quality studies. First, some software applications are now so large that normal function point analysis is too slow and too expensive to be used. Second, the success of function points has triggered an explosion of function point clones, and as of 2008, there are at least 24 function point variations. The number of variations tends to make baseline studies difficult because there are very few conversion rules from one variation to another.

17.2.3 Conditional Complexity[5]

Conditional complexity also can be called cyclomatic complexity. Conditional complexity was developed in the mid-1970s by Thomas McCabe and is used to measure the complexity of a program.[6] Cyclomatic complexity sometimes is referred to as McCabe's complexity as well. This metric has two primary uses:

1. To indicate escalating complexity in a module as coded and assisting programmers in determining the size of a module
2. To determine the upper bound on the number of tests that must be run (LaPlante, 2005)

The complexity of a section of code is the count of the number of linearly independent paths through the source code. To compute conditional complexity, Equation (5.1) is used:

$$C = e - n + 2$$

[5] See Chapter 5.

[6] http://en.wikipedia.org/wiki/Cyclomatic_complexity

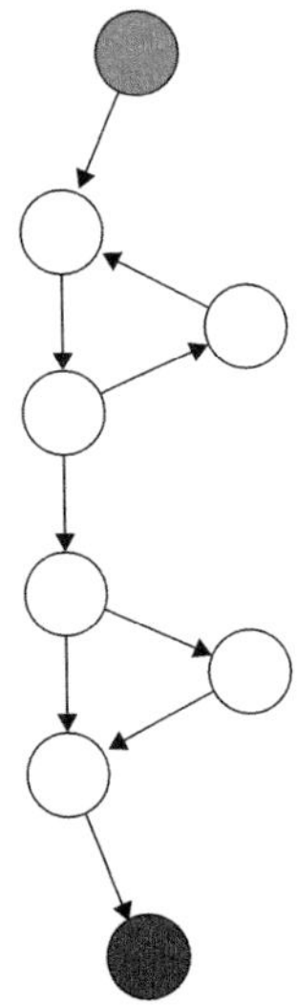

FIGURE 17.1 A control flow graph.

where, if a flow graph is provided, the nodes represent program segments and edges represent independent paths. In this situation, e is the number of edges, n is the number of nodes, and C is the conditional complexity. Using this equation, a conditional complexity with a higher number is more complex.

In Figure 17.1, the program begins at the red node and enters the loop with three nodes grouped immediately below the red node. There is a conditional statement located at the group below the loop, and the program exits at the blue node. For this graph, $e = 9$, $n = 8$, and $P = 1$, so the complexity of the program is 3.[7]

It often is desirable to limit the complexity. This is because complex modules are more error prone, harder to understand, harder to test, and harder to modify (McCabe, 1996). Limiting the complexity may help avoid some issues are associated with high-complexity software. It should be noted that many organizations successfully have implemented complexity limits, but the precise number to use as a limit remains up in the air. The original limit is 10 and was proposed by McCabe himself. This limit of 10 has significant supporting evidence; however, limits as high as 15 have been used as well.

Limits greater than 10 typically are used for projects that have several operational advantages over typical projects, for example, experienced staff, formal design, a modern programming language, structured programming, code walkthroughs, and a comprehensive test plan. This means that an organization can select a complexity limit greater than 10 but only if the organization has the resources. Specially, if a limit greater than 120 is used, then the organization should be willing to devote the additional testing effort required by more complex modules. There are exceptions to

[7] http://en.wikipedia.org/wiki/Cyclomatic_complexity

the complexity limit as well. McCabe originally recommended exempting modules including single multiway decision statements from the complexity limit.

Cyclomatic complexity has its own drawbacks as well. One drawback is that it only measures complexity as a function of control flow. However, complexity also can exist internally in the way that a programming language is used. Halstead's metrics are suitable for measuring how intensely the programming language is used.

17.2.4 Halstead's Metric[8]

The Halstead metric bases its approach on the mathematical relationships among the number of variables, the complexity of the code, and the type of programming language statements. The Halstead metric has been criticized for its difficult computations as well as its questionable methodology for obtaining some mathematical relationships.[9]

Some of Halstead's metrics can be computed using Section 5.3.2. Another metric is the amount of mental effort used to develop the code, which is E and is defined as $E = V/L$. Decreasing the effort will increase the reliability and implementation (LaPlante, 2005).

17.2.5 Cohesion

Cohesion is the measure of the extent to which related aspects of a system are kept together in the same module and unrelated aspects are kept out.[10] High cohesion implies that each module represents a single part of the problem solution; thus, if the system ever needs to be modified, then the part that needs to be modified exists in a single place, making it easier to change (LaPlante, 2002). In contrast, low cohesion typically means that the software is difficult to maintain, test, reuse, and understand. Coupling, which is discussed in greater detail in the next section, is related to cohesion. Specifically, a low coupling and a high cohesion are desired in a system and not a high coupling and a low cohesion.

LaPlante has identified seven levels of cohesion, and they are listed in order of strength:

1. Coincidental—parts of the module are not related but are bundled in the module
2. Logical—parts that perform similar tasks are put together in a module
3. Temporal—tasks that execute within the same time span are brought together
4. Procedural—the elements of a module make up a single control sequence
5. Communicational—all elements of a module act on the same area of a data structure

[8] See Chapter 5.
[9] http://cispom.boisestate.edu/cis320emaxson/metrics.htm
[10] http://www.site.uottawa.ca:4321/oose/index.html#cohesion

6. Sequential—the output of one part in a module serves as the input for an other part
7. Functional—each part of the module is necessary for the execution of a function

17.2.6 Coupling[11]

Coupling can be defined as the degree each program module relies on another program module. It is in a programmer's best interest to reduce coupling so that changes to one unit of code do not affect another. A program is considered to be modular if it is decomposed into several small, manageable parts.[12] The following is a list of factors in defining a manageable module: the modules must be independent of each other, the module implements an indivisible function, and the module should have only one entrance and one exit. In addition to this list, the function of a module should be unaffected by: the source of its input, the destination of its output, and the history of the module. Modules also should be small, which means that they should have less than one page of source code, less than one page of flowchart, and less than 10 decision statements.

Coupling also has been characterized in increasing levels, starting with:

1. *No direct coupling*—where all modules are unrelated
2. *Data*—when all arguments are homogenous data items
3. *Stamp*—when a data structure is passed from one module to another, but that module operates on only some data elements of the structure
4. *Control*—one module passes an element of control to another
5. *Common*—if two modules have access to the same global data
6. *Content*—one module directly references the contents of another

17.3 COMPARING SOFTWARE OPTIMIZATION METRICS

Some of the most effective response time techniques are probably cohesion and coupling. As discussed, the LOC metric is rather outdated and is usually not that effective anymore. This metric was used commonly in the 1960s but is not used much today. In fact, it could be viewed as a professional malpractice to use LOC as a metric if more than one programming language is part of the study or the study seeks to measure real economic productivity. Instead of using the LOC metric, some organizations look to use function point analysis.

Function point analysis was introduced in the late 1970s as an alternative to the LOC metric. Function point metrics have become the dominant metric for some types of economic and quality studies; however, there are several issues that have kept function point metrics from becoming the industry standard. As discussed,

[11] See Chapter 13.
[12] http://www.jodypaul.com/SWE/HAL/hal.html

TABLE 17.1 Summary of Optimization Metrics

Type of Metric	Comments	Ranking
Cohesion	High cohesion is an indication of a well-designed system	1
Coupling	Low coupling is an indication of a well-designed system	1
Cyclomatic Complexity	Only measures complexity as a function of control flow	2
Halstead's Metric	Difficult computations as well as questionable methodology for obtaining some mathematical relationships	2
Function Point	Dominant metric for some types of economic and quality studies. However, some software applications so large that normal function point analysis is too slow	2
Lines of Code	Outdated, has not been used widely since the mid-1980s	3

some software applications are now so large that normal function point analysis is too slow and too expensive to be used. Second, as of 2008, there are at least 24 function point variations in which the number of variations tends to make baseline studies difficult.

The next optimization metric, cyclomatic complexity, also has drawbacks, as it only measures complexity as a function of control flow. Instead, Halstead's metrics are suitable for measuring how intensely the programming language is used. However, the Halstead metric has been criticized for its difficult computations as well as its questionable methodology for obtaining some mathematical relationships.

In contrast to LOC, function point, cyclomatic complexity, and Halstead's metric, some simpler metrics to use are cohesion and coupling. Indeed, high cohesion combined with low coupling is a sign of a well-structured computer system and a good design. Such a system supports the goals of high readability and high maintainability. Table 17.1 is a table summarizing each optimization metric, comments, and a ranking in which 1 is the best, 2 is average, and 3 is worst. As seen in Table 17.1, both cohesion and coupling rank the highest, followed by cyclomatic complexity, Halstead's metric, function point analysis, and LOC.

Therefore, although there are many types of optimization techniques on the market today, some of the best optimization techniques are probably cohesion and coupling.

17.3.1 Response Time Techniques

Response time is the presentation of an input to a system and the realization of the required behavior including the availability of all associated outputs (LaPlante, 2005). Response time is important in real-time applications because it estimates the maximum amount of time until an event, such as when a communication from another task or an external input is serviced in the system (Na'Cul & Givargis, 1997). In a system with cyclic tasks and different task priorities, the response time determines the wait time of the tasks until they are granted access to the processor and put into a running state.

The response time for an embedded system usually will include three components, and the sum of these three components is the overall response time of the embedded system.[13] The components are:

1. The time between when a physical interrupt occurs and when the interrupt service routine begins. This is commonly known as the interrupt latency or the hardware interrupt latency
2. The time between when the interrupt service routine begins to run and when the operating system switches the tasks to the interrupt service thread (IST) that services the interrupt, known as scheduling latency
3. The time required for the high-priority interrupt to perform its tasks. This period is the easiest to control

Almost all real time operating systems employ a priority-based preemptive scheduler.[14] This exists despite the fact that real-time systems vary in their requirements. Although there are good reasons to use priority-based preemption in some applications, preemption also creates several problems for embedded software developers as well. For example, preemption creates excess complexity when the application is not well suited to being coded as a set of tasks that can preempt each other and may result in system failures. However, preemption is beneficial to task responsiveness. This is because a preemptive priority-based scheduler treats software tasks as hardware treats an Interrupt Service Routine (ISR). This means that as soon as the highest priority task ISR is ready to use the central processing unit (CPU), the scheduler (interrupt controller) makes it so. Thus, the latency in response time for the highest priority-ready task is minimized to the context switch time.

Specifically, most real-time operating systems use a fixed-priority preemptive system in which schedulability analysis is used to determine whether a set of tasks are guaranteed to meet their deadlines (Davis et al., 2008). A schedulability test is considered sufficient if all task-sets deemed to be schedulable by the test are, in fact, schedulable. A schedulability test is considered necessary if all task sets that are considered unschedulable actually are. Tests that are both sufficient and necessary are considered to be exact. Efficient exact schedulability tests are required for the admission of applications to dynamic systems at the runtime and the design of complex real-time systems. One of the most common fixed-priority assignments follows the rate monotonic algorithm (RMA). This is where the tasks' priorities are ordered based on activation rates. This means that the task with the shortest period has the highest priority.

[13] www.tmworld.com/article/CA1187159.html
[14] http://www.embedded.com/columns/technicalinsights/192701173?_requestid=343970

17.3.2 Interrupt Latency

As discussed in the previous section, interrupt latency is a component of response time and is the period of time between when a device requests the interrupt and when the first instruction for the hardware Interrupt Service Routine executes (LaPlante, 2005). In regard to real-time systems, it is important to calculate the worst-case interrupt latency of a system. Real-time systems usually have to disable interrupts while the system processes waiting threads.

An interrupt fires only when all of the following conditions are true:

1. The interrupt is pending.
2. The processor's master interrupt enable bit is set.
3. The individual enable bit for the interrupt is set.
4. The processor is in between executing instructions or else is in the middle of executing an interruptible instruction.
5. No higher priority interrupt meets conditions 1–4 (Regehr, 2008).

Because an interrupt only fires when all five of the conditions are met, all five factors can contribute to interrupt latency. The worst-case interrupt latency is the longest possible latency of a system. The worst-case latency usually is determined by static analysis of an embedded system's object code.

If the embedded system does not react in time, then degradation or failure of the operating system may occur, depending on whether it is a hard or soft real-time system.[15] Real-time capability generally is defined by interrupt latency and context switch time. Interrupts typically are prioritized and are nested. Thus, the latency of the highest priority interrupt usually is examined. Once the latency is known, it can be determined whether it is tolerable for a particular application. As a result, a real-time application will mandate certain maximum latencies to avoid failure or degradation of the system. If a system's worst-case interrupt latency is less than the application's maximum tolerable latency, then the design can work. Interrupt latency may be affected by several factors, including interrupt controllers, interrupt masking, and the operating system's interrupt handling methods.

In addition to other factors such as context switch time, interrupt latency is probably the most often analyzed and benchmarked measurement for embedded real-time systems.[16] Software actually can increase interrupt latency by deferring interrupt processing during certain types of critical operating system operations. The operating system does this by disabling interrupts while it performs critical sequences of instructions. The major component of worst-case interrupt latency is the number and length of these sequences. If an interrupt occurs during a period of time in which the operating system has disabled interrupts, then the interrupt will remain pending until software reenables interrupts which is illustrated in Figure 17.2.

[15] http://www.rtcmagazine.com/articles/view/100152

[16] http://www.cotsjournalonline.com/articles/view/100129

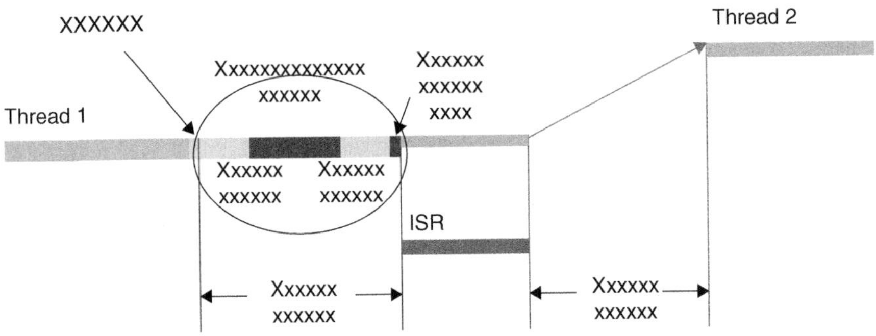

FIGURE 17.2 Interrupt events.

It is important to understand the worst-case interrupt disabling sequence, as a real-time system depends on the critical events in the system being executed within the required time frame.

17.3.3 Time Loading

The CPU utilization or time-loading factor U is a measure of the percentage of nonidle processing in a computer. A system is considered time overloaded if the CPU utilization is more than 100%. Figure 17.3[17] is an illustration of the typical CPU utilization zones and typical applications and recommendations.

A utilization of about 50% is common for new products; however, a CPU utilization of up to about 80% may be acceptable for a system that does not anticipate growth (LaPlante, 2005). A CPU utilization of about 70% is probably the most common and most recommended CPU utilization for a real-time system. However, there are several different opinions available. For example, one study indicates that system designers should strive to keep CPU use below 50%, as a CPU with a high utilization will lead to unpredictable real-time behavior. Also, it is possible that the high-priority tasks in the system will starve the low-priority tasks of any CPU time. This can cause the low-priority tasks to misbehave (Eventhelix.com, 2001).

CPU utilization, U, can be defined by the following:

$$U = 100\% - (\text{time spent in a idle task})$$

where the idle task is the task with the absolute lowest priority in a multitasking system.[18] This task also sometimes is called the background task or background loop. This logic traditionally has a while(1) type of loop in which an infinite loop spins the CPU waiting for an indication that critical work needs to be done.

[17] www.cse.buffalo.edu/~bina/cse321/fall2007/IntroRTSAug30.ppt

[18] http://www.design-reuse.com/articles/8289/how-to-calculate-cpu-utilization.html

Utilization %	Zone Type	Type of Application
0–25	CPU under utilized	General purpose
25–50	Very safe	,,
51–68	Safe	,,
69	Theoretical limit	Embedded system
700–82	Questionable	Embedded system
83–99	Dangerous	Embedded system

FIGURE 17.3 Typical CPU utilization zones and typical applications and recommendations.

The following is a simple example of a background loop:

```
int main( void )
{
 SetupInterrupts();
 InitializeModules();
 EnableInterrupts();
 while(1) /* endless loop - spin in the background */
 {
 CheckCRC();
 MonitorStack();
 ... do other non time critical logic here.
 }
}
```

This depiction is an oversimplification, as some kind of work often is done in the background task. However, the logic coded for execution during the idle task must have no hard real-time requirements. In fact, one technique that may be used in an overloaded system is to move some of the logic with less strict timing requirements out of the hard real-time tasks and into the idle task.

17.3.4 Memory Requirements

A system's memory can be important for a real-time system, as a computer's memory directly can influence the performance of a real-time system. In particular, a system's memory can affect access time. Access time is defined as the interval between when a datum is requested and when it is available to the CPU. The effective access time may depend on the memory type as well as the memory technology, the memory layout, and other various factors. In the last few years, memory has become cheaper and more plentiful. Thus, memory has become less of an issue than it was a decade or two ago. However, embedded real-time systems must be small, inexpensive, and efficient.

Moreover, embedded systems are used in smaller and more portable applications, making memory space smaller and at a premium. As a result, memory is still an issue.

One way to classify memory is through the term coined volatility. Volatile memories only hold their contents while power is applied to the memory device, and as power is removed, the memories lose their contents.[19] Volatile memories are unacceptable if data must be retained when the memory is switched off. Some examples of volatile memories include static random access memory (SRAM), and synchronous dynamic random access memory (SDRAM), which are discussed in greater detail subsequently.

In contrast, nonvolatile memories retain their contents when power is switched off. Items such as CPU boot-code typically are stored in nonvolatile memory. Although nonvolatile memory has the advantage of retaining its data when power is removed, it is typically much slower to write to than volatile memory and often has more complex writing and erasing procedures. Moreover, nonvolatile memory is also usually only erasable for a given number of times. Some types of nonvolatile memories include flash memory, erasable programmable read only memory (EPROM), and electrically erasable programmable read only memory (EEPROM), which also are discussed in greater detail subsequently. Most types of embedded systems available today use some type of flash memory for nonvolatile storage. Many embedded applications require both volatile and nonvolatile memories because the two memory types serve unique and exclusive purposes.

The main types of memory are random access memory (RAM), read only memory (ROM), and a hybrid of the two different types. The RAM family includes two important memory devices: static RAM (SRAM) and dynamic RAM (DRAM).[20] SRAM is retained as long as electrical power is applied to the chip, and DRAM has a short data lifetime of a few milliseconds. When deciding which type of RAM to use, a system designer must consider access time and cost. SRAM offers fast access times but are much more expensive to produce. DRAM can be used when large amounts of RAM are required. Most types of embedded systems include both types of memory in which a small block of SRAM and a large block of DRAM is used for everything else.

ROM memory can have new data written. Some types of ROM rewritten, reflect the evolution of ROM devices from hardwired to programmable to erasable and programmable. However, all ROM devices are capable of retaining data and programs forever. The first ROMs contained a preprogrammed set of data or instructions in which the contents of the ROM had to be specified before chip production. Hardwired memories still can be used, and are called masked ROM. The primary advantage of a masked ROM is its low production cost. PROM (programmable ROM or a one-time programmable device) is purchased in an unprogrammed state. A device programmer writes data to the PROM one word at a time by applying an electrical charge to the input pins of the chip. Once a PROM has been programmed in this way,

[19] http://www.altera.com/literature/hb/nios2/edh_ed51008.pdf

[20] http://www.netrino.com/Embedded-Systems/How-To/Memory-Types-RAM-ROM-Flash

its contents never can be changed. An erasable-and-programmable ROM (EPROM) is programmed in exactly the same manner as a PROM but can be erased and reprogrammed repeatedly. To erase an EPROM, the device should be exposed to a strong source of ultraviolet light.

Nowadays, several types of memory combine features of both RAM and ROM. These devices do not belong to either group and can be referred to collectively as hybrid memory devices, which include EEPROM and flash, and nonvolatile random access memory (NVRAM). EEPROMs are electrically erasable and programmable, but the erase operation is accomplished electrically, rather than by exposure to ultraviolet light, like EPROM. Any byte within an EEPROM may be erased and rewritten.

Flash memory originally was created as a replacement for mass storage media such as floppy and hard disks and is designed for maximum capacity and density, minimum power consumption, and a high number of write cycles.[21] However, it should be noted that all nonvolatile solid-state memory can endure a limited number of write cycles. Information stored in flash memory usually is written in blocks rather than one byte or word at a time. Despite this, flash memory is still more preferred than EEPROM and is rapidly displacing many of the ROM devices as well.[22]

There are generally two main types of flash memory—linear flash and advanced technology attachment (ATA) flash.[23] Linear flash is laid out and addressed linearly, in blocks, where the same address always maps to the same physical block of memory, and the chips and modules contain only memory with address decoding and buffer circuits. This makes linear memory relatively simple and energy-efficient. This type of memory typically is used for nonvolatile memory that is permanently part of an embedded system. The ATA flash memory module interfaces with the rest of the system using the AT Attachment standard in which the memory seems as if it were sectors on a hard disk. The main advantages of an ATA flash are flexibility and interchangeability with hard disks, as linear flash modules are not completely interchangeable between devices that accept removable memory modules.

The third member of the hybrid memory class is NVRAM (nonvolatile RAM). An NVRAM is basically an SRAM with battery backup, and when power is supplied, the NVRAM operates just like SRAM. When the power is turned off, the NVRAM draws just enough power from the battery to retain its data. NVRAM is fairly common in embedded systems but is even more expensive than SRAM because of the battery.[24]

Figure 17.4 is a useful illustration of the different classifications of memory that typically are used in embedded systems.

Table 17.2 (LaPlante, 2005) is a summary of the memory discussed; however, different memory types serve different purposes, and each type of memory has its own strengths and weaknesses.

[21] http://www.embedded.com/98/9801spec.htm

[22] http://www.netrino.com/Embedded-Systems/How-To/Memory-Types-RAM-ROM-Flas

[23] http://www.embedded.com/98/9801spec.htm

[24] http://www.netrino.com/Embedded-Systems/How-To/Memory-Types-RAM-ROM-Flas

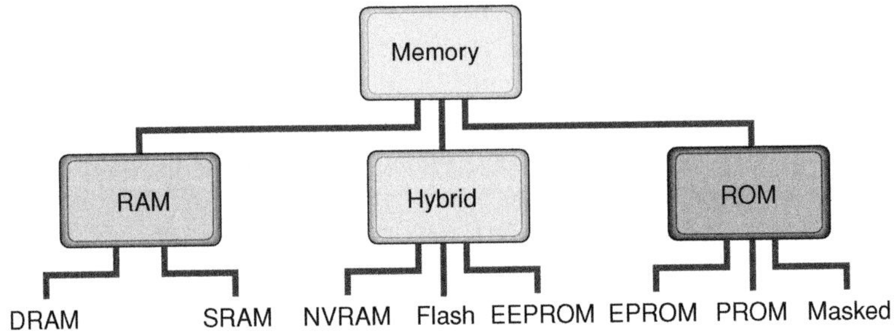

FIGURE 17.4 Memory classification in embedded systems.

The fastest possible memory that is available is desired for a real-time system; however, cost should be considered as well. The following is a list of memory in order of fastest to slowest while still considering cost:

1. Internal CPU memory
2. Registers
3. Cache
4. Main memory
5. Memory on board external devices

TABLE 17.2 Memory Types and Attributes

Type	Volatile?	Writeable?	Erase Size	Max Erase Cycles	Cost (per Byte)	Speed
SRAM	Yes	Yes	Byte	Unlimited	Expensive	Fast
DRAM	Yes	Yes	Byte	Unlimited	Moderate	Moderate
Masked ROM	No	No	n/a	n/a	Inexpensive	Fast
PROM	No	Once, with a device programmer	n/a	n/a	Moderate	Fast
EPROM	No	Yes, with a device programmer	Entire Chip	Limited (consult datasheet)	Moderate	Fast
EEPROM	No	Yes	Byte	Limited (consult datasheet)	Expensive	Fast to read, slow to erase/write
Flash	No	Yes	Sector	Limited (consult datasheet)	Moderate	Fast to read, slow to erase/write
NVRAM	No	Yes	Byte	Unlimited	Expensive (SRAM + battery)	Fast

In general, the closer the memory is to the CPU, the more expensive it tends to be.

The main memory holds temporary data and programs for execution by the CPU. Cache memory is a type of memory designed to provide the most frequently and recently used instructions and data for the processor, and it can be accessed at rates many times faster than the main memory can.[25] The processor first looks at cache memory to find needed data and instructions. There are two levels of cache memory—internal cache memory and external cache memory. Internal cache memory is called level 1 and is located inside the CPU chip. Internal cache memory ranges from 1KB to 32KB. External cache memory is called level 2 and is located on the system board between the processor and RAM. It is SRAM memory, which can provide much more speed than main memory.

The registers are for temporary storage for the current instructions, address of the next instruction, and storage for the intermediate results of execution and are not a part of main memory. They are under the direction of the control unit to accept, store, and transfer data and instructions and perform at a very high speed. Earlier models of computers, such as the Intel 286, had eight general-purpose registers. Some types of registers have special assignments such as the accumulator register, which holds the results of execution; the address register, which keeps address of the next instruction; the storage register, which temporarily keeps instruction from memory and general-purpose registers, which are used for operations.

The part of the system that manages memory is called the memory manager. Memory management primarily deals with space multiplexing (Sobh & Tibrewal, 2006). Spooling enables the transfer of a process while another process is in execution. The job of the memory manager is to keep track of which parts of memory are in use and which parts are not, to allocate memory to processes when they need it and to deallocate it when they are done, and to manage swapping between main memory and disc when the main memory is not big enough to hold all the processes. However, the three disadvantages related to memory management are synchronization, redundancy, and fragmentation. Memory fragmentation does not affect memory utilization; however, it can degrade a system's response, which gives the impression of an overloaded memory.

Spooling allows the transfer of one or more processes while another process is in execution. When trying to transfer a very big process, it is possible that the transfer time exceeds the combined execution time of the processes in the RAM and results in the CPU being idle, which was the problem for which spooling was invented. This problem is termed as the synchronization problem. The combined size of all processes is usually much bigger than the RAM size, and for this reason, processes are swapped in and out continuously. The issue regarding this is the transfer of the entire process when only part of the code is executed in a given time slot. This problem is termed as the redundancy problem. Fragmentation is when free memory space is broken into pieces as processes are loaded and removed from memory. External fragmentation exists when enough total memory space exists to satisfy a request, but it is not continuous.

[25]http://www.bsu.edu/classes/nasseh/cs276/module2.html

17.3.5 Queuing Theory

Queuing theory is the study of waiting lines and analyzes several related processes, including arriving at the queue, waiting in the queue, and being served by the server at the front of the queue.[26] Queuing theory calculates performance measures including the average waiting time in the queue or the system, the expected number waiting or receiving service, and the probability of encountering the system in certain states, such as empty, full, having an available server, or having to wait a certain time to be served. Some different types of queuing theories include first-in-first out, last-in-first-out, processor sharing, and priority.

A queuing model can be characterized by several different factors. Some of them are: the arrival process of customers, the behavior of customers, the service times, the service discipline, and the service capacity (Adan & Resing, 2002). Kendall introduced a shorthand notation to characterize a range of queuing models, that is, a three-part code a = b = c. The first letter specifies the interarrival time distribution, and the second one specifies the service time distribution. For example, for a general distribution, the letter G is used, M is used for the exponential distribution (M stands for memoryless), and D is used for the deterministic times. The third letter specifies the number of servers. Some examples are M = M = 1, M = M = c, M = G = 1, G = M = 1, and M = D = 1. This notation can be extended with an extra letter to cover other types of models as well.

One of the simplest queuing models is the M/M/1 model, which is the single-server model. Letting $\rho = \lambda / \mu$, the average number of customers in a queue can be calculated by:

$$\bar{N} = \frac{\rho}{1-\rho} \tag{17.1}$$

Although the system's variance can be calculated by the following:

$$\sigma_N^2 = \frac{\rho}{(1-\rho)^2} \tag{17.2}$$

The expected number of requests in the server is:

$$\bar{N}_S = \lambda \bar{x} = \rho \tag{17.3}$$

The expected number of requests in the queue[27] is

$$\bar{N}_Q = \frac{\rho^2}{1-\rho} \tag{17.4}$$

[26] http://en.wikipedia.org/wiki/Queuing_theory
[27] http://en.wikipedia.org/wiki/M/M/1_model

M/M/1 queuing systems assume a Poisson arrival process. This is a very good approximation for the arrival process in real systems that meet the following rules:

1. The number of customers in the system is very large.
2. The impact of a single customer on the performance of the system is very small.
3. All customers are independent.[28]

In the M/M/1 model, the probability of exceeding a particular number of customers in the system decreases geometrically, and if interrupt requests are considered customers, then the two such requests in the system have a far greater probability than three or more such requests (LaPlante, 2002). This means that a system that can tolerate a single time overload should be able to contribute to the system's reliability.

Another type of queuing model is the M/M/c model. This type of model is a multiserver model. Another useful queuing theory is Erlang's formula. If there are m servers, then each newly arriving interrupt is serviced by a process, unless all servers are busy. In this instance, the customer or interrupt is lost. The Erlang distribution can be used to model service times with a low coefficient of variation (less than one), but it also can develop naturally. For instance, if a job has to pass, stage by stage, through a series of r independent production stages, where each stage takes a exponentially distributed time, then the analysis of the $M = Er = 1$ queue is similar to that of the $M = M = 1$ queue (Adan & Resing, 2002).

17.4 PERFORMANCE ANALYSIS

Performance analysis is the study of a system, especially a real-time system, to see if it will meet its deadlines. The first step in performing this type of analysis involves determining the execution of code units. The ability to calculate the execution time for a specific real-time system can be critical because the system's parameters, such as CPU utilization requirements, are calculated beforehand. With this information, the hardware and the software of a system is selected as well. There are several methods available to conduct performance analysis.

One way to measure real-time performance estimate is though an execution time estimate in which the execution time is calculated by the following:

execution time = program path + instruction timing[29]

The path is the sequence of instructions executed by the program, and the instruction timing is determined based on the sequence of instructions traced by the program path, which takes into account data dependencies, pipeline behavior, and caching. The execution path of a program can be traced through a high-level language specification; however, it may be difficult to obtain accurate estimates of total execution time from a high-level language program, as there is not a direct correspondence between

[28] http://www.eventhelix.com/realtimemantra/CongestionControl/m_m_1_queue.htm

[29] http://www.embedded.com/design/multicore/201802850

program statements and instructions. The number of memory locations and variables must be estimated. These problems become more challenging as the compiler puts more and more effort into optimizing the program.

Some aspects of program performance can be estimated by looking directly at the program. For example, if a program contains a loop with a large, fixed iteration bound, or if one branch of a conditional is much longer than another, then we can get at least a rough idea that these are more time-consuming segments of the program. However, a precise estimate of performance also relies on the instructions to be executed because different instructions take different amounts of time. The following snippet of code[30] is a data-dependent program path with a pair of nested if statements:

```
if (a || b){/ * test 1 * /
if (c) / * test 2 * /
{x = r * s + t; / * assignment 1 * /}
else {y = r + s; / * assignment 2 * /}
z = r + s + u; / * assignment 3 * /
} else {
if (c)/ * test 3 * /
{y = r't; / * assignment 4 * /}
}
```

One way to enumerate all the paths is to create a truth table structure in which the paths are controlled by the variables in the if-conditions, namely, a, b, and c.

Results for all controlling variable values follow:

a	b	c	Path
0	0	0	test 1 false, test 3 false: no assignments
0	0	1	test 1 false, test 3 true: assignment 4
0	1	0	test 1 true, test 2 false: assignments 2, 3
0	1	1	test 1 true, test 2 true: assignments 1, 3
1	0	0	test 1 true, test 2 false: assignments 2, 3
1	0	1	test 1 true, test 2 true: assignments 1, 3
1	1	0	test 1 true, test 2 false: assignments 2, 3
1	1	1	test 1 true, test 2 true: assignments 1, 3

Notice that there are only four distinct cases: no assignment, assignment 4, assignments 2 and 3, or assignments 1 and 3. These correspond to the possible paths through the nested ifs; the table adds value by telling us which variable values exercise each of these paths. Enumerating the paths through a fixed-iteration for a loop is seemingly simple.

After the execution path of the program is calculated, the execution time of the instructions executed along the path must be measured. The simplest estimate is to

[30] http://www.embedded.com/design/multicore/201802850

assume that every instruction takes the same number of clock cycles. However, even ignoring cache effects, this technique is unrealistic for several reasons. First, not all instructions take the same amount of time. Second, it is important to note that execution times of instructions are not independent. This means that the execution time of one instruction depends on the instructions around it. For example, many CPUs use register bypassing to speed up instruction sequences when the result of one instruction is used in the next instruction. As a result, the execution time of an instruction may depend on whether its destination register is used as a source for the next operation. Third, the execution time of an instruction may depend on operand values. This is true of floating-point instructions in which a different number of iterations may be required to calculate the result. The first two problems can be addressed more easily than the third.

17.5 SYNCHRONIZATION AND DEADLOCK HANDLING

To ensure the orderly execution of processes, jobs should not get stuck in a deadlock, forever waiting for each other (Sobh & Tibrewal, 2006). Synchronization problems develop because sections of code that constitutes the critical sections overlap and do not run atomically. A critical section of code is a part of a process that accesses shared resources. Two processes should not enter their critical sections at the same time. Synchronization can be implemented by using semaphores, monitors, and message passing.

Semaphores are either locked or unlocked. When locked, a queue of tasks wait for the semaphore. Problems with semaphore designs are priority inversion and deadlocks.[31] In priority inversion, a high-priority task waits because a low-priority task has a semaphore. A typical solution is to have the task that has a semaphore run at the priority of the highest waiting task. Another solution is to have tasks send messages to each other. These have exactly the same problems, as priority inversion occurs when a task is working on a low-priority message and ignores a higher priority message in its inbox. Deadlocks happen when two tasks wait for the other to respond. Although their real-time behavior is less crisp than semaphore systems, message-based systems are generally better behaved than semaphore systems. Figure 17.5 shows a comparison between the three synchronization methods (Sobh & Tibrewal, 2006).

A set of processes or threads is deadlocked when each process or thread is waiting for a resource to be freed that is controlled by another process.[32] For deadlock to occur, four separate conditions must be met. They are:

1. Mutual exclusion
2. Circular wait

[31] http://www.webcomtechnologiesusa.com/embeddedeng.htm

[32] http://www.cs.rpi.edu/academics/courses/fall04/os/c10/index.html

Implementation	Synchronization	Mutual Exclusion	Advantages	Disadvantages
Semaphores	√	√		Low-level implementation, Can cause deadlock
Monitors	√	√	High-level implementation	
Message Passing	√	√		

FIGURE 17.5 A comparison between the three synchronization methods.

3. Hold and wait
4. No preemption

Eliminating any of these four conditions will eliminate deadlock. Mutual exclusion applies to those resources that possibly no longer can be shared, such as printers, disk drives, and so on (LaPlante, 2002). The circular wait condition will occur when a chain of processes exist that holds resources needed by an other process. Circular wait can be eliminated by imposing an explicit order on the resources and forcing all processes to request all the resources listed. The hold and wait condition occurs when processes request a resource and then lock it until that resource is filled. Also, eliminating preemption will eliminate deadlock. This means that if a low-priority task holds a resource protected by semaphore S, and if a higher priority interrupts, then the lower priority task will cause the high-priority task to wait forever.

Once deadlock in the system has been detected, there are several ways to deal with the problem. Some strategies include:

1. Preemption—take an already allocated resource away from a process and give it to another process. This can present problems. Suppose the resource is a printer and a print job is half completed. It is often difficult to restart such a job without completely starting over.
2. Rollback—In situations in which deadlock is a real possibility, the system periodically can make a record of the state of each process, and when deadlock occurs, it can roll everything back to the last checkpoint and restart but allocating resources differently so that deadlock does not occur. This means that all work done after the checkpoint is lost and will have to be redone.
3. Kill one or more processes—this solution is the simplest and the crudest but is also effective.[33]

[33] http://www.cs.rpi.edu/academics/courses/fall04/os/c10/index.html

Another approach is to avoid deadlock by only granting resources if granting them cannot result in a deadlock situation later on, but this works only if the system knows what requests for resources a process will be making in the future, which is an unrealistic assumption. Yet another approach is deadlock avoidance. Deadlock avoidance is a strategy in which whenever a resource is requested, it is only granted if it cannot result in deadlock. Deadlock prevention strategies involve changing the rules so that processes will not make requests that could result in deadlock.

A variant of deadlock called live-lock is a situation in which two or more processes continuously change their state in response to changes in the other processes without doing any useful work. This is similar to deadlock in that no progress is made but differs in that neither process is blocked or waiting for anything. Some specialized systems have deadlock avoidance/prevention mechanisms. For example, many database operations involve locking several records, which can result in deadlock, so database software often has a deadlock-prevention algorithm.

17.6 PERFORMANCE OPTIMIZATION

There are many approaches available to optimize performance. Indeed, identifying sections of wasteful or unneeded code is probably the first step in optimizing a real-time system. Several approaches are available today to optimize software; however, this book concentrates more on the approaches that are most effective in real-time systems. It should be noted that all processing should be done at the slowest rate that can possible be tolerated by the system (LaPlante, 2002).

17.6.1 Look-Up Tables

A look-up table is a technique used to speed up computation time and especially is used in an application such as a real-time system, where time is of the essence. Look-up tables are used particularly in the implementation of continuous functions such as exponential sine, cosine, and tangent.

A look-up table can be defined as an array that holds a set of precomputed results for a given operation.[34] The array provides access to the results that is faster than computing each time of the result of the given operation. For this reason, look-up tables typically are used in real-time data acquisition and in processing systems, especially embedded systems, because of their demanding and strict timing restriction. However, look-up tables require a considerable amount of execution time to initialize the array, but in real-time systems, it is in general acceptable to have a delay during the initialization of the application.

The snippet of code below represents a real-time data acquisition and processing system in which data are sampled as eight-bit numbers, representing positive values from 0 to 255. In this example, the required processing involves computing the square

[34] http://www.mochima.com/articles/LUT/LUT.html

root of every sample. The use of a look-up table to compute the square root would look as follows:

```
double LUT_sqrt [256]; /* presumably declared globally */
 /* Somewhere in the Initialization code */
 for (i = 0; i < 256; i++)
 {
    LUT_sqrt[i] = sqrt(i);/* provided that <math.h/>
    was #included */
 }
/* During the normal execution of the application */
result = LUT_sqrt[sample]; /* instead of
result = sqrt(sample); */
```

During the initialization phase, the application sacrifices a certain amount of time to compute all 256 possible results, but after that, when the system starts to read data in real time, the system can complete the processing required in the time available.

17.6.2 Scaled Numbers

In almost all types of computing systems, integer operations are faster than floating point operations. Because of this, floating point algorithms often are converted into scaled integer algorithms. In this example, the least significant bit of an integer variable is assigned a real-number scale factor. These scaled numbers then can be added, subtracted, multiplied, or divided, and then converted back to floating point numbers. However, it should be noted that accuracy may be sacrificed by excessive use of scaled numbers.

17.7 COMPILER OPTIMIZATION TOOLS

Most types of code optimization techniques can be used in an effort to improve the real-time performance of an embedded system. In this section, several optimization techniques and their impact on real-time performance will be discussed. These techniques are:

1. Reduction in strength
2. Common subexpression elimination
3. Constant folding
4. Loop invariant removal
5. Loop induction elimination
6. Dead code removal
7. Flow of control

8. Loop unrolling
9. Loop jamming

17.7.1 Reduction in Strength

A reduction in strength is a type of compiler optimization in which an expensive operation is combined with a less expensive one. More specifically, strength reduction is a transformation that a compiler uses to replace strong, costly instructions with cheaper and weaker instructions (Cooper et al., 2005). For example, a weak form of strength reduction replaces 2 x x with either x + x, or $x \gg 1$.

Another type of strength reduction replaces an iterated series of strong computations with an equivalent series of weaker computations. For example, the reduction replaces certain multiplications inside a loop with repeated additions, which results in loop nests that manipulate arrays. These resulting additions are usually cheaper than the multiplications they replace. It should be noted that many operations, other than multiplication, also can be reduced in this manner.

Strength reduction is important for two reasons. First, multiplying integers usually has taken longer than adding them. This makes strength reduction profitable, as the amount of improvement varies with the relative costs of addition and multiplication. Second, strength reduction decreases the "overhead" introduced by translation from a higher level language down to assembly code. Strength reduction often decreases the total number of operations in a loop. Smaller operations lead to faster code. The shorter sequences used to generate addresses may lead to tighter schedules as well.

17.7.2 Common Subexpression Elimination

Repeated calculations of the same subexpression in two different equations should be avoided in code. The following is an example of subexpression elimination:

$$x = 6 + a \times b;$$

$$y = a \times b + z;$$

could be replaced with

$$t = a \times b;$$

$$x = y + t;$$

$$y = t + z;$$

The purpose of common subexpression elimination is to reduce the runtime of a program by avoiding the repetition of the same computation (Chiil, 1997).

The transformation statically identifies a repeated computation by locating multiple occurrences of the same expression. These repeated computations are eliminated by storing the result of evaluating the expression in a variable and accessing this variable instead of reevaluating the expression.

17.7.3 Constant Folding

Constant folding is the process of simplifying a group of constants in a program. One good example of constant folding is as follows:

$$x = y \times 2.0 \times 2.0$$

which could be simplified to

$$x = y \times 4.0$$

In other words, x has been optimized by combing the 2.0 and the 2.0, and using 4.0 instead.

In some cases, constant folding is similar to a reduction in strength optimizations and is most easily implemented on a directed acyclic graph (DAG) intermediate representation.[35] However, it can be performed in almost any stage of compilation. The compiler seeks any operation that has constant operands and, without side effects, computes the result replacing the entire expression with instructions to load the result.

In another example of constant folding, if the program uses the expression $\Pi/2$, then this value should be precalculated during initialization be a and stored value, such as pi_div_2. This typically saves one floating point load and one floating point divide instruction, which translates into a time savings of a few microseconds.

17.7.4 Loop Invariant Optimization

If a computation is calculated within a loop but does not need to be, then the calculation can be moved outside of the loop instead. When a computation in a loop does not change during the dynamic execution of the loop, this computation can be removed from the loop to improve execution time performance (Song, et al., 2003). In one example illustrated below in Figure 17.6, the evaluation of expression ax100 is loop invariant in (a), and (b) shows a more efficient version of the loop in which the loop invariant code has been removed from the loop.

17.7.5 Loop Induction Elimination

Some types of code optimizations, such as dead code elimination and common subexpresssion elimination reduce the execution time by removing redundant computation.

[35] http://en.citizendium.org/wiki/Constant_folding

for $(i = 1; i <= 100; i++)$ $\{x = a \times 100; y = y + i;\}$

(a) A source loop

$t = a \times 100;$ for $(i = 1; i <= 100; i++)$ $\{x = t; y = y + i;\}$

(b) The resulting code

FIGURE 17.6 Loop invariant example.

However, a loop induction variable elimination reduces the execution time by moving instructions from frequently executed program regions to infrequently executed program regions (Chang, et al., 2006).

Induction variables are variables in a loop incremented by a constant amount each time the loop iterates, and replaces the uses of an induction variable by another induction variable, thereby eliminating the need to increment the variable on each iteration of the loop. If the induction variable eliminated is needed after the loop is exited, then its value can be derived from one of the remaining induction variables. The following is an example of loop induction elimination in which the variable i is the induction variable of the loop:

```
for (i=1,i<=10;i++)
        a [i + 1] = 1;
```

an optimized version is

```
for (i=2,i<=11;i++)
        a[j] = 1;
```

17.7.6 Removal of Dead Code

Another simple and easy way to decrease memory requirements in a system is through the removal of dead code. Dead code is unnecessary, inoperative code.[36] Some types of dead code are dead procedures, dead variables, dead parameters, dead return values, dead event declarations, dead enumeration values, dead user-defined types, dead classes, dead modules, and dead control.

A dead procedure is not called by any other procedure, which means that it is never executed and is not required for any purpose. A dead variable is not read nor written. It is completely useless, only taking up memory and a line or two of code. A dead parameter is passed to a procedure but is not used by it. Passing parameters takes a little bit of time during each call. Dead parameters can make a program slower. A dead return value of a function is not stored or used by any of the callers. A dead event does not fire, and a semidead event does not fire and its handlers are not executed. A dead enumeration or constant value is not required by the program. A user-defined type is a structure or record of one that is not used anywhere. A dead class is not used

[36] http://www.aivosto.com/vbtips/deadcode.html

anywhere but still may be compiled in the executable and even published as a part of the library interface. This bloats the executable and makes the library unnecessarily complex. A dead module or file is with contents not used for any purpose. They are only making the program more complex, more bloated, and harder to understand.

17.7.7 Flow of Control Optimization

Program flow of control is essentially the order in which instructions in the program are executed. In flow of control optimization, unnecessary jump statements can be removed and replaced with a single jump statement (LaPlante, 2002). The following is an example of flow of control optimization:

```
(n)Goto L1
 .
 .
(n+k)L1  : goto L2
```

This code[37] can be replaced by the following:

```
(n)Goto L2
 .
 .
(n+k) goto L2
```

17.7.8 Loop Unrolling

Loop unrolling duplicates statements that are executed in a loop to reduce the number of operations. The following is an example of loop unrolling:

```
for (i = 1; i <  =  60; i +  +) a[i]} = a[i] *  b + c;
```

This loop can be transformed into the following equivalent loop consisting of multiple copies of the original loop body[38]:

```
for (i = 1; i < = 60; i+ = 3)
{
a[i] = a[i] * b + c;
a[i+1] = a[i+1] * b + c;
a[i+2] = a[i+2] * b + c;
}
```

[37] www.facweb.iitkgp.ernet.in/~niloy/Compiler/notes/TCheck1.doc
[38] http://www2.cs.uh.edu/~jhuang/JCH/JC/loop.pdf

The loop is said to have been unrolled twice, and the unrolled loop should run faster because of reduction in loop overhead. Loop unrolling initially was developed for reducing loop overhead and for exposing instruction-level parallelism for machines with multiple functional units.

17.7.9 Loop Jamming

Loop jamming is a technique in which two loops essentially can be combined into a single loop. The advantages of loop jamming are that loop overhead is reduced, which results in a speed-up in execution as well as a reduction in code space. The following is an example of loop jamming:

```
LOOP I = 1 to 100

     A(I) := 0

ENDLOOP
LOOP I := 1 to 100

     B(I) = X(I) + Y

ENDLOOP
```

These two loops can be combined together to produce a single loop[39]:

```
LOOP I = 1 to 100

     A(I) = 0
     B(I) = X(I) + Y

ENDLOOP
```

The conditions for performing this optimization are that the loop indices be the same, and the computations in one loop cannot depend on the computations in the other loop.

17.7.10 Other Techniques

There are other optimization techniques available as well, which will be discussed briefly.

[39] http://web.cs.wpi.edu/~kal/PLT/PLT10.2.5.html

- Optimize the common case. The most frequently used path also should be the most efficient.
- Arrange table entries so that the most frequently used value is the first to be compared.
- Replace threshold tests on monotone with tests on their parameters.
- Link the most frequently used procedures together.
- Store redundant data elements to increase the locality of a reference.
- Store procedures in memory in sequence so that calling and called subroutines can be loaded together (LaPlante, 2002).

17.8 CONCLUSION

This chapter has attempted to explain, compare, and contrast different types of software optimization methods and analysis. It is important to note different metrics and techniques often serve different purposes. Thus, each type of technique or approach usually has its own strengths and weaknesses. Indeed, in any system, but especially a real-time system, it is important to maintain control and to ensure that the system works properly.

REFERENCES

Adan, Ivo and Resing, Jaques (2002), *Queuing Theory*, Department of Mathematics and Computing Science, Eindhoven University of Technology. http://www.cs.duke.edu/~fishhai/misc/queue.pdf.

Capers Jones & Associates LLC (2008), *A Short History of Lines of Code Metric.* http://www.itmpi.org/assets/base/images/itmpi/privaterooms/capersjones/LinesofCode2008.pdf.

Chang, Pohua P., Mahlke, Scott A. and Hu, Wen-Mei W. (1991), "Using profile information to assist classic code optimizations" software—practice and experience, Volume *21* #12, pp.1301–1321.

Chiil, Olaf (1997), "Common Subexpression Elimination in a Lazy Functional Language," Draft Proceedings of the 9th International Workshop on Implementation of Functional Languages, St Andrews, Scotland, Sept., pp. 501–516.

Cooper, Keith D., Simpson, L. Taylor, and Vick, Christopher A. (2001), "Operator strength reduction." *ACM Transactions on Programming Languages and Systems*, Volume *23*, #5, p. 603.

Davis, Robert I., Zabos, Attila, and Burns, Alan (2008), "Efficient exact schedulability tests for fixed priority real-time systems." *IEEE Transactions on Computers*, Volume *57*, #9.

El-Haik, and Mekki (2008).

Eventhelix.com (2001), *Issues in Real Time System Design, 2000.* http://www.eventhelix.com/realtimemantra/ issues in Realtime System Design.htm.

LaPlante, Phillip (2005), *Real Time Systems Design and Analysis*, 3rd ed., IEEE Press, Washington, DC.

Watson, McCabe (1996), *Structured Testing: A Testing Methodology Using the Cyclomatic Complexity Metric.* http://hissa.nist.gov/HHRFdata/Artifacts/ITLdoc/235/title.htm.

Na'Cul, Andre'C. and Givargis, Tony (2006), "Synthesis of time constrained multitasking embedded software." *ACM Transactions on Design Automation of Electronic Systems*, Volume *11*, #4, pp. 827–828.

Regehr, John, (2006), *Safe and Structured Use of Interrupts in Real-Time and Embedded Software.* http://www.cs.utah.edu/~regehr/papers/interrupt_chapter.pdf.

Sobh, Tarek M, and Abhilasha, Tibrewal (2006), "*Parametric Optimization of Some Critical Operating System Functions—An Alternative Approach to the Study of Operating Systems Design*," AEEE Conference Paper. www.bridgeport.edu.

Song, Litong, Kavi, Krishna, and Cytron, Ron (2003), "An Unfolding-Based Loop Optimization Technique," International Workshop on Software Compilers for Embedded Systems N°7, Vienna, Austria, Sept.

CHAPTER 18

ROBUST DESIGN FOR SOFTWARE DEVELOPMENT

18.1 INTRODUCTION

In the context of this book, the terms "quality" and "robustness" can be used interchangeably. Robustness is an important dimension of software quality, and it is a hallmark of the software Design for Six Sigma (DFSS) process. The subject is not familiar in mainstream software professionals, despite the ample opportunity for application. This chapter will explore the application of the Taguchi robustness techniques in software DFSS, introducing concepts, developing basic knowledge, and formulating for application.[1]

In general, robustness is defined as a design attribute that represents the reduction of the variation of the functional requirements (FRs) or design parameters (DPs) of a software and having them on target as defined by the customer (Taguchi, 1986), (Taguchi & Wu, 1986), (Phadke, 1989), (Taguchi et al., 1989), and (Taguchi et al., 1999).

Variability reduction has been the subject of robust design (Taguchi, 1986) through methods such as parameter design and tolerance design. The principal idea of robust design is that statistical testing of a product or a process should be carried out at the developmental stage, also referred to as the "offline stage." To make the software *robust* against the effects of variation sources in the development, production, and use environments the software entity is viewed from the point of view of quality and

[1]Contact Six Sigma Professionals, Inc. at www.sixsigmapi.com for further details.

Software Design for Six Sigma: A Roadmap for Excellence, By Basem El-Haik and Adnan Shaout

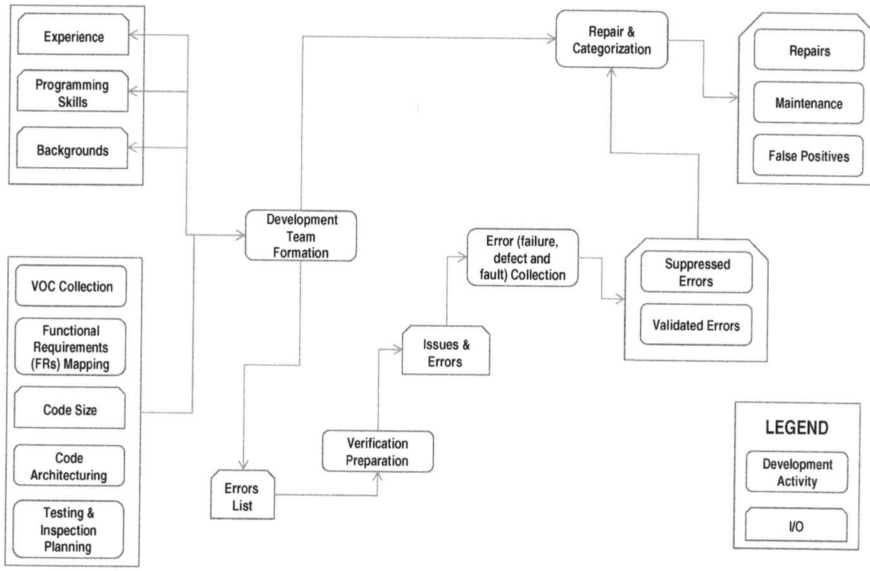

FIGURE 18.1 Software developmental activities are sources of variation.

cost (Taguchi, 1986), (Taguchi & Wu, 1986), (Taguchi et al., 1989), (Taguchi et al., 1999), and (Nair, 1992).

Quality is measured by quantifying statistical variability through measures such as standard deviation or mean square error. The main performance criterion is to achieve an on-target performance metric on average while simultaneously minimizing variability around this target. Robustness means that a software performs its intended functions under all operating conditions (different causes of variations) throughout its intended life. The undesirable and uncontrollable factors that cause a software code under consideration to deviate from target value are called "noise factors." Noise factors adversely affect quality, and ignoring them will result in software not optimized for conditions of use and possibly in failure. Eliminating noise factors may be expensive (e.g., programming languages, programming skill levels, operating systems bugs, etc.). Many sources of variation can contribute negatively to software quality level. All developmental activities in a typical process similar to the one depicted in Figure 18.1 can be considered rich sources of variation that will affect the software product. Instead, the DFSS team seeks to reduce the effect of the noise factors on performance by choosing design parameters and their settings that are *insensitive* to the noise.

In software DFSS, robust design is a disciplined methodology that seeks to find the *best* expression of a software design. "Best" is defined carefully to mean that the design is the lowest cost solution to the specification, which itself is based on the identified customer needs. Dr. Taguchi has included design quality as one more dimension of product cost. High-quality software minimizes these costs by performing consistently at targets specified by the customer. Taguchi's philosophy of robust design is aimed at reducing the loss caused by a variation of performance from the target value based on a portfolio of concepts and measures such as *quality loss function* (QLF), *signal-to-noise* (SN) *ratio*, optimization, and experimental design. *Quality loss* is the loss experienced by customers and society and is a function of how far performance deviates from the target. The QLF relates quality to cost and is considered a better evaluation system than the traditional binary treatment of quality (i.e., within/outside specifications). The QLF of a functional requirement, a design parameter, or a process variable (generically denoted as response y) has two components: mean (μ_y) deviation from targeted performance value (T_y) and variance (σ_y^2). It can be approximated by a quadratic polynomial of the response of interest.

18.2 ROBUST DESIGN OVERVIEW

In Taguchi's philosophy, robust design consists of three phases (Figure 18.2). It begins with the *concept design phase* followed by the *parameter design* and *tolerance design* phases. It is unfortunate to note that the concept phase did not receive the attention it deserves in the quality engineering community, hence, the focus on it in this book.

The goal of parameter design is to minimize the expected quality loss by selecting design parameters settings. The tools used are quality loss function, design of experiment, statistics, and optimization. Parameter design optimization is carried out in two sequential steps: variability minimization of σ_y^2 and mean (μ_y) adjustment to target T_y. The first step is conducted using the mapping parameters or variables (x's) **(in the context of Figure 13.1) that** affect variability, whereas the second step is accomplished via the design parameters that affect the mean but do not adversely influence variability. The objective is to carry out both steps at a low cost by exploring the opportunities in the design space.

18.2.1 The Relationship of Robust Design to DFSS

Let us discuss this relationship through an example. Consider a digital camera application in which the central processing unit (CPU) unit uses illumination levels to produce images of a specified quality. A measurement system measures illumination levels and feeds it into the CPU. The system measures the performance of a camera in four dimensions: luminance, black level, signal-to-noise level, and resolution.

For each dimension, an illumination level is defined at which the camera fails. The highest of these (i.e., the worst performance dimension) is defined as the minimum

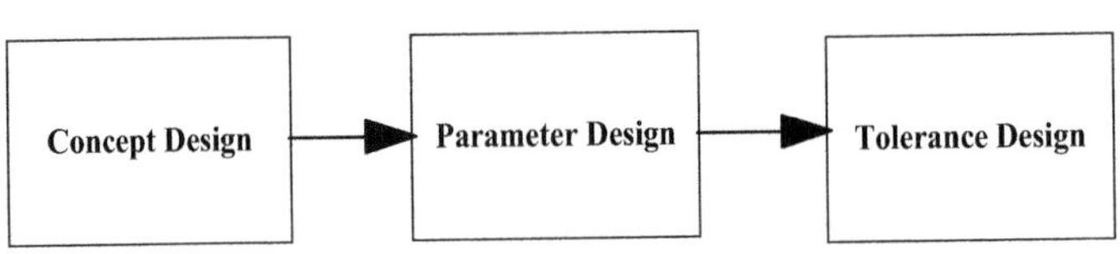

FIGURE 18.2 Taguchi's robust design.

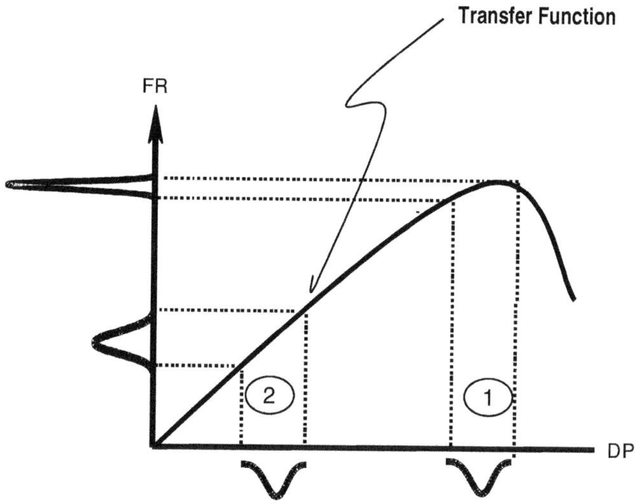

FIGURE 18.3 Robustness optimization definition.

illumination required by the camera. This value represents the lowest illumination that the camera can operate under with acceptable image quality, as defined by this method.

The light sensitivity of a camera is affected by many design parameters and their variation (noise). These include the aperture, the quality of the lens, the size and quality of the sensor, the gain, the exposure time, and image processing. When using several criteria, it is difficult to compensate for the quality of the camera with gain and image processing. For example, increasing the gain level may provide better luminance, but it also may increase the noise in the image. Illumination measures [in lux (lx)] the visible light falling per unit area in a given position. It is important to note that illumination concerns the spectral sensitivity of the human eye, so that electromagnetic energy in the infrared and ultraviolet ranges contributes nothing to illumination. Illumination also can be measured in foot-candles (fc).[2]

Consider two settings or means of the minimum illumination parameter (DP)—setting 1 (DP*) and setting 2 (DP**)—having the same variance and probability density function (statistical distribution) as depicted in Figure 18.3. Consider, also, the given curve of a hypothetical transfer function relating illumination to image quality, an FR,[3] which in this case is a nonlinear function in the DP. It is obvious that setting 1 produces less variation in the FR than setting 2 by capitalizing on nonlinearity.[4] This also implies a lower information content and, thus, a lower degree

[2] 1 lx = 10.76 fc. Note that 1 lx can be interpreted as 1 "meter-candle."

[3] A mathematical form of the design mapping. See Chapter 13.

[4] In addition to nonlinearity, leveraging the interactions between the noise factors and the design parameters is another popular empirical parameter design approach.

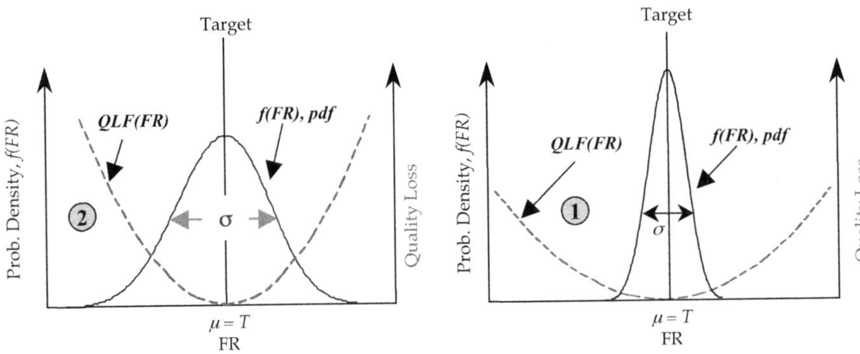

FIGURE 18.4 The quality loss function scenarios of figure 18.3.

of complexity based on axiom 2.[5] Setting 1 (DP*) also will produce a lower quality loss similar to the scenario on the right of Figure 18.4. In other words, the design produced by setting 1 (DP*) is more robust than that produced by setting 2. Setting 1 (DP*) robustness is evident in the amount of transferred variation through the transfer function to the FR response of Figure 18.2 and the flatter quadratic quality loss function in Figure 18.3. When the distance between the specification limits is six times the standard deviation (6 σ_{FR}), a Six Sigma level optimized FR is achieved. When all design FRs are released at this level, a Six Sigma design is obtained.

The important contribution of robust design is the systematic inclusion into experimental design of noise variables, that is, the variables over which the designer has little or no control. A distinction also is made between internal noise (such as dimensional variation in aperture, the size and quality of the sensor, the gain, and the exposure time), and environmental noise, which the DFSS team cannot control (e.g., humidity and temperature). The robust design's objective is to suppress, as far as possible, the effect of noise by exploring the levels of the factors to determine their potential for making the software insensitive to these sources of variation in the respective responses of interest (e.g., FRs).

The noise factors affect the FRs at different segments in the life cycle. As a result, they can cause a dramatic reduction in product reliability, as indicated by the failure rate. The bath tub curve in Figure 18.5 implies that robustness can be defined as reliability throughout time. Reliability is defined as the probability that the design will perform as intended (i.e., deliver the FRs to satisfy the customer attributes (CAs) [(Figure 13.1)], throughout a specified time period when operated under some stated conditions). The random failure rate of the DPs that characterizes most of the life is the performance of the design subject with external noise. Notice that the coupling vulnerability contributes to unreliability of the design in customer hands. Therefore, a product is said to be robust (and, therefore, reliable) when it is insensitive to the effect of noise factors, even though the sources themselves have not been eliminated.

[5] See Chapter 13.

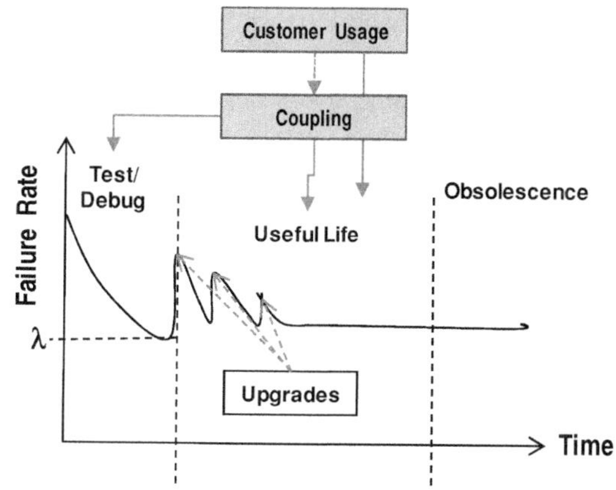

FIGURE 18.5 The effect of noise factors during the software life cycle.[6]

Parameter design is the most used phase in the robust design method. The objective is to design a solution entity by making the functional requirement insensitive to the variation. This is accomplished by selecting the optimal levels of design parameters based on testing and using an optimization criterion. Parameter design optimization criteria include both quality loss function and SN. The optimum levels of the *x's* or the design parameters are the levels that maximize the SN and are determined in an experimental setup from a pool of economic alternatives. These alternatives assume the testing levels in search for the optimum.

Several robust design concepts are presented as they apply to software and product development in general. We discuss them in the following sections.

18.3 ROBUST DESIGN CONCEPT #1: OUTPUT CLASSIFICATION

An output response of software can be classified as static or dynamic from a robustness perspective. A static entity has a fixed target value. The parameter design phase in the case of the static solution entity is to bring the *response (y)* mean, μ_y, to the target, T_y. For example, we may want to maximize the quality of programmer ability or streamlining software debugging. However, the dynamic response expresses a variable target depending on the customer intent. In this case, the DFSS optimization phase is carried across a range of the useful customer applications, called the signal factor. The signal factor can be used to set the y to an intended value.

Parameter design optimization requires the classification of the output responses (depending on the mapping of interest in the context of Figure 13.1) as *smaller-the-better* (e.g., minimize coding man-hour), *larger-the-better* (e.g., increase

[6]Modified graph. The original can be found at http://www.ece.cmu.edu/~koopman/des_s99/sw_reliability/presentation.pdf.

effectiveness), *nominal-the-best* (keeping the software on a single performance objective is the main concern, (e.g., produce correct results for a test case), and *dynamic* (energy-related functional performance across a prescribed dynamic range of usage is the perspective, (e.g., produce correct results for a range of inputs).

When robustness cannot be assured by parameter design, we resort to the tolerance design phase. Tolerance design is the last phase of robust design. The practice is to upgrade or tighten tolerances of some design parameters so that quality loss can be reduced. However, lightening tolerance practice usually will add cost to the process of controlling tolerance. El-Haik (2005) formulated the problem of finding the optimum tolerance of the design parameters that minimizes both quality loss and tolerance control costs (Chapter 16).

The important contribution of robust design is the systematic inclusion of the experimental design of noise variables, that is, the variables over which the designer has little or no control. Robust design's objective is to suppress, as much as possible, the effect of noise by exploring the levels of the factors to determine their potential for making the software insensitive to these sources of variation.

18.4 ROBUST DESIGN CONCEPT #2: QUALITY LOSS FUNCTION

Traditional inspection schemes represent the heart of online quality control. Inspection schemes depend on the binary characterization of design parameters (i.e., being within or outside the specification limits). A process is conforming if all its inspected design parameters are within their respective specification limits; otherwise, it is nonconforming. This binary representation of the acceptance criteria per design parameter, for example, is not realistic because it characterizes, equally, entities that are marginally off these specification limits and entities that are marginally within these limits. In addition, this characterization also does not discriminate the marginally off entities with those that are significantly off. The point here is that it is not realistic to assume that, as we move away from the nominal specification in software, the quality loss is zero as long as you stay within the set tolerance limits. Rather, if the software functional requirement is not exactly "on target," then loss will result, for example, in terms of customer satisfaction. Moreover, this loss is probably not a linear function of the deviation from nominal specifications but rather a quadratic function similar to what you see in Figure 18.4. Taguchi and Wu (1980) proposed a continuous and better representation than this dichotomous characterization—the quality loss function. The loss function provides a better estimate of the monetary loss incurred by production and customers as an output response, y, deviating from its targeted performance value, T_y. The determination of the target T_y implies the nominal-the-best and dynamic classifications.

A quality loss function can be interpreted as a means to translate variation and target adjustment to a monetary value. It allows the design teams to perform a detailed optimization of cost by relating technical terminology to economical measures. In its quadratic form (Figure 18.6), quality loss is determined by first finding

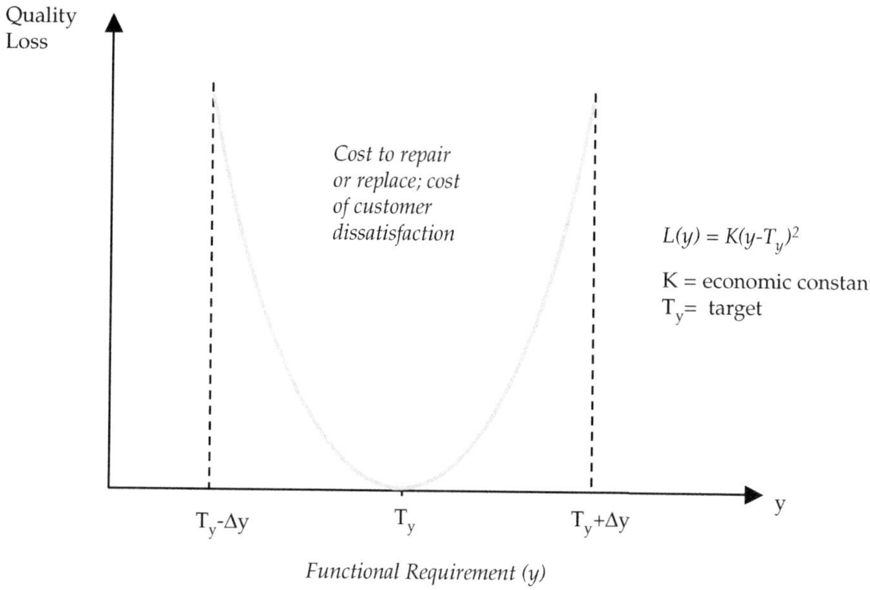

FIGURE 18.6 Quality loss function.

the functional limits,[7] $T_y \pm \Delta y$, of the concerned response. The functional limits are the points at which the process would fail (i.e., produces unacceptable performance in approximately half of the customer applications). In a sense, these limits represent performance levels that are equivalent to average customer tolerance. Kapur (1988) continued with this path of thinking and illustrated the derivation of specification limits using Taguchi's quality loss function. A quality loss is incurred as a result of the deviation in the response (y or FR), as caused by the noise factors, from their intended targeted performance, T_y. Let "L" denote the QLF, taking the numerical value of the FR and the targeted value as arguments. By Taylor series expansion[8] at FR $= T$ and with some assumption about the significant of the expansion terms we have:

$$L(\mathrm{FR}, T) \cong K(\mathrm{FR} - T_{\mathrm{FR}})^2 \tag{18.1}$$

Let FR $\in [T_y - \Delta y, T_y + \Delta y]$, where T_y is the target value and Δy is the functional deviation from the target (see Figure 18.2). Let A_Δ be the quality loss incurred as a result of the symmetrical deviation, Δy, then by substitution into Equation 18.1 and

[7]Functional limits or customer tolerance in robust design terminology is synonymous with design range, (DR) in axiomatic design approach terminology. See Chapter 13.
[8]The assumption here is that L is a higher order continuous function such that derivatives exist and is symmetrical around $y = T$

solving for K:

$$K = \frac{A_\Delta}{(\Delta y)^2} \tag{18.2}$$

In the Taguchi tolerance design method, the quality loss coefficient K can be determined based on losses in monetary terms by falling outside the *customer tolerance limits (design range)* instead of the specification limits usually used in process capability studies, for example, or the *producer limits*. The specification limits most often are associated with the design parameters. Customer tolerance limits are used to estimate the loss from customer perspective or the quality loss to society as proposed by Taguchi. Usually, the customer tolerance is wider than manufacturer tolerance. In this chapter, we will side with the design range limits terminology. Deviation from this practice will be noted where needed.

Let *f(y)* be the probability density function (pdf) of the y, then via the expectation operator, E, we have the following:

$$E\left[L(y,T)\right] = K\left[\sigma_y^2 + (\mu_y - T_y)^2\right] \tag{18.3}$$

Equation (18.3) is fundamental. Quality loss has two ingredients: loss incurred as a result of variability σ_y^2 and loss incurred as a result of mean deviation from target $(\mu_y - T_y)^2$. Usually the second term is minimized by adjusting the mean of the critical few *design parameters*—the affecting x's.

The derivation in (18.3) suits the nominal-is-best classification. Other quality loss function mathematical forms may be found in El-Haik (2005). The following forms of loss function were borrowed from their esteemed paper.

18.4.1 Larger-the-Better Loss Function

For functions like "increase yield" (y = yield), we would like a very large target, ideally $T_y \rightarrow \infty$. The requirement (output y) is bounded by a lower functional specifications limit y_l. The loss function then is given by

$$L(y, T_y) = \frac{K}{y^2} \quad \text{where} \quad y \geq y_l \tag{18.4}$$

Let μ_y be the average y numerical value of the software range (i.e., the average around which performance delivery is expected). Then by Taylor series expansion around $y = \mu_y$, we have

$$E\left[L(y, T_y)\right] = K\left[\frac{1}{\mu_y^2} + \frac{3}{\mu_y^4}\sigma_y^2\right] \tag{18.5}$$

18.4.2 Smaller-the-Better Loss Function

Functions like "reduce audible noise" would like to have zero as their target value. The loss function in this category and its expected values are given in (18.6) and (18.7), respectively.

$$L(y, T) = Ky^2 \tag{18.6}$$

$$E[L(y, T)] = K\left(\sigma_y^2 + \mu_y^2\right) \tag{18.7}$$

In this development as well as in the next sections, the average loss can be estimated from a parameter design or even a tolerance design experiment by substituting the experiment variance S^2 and average $\bar{y}$ as estimates for σ_y^2 and μ_y into Equations (18.6) and (18.7).

Recall the example of two settings in Figure 18.3. It was obvious that setting 1 was more robust (i.e., produced less variation in the functional requirement [y] than setting 2 by capitalizing on nonlinearity as well as on lower quality loss similar to the scenario on the right of Figure 18.4). Setting 1 (DP*) robustness is even more evident in the flatter quadratic quality loss function.

Because quality loss is a quadratic function of the deviation from a nominal value, the goal of the DFSS project should be to minimize the squared deviations or variance of a requirement around nominal (ideal) specifications, rather than the number of units within specification limits (as is done in traditional statistical process control (SPC) procedures).

Several books recently have been published on these methods, for example, Phadke (1989), Ross (1988), and—within the context of product DFSS—Yang and El-Haik (2008), El-Haik (2005), and El-Haik (2008) to name a few, and it is recommended that the reader refer to these books for further specialized discussions. Introductory overviews of Taguchi's ideas about quality and quality improvement also can be found in Kackar (1985).

18.5 ROBUST DESIGN CONCEPT #3: SIGNAL, NOISE, AND CONTROL FACTORS

Software that is designed with Six Sigma quality always should respond in exactly the same manner to the signals provided by the customer. When you press the ON button of a television remote control you expect the television to switch on. In a DFSS-designed television, the starting process always would proceed in exactly the same manner; for example, after three seconds of the remote pressing action, the television comes to life. If, in response to the same signal (pressing the ON button) there is random variability in this process, then you have less-than-ideal quality. For example, because of such uncontrollable factors such as speaker conditions, weather conditions, battery voltage level, television wear, and so on, the television sometimes

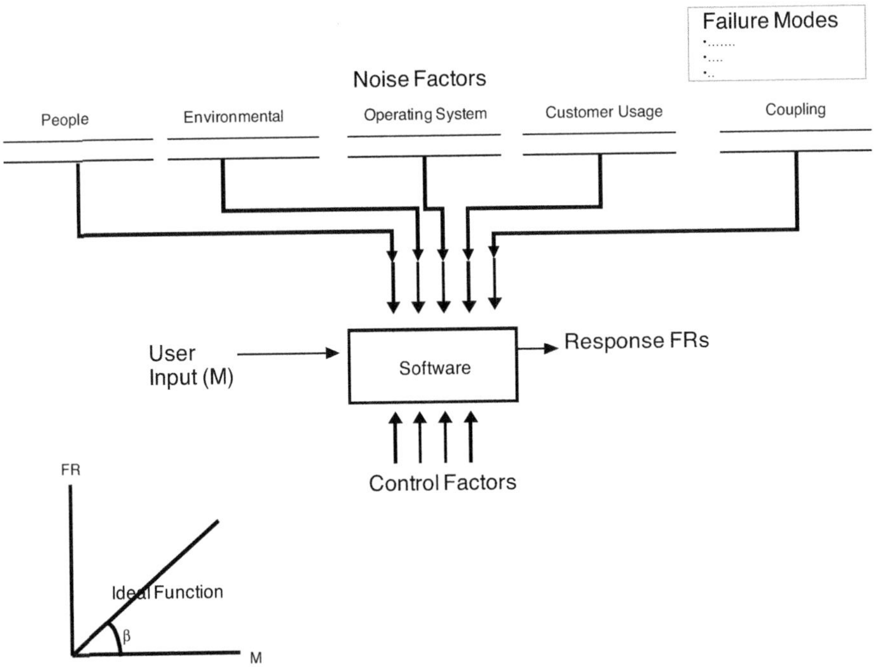

FIGURE 18.7 P-diagram.

may start only after 10 seconds and, finally, may not start at all. We want to minimize the variability in the output response to noise factors while maximizing the response to signal factors.

Noise factors are those factors that are not under the control of the software design team. In this television example, those factors include speaker conditions, weather conditions, battery voltage level, and television wear. Signal factors are those factors that are set or controlled by the customer (end user) of the software to make use of its intended functions.

The goal of a DFSS optimize phase is to find the best experimental settings of factors under the team's control involved in the design to minimize quality loss; thus, the factors in the experiment represent control factors. Signal, noise and control factors (design parameters) usually are summarized in a P-diagram similar to the one in Figure 18.7.

18.5.1 Usage Profile: The Major Source of Noise

A software profile is the set of operations that software can execute along with the probability with which they will occur (Halstead, 1977). Software design teams know that there are two types of features: tried-and-true features that usually work and fancy features that cause trouble. The later is a big source of frustration. End users will not stay on the prescribed usage catalogs, even when highly constrained by user interfaces. The software robustness argument against this concern is one for which no counter argument can prevail; certified robustness and reliability is valid only for the profile used in testing.

The operational profile includes the operating environment or system, third-party application programming interfaces, language-specific run-time libraries, and external data files that the tested software accesses. The state of each of these other users can determine the software robustness. If an e-mail program cannot access its database, then it is an environmental problem that the team should incorporate into the definition of an operational profile.

To specify an operational profile, the software DFSS team must account for more than just the primary users. The operating system and other applications competing for resources can cause an application to fail even under gentle uses. Software operating environments are extremely diverse and complex. For example, the smooth use of a word processor can elicit failure when the word processor is put in a stressed operating environment. What if the document gently being edited is marked read-only by a privileged user? What if the operating system denies additional memory? What if the document autobackup feature writes to a bad storage sector? All these aspects of a software environment can cause failures even though the user follows a prescribed usage profile. Most software applications have multiple users. At the very least, an application has one primary user and the operating system. Singling out the end user and proclaiming a user profile that represents only that user is naive. That user is affected by the operating environment and by other system users. A specified operational profile should include all users and all operating conditions that can affect the system under test (Whittaker & Voas, 2000).

18.5.2 Software Environment: A Major Source of Noise

The fundamental problem with trying to correlate code metrics to software quality such as robustness is that quality is a behavioral characteristic, not a structural one. A perfect reliable system can suffer from terrible spaghetti logic. Although spaghetti logic may be difficult to test thoroughly using coverage techniques, and although it will be hard to debug, it still can be correct. A simple straight-line chunk of code (without decision points) can be totally unreliable. These inconsistencies make generalizing code metrics impossible. Consider the Centaur rocket and Ariane 5 problems. Aviation Week and Space Technology announced that the Centaur upper stage failure was caused by a simple programming error. A constant was set to one-tenth of its proper value (−0.1992476 instead of −1.992476). This tiny miscoding of a constant caused the rocket failure. Code complexity was not the issue. The Ariane 5 disaster was caused by failing to consider the environment in which a software component was being reused (Lions, 1996). That is, the complexity or lack thereof had nothing to do with the resulting software failure.

Although it is true that metrics correlate to characteristics like readability or maintainability, they do not correlate well to robustness. Therefore, design teams need complexity metrics with respect to the environment in which the code will reside.

Propagation, infection, and execution (PIE) provides a behavioral set of measures that assess how the structure and semantics of the software interact with its environment (Voas, 1992). The code complexity metrics must be a function of the software's semantics and environment in a robustness study. If they are, then they will be useful for creating a more universally applicable robustness theory.

It is possible to use the three algorithms of the PIE model (Voas, 1996)—propagation analysis, infection analysis, and execution analysis in a robustness study. Execution analysis provides a quantitative assessment of how frequently a piece of code actually is executed with respect to the environment. For example, a deeply nested piece of code, if viewed only statically, seems hard to reach. This assumption could be false. If the environment contains many test vectors that toggle branch outcomes in ways that reach the nested code, then executing this code will not be difficult. Similarly, infection analysis and propagation analysis also quantitatively assess the software semantics in the context of the internal states that are created at runtime (Whittaker & Voas, 2000).

Software does not execute in isolation; it resides on hardware. Operating systems are the lowest level software programs we can deal with, and they operate with privileged access to hardware memory. Application software cannot touch memory without going through the operating system kernel. Device drivers are the next highest level. Although they must access hardware memory through an operating system kernel, device drivers can interact directly with other types of hardware, such as modems and keyboards.

Application software communicates with either device drivers or an operating system. In other words, most software does not interact directly with humans; instead, all inputs come from an operating system, and all outputs go to an operating system. Too often, developers perceive humans as the only user of software. This misconception fools testers into defining operational profiles based on how human users interact with software. In reality, humans interact only with the drivers that control input devices.

The current practice for specifying an operational profile is to enumerate input only from human users and lump all other input under abstractions called environment variables. For example, you might submit inputs in a normal environment and then apply the same inputs in an overloaded environment. Such abstractions greatly oversimplify the complex and diverse environment in which the software operates. The industry must recognize not only that humans are not the primary users of software but also that they often are not users at all. Most software receives input only from other software. Recognizing this fact and testing accordingly will ease debugging and make operational profiles more accurate and meaningful. Operational profiles must encompass every external resource and the entire domain of inputs available to the software being tested. One pragmatic problem is that current software testing tools are equipped to handle only human-induced noise (Whittaker & Voas, 2000).

Sophisticated and easy-to-use tools to manipulate graphical user interface (GUIs) and type keystrokes are abundant. Tools capable of intercepting and manipulating software-to-software communication fall into the realm of hard-to-use system-level debuggers. It is difficult to stage an overloaded system in all its many variations, but it is important to understand the realistic failure situations that may result.

18.6 ROBUSTNESS CONCEPT #4: SIGNAL–TO-NOISE RATIOS

A conclusion of the previous sections is that quality can be quantified in terms of the respective software response to noise and signal factors. The ideal software only will respond to the customer signals and will be unaffected by random noise factors. Therefore, the goal of the DFSS project can be stated as attempting to maximize the SN ratio for the respective software. The SN ratios described in the following paragraphs have been proposed by Taguchi (1987).

- *Smaller-is-better*. For cases in which the DFSS team wants to minimize the occurrences of some undesirable software responses, you would compute the following SN ratio:

$$SN = -10\log_{10}\left(\frac{1}{N}\sum_{n=1}^{N} y_i^2\right) \tag{18.8}$$

  The constant, N, represents the number of observations (that has y_i as their response) measured in an experiment or in a sample. Experiments are conducted, and the y measurements are collected. Note how this SN ratio is an expression of the assumed quadratic nature of the loss function.

- *Nominal-the-best*. Here, the DFSS team has a fixed signal value (nominal value), and the variance around this value can be considered the result of noise factors:

$$SN = -10\log_{10}\left(\frac{\mu^2}{\sigma^2}\right) \tag{18.9}$$

  This signal-to-noise ratio could be used whenever ideal quality is equated with a particular nominal value. The effect of the signal factors is zero because the target date is the only intended or desired state of the process.

- *Larger-is-better*. Examples of this type of software requirement are therapy software yield, purity, and so on. The following SN ratio should be used:

$$SN = -10\log_{10}\left(\frac{1}{N}\sum_{n=1}^{N} \frac{1}{y_i^2}\right) \tag{18.10}$$

- *Fraction defective* (p). This SN ratio is useful for minimizing a requirement's defects (i.e., values outside the specification limits or minimizing the percent of software error states, for example).

$$SN = 10\log_{10}\left(\frac{p}{1-p}\right) \tag{18.11}$$

  where p is the proportion defective.

18.7 ROBUSTNESS CONCEPT #5: ORTHOGONAL ARRAYS

This aspect of Taguchi robust design methods is the one most similar to traditional design of experience (DOE) technique. Taguchi has developed a system of tabulated designs (arrays) that allow for the maximum number of main effects to be estimated in an unbiased (orthogonal) manner, with a minimum number of runs in the experiment. Latin square designs, 2^{k-p} designs (Plackett–Burman designs, in particular) and Box–Behnken designs also are aimed at accomplishing this goal. In fact, many standard orthogonal arrays tabulated by Taguchi are identical to fractional two-level factorials, Plackett–Burman designs, Box–Behnken designs, Latin square, Greco–Latin squares, and so on.

Orthogonal arrays provide an approach to design efficiently experiments that will improve the understanding of the relationship between software control factors and the desired output performance (functional requirements and responses). This efficient design of experiments is based on a fractional factorial experiment, which allows an experiment to be conducted with only a fraction of all possible experimental combinations of factorial values. Orthogonal arrays are used to aid in the design of an experiment. The orthogonal array will specify the test cases to conduct the experiment. Frequently, two orthogonal arrays are used: a control factor array; and a noise factor array; the latter used to conduct the experiment in the presence of difficult-to-control variation so as to develop robust software.

In Taguchi's experimental design system, all experimental layouts will be derived from about 18 standard "orthogonal arrays." Let us look at the simplest orthogonal array, L_4 array, (Table 18.1).

The values inside the array, that is, 1 and 2, represent two different levels of a factor. By simply use "-1" to substitute "1," and "$+1$" to substitute "2," we find that this L_4 array becomes Table 18.2.

Clearly, this is a 2^{3-1} fractional factorial design, with the defining relation,[9] $I = -ABC$ Where "column 2" of L_4 is equivalent to the "A column" of the 2^{3-1} design, "column 1" is equivalent to the "B column" of the 2^{3-1} design, and "column 3" is equivalent to "C column" of the 2^{3-1} design, with $C = -AB$.

In each of Taguchi's orthogonal array, there are "linear graph"(s) to go with it. A linear graph is used to illustrate the interaction relationships in the orthogonal array, for example, the L_4 array linear graph is given in Figure 18.8. The numbers "1" and "2" represent column 1 and column 2 of the L_4 array, respectively, "3" is above the line segment connecting "1" and "2," which means that 'the interaction of column 1 and column 2 is confounded with column "3," which is perfectly consistent with $C = -AB$ in the 2^{3-1} fractional factorial design.

For larger orthogonal arrays, not only are there linear graphs but there are also interaction tables to explain interaction relationships among columns. For example, The L_8 array in Table 18.3 has the linear graph and table shown in Figure 18.9.

This approach to designing and conducting an experiment to determine the effect of control factors (design parameters) and noise factors on a performance characteristic is represented in Figure 18.10.

[9]The defining relation is covered in Chapter 12 of El-Haik and Roy (2005).

TABLE 18.1 $L_4(2^3)$ Orthogonal Array

	Column		
No.	1	2	3
1	1	1	1
2	1	2	2
3	2	1	2
4	2	2	1

TABLE 18.2 L_4 Using "–1" and "+1" Level Notation

	Column		
No.	1	2	3
1	−1	−1	−1
2	−1	1	1
3	1	−1	1
4	1	1	−1

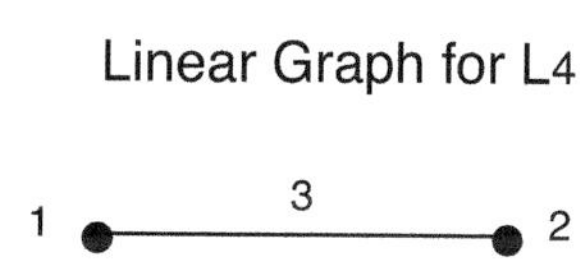

FIGURE 18.8 L_4 linear graph.

TABLE 18.3 $L_8(2^7)$ Orthogonal Array

	Column						
No	1	2	3	4	5	6	7
1	1	1	1	1	1	1	1
2	1	1	1	2	2	2	2
3	1	2	2	1	1	2	2
4	1	2	2	2	2	1	1
5	2	1	2	1	2	1	2
6	2	1	2	2	1	2	1
7	2	2	1	1	2	2	1
8	2	2	1	2	1	1	2

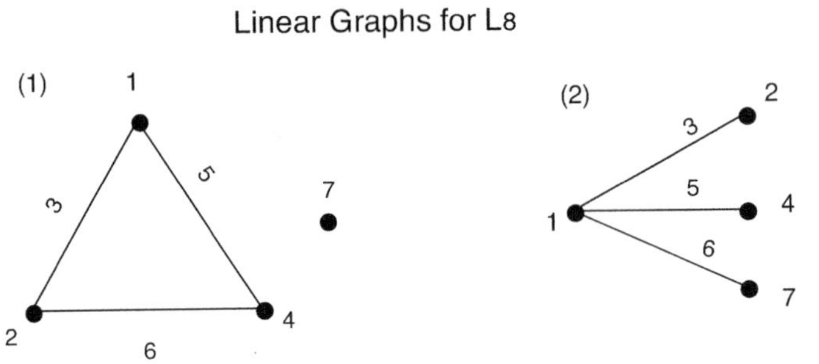

Column	Column						
	1	2	3	4	5	6	7
1	(1)	3	2	5	4	7	6
2		(2)	1	6	7	4	5
3			(3)	7	6	5	4
4				(4)	1	2	3
5					(5)	3	2
6						(6)	1
7							(7)

FIGURE 18.9 Interaction table and linear graph of L_8.

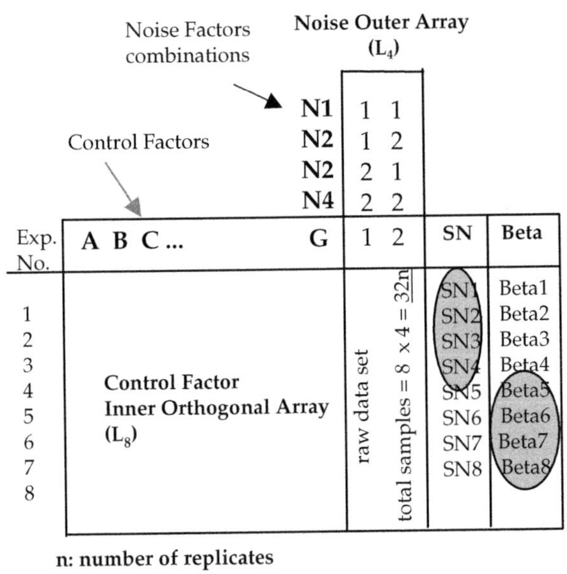

FIGURE 18.10 Parameter design orthogonal array experiment.

Control Factors	A	B	C	D
Level 1	0.62	1.82	3.15	0.10
Level 2	3.14	1.50	0.12	2.17
Gain (meas. in dB)	2.52	0.32	3.03	2.07

FIGURE 18.11 Signal-to-noise ration response table example.

The factors of concern are identified in an inner array, or control factor array, which specifies the factorial levels. The outer array, or noise factor array, specifies the noise factors or the range of variation the software possibly will be exposed to in its life cycle. This experimental setup allows the identification of the control factor values or levels that will produce the best performing, most reliable, or most satisfactory software across the expected range of noise factors.

18.8 ROBUSTNESS CONCEPT #6: PARAMETER DESIGN ANALYSIS

After the experiments are conducted and the signal-to-noise ratio is determined for each run, a mean signal-to-noise ratio value is calculated for each factor level. This data is analyzed statistically using analysis of variance (ANOVA) techniques (El-Haik & Roy, 2005).[10] Very simply, a control factor with a large difference in the signal noise ratio from one factor setting to another indicates that the factor is a significant contributor to the achievement of the software performance response. When there is little difference in the signal-to-noise ratio from one factor setting to another, it indicates that the factor is insignificant with respect to the response. With the resulting understanding from the experiments and subsequent analysis, the design team can:

- Identify control factors levels that maximize output response in the direction of goodness and minimize the effect of noise, thereby achieving a more robust design.
- Perform the two-step robustness optimization[11]:
  - Step 1: Choose factor levels to reduce variability by improving the SN ratio. This is robustness optimization step 1. The level for each control factor with the highest SN ratio is selected as the parameter's best target value. All these best levels will be selected to produce the "robust design levels" or the "optimum levels" of design combination. A response table summarizing SN gain usually is used similar to Figure 18.11. Control factor level effects are calculated by averaging SN ratios that correspond to the individual control factor levels as depicted by the orthogonal array diagram. In this example, the

[10] See Appendix 18.A.

[11] Notice that the robustness two-step optimization can be viewed as a two-response optimization of the functional requirement (y) as follows: Step 1 targets optimizing the variation (σ_y), and step 2 targets shifting the mean (μ_y) to target T_y. For more than two functional requirements, the optimization problem is called multiresponse optimization.

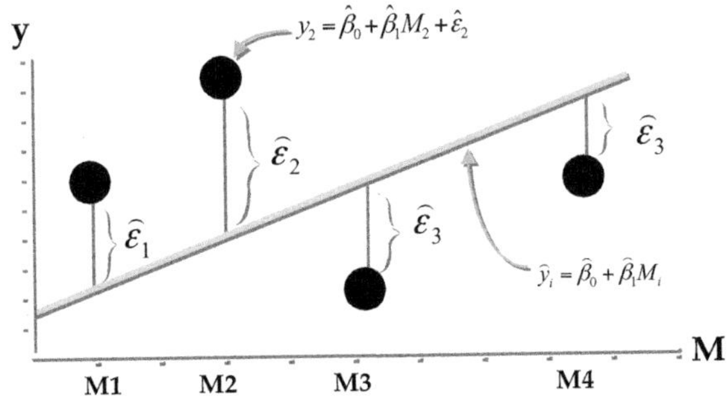

FIGURE 18.12 Best-fit line of a dynamic robust design DOE.

robust design levels are as follows: factor A at level 2, factor C at level 1 and factor D at level 2, or simply A2C1D2.

Identify control factors levels that have no significant effect on the functional response mean or variation. In these cases, tolerances can be relaxed and cost reduced. This is the case for Factor B of Figure 18.11.

- Step 2: Select factor levels to adjust mean performance. This is the robustness optimization step 2. This is more suited for dynamic characteristic robustness formulation, with sensitivity defined as Beta (β). In a robust design, the individual values for$\beta$are calculated using the same data from each experimental run as in Figure 18.10. The purpose of determining the Beta values is to characterize the ability of control factors to change the average value of the functional requirement (y) across a specified dynamic signal range as in Figure 18.12. The resulting Beta performance of a functional requirement (y) is illustrated by the slope of a best-fit line in the form of $y = \beta_0 + \beta_1 M$, where $\beta_1$is the slope and β_0 is the intercept of the functional requirement data that is compared with the slope of an ideal function line. A best-fit line is obtained by minimizing the squared sum of error (ε) terms.

In dynamic systems, a control factor's importance for decreasing sensitivity is determined by comparing the gain in SN ratio from level to level for each factor, comparing relative performance gains between each control factor, and then selecting which ones produce the largest gains.

That is, the same analysis and selection process is used to determine control factors that can best used to adjust the mean functional requirement. These factors may be the same ones that have been chosen based on SN improvement, or they may be factors that do not affect the optimization of SN. Most analyses of robust design experiments amount to a standard ANOVA[12] of the respective SN ratios, ignoring two-way or higher order interactions. However, when estimating error variances, one customarily pools together main effects of negligible size. It should be noted at this

[12]See Appendix 18.A.

point that, of course, all experimental designs (e.g., 2^k, 2^{k-p}, 3^{k-p}, etc.) can be used to analyze SN ratios that you computed. In fact, the many additional diagnostic plots and other options available for those designs (e.g., estimation of quadratic components, etc.) may prove very useful when analyzing the variability (SN ratios) in the design. As a visual summary, an SN ratio plot usually is displayed using the experiment average SN ratio by factor levels. In this plot, the optimum settings (largest SN ratio) for each factor easily can be identified.

For prediction purposes, the DFSS team can compute the expected SN ratio given optimum settings of factors (ignoring factors that were pooled into the error term). These predicted SN ratios then can be used in a verification experiment in which the design team actually sets the process accordingly and compares the resulting observed SN ratio with the predicted SN ratio from the experiment. If major deviations occur, then one must conclude that the simple main effect model is not appropriate. In those cases, Taguchi (1987) recommends transforming the dependent variable to accomplish additivity of factors, that is, to make the main effects model fit. Phadke (1989: Chapter 6) also discusses, in detail, methods for achieving additivity of factors.

A robustness case study is provided in the following section.

18.9 ROBUST DESIGN CASE STUDY NO. 1: STREAMLINING OF DEBUGGING SOFTWARE USING AN ORTHOGONAL ARRAY[13]

Debugging is a methodical process of finding and reducing the number of bugs, or defects, in a computer program or a piece of electronic hardware, thus making it behave as expected. Debugging tends to be harder when various subsystems are tightly coupled, as changes in one may cause bugs to emerge in another. Debuggers are software tools that enable the programmer to monitor the execution of a program, stop it, restart it, set breakpoints, change values in memory, and even in some cases, go back in time. The term debugger also can refer to the person who is doing the debugging.

Generally, high-level programming languages, such as Java, make debugging easier because they have features such as exception handling that make real sources of erratic behavior easier to spot. In lower level programming languages such as C or assembly, bugs may cause silent problems such as memory corruption, and it is often difficult to see where the initial problem happened. In those cases, memory debugger tools may be needed.[14]

Software debugging is the process by which DFSS teams attempt to remove coding defects from a computer program. It is not untypical for the debugging of software development to take 40%–60% of the overall development time. Ultimately, a great amount of difficulty and uncertainty surround the crucial process of software

[13]Reprinted with permission of John Wiley & Sons, Inc. from Taguchi et al. (2005).
[14]http://en.wikipedia.org/wiki/Debugging.

debugging. It is difficult to determine how long it will take to find and fix an error, not to mention whether the fix actually will be effective. To remove bugs from the software, the team first must discover that a problem exists, then classify the error, locate where the problem actually lies in the code, and finally, create a solution that will remedy the situation (without introducing other problems!). Software professionals constantly are searching for ways to improve and streamline the debugging process. At the same time, they have been attempting to automate techniques used in error detection.

Where bugs are found by users after shipment, not only the software per se but also the company's reputation will be damaged. However, thanks to the widely spreading Internet technology, even if software contain bugs, it is now easy to distribute bug-fix software to users. Possibly because of this trend, the issue of whether there are bugs seems to become of less interest. However, it is still difficult to correct bugs after shipping in computerized applications (e.g., automation). This case study establishes a method of removing bugs within a limited period before shipping, using an orthogonal array.

This case study is based on the work of Dr. G. Taguchi in (Taguchi, 1999a) and (Taguchi, 1999b). The method was conducted by Takada et al. (2000). They allocated items selected by users (signal factors) to L_{36} or L_{18} orthogonal arrays, ran software in accordance to the combination in each row, and judged using binary output (0 or 1) whether an output was normal. Subsequently, and using the output obtained, the authors calculated the variance of interaction to identify bugging root cause factors in the experiment. Through this process, the authors found almost all bugs caused by combinations of factors on the beta version (contains numerous recognized bugs) of their company software. Therefore, the effectiveness of this experiment easily can be confirmed. However, because bugs detected cannot be corrected, they cannot check whether the trend in regard to the number of bugs is decreasing. As signal factors, they selected eight items that frequently can be set up by users, allocating them to an L_{18} orthogonal array. When a signal factor has four or more levels, for example, continuous values ranging from 0 to 100, they selected 0, 50, and 100.

When dealing with a factor that can be selected, such as patterns 1 to 5, three of the levels that are used most commonly by users were selected. Once they assigned these factors to an orthogonal array, they noticed that there were quite a few two-level factors. In this case, they allocated a dummy level to level 3. For the output, they used a rule of normal = 0 and abnormal = 1, based on whether the result was what they wanted. In some cases, "no output" was the right output. Therefore, normal or abnormal was determined by referring to the specifications. Signal factors and levels are shown in Table 18.4.

From the results of Table 18.5, they created approximate two-way tables for all combinations. The upper left part of Table 18.6 shows the number of each combination of A and B: A_1B_1, A_1B_2, A_1B_3, A_2B_1, A_2B_2, and A_2B_3. Similarly, they created a table for all combinations.

Where many bugs occur on one side of this table was regarded as a location with bugs. Looking at the overall result, they can see that bugs occur at H_3. After

TABLE 18.4 Signal Factors & Levels[15]

Factor	Level 1	Level 2	Level 3
A	A_1	A_2	—
B	B_1	B_2	B_3
C	C_1	C_2	C_3
D	D_1	D_2	D_3
E	E_1	E_2	E_2'
F	F_1	F_2	F_1'
G	G_1	G_2	G_1'
H	H_1	H_2	H_3

investigation, it was found that bugs did not occur in the on-factor test of H, but occur with its combination with G) (= G', the same level because of the dummy treatment used) and B_1 or B_2. Because B_3 is a factor level whose selection blocks us from choosing (or annuls) factor levels of H and has interactions among signal factors, it was considered to be the reason this result was obtained.

TABLE 18.5 L_{18} Orthogonal Array and Response (y)

No	A	B	C	D	E	F	G	H	y
1	1	1	1	1	1	1	1	1	0
2	1	1	2	2	2	2	2	2	0
3	1	1	3	3	2'	1'	1'	3	1
4	1	2	1	1	2	2	1'	3	1
5	1	2	2	2	2'	1'	1	1	0
6	1	2	3	3	1	1	2	2	0
7	1	3	1	2	1	1'	2	3	0
8	1	3	2	3	2	1	1'	1	0
9	1	3	3	1	2'	2	1	2	0
10	2	1	1	3	2'	2	2	1	0
11	2	1	2	1	1	1'	1'	2	0
12	2	1	3	2	2	1	1	3	1
13	2	2	1	2	2'	1	1'	2	0
14	2	2	2	3	1	2	1	3	1
15	2	2	3	1	2	1'	2	1	0
16	2	3	1	3	2	1'	1	2	0
17	2	3	2	1	2'	1	2	3	0
18	2	3	3	2	1	2	1'	1	0

[15]Because of author's company confidentiality policy, they have left out the details about signal factors and levels.

TABLE 18.6 Binary Table Created From L_{18} Orthogonal Array

B_1	B_2	B_3	C_1	C_2	C_3	D_1	D_2	D_3	E_1	E_2	E_3	F_1	F_2	F_3	G_1	G_2	G_3	H_1	H_2	H_3
1	1	0	1	0	1	1	0	1	0	1	1	0	1	1	0	0	2	0	0	2
1	1	0	0	1	1	0	1	1	1	1	0	1	1	0	2	0	0	0	0	2
		B_1	0	0	2	0	1	1	0	1	1	1	0	1	1	0	1	0	0	2
		B_2	1	1	0	1	0	1	1	1	0	0	2	0	1	0	1	0	0	2
		B_3	0	0	0	0	0	0	0	0	0	0	0	0	0	0	0	0	0	0
					C_1	1	0	0	0	1	0	0	1	0	0	0	1	0	0	1
					C_2	0	0	1	1	0	0	0	1	0	1	0	0	0	0	1
					C_3	0	1	1	0	1	1	1	0	1	1	0	1	0	0	2
								D_1	0	1	0	0	1	0	0	0	1	0	0	1
								D_2	0	1	0	1	0	0	1	0	0	0	0	1
								D_3	1	0	1	0	1	1	1	0	1	0	0	2
											E_1	0	1	0	1	0	0	0	0	1
											E_2	1	1	0	1	0	1	0	0	2
											E_2’	0	0	1	0	0	1	0	0	1
														F_1	1	0	0	0	0	1
														F_2	1	0	1	0	0	2
														F_1’	0	0	1	0	0	1
																	G_1	0	0	2
																	G_2	0	0	0
																	G_1’	0	0	2

Now the calculated variance and interaction were as follows:

- Variation between A and B combinations with 5 degrees of freedom:

$$S_{AB} = \frac{1^2 + 1^2 + 0 + 1^2 + 1^2 + 0^2}{3} - \frac{4^2}{18}$$
$$= 0.44$$

- Variation of A with 1 degree of freedom:

$$S_A = \frac{2^2 + 2^2}{9} - \frac{4^2}{18}$$
$$= 0.00$$

- Variation of B with 1 degree of freedom:

$$S_B = \frac{2^2 + 2^2 + 0}{6} - \frac{4^2}{18}$$
$$= 0.44$$

TABLE 18.7 Main Effect

Factor	Main Effect
A	0.00
B	0.44
C	0.11
D	0.11
E	0.03
F	003
G	0.44
H	1.77

A summary of all main effects is shown in Table 18.7.

- Variation of AB interaction with 2 degrees of freedom:

$$\begin{aligned} S_{AxB} &= S_{AB} - S_A - S_B \\ &= 0.44 - 0.00 - 0.44 \\ &= 0.00 \end{aligned}$$

In the next step, they divided the combinational effect, S_{AB}, and interaction effect, $S_{A\times B}$, by each corresponding degree of freedom:

$$\text{Combination effect} = \frac{S_{AB}}{5} = 0.09$$

$$\text{Interaction effect} = \frac{S_{A\times B}}{5} = 0.00$$

Because these results are computed from the approximate two-way tables, they considered such results to be a clue for debugging in particular if the occurrence of bugs is infrequent. When there are more bugs or when a large-scale orthogonal array is used, they used these values for finding bug locations.

The authors succeeded in finding bugs by taking advantage of each combination of factors (Table 18.8). As is shown, using the method as described, the bugs can be found from an observation of specific combinations.

Following are the differences between our current debugging process and the method using an orthogonal array:

1. Efficiency of finding bugs
   a. Current process: What can be found through numerous tests are mainly independent bugs. To find bugs caused by a combination of factors, many repeated tests need to be performed.
   b. Orthogonal array: Through a few experiments, they can find independent bugs and bugs generated by a combination of factors. However, for a multiple-level factor, they need to conduct one-factor tests later on.

TABLE 18.8 Combinational and Interaction Effects

Factor	Combination	Interaction
AB	0.09	0.00
AC	0.09	0.17
AD	0.09	0.17
AE	0.09	0.25
AF	0.04	0.00
AG	0.15	0.00
AH	0.36	0.00
BC	0.26	0.39
BD	0.14	0.14
BE	0.17	0.19
BF	0.42	0.78
BG	0.22	0.11
BH	0.39	0.22
CD	0.14	0.22
CE	0.17	0.36
CF	0.22	0.44
CG	0.12	0.03
CH	0.26	0.06
DE	0.07	0.11
DF	0.12	0.19
DG	0.12	0.03
DH	0.26	0.11
EF	0.12	0.22
EG	0.16	0.01
EH	0.23	0.01
FG	0.20	0.06
FH	0.42	0.11
GH	0.62	0.44

2. Combination of signal factors
   a. Current process: DFSS team tend to check only where the bug may exist and unconsciously neglect the combinations that users probably do not use.
   b. Orthogonal array: This method is regarded as systematic. Through nonsubjective combinations that do not include debug engineers' presuppositions, a well-balanced and broadband checkup can be performed.
3. Labor required
   a. Current process: After preparing a several-dozen-page checksheet, they have to investigate all its checkpoints.
   b. Orthogonal array: The only task they need to do is to determine signal factors and levels. Each combination is generated automatically. The number of checkups required is much smaller, considering the number of signal factors.

4. Location of bugs
   a. Current process: Because they need to change only a single parameter for each test, they can easily notice whether changed items or parameters involve bugs.
   b. Orthogonal array: Locations of bugs are identified by looking at the numbers after the analysis.
5. Judgment of bugs or normal outputs
   a. Current process: They easily can judge whether a certain output is normal or abnormal only by looking at one factor changed for the test.
   b. Orthogonal array: Because they need to check the validity for all signal factors for each output, it is considered cumbersome in some cases.
6. When there are combinational interactions among signal factors
   a. Current process: Nothing in particular.
   b. Orthogonal array: they cannot perform an experiment following combinations determined in an orthogonal array.

Although several problems remain before they can conduct actual tests, they believe that through the use of our method, the debugging process can be streamlined. In addition, because this method can be employed relatively easily by users, they can assess newly developed software in terms of bugs. In fact, as a result of applying this method to software developed by outside companies, they have found a certain number of bugs.

18.10 SUMMARY

To briefly summarize, when using robustness methods, the DFSS team first needs to determine the design or control factors that can be controlled. These are the factors in the DFSS team for which the team will try different levels. Next, they decide on an appropriate orthogonal array for the experiment. Then, they need to decide how to measure the design requirement of interest. Most SN ratios require that multiple measurements be taken in each run of the experiment; so that the variability around the nominal value otherwise cannot be assessed. Finally, they conduct the experiment and identify the factors that most strongly affect the chosen SN ratio, and they reset the process parameters accordingly.

APPENDIX 18.A

Analysis of Variance (ANOVA)

Analysis of variance (ANOVA)[16] is used to investigate and model the relationship between a response variable (y) and one or more independent factors. In effect,

[16] ANOVA differs from regression in two ways; the independent variables are qualitative (categorical), and no assumption is made about the nature of the relationship (i.e., the model does not include coefficients for variables).

analysis of variance extends the two-sample t test for testing the equality of two population means to a more general null hypothesis of comparing the equality of more than two means versus them not all being equal. ANOVA includes procedures for fitting ANOVA models to data collected from several different designs and graphical analysis for testing equal variances assumption, for confidence interval plots as well as graphs of main effects and interactions.

For a set of experimental data, most likely the data varies as a result of changing experimental factors, whereas some variation might be caused by unknown or unaccounted for factors, experimental measurement errors, or variation within the controlled factors themselves.

Several assumptions need to be satisfied for ANOVA to be credible, which are as follows:

1. The probability distributions of the response (y) for each factor-level combination (treatment) is normal.
2. The response (y) variance is constant for all treatments.
3. The samples of experimental units selected for the treatments must be random and independent.

The ANOVA method produces the following:

1. A decomposition of the total variation of the experimental data to its possible sources (the main effect, interaction, or experimental error);
2. A quantification of the variation caused by each source;
3. Calculation of significance (i.e., which main effects and interactions have significant effects on response (y) data variation).
4. Transfer function when the factors are continuous variables (noncategorical in nature).

18.A.1 ANOVA STEPS FOR TWO FACTORS COMPLETELY RANDOMIZED EXPERIMENT[17]

1. Decompose the total variation in the DOE response (y) data to its sources (treatment sources: factor A; factor B; factor A $\times$ factor B interaction, and error). The first step of ANOVA is the "sum of squares" calculation that produces the variation decomposition. The following mathematical equations are needed:

$$\overline{y}_{i..} = \frac{\sum_{j=1}^{b}\sum_{k=1}^{n} y_{ijk}}{bn} \text{(Row average)} \qquad (18.A.1)$$

[17]See Yang and El-Haik (2008).

$$\overline{y}_{.j.} = \frac{\sum_{i=1}^{a}\sum_{k=1}^{n} y_{ijk}}{an} \text{(Column average)} \tag{18.A.2}$$

$$\overline{y}_{ij.} = \frac{\sum_{k=1} y_{ijk}}{n} \text{(Treatment or cell average)} \tag{18.A.3}$$

$$\overline{y}_{...} = \frac{\sum_{i=1}^{a}\sum_{j=1}^{b}\sum_{k=1}^{n} y_{ijk}}{abn} \text{(Overall average)} \tag{18.A.4}$$

It can be shown that:

$$\underbrace{\sum_{i=1}^{a}\sum_{j=1}^{b}\sum_{k=1}^{n}\left(y_{ijk} - \overline{y}_{...}\right)^2}_{SS_T} = \underbrace{bn\sum_{i=1}^{a}\left(\overline{y}_{i..} - \overline{y}_{...}\right)^2}_{SS_A} + \underbrace{an\sum_{j=1}^{b}\left(\overline{y}_{.j.} - \overline{y}_{...}\right)^2}_{SS_B}$$
$$+ \underbrace{n\sum_{i=1}^{a}\sum_{j=1}^{b}\left(\overline{y}_{ij.} - \overline{y}_{i..} - \overline{y}_{.j.} + \overline{y}_{...}\right)^2}_{SS_{AB}} + \underbrace{\sum_{i=1}^{a}\sum_{j=1}^{b}\sum_{k=1}^{n}\left(y_{ijk} - \overline{y}_{ij.}\right)^2}_{SS_E} \tag{18.A.5}$$

Or simply:

$$SS_T = SS_A + SS_B + SS_{AB} + SS_E \tag{18.A.6}$$

As depicted in Figure 18.A.1, SS_T denotes the "total sum of squares," which is a measure for the "total variation" in the whole data set. SS_A is the "sum of squares" because of factor A, which is a measure of the total variation caused by the main effect of A. SS_B is the sum of squares because of factor B, which is a measure of the total variation caused by the main effect of B. SS_{AB} is the sum of squares because of factor A and factor B interaction (denoted as AB) as a measure of variation caused by interaction. SSE is the sum of squares because of error, which is the measure of total variation resulting from error.

2. Test the null hypothesis toward the significance of the factor A mean effect and the factor B mean effect as well as their interaction. The test vehicle is the mean square calculations. The mean square of a source of variation is calculated by dividing the source of the variation sum of squares by its degrees of freedom.

   The actual amount of variability in the response data depends on the data size. A convenient way to express this dependence is to say that the sum of square has degrees of freedom (DF) equal to its corresponding variability source data size reduced by one. Based on statistics, the number of degree of freedom associated with each sum of squares is shown in Table 18.A.1.

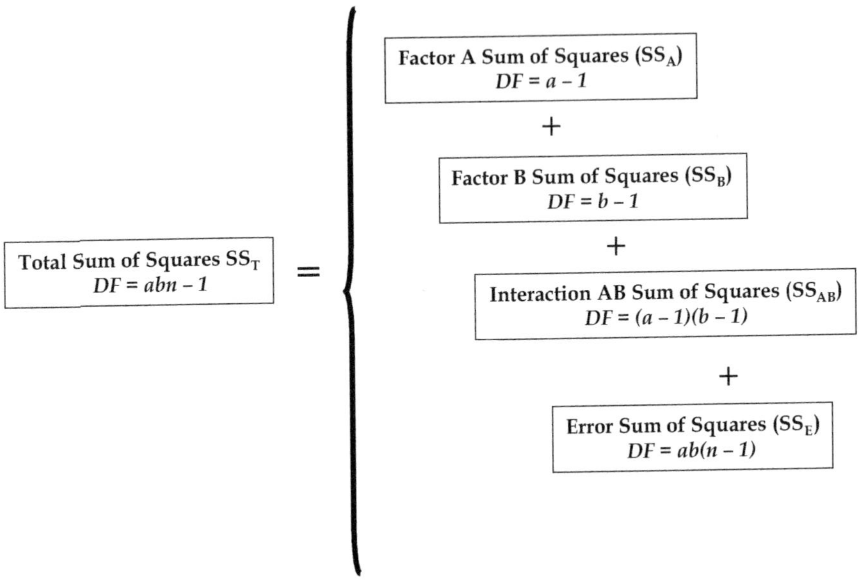

FIGURE 18.A.1 ANOVA variation decomposition.

Test for Main Effect of Factor A

H_0: No difference among the a mean levels of factor A ($\mu_{A1} = \mu_{A2} = \ldots = \mu_{Aa}$)

H_a: At least two factor A mean levels differ

Test for Main Effect of Factor B

H_0: No difference among the a mean levels of factor B ($\mu_{B1} = \mu_{B2} = \ldots = \mu_{Ba}$)

H_a: At least two factor B mean levels differ

Test for Main Effect of Factor A × Factor B Interaction

H_0: Factor A and factor B do not interact in the response mean

H_a: Factor A and factor B interact in the response mean

TABLE 18.A.1 Degree of Freedom for Two Factor Factorial Design

Effect	Degree of Freedom
A	$a - 1$
B	$b - 1$
AB interaction	$(a - 1)(b - 1)$
Error	$ab(n - 1)$
Total	$abn - 1$

TABLE 18.A.2 ANOVA Table

Source of Variation	Sum of Squares	Degree of Freedom	Mean Squares	F_0
A	SS_A	a − 1	$MS_A = \frac{SS_A}{a-1}$	$F_0 = \frac{MS_A}{MS_E}$
B	SS_B	b − 1	$MS_B = \frac{SS_B}{b-1}$	$F_0 = \frac{MS_B}{MS_E}$
AB	SS_{AB}	(a − 1)(b − 1)	$MS_{AB} = \frac{SS_{AB}}{(a-1)(b-1)}$	$F_0 = \frac{MS_{AB}}{MS_E}$
Error	SS_E	ab(n − 1)		
Total	SS_T	abn − 1		

3. Compare the Fisher *F* test of the mean square of the experimental treatment sources with the error to test the null hypothesis that the treatment means are equal.

   - If the test results are in non-rejection region of the null hypothesis, then refine the experiment by increasing the number of replicates, *n*, or by adding other factors; otherwise, the response is unrelated to the two factors.

     In the Fisher *F* test, the F_0 will be compared with the F-critical defining the null hypothesis rejection region values with appropriate degrees of freedom; if F_0 is larger than the critical value, then the corresponding effect is statistically significant. Several statistical software packages, such as MINITAB (Pensylvania State University, University Park, PA), can be used to analyze DOE data conveniently, otherwise spreadsheet packages like Excel (Microsoft, Redmond, WA) also can be used.

     In ANOVA, a sum of squares is divided by its corresponding degree of freedom to produce a statistic called the "mean square" that is used in the Fisher *F* test to see whether the corresponding effect is statistically significant. An ANOVA often is summarized in a table similar to Table 18.A.2.

Test for Main Effect of Factor A

Test statistic: $F_{0,a-1,ab(n-1)} = \frac{MS_A}{MS_E}$ with a numerator degree of freedom equal to $(a-1)$ and denominator degree of freedom equal $ab(n-1)$.

H_0 hypothesis rejection region: $F_{0,a-1,ab(n-1)} \geq F_{\alpha,a-1,ab(n-1)}$ with a numerator degree of freedom equal to $(a-1)$ and denominator degree of freedom equal to $ab(n-1)$

Test for Main Effect of Factor B

Test statistic: $F_{0,b-1,ab(n-1)} = \frac{MS_B}{MS_E}$ with a numerator degree of freedom equal to $(b-1)$ and a denominator degree of freedom equal to $ab(n-1)$.

H_0 hypothesis rejection region: $F_{0,b-1,ab(n-1)} \geq F_{\alpha,b-1,ab(n-1)}$ with a numerator degree of freedom equal to $(b-1)$ and a denominator degree of freedom equal to $ab(n-1)$

Test for Main Effect of Factor A × Factor B Interaction

Test statistic: $F_{0,(a-1)(b-1),ab(n-1)} = \frac{MS_{AB}}{MS_E}$ with a numerator degree of freedom equal to $(a-1)(b-1)$ and a denominator degree of freedom equal to $ab(n-1)$.

H_0 hypothesis rejection region: $F_{0,(a-1)(b-1),ab(n-1)} \geq F_{\alpha,(a-1)(b-1),ab(n-1)}$ with a numerator degree of freedom equal to $(a-1)(b-1)$ and a denominator degree of freedom equal to $ab(n-1)$

The interaction null hypothesis is tested first by computing the Fisher F test of the mean square for interaction with the mean square for error. If the test results in nonrejection of the null hypothesis, then proceed to test the main effects of the factors. If the test results in a rejection of the null hypothesis, then we conclude that the two factors interact in the mean response (y). If the test of interaction is significant, then a multiple comparison method such as Tukey's grouping procedure can be used to compare any or all pairs of the treatment means.

Next, test the two null hypotheses that the mean response is the same at each level of factor A and factor B by computing the Fisher F test of the mean square for each factor main effect of the mean square for error. If one or both tests result in rejection of the null hypothesis, then we conclude that the factor affects the mean response (y). If both tests result in nonrejection, then an apparent contradiction has occurred. Although the treatment means apparently differ, the interaction and main effect tests have not supported that result. Further experimentation is advised. If the test for one or both main effects is significant, then a multiple comparison is needed, such as the Tukey grouping procedure, to compare the pairs of the means corresponding with the levels of the significant factor(s).

The results and data analysis methods discussed can be extended to the general case in which there are a levels of factor A, b levels of factor B, c levels of factor C, and so on arranged in a factorial experiment. There will be $abc \ldots n$ total number of trials if there are n replicates. Clearly, the number of trials needed to run the experiment will increase quickly with the increase in the number of factors and the number of levels. In practical application, we rarely use a general full factorial experiment for more than two factors. Two-level factorial experiments are the most popular experimental methods.

REFERENCES

El-Haik, Basem, S. (2005), *Axiomatic Quality: Integrating Axiomatic Design with Six-Sigma, Reliability, and Quality*, Wiley-Interscience, New York.

El-Haik, Basem S., and Mekki, K (2008), *Medical Device Design for Six Sigma: A Road Map for Safety and Effectiveness*, 1st Ed., Wiley-Interscience, New York.

El-Haik, Basem S., and Roy, D. (2005), *Service Design for Six Sigma: A Roadmap for Excellence*, Wiley-Interscience, New York.

Halstead, M. H. (1977), *Elements of Software Science*, Elsevier, Amsterdam, The Netherlands.

Kacker, R. N. (1985), "Off-line quality control, parameter design, and the Taguchi method," *Journal of Quality Technology*, Volume *17*, #4, pp. 176–188.

Kapur, K. C. (1988), "An approach for the development for specifications for quality improvement," *Quality Engineering*, Volume *1*, #1, pp. 63–77.

Lions, J. L. (1996), *Ariane 5 Flight 501 Failure*, Report of the Inquiry Board, Pairs, France. http://www.esrin.esa.it/htdocs/tidc/Press/Press96/ariane5rep.html.

Nair, V. N. (1992), "Taguchi's parameter design: a panel discussion", *Econometrics*, Volume *34*, #2, pp. 127–161.

Phadke, M. S. (1989), "*Quality Engineering Using Robust Design*," Prentice-Hall, Englewood Cliffs, NJ.

Ross, P. J. (1988), "*Taguchi Techniques for Quality Engineering*," McGraw-Hill, New York.

Taguchi, G. (1986), "*Introduction to Quality Engineering*," UNIPUB/Kraus International Publications, White Plains, NY.

Taguchi, G. (1987), "*System of Experimental Design: Engineering Methods to Optimize Quality and Minimize Costs*," Kraus International Publications, NY.

Taguchi, G. (1999a), "Evaluation of objective function for signal factor—part 1." *Standardization and Quality Control*, Volume *52*, #3, pp. 62–68.

Taguchi, G. (1999b), "Evaluation of objective function for signal factor—part 2." *Standardization and Quality Control*, Volume *52*, #4, pp. 97–103.

Taguchi, G. and Wu, Y. (1980), *Introduction to Off-line Quality Control*, Central Japan Quality Control Association, Nagoya.

Taguchi, G. (1986), "*Introduction to Quality Engineering*," Kraus International Publications, NY.

Taguchi, G., Elsayed, E., and Hsiang, T. (1989), "*Quality Engineering in Production Systems*," McGraw-Hill, NY.

Taguchi, G., Chowdhury, S., and Taguchi, S. (1999), "*Robust Engineering: Learn How to Boost Quality While Reducing Costs and Time to Market*," 1st Ed., McGraw-Hill Professional, New York.

Taguchi, G., Chowdhury, S., and Taguchi, Wu, Y. (2005), *Quality Engineering Handbook*, John Wiley & Sons, Hoboken, NJ.

Takada, K., Uchikawa, M., Kajimoto, K., and Deguchi, J. (2000), "Efficient debugging of a software using and orthogonal array." *Journal of Quality Engineering Society*, Volume *8*, #1, pp. 60–64.

Voas, J. (1992), "PIE: A dynamic failure-based technique." *IEEE Transactions of Software Engineering*, Volume *18*, #8, pp. 717–727.

Whitttaker, J. A. and Voas, J. (2000), "Toward a more reliable theory of software reliability." *IEEE Computer*, Volume *33*, #12, pp. 36–42.

Yang, K. and El-Haik, Basem. (2008), *Design for Six Sigma: A Roadmap for Product Development*, 2nd Ed., McGraw-Hill Professional, New York.

CHAPTER 19

SOFTWARE DESIGN VERIFICATION AND VALIDATION

19.1 INTRODUCTION

The final aspect of DFSS methodology that differentiates it from the prevalent "launch and learn" method is design verification and design validation. This chapter covers in detail the Verify/Validate phase of the Design for Six Sigma (DFSS), (Identify, conceptualize, optimize, and verify/validate [ICOV]) project road map (Figure 11.1). Design verification, process validation, and design validation help identify the unintended consequences and effects of software, develop plans, and reduce risk for full-scale commercialization to all stakeholders, including all customer segments.

At this final stage before the release stage, we want to verify that software product performance is capable of achieving the requirements specified, and we also want to validate that it met the expectations of customers and stakeholders at Six Sigma performance levels. We need to accomplish this assessment in a low-risk, cost-effective manner. This chapter will cover the software relevant aspects of DFSS design verification and design validation.

Software companies still are finding it somewhat difficult to meet the requirements of both verification and validation activities. Some still confound both processes today and are struggling to distinguish between them. Many literatures do not prescribe how companies should conduct software verification and validation activities because so many ways to go about it were accumulated through mechanisms such as in-house tribal knowledge. The intent in this chapter is not to constrain the manufacturers and to allow them to adopt definitions that satisfy verification and validation terms that

Software Design for Six Sigma: A Roadmap for Excellence, By Basem El-Haik and Adnan Shaout

they can implement with their particular design processes. In this chapter, we provide a DFSS recipe for device verification and validation. Customization is warranted by an industry segment and by application.

The complexities of risk management and software make it harder for researchers to uncover deficiencies, and thus, produce fewer defects, faults, and failures. In addition, because many companies are often under budget pressure and schedule deadlines, there is always a motivation to compress that schedule, sacrificing verification and validation more than any other activities in the development process.

Verification can be performed at all stages of the ICOV DFSS process. The requirement instructs firms to review, inspect, test, check, audit, or otherwise, establish whether components, subsystems, systems, the final software product, and documents conform to requirements or design inputs. Typical verification tests may include risk analysis, integrity testing, testing for conformance to standards, and reliability. Validation ensures that a software meets defined user needs and intended uses. Validation includes testing under simulated and/or actual use conditions. Validation is, basically, the culmination of risk management, the software, and proving the user needs and intended uses is usually more difficult than verification. As the DFSS team goes upstream and links to the abstract world of customer and regulations domains—the validation domain—things are not in black-and-white, as in the engineering domain—the verification domain.

Human existence is defined in part by the need for mobility. In modern times, such need is luxuriated and partially fulfilled by commercial interests of automotive and aerospace/avionic companies. In the terrestrial and aeronautic forms of personal and mass transportation, safety is a critical issue. Where human error or negligence in a real-time setting can result in human fatality on a growing scale, the reliance on machines to perform basic, repetitive, and critical tasks grows in correlation to the consumer confidence in that technology. The more a technology's reliability is proven, the more acceptable and trusted that technology becomes. In systems delivered by the transportation industry—buses, trains, planes, trucks, automobiles—as well as in systems that are so remote that humans can play little or no role in control of those systems such as satellites and space stations, computerized systems become the control mechanism of choice. In efforts to implement safety and redundancy features in larger commercial transportation vehicles such as airplanes, this same x-by-wire (brake by wire, steer by wire, drive-by wire, etc.) concept is now being explored and implemented in aerospace companies that make or supply avionic systems into a fly-by-wire paradigm, that is, the proliferation of electronic control by-wire over the mechanical aspects of the system.

In the critical industries, automobile or aircraft development processes endure a time to market that is rarely measured in months but instead in years to tens of years, yet the speed of development and the time to market are every bit as critical as for small-scale electronics items. Product and process verification and validation, including end-of-line testing, contribute to longer time to market at the cost of providing quality assurances that are necessary to product development. In the industry of small-scale or personal electronics where time-to market literally can be the life or death of a product, validation, verification, and testing processes are less likely to receive the level of attention that they do in mission-critical or life-critical industries. An integrated software solution for a validation and verification procedure, therefore, can provide utility for systems of any size or scale.

The software development process encounters various changes in requirements and specifications. These changes could be a result of the customer who requires the software application, the developer, or any other person involved in the software development process. Design verification and validation is there to reduce the effects of continually having to keep changing the requirements and specifications in the software development process, to increase the performance, and to achieve software quality and improvement. Software verification and validation (V&V) is one of the important processes during the software development cycle. No software application should be released before passing this process. Design verification is to ensure that the software has implemented the system correctly, whereas design validation is to ensure that the software has been built according to the customer requirements.

Software testing strategy is another goal of this chapter. Pressman (1997) defined software testing as a process of exercising a program with the specific intent of finding errors prior to delivery to the end user. Testing is the software development process to ensure the quality and performance by uncovering the errors and find relative solutions; also, testing plays an important role before releasing the application in the software deployment phase.

The problem of using a certain verification method exists with most software or system engineers because of the many verification methods available with specific software environment needed. For that reason, we will be discussing three types of verification methods, which are: the hybrid verification method, basic functional verification method, and verification method using Petri Net. We also will be discussing several types of testing strategies, which are: test data generation testing, traditional manual testing, proof of correctness. Finally, we will be presenting some V&V standards with their domain of use.

19.2 THE STATE OF V&V TOOLS FOR SOFTWARE DFSS PROCESS

A development process provides a developer (individual, group, or enterprise) with some degree of traceability and repeatability. There are several useful process paradigms commonly used, including the waterfall model of which the V design paradigm is a derivative process (for more on software process methods, see Chapter 2). The V development process depicts a modifiable paradigm in which the design process flows down one branch of the V from a high level (abstract) to a low level (detailed) and up the opposing branch of the V to a high level, with each successive level or phase relying on the accumulated framework of the previous phase for its developmental foundation.

Figure 19.1 depicts a V model software development process that has been modified to include concurrent test design phases that are derived from the fundamental stages of the design process. A key feature of Figure 19.1 is that test designs are linked to phases in both the design arm (the left-most process from a high level to a low

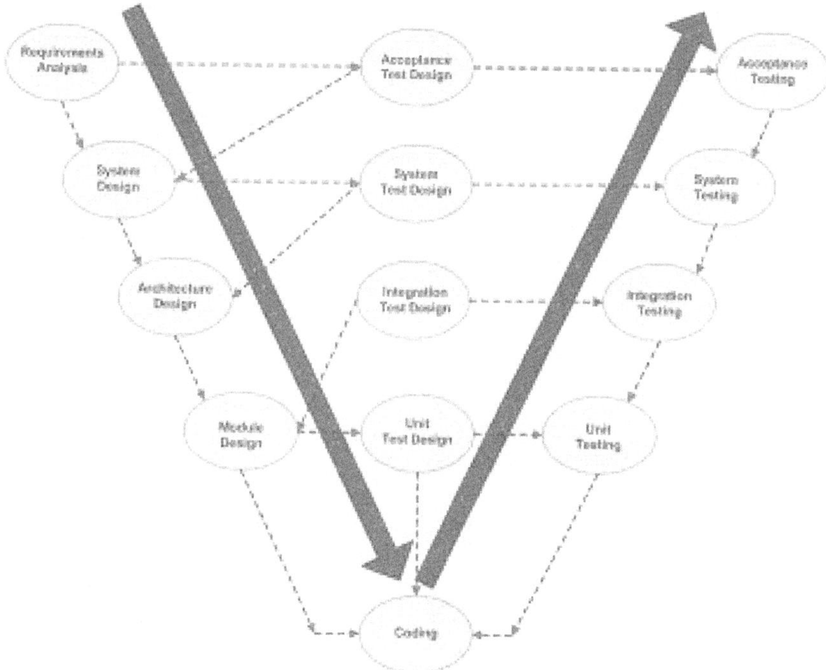

FIGURE 19.1 Modified V development process.[1]

level) of the V and the verification and validation arm (the right-most branch of the V from a low level back up to a high level). Although this process suggests that a testing suite for complete verification and validation is linked easily throughout the design process; in practice, this is far from the case. An Internet survey for verification, validation, and testing process software application tools quickly revealed the absence of unifying tool support for conjoining the terminal phase to support predesign and postdesign documentation, which is vastly used in the design branch of the process. There is a variety of commercial software tools available that provide process stages; some of which singularly or in tandem incrementally approach a complete solution; however, nothing currently fills 100% of the void.

Figure 19.2 depicts a flow diagram of product–process development based on the V model. The intention of this diagram is the graphical representation of series and parallel activities at several levels of detail throughout the development of a product. The subset of phases along the bottom of the diagram effectively represent the required activities that are not necessarily associated with particular phases connected to the V path. This set, however, constitutes interdependent phases that may occur along the timeline as the process is implemented from the farleft to the farright phase of the V-represented phases. This diagram is color coded such that

[1] http://en.wikipedia.org/wiki/V-Model_(software development).

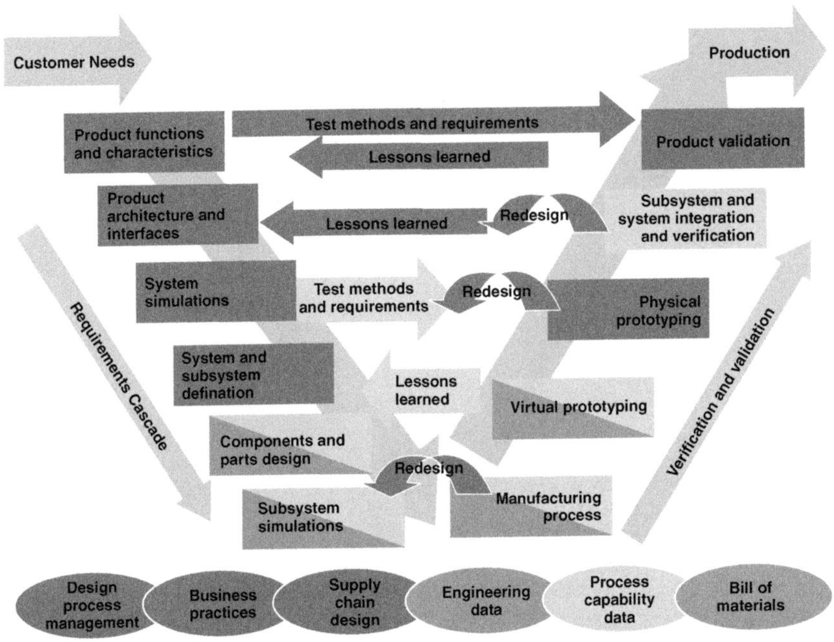

FIGURE 19.2 V process model Modified to Indicate the Potential for a Completely Unified Process Approach.[2]

commonly colored stages represent a known and/or proven process trace among like-colored phases. For the red phases and chains of phases, software tools are not available. The yellow phases represent emerging software tool developments. For the green colored components, there may be one or more well-known or proven software tools however these may not necessarily be "interoperable or are used inefficiently.[3] Figure 19.2 plainly indicates the lack of conjunctive applications to facilitate a unified process flow conjoining requirements and design to validation and verification phases at all levels of development. Some design/development platforms such as MATLAB/Simulink (The MathWorks, Inc., MA, USA) offer solutions that partially bridge the gap. Some source integrity and requirements management tools provide solutions that may involve complex configuration management or require process users to learn additional program languages or syntaxes.

19.3 INTEGRATING DESIGN PROCESS WITH VALIDATION/VERIFICATION PROCESS

Testing, debugging, verification, and validation are inarguably essential tasks of system development, regardless of the process adopted. By extending the abilities of

[2]http://en.wikipedia.org/wiki/V-Model_(software development).
[3]http://www.nap.edu/catalog/11049.html.

existing tools to enable the addition of an integrated test procedure suite and making it applicable at the earliest development stages, the ability to move from left to right across the V gap can become greatly enhanced. What can be gained from this approach is a very high degree of concurrent development, bolstering early fault detection, design enhancement, and the potential shortening of overall development time. Although the diagram in Figure 19.2 is somewhat outdated, it clearly depicts deficiencies in current software tools and tool availability to meet well-defined tasks and requirements. Software companies are making inroads to these territories, but it is also evident that gross discontinuities exist between the conceptual framework of a development process and a real-world ability to implement such a process economically. Furthermore, Figure 19.2 makes clear the need for unification of the subprocesses that can lead to the unification of an entire system process that allows real-world implementation of the theoretical model. The color coding in Figure 19.2 represents a "bridging" process applicable to components of an overall system development process considered analogous with concurrent engineering design and design for manufacturing and assembly, which also are accepted development processes. It is evident that an evolution toward the integration of these subprocesses can increase oversight and concurrency at all levels of development.

Typical engineering projects of systems with even low-to-moderate complexity can become overly convoluted when multiple tools are required to complete the various aspects of the overall system tasks. It is typical for a modern software engineering project to have multiple resource databases for specifications, requirements, project files, design and testing tools, and reporting formats. A fully integrated and unified process that bridges the V gap would solve the problem of configuring multiple tools and databases to meet the needs of a single project. Furthermore, such an approach can simplify a process that follows recent developmental trends of increased use of a model-based design paradigm.

Potential benefits of an integrated development process include

- High degree of traceability, resulting in ease of project navigation at all levels of engineering/management
- High degree of concurrent development, resulting in a reduction of overall project development time/time to market
- Testing at early/all stages enabled, resulting in potential for improved product and reduced debugging costs

These benefits alone address several of the largest issues faced by developers to improve quality and reduce costs and, therefore, remain competitive in the global marketplace. Benefits also apply to other developmental practices adapted to enhance recent trends in design and quality processes such as the model-based design paradigm in which testing can be done iteratively throughout the entire development process. An integrated process also can add utility to reiterative process structures such as Capability Maturity Model Integration (CMMI) and Six Sigma/DFSS, which have become instrumental practices for quality assurance (see Chapter 11).

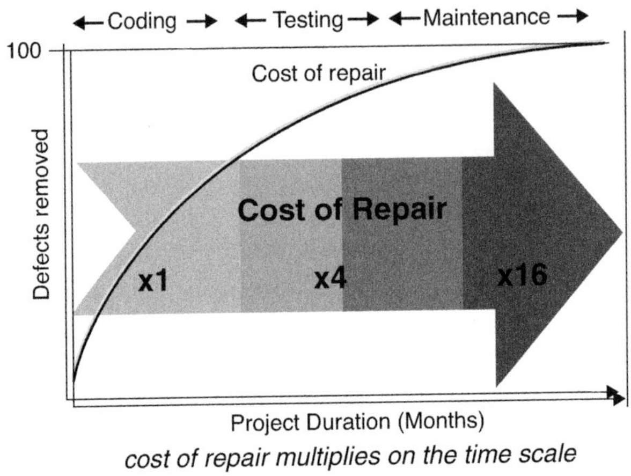

FIGURE 19.3 Cost of repair increase estimated against development time scale (Quality Management and Assessment).[4]

In model-based design testing, an integrated development process will enhance and simplify the procedures on multiple levels. Because model-based design lends itself to modularization of components using serial systems, parallel systems, and subsystems, testing can begin near the beginning of the design process as opposed to a postintegration phase such as in legacy testing paradigms. Component- and system-level software and hardware testing can be increased, and testing can begin at earlier stages in the design. Testing additionally can occur with more concurrency as a result of the modular nature of the newer design paradigms, decreasing the time to market via a parallel/pipeline type of approach. As indicated in Figure 19.3, changes and improvements made later in the design process are far more costly than those made in the earlier stages.

Validation and verification procedures are a certain means of improving product quality and customer satisfaction. By applying such procedures at every step of development, an enormous cost savings can be realized by iterating improvements at the earliest possible stage of the development, when integration is considerably less complex.

19.4 VALIDATION AND VERIFICATION METHODS

Model-Based development/design has become a standard engineering industry operating procedure in computer aided design (CAD) and computer-aided engineering (CAE). There are several proven capable and trusted tools for graphical modeling and

[4]Quality Management and Assessment: Renate Stuecka, Telelogic Deutschland GmbH Otto-Brenner-Strasse 247, 33604 Bielefeld, Germany http://download.telelogic.com/download/paper/qualitymanagement.pdf.

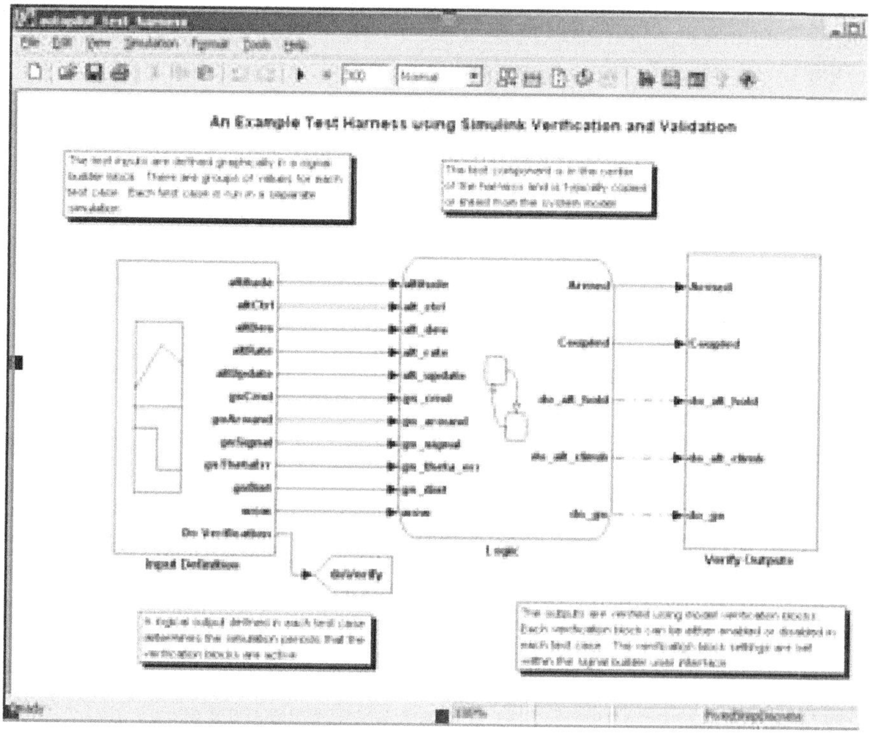

FIGURE 19.4 MATLAB/Simulink's verification and validation platform for logic testing application extension; an example of a model-based development platform (Wakefield, 2008).

simulation of commonly engineered systems such as manufacturing, electrical, medical, computational, mechanical, and communications. Commercial software simulation tools are presently in a highly advanced state of development, having long since proven their usefulness and reliability in many engineering fields in the global marketplace. An example of a model-based development platform is shown in Figure 19.4. Figure 19.4 shows MATLAB/Simulink's Verification and Validation platform for logic testing application extension (Wakefield, 2008).

Extensive efforts are made by simulation suppliers to upgrade continually and extend the application potential for their products. The concept of graphical modeling is a simple representation of any physical system by its inputs—a "black box" containing functional logic, and outputs. The approach can be a top-down or bottom-up hierarchical structure within which each black box may contain multiple subsystems, with the lowest level containing the basic logic and arithmetic operations. The architecture of a given physical system is graphically arranged or designed graphically to simplify conceptualization and understanding of underlying logic. This has tremendous advantages over the interpretation of a system by analysis of potentially hundreds of thousands of lines of code.

19.4.1 The Successive Processes of Model in the Loop (MIL), Software in the Loop (SIL), Processor in the Loop (PIL), and Hardware in the Loop (HIL) Testing Approaches

The usefulness of this approach is that in a well-modeled system using the proper software tools, the software that commands the model can be the same control software for microprocessors and microcontrollers in the physical system. This approach also lends itself to an iterative validation or verification approach in that the custom hand written software or computer-generated code can be tested at multiple levels and with several interfaces that progressively approach integration into the final physical system. This iterative approach commonly is recognized in modern design engineering as the successive processes of model in the loop (MIL), software in the loop (SIL), processor in the loop (PIL), and hardware in the loop (HIL) testing.

19.4.1.1 MIL. MIL occurs when model components interface with logical models for model-level correctness testing. Figure 19.5 shows an example of MIL process from dSPACE (Wixom, MI).[5]

Consider a software system that is ready for testing. Within the same paradigm that the software itself represents a physical reality, it is reasonable to expect that a software representation of inputs to the system can achieve the desired validation results. As the system has been designed and is ready for testing, so can test software be designed to represent real-world input to the system. A reasonable validation procedure can be undertaken by replacing inputs with data sources that are expected, calculable, or otherwise predefined, and monitoring the output for expected results is an ordinary means of simulating real system behavior.

19.4.1.2 SIL. SIL occurs after code has been generated from a model and run as an executable file that is configured to interact with the model software. Figure 19.6 shows an example of an SIL process from dSPACE. This midway point in the V design methodology is perhaps the most important stage of testing, as the progression will begin at this point to lead into hardware testing. This is the optimal stage at which code optimization for hardware should be considered and before the configuration grows in complexity. Code optimization is dependent on the constraints of the design under test (DUT). For example, it may be necessary to minimize the line of code count to not exceed read only memory (ROM) limitations based on particular microprocessor architecture. Other code optimizations can include loop unravelling.

19.4.1.3 PIL. From this point, PIL testing is undertaken for proof that the generated code can run on a hardware platform such as microcontroller, electrically erasable programmable read-only memory (E/EE/PROM) or field programmable gate array (FPGA). Figure 19.7 shows an example of a PIL process from dSPACE.

[5]http://www.dSPACE.de.

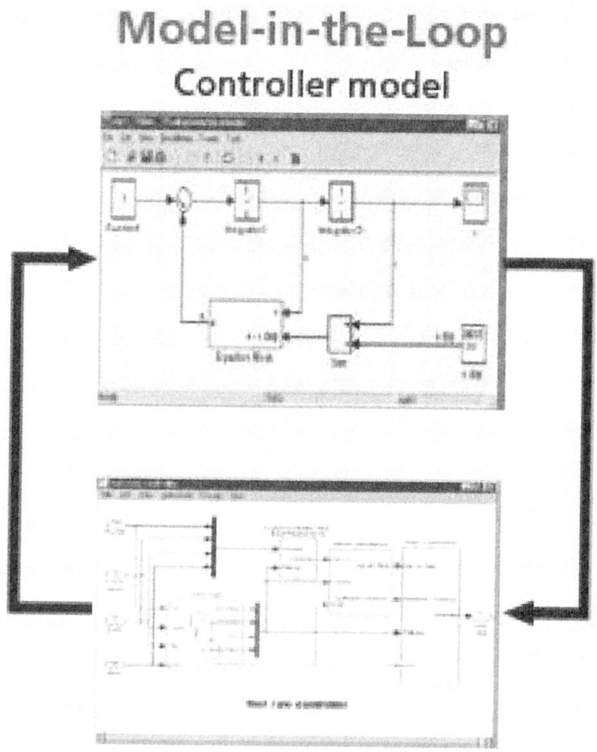

FIGURE 19.5 MIL process from dSPACE catalog (Wixom, MI) dSPACE 2008.[6]

19.4.1.4 HIL. Once it is certain that the software performs as intended and that there are no defects in the hardware, the final in-the-loop stage, HIL is undertaken to prove that the control mechanism can perform its intended functionality of operating on a hardware system. At this point, a test procedure such as joint test action group (JTAG)/boundary scan may be considered for hardware testing prior to implementing a system under test (SUT) scheme. JTAG boundary scan specification outlines a method to test input and output connection, memory hardware, and other logical subcomponents that reside within the controller module or the printed circuit board. The JTAG specification makes it possible to access transparently structural areas of the board under test using a software controlled approach.

According to joint open source initiative UNISIM[7] (Houston, TX), "Simulation is a solution to the test needs of both microprocessors and software running on microprocessors." "A silicon implementation of these microprocessors usually is

[6]http://www.dSPACE.de.
[7](www.unisim.org).

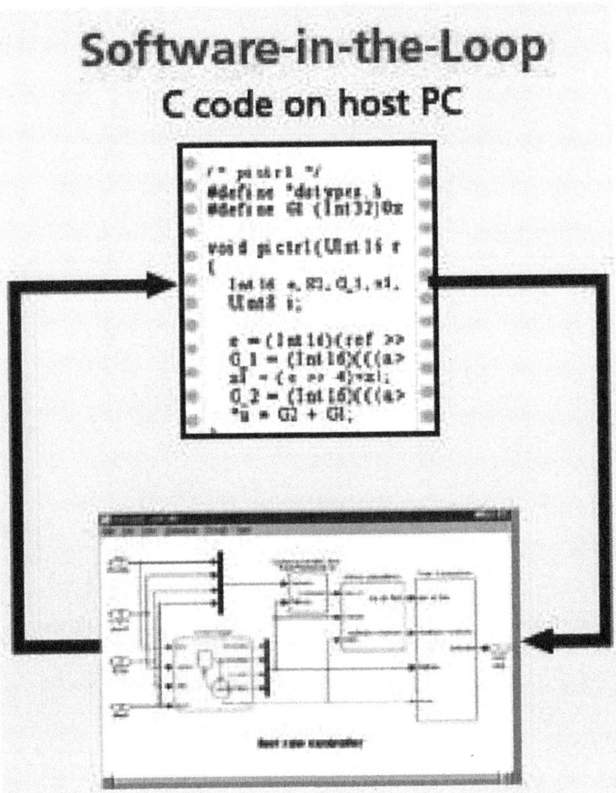

FIGURE 19.6 SIL process from dSPACE catalog (Wixom, MI) dSPACE 2008.[8]

not available before the end of the architecture design flow, essentially for cost reasons. The sooner these simulation models are available, the sooner the compilers, the operating system, and the applications can be designed while meeting a good integration with the architecture".[9]

19.4.2 Design Verification Process Validation (DVPV) Testing

Design verification process validation (DVPV) testing has become overwhelmingly reliant on graphic design with simulation tools such MATLABs Simulink, dSPACEs Targetlink, Simplorer (Wixom, MI), and IBMs Telelogic Statemate (North Castle, NY) is in part because of their sophistication and versatility and largely in part of their added functionality of automated code generation. Simulation tools make for excellent testing tools because they provide instantaneous feedback to system or

[8]http://www.dSPACE.de.
[9](www.unisim.org).

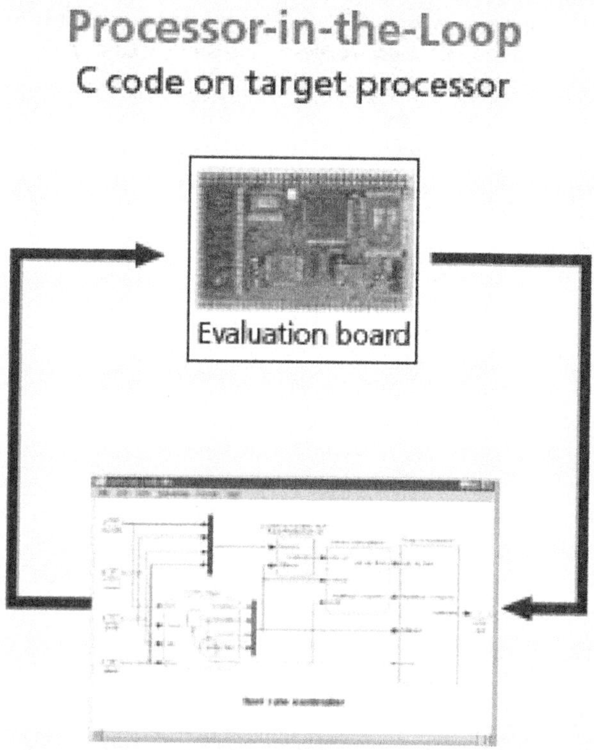

FIGURE 19.7 PIL process from dSPACE catalog (Wixom, MI) dSPACE 2008.[10]

subsystem designs. Test input may be provided internally, modularly, and/or from external scripts/application resources. Simulation allows developers to test designs quickly for completeness and correctness, and many tools also offer autogenerated test reports such as for code coverage and reachability of generated code.[10]

Computer-aided simulation offers a means of interpretive modeling of real systems. Sophisticated software applications allow increasingly larger phases of design and development to remain in a unified development environment. Such an environment may include single-or multiple-tool custom tool chains where the software applications required are correlated to the choice of hardware (microcontrollers, communication networks, etc.). For example, a configuration of software tools to support the Motorola MPC555 (Schaumburg, IL) can be implemented with a particular MATLAB configuration. Support to develop a system using a Fujitsu (Melborne, Australia) microcontroller could include MATLAB but additionally may require dSPACE Targetlink and the Green Hills MULTIintegration development environment

[10]http://www.dSPACE.de.

tool (Santa Baibara, A). Although there may be redundancy among some software applications in the supported hardware, there is presently no single tool that easily configures to a broadbase hardware support. The ongoing development of many of these tools largely is relegated to the logical and functional domains, whereas the needle barely has moved in the domain of external interface configuration. Although simulation remains the most common verification method for software system design, there is room for vast improvement in a move toward a unified integrated development environment.

19.4.3 Process Knowledge Verification Method Based on Petri Net

This method of modeling and verifying a process model is based on Petri Net. This method solves issues such as problems of weak accessibility, deadlock, and dead circle within a software system.

Domain knowledge structure is a complicated system, and its base contains rich knowledge to fulfill application needs according to certain facts, rules, cases, and process knowledge. Process knowledge is the key element of domain knowledge, and it is the representation of all kinds of process, flow, logic, and situational knowledge. Knowledge base, as an important kind of knowledge representation form, receiving more and more attention lately (Kovalyov & Mcleod, 1998).

A process model is an abstract description of enterprise operation mechanism and operating process. Previous process models are described by nonformalized language, whereas to manage and reuse the knowledge requires that models not only should describe relations among activities but also have much other incidental information to facilitate the explanation and implementation. This process modeling mainly aims at solving the problem of how properly to describe the system behavior state according to process objectives and system constraint conditions. The common characteristics are difficulties in modeling and the formalization of algorithms and tools abroad.

19.4.3.1 Process Model Verification Based on Petri Net. The theory foundation and development experience of Petri Net make it suitable for domain knowledge base analysis (Civera et al., 1987). It is necessary to specify corresponding relations between Petri Net and an enterprise process model to verify a process model by Petri Net. Petri Net can be defined as a Quadruple (P, T, I, O) (Daliang et al., 2008), therein:

$P = \{p1, p2..., pn\}$ is finite set of place;

$T = \{t1, t2..., tm\}$ is definite set of transition, and, T and P are disjoint;

I is the input function, mapping of transition T to a place. For each tk to belong to T, relevant results can be determined as follows: $I(tk) = \{pi, pj \ldots\}$;

O is the output function, also mapping of a transition to a place. For each tk to belong to T, relevant results can be determined as follows: $O(tk) = \{pr, ps \ldots\}$.

TABLE 19.1 Relation between Petri Net and Business Process

Petri Net	Process Model
Store-place	Resources (like employees, warehouse), resource state (like busy idle) process
Transition	Beginning or ending of operation, events, process and time
Token	Resource, resource amount
Label	System state
Accessibility	Whether system can reach certain state

Table 19.1 shows the relation between Petri Net and business process.

In the Petri Net flow diagram, we usually have a preference relation, parallel relation, conditional branch relation, circular relation, and other basic relations.

19.4.3.1.1 Process Knowledge Verification Technologies Based on Petri Net. A process model is the result of a business process design. Thus, its construction is a complex process. In a system, business models of different enterprises are mapped to the Petri Net model according to the interface technology, then relative data output, and finally the process model is verified according to relevant theories and methods.

A good summary that summarized the properties needed to be verified in process model verification is Commoner's Deadlocks in Petri Nets (1972). The basic properties of Petri Net are classified into two major kinds: one is dynamic properties that depend on initial marking, such as accessibility, boundedness, reliability, activity, coverability, continuity, fairness, and so on; the other is structure properties that do not depend on initial marking, such as structural liveness, repeatability, and consistency. The mentioned properties all are consistent with basic properties of Petri Net, and Petri Net has remarkable advantages in process model verification. Reachability is the most basic dynamic properties of Petri Net, and the rest of the properties can be defined by it. However, Boundedness reflects the demand for resource volumes while the system is running.

19.4.3.1.2 Deadlock Issue. If transition from state A to state B is impossible (directly or indirectly), then the transition is said to be unreachable. If it is unreachable from the initial state to a certain state, then it demonstrates that there are mistakes in the workflow (Girault & Valk, 2003).

All entities in a process are in a waiting state. The entities can change its state when an event occurs. If this certain event is impossible in the state, then it state is called a deadlock state. The other form of deadlock that can occur in the system is caused by an endless loop that no other events can help get rid of. This kind of deadlock is called livelock, which means that the overall state is changing, but it cannot get rid of the dead circle.

There are many analysis methods used with Petri Nets, among them are the reachability tree (Kovalyov et al., 2000), coverability trees, and incidence matrices with state equation.

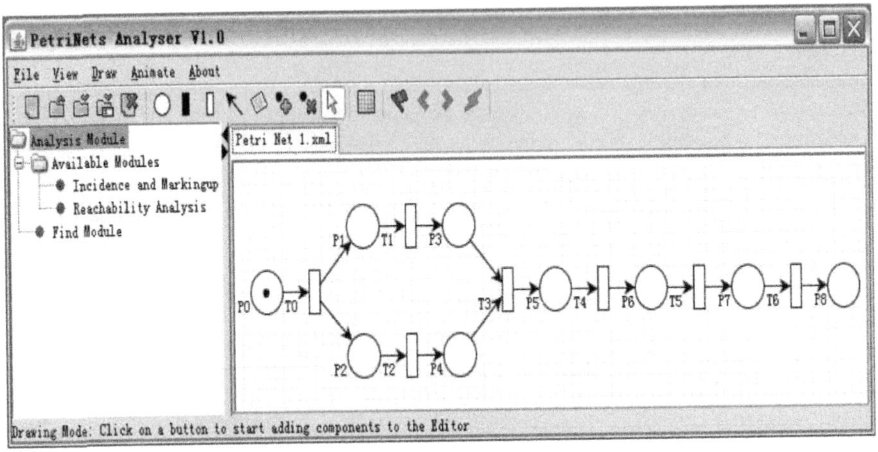

FIGURE 19.8 A pertri net model for a process model.

19.4.3.1.3 Verification Using Petri Net Tool: Petri Net Analyzer Version 1.0. This tool uses the process model that is constructed by Petri Net, then a performance evaluation is performed. Conclusions then are drawn on the feasibility of a process model verification by Petri Net, as shown in Figure 19.8. Figure 19.8 shows an example representation of a Petri Net model for a process model using this analyzer version 1.0:

Figure 19.9 shows the analysis result of a process model that will help in drawing some conclusions and analysis about a certain process model.

Furthermore, the analysis shows the reachability tree of the Petri Net model result. The first row shows that the model is bounded, which indicates that there is not a new token in the changing state of Petri Net, and there is no generation of new

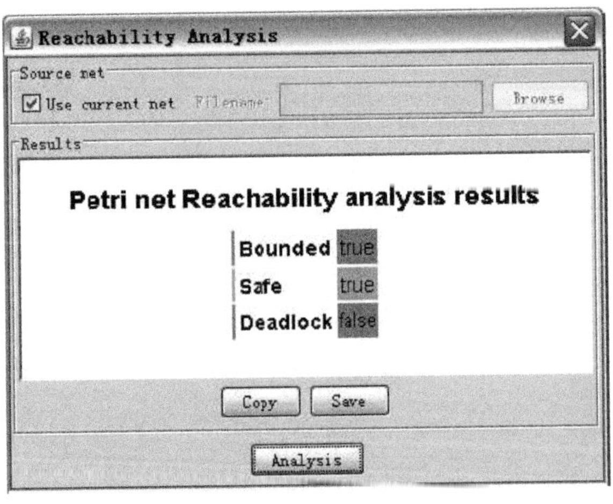

FIGURE 19.9 Analysis result of the process model.

resources in the transition process. The second row shows that the model is safe, which indicates that the token number of all places in the model is no more than one token. The third row shows that there is no deadlock in the model, which shows that deadlock is impossible for resource competition.

19.4.3.1.4 Evaluating Verification approach using Petri Net. When using Petri Nets, a process model gains some effective verification methods. It ensures the correctness and effectiveness of the process model. This method provides practica and effective means for the management and maintenance of the domain knowledge system.

19.4.4 A Hybrid Verification Approach

Hybrid verification technology, which has been tested on current Intel designs, combines symbolic trajectory evaluation (STE) with either symbolic model checking (SMC) or SAT[11]-based model checking. Both human and computing costs are reduced by this hybrid approach.

STE deals with much larger circuits than SMC. It computes characteristic or parametric representation sequence of states as initial values (Hazelhurst, 2002) and then executes the circuit and checks to see that the consequent is satisfied. STE complements SMC, in which SMC is one of the more automated techniques but requires human intervention in modeling. It deals more with commercial sequential designs, and it is limited with respect to the size of verifiable designs (McMillan, 1993).

In the hybrid approach (which we call MIST), the user of the model provides a design being tested and the specification just as if using a classic model checker. However, the user specifies the design behavior being tested or the initial state(s) for the model checker.

The hybrid verification flow consists of two phases:

1. STE performs the initializing computation and calculates the state (or set of states) that the design would be in at the end of this initialization process.
2. Using the set of states in the previous step as the starting point, a SAT/BDD-based model checker completes the verification (Hazelhurst & Seger, 1997).

19.4.4.1 Hybrid Method Workflow. The hybrid approach works as follows (Hazelhurst et al., 2002):

1. Build M'—a pruned model of M (automated step). The initializing behavior and inputs of M are given by an STE antecedent, A.
2. Use STE to exercise the *unpruned* model M under the influence of A. STE's ability to deal with the large unpruned model easily provides the key benefits of enhancements of performance and simplification of modeling.

[11] SATisfiability: Given a propositional formula, find if there exists an assignment to Boolean variables that makes the formula true.

3. The run computes a symbolic set of states that gives, *S*, the set of states of the machine after initialization.
4. Proof of *M'* using SMC/BMC (base model checking) starting from the state set *S*.

In principle, STE's computation to find the set of states after the initialization is complete is the same as the SMC computation.

The workflow of the MIST approach passes through the following steps:

1. Generating the initializing behavior so that MIST requires the initializing sequence of the model *M* as circuit's reset behavior or any user's behavior request.
2. Specifying external stimulus for initializing; in this mode, the cost of modeling the environment is reduced. Computation is done by an STE model. Providing external stimulus is particularly useful when the circuit has a relatively long reset behavior. Here a significant reduction in computation times will be seen, too, because the computation of the reset behavior by STE is extremely efficient compared with SMC. The longer the reset sequence which is the greater the savings (Clarke et al., 1995).
3. Providing an initializing sequence; which is very useful in specification debugging where we can use the same computation several times to find specification errors. A typical use of providing initialization sequences is finding multiple counterexamples (MCEs) (Clarke et al., 1995). The set of MCEs often forms a tree structure that shares a long common prefix. So, before switching to SMC/BMC-based approaches to find MCEs, the first part of the counterexample can be skipped by replaying it with STE to get to the interesting part.
4. The counterexample using BMC depends on the bound chosen; SMC always finds the shortest counterexample, so replaying the prefix always will lead to the same counterexample. Then we can reuse the result of one STE run in many SMC verifications.

19.4.4.2 A Hybrid Verification Tool. The tool prototype is based on forecast and thunder that support both STE and SMC.

Using this tool, the user provides:

- The model description in register transfer level (RTL) format[12]
- Pruning directives that indicate which part of the RTL model should be pruned for SMC
- The initialization information
- The properties to be proved

[12]The RTL format is designed as an extension of the international symposium of circuits and systems (ISCAS) format. Its difference from the ISCAS format is in the possibility to work with multi-bit variables. http://logic.pdmi.ras.ru/~basolver/rtl.html

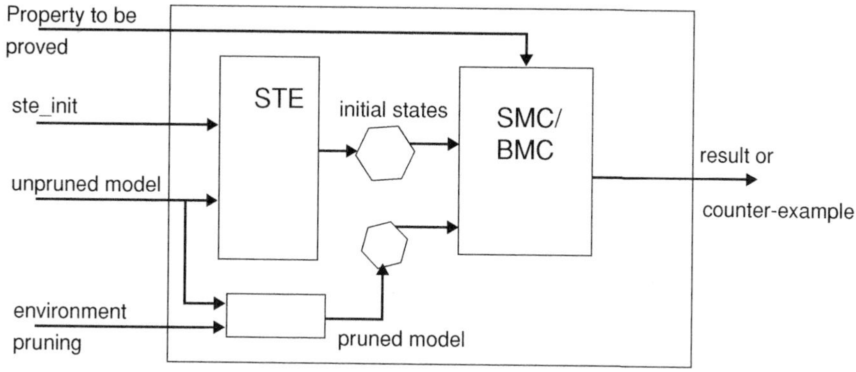

FIGURE 19.10 Overview of MIST prototype system.

The verification process starts by running the original model using STE and computing the initial states for SMC. After that, the parametric representations are converted to characteristic representations. The SMC tool then is invoked. First, the large model is pruned automatically using the pruning directives. The resultant model then is model checked, taking into account the starting state. Although, in MIST, we must provide some additional information, but the benefit is the reduced cost in modeling the environment and the performance improvements. Figure 19.10 shows the MIST steps.

19.4.4.3 Evaluating MIST: A Hybrid Verification Approach. MIST is a hybrid method of verification using STE, BMC, and SMC. We can to use the power of STE to deal with large circuits directly to improve ease of specification, performance, and productivity. It reduces the work required by the user. The user also can have much higher confidence in the result, as the validity of the result will not be affected by the modeling of the environment.

The application of this hybrid approach on real-life industrial test cases shows that MIST significantly can boost the performance and capacity of SAT/BDD-based symbolic model checking. Moreover, this methodology enables the verification engineer to have much more control over the verification process, facilitating a better debugging environment (Yuan et al., 1997).

The insight that initialization is a very important part of the verification of many systems may be helpful in other flows and verification methodologies.

19.5 BASIC FUNCTIONAL VERIFICATION STRATEGY

The functional verification is based on the idea that some specification implemented at two different levels of abstraction may have its behaviors compared automatically by a tool called the test bench.

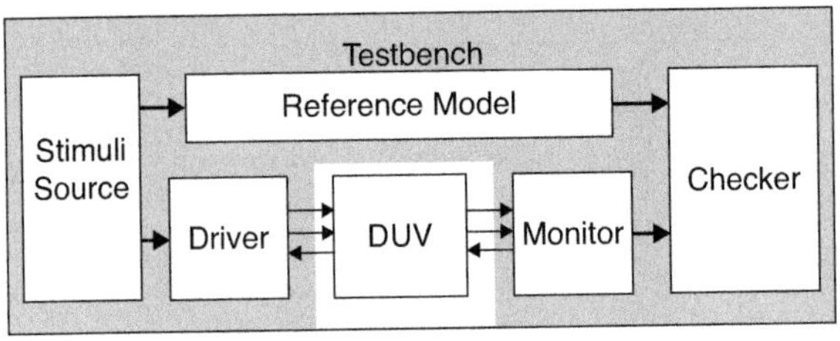

FIGURE 19.11 Basic test bench model.

The test bench architecture used in this verification method is characterized for modularity and reusability of its components. The test bench model comprises all elements required to stimulate and check the proper operation of the design under verification (DUV); the DUV is an RTL description.

Figure 19.11 shows the basic test bench model in which the stimuli source is based on aid tools and applies pseudorandom-generated test cases to both the DUV and the reference model, a module with a behavioral description at a higher level of abstraction. The driver and monitor are blocks aimed to convert the transaction-level data to RTL signals and vice versa. Outputs from the simulation performed on both the reference model and the RTL modules are compared, and outcomes on coverage are computed and presented in the checker.

The designer must carefully plan aspects of the coverage model and the stimuli source. The stimuli can be classified in the following categories:

- Directed cases, whose responses previously are known (e.g., compliance test)
- Real cases dealing with expected stimuli for the system under normal conditions of operation
- Corner cases, aimed to put the system on additional stress (e.g., boundary conditions, design discontinuities, etc.)
- Random stimuli, determined by using probability functions (Bergeron, 2003)

Moving to the coverage related to the strategy, the coverage is an aspect that represents the completeness of the simulation, being particularly important when random stimuli are applied. Functional coverage usually is considered the most relevant type because it directly represents the objectives of the verification process, and it is limited by project deadlines.

Each engineer has his own verification coverage measurement metrics. Thus, to deal with the complexity of a problem, the engineer follows some generic steps for functional coverage. The steps are as follows:

A judicious selection must be made on a set of parameters associated with input and output data, for instance, the size of packets (words with specific meaning) as keys, passwords, and so on.

For every selected parameter, the designer must form groups defined by ranges of values it may assume, following a distribution considered relevant.

The 100% coverage level is established by a sufficient amount of items per group (i.e., test cases) whose corresponding applied stimuli and observed responses match the parameter group characteristics. The larger the number of items is considered, the stronger the functional verification process will be (Tasiran & Keutzer, 2001).

19.5.1 Coverage Analysis

Hekmatpour suggests that the functional verification should be carried out by following several phases such as planning, reference model implementation, coverage analysis, and others (Hekmatmpour & Coulter, 2003).

The coverage analysis is an important phase for certifying the test bench robustness. After the test bench application, in case of evidence of coverage holes, the stimuli generation should be redirected, and the verification should restart until no missing coverage aspects are found.

Under random stimuli, the coverage evolution, in terms of time, presents a fast growth in the initial phase of the test bench application, and then it follows a saturation tendency if higher levels of coverage are reached as a result of an increased occurrence of redundant stimuli.

The functional coverage saturation effect has motivated two types of techniques known as closed-loop test benches and reactive test benches. One example of a closed-loop test bench technique, is the stimuli filtering technique, which is based on the observation that simulating the reference model is much faster than performing it on the RTL model of the DUV, and it shows that important time can be saved without much computational expense or development effort. Although a basic functional technique is a good example for a reactive test bench technique.

19.5.2 Evaluating Functional Approach

The importance of the functional verification strategy is shown in the success of the verification process. With respect to coverage point of view, random stimulation is a big source of redundant cases (i.e., stimuli that do not increase coverage). Consequently, the effective and appropriate use of random stimulation requires using techniques to modify the generation patterns according to the desired coverage.

19.5.3 Verification Methods Summary Table

Table 19.2 summarizes and compares some of the verification methods.

19.6 COMPARISON OF COMMERCIALLY AVAILABLE VERIFICATION AND VALIDATION TOOLS

Many code generation, partial process automation, or test bench generation tools require the use of additional software tools for patching through to another root software platform to complete an uninterrupted tool chain. In this section, a brief

TABLE 19.2 Summarizes and Compares Some of the Verification Methods.

Verification Method	Include	Complexity	Most Applied to	Tool
Hybrid	Symbolic checking model	Invoking RTL document, initializing computation, and calculating the state	Large circuits, systems support SMC and STE, such as thunder forecast systems	Hybrid verification tool
Basic Functional	Test bench model	Coverage measurements metrics	Parameters associated to input and output data	Stimuli filtering and reactive test bench
Using Petri Net	Petri Net state flows	3 model metrics: bounded, safe, and deadlock	Domain knowledge systems	Petri Net Analyzer Version 1.0

overview of commercially available tools that are integral pieces, providing essential large-stage or small-step additions to a movement toward a universal all-in-one verification and validation tool paradigm.

19.6.1 MATHWORKS HDL Coder

MATHWORKS (Natick, MA) provides a widely used and highly sophisticated tool set for model-based design and a variety of code generation utilities for application software development. One example of an extension product that does not fulfill the implications of its utility is the Simulink hardware descriptive language (HDL) Coder. HDL coder is an extension of the model-based development package whose intended use is the autocreation of HDL code for use in a third-party synthesis tool. Although the HDL coder offers the automation of a test bench and saves the user from learning additional software programming languages (VHDL or Verilog-HDL), it still lacks a complete solution because the tool requires the acquisition of other tool sets: synthesizers such as Synplify (San Jose, CA) and simulation tools Mentor Graphics (Wilsonville, OB) ModelSim (Wilsonville, OR) simulator or Cadence Incisive (Natick, MA) to affect code instrumentation (device programming with production ready microcontroller code). The end result is that this tool is really just a conversion instrument from one type of simulation (model-based) to another (hardware description), prompting third-party providers such as Impulse C (San Jose, CA) to develop a code optimization extension tool, again with the impetus landing on the engineer to learn to navigate an additional development platform, complete with new flavors of proprietary C-code syntax.

19.6.2 dSPACE Targetlink Production Code Generator

dSPACE is an internationally reputable provider of complete system development software (control desk and automation desk integreated development environment (IDEs), targetlink model simulator, etc.), hardware (HIL testers, load-boxes), and staff solutions for automotive and aerospace industries. Basing observations on page 213 of the 2008 dSPACE Product Catalog[13] illustrating the "workflow between model design and code implementation," dSPACE offers a complete MIL/SIL/PIL integrated environment. The dSPACE design workflow, "modeling simulation and code specification," encompasses several tasks including behavioral validity checking that can be tested in model in the loop paradigm. This stage is related to reference checking. The next stage of the design process testing is software in the loop in which production code is hosted in the simulation environment. This stage is related to precision checking. The next stage of the design process facilitates "production code target simulation," which encompasses several tasks including target code verification testing with processor in the loop evaluation. This stage is related to the optimization of production code. What the architecture of this environment lacks is clarity as to the configuration and interface needs and complexity required to link Simulink modeling, third-party calibration tools and engine control unit (ECU) programmers. Again, the onus is left to engineering for configuration and interface.

19.6.3 MxVDEV—Unit/System Test Tool Solution

Micro-Max Technology (Bologna, Italy) offers a foundation development environment, a virtual development bench of programs used for requirements capture, design, development, and test of real-time control systems. Some components of this suite include:

- Mx-VDev unit/system test tool
- Mx-Sim system simulator
- Mx-automate continuous integration tool
- Mx-Xchange test data interchange tool

Again, the configuration given in Figure 19.12 implies a complete development environment yet remains vague in the areas of model development, integration, and especially verification, validation, and test reporting as the project moves toward completion along the right arm of the V. Alternatively, this figure indicates the extensibility requirements of tool chains using the V design in which the Mx-VDev unit test tool can provide an integral component for integration of test development. Other issues involved with this and other tools include server/resource repository storage, mapping, and configuration.

[13]http://www.dSPACE.de.

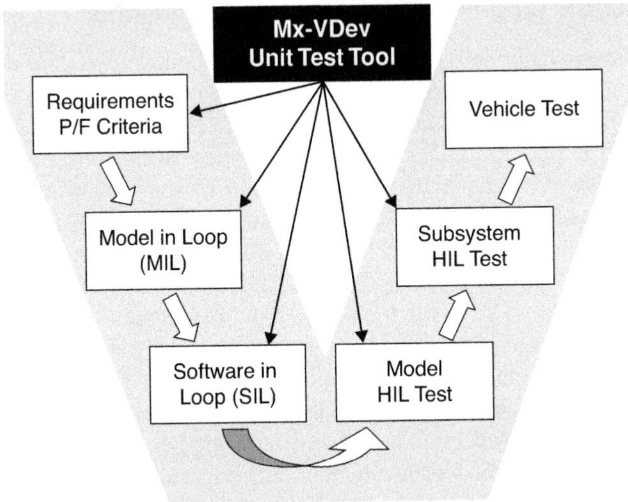

FIGURE 19.12 Micro-max technology's Mx-VDev unit/system test tool (Adrion, et al., 1982).

19.6.4 Hewlett-Packard Quality Center Solution

Adding more quality management concerns into the mix, Hewlett-Packard (Palo, Alto, CA) provides their quality center solution tool boasting an enterprise-ready integration approach to quality management that extends visibility and control up to a generalized-project-management level, including out-of-the box capabilities for SAP, SOP, and Oracle quality management (Paradkar, 2000) environments. This high-level control environment empowers the strictest management up to and including purchasing, billing, and individual time-management control over all aspects of a project.

19.7 SOFTWARE TESTING STRATEGIES

Software testing is an important and huge process that is present in every phase of the software development cycle. Testing the software helps in generating error reports with their solutions to increase the software quality and assurance or to achieve software improvement. Testing might seem to be a certain phase before software release or deployment, maybe because of its great importance before delivering the software to the customer. In fact, software verification and validation testing moves with the software at each single phase, each software iteration, and after finishing a certain step in the software development process.

Validation and verification testing will be the focus of this section. Testing will be covered in the order of use during the software development cycle. Figure 19.13

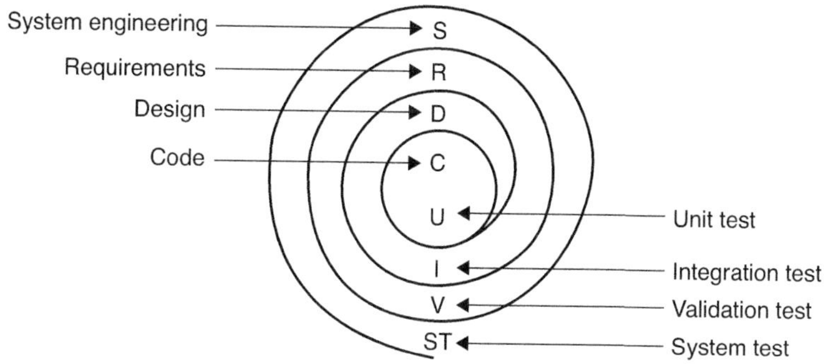

FIGURE 19.13 Testing strategies during software cycle.

shows the testing strategies during a typical software cycle. The testing types that will be conducted are as follows:

- Unit testing
- Integration testing
- Validation testing
- System testing

The development cycle deals with different kinds of V&V testing according to the development phase.

At the very beginning and during the requirement phase, reviews and inspections tests are used to assure the sufficiency of the requirements, the software correctness, completeness, and consistency and must be analyzed carefully, and initial test cases with the correct expected responses must be created.

During the design phase, validation tools should be developed, and test procedures should be produced. Test data to exercise the functions introduced during the design process as well as test cases should be generated based on the structure of the application system. Simulation can be used here to verify specifications of the system structures and subsystem interaction; also, a design walk-through can be used by the developers to verify the flow and logical structure of the system. Furthermore, design inspection and analysis should be performed by the test team to discover missing cases, some logical errors/faults, faulty logic, I/O assumptions, and many other fault issues to assure the software consistency.

Many test strategies are in the implementation phase applied. Static analysis is used to detect errors by analyzing program characteristics. Dynamic analysis is performed as the code actually executes, and it is used to determine test coverage through various instrumentation techniques. Formal verification or proof techniques are used on selected code to provide quality assurance.

At the deployment phase, and before delivering the software, maintenance costs are expensive, especially if certain requirement changes are or a necessary certain upgrade occurs. Regression testing is applied here so that test cases generated during system development are reused or used after any modifications.

19.7.1 Test Data-Generation Testing

This type of testing exercises the software input and provides the expected correct output. It can be done by two popular approaches: the black box strategy and the white box strategy.

The black box, which is classified as a functional analysis for the software, only considers the external specifications of the software without any consideration of its logic, control, or data flow. It mainly concerns the selection of appropriate data as per functionality and testing it against the functional specifications to check for normal and abnormal behavior of the system. The tester is needed to be thorough with the requirement specifications of the system, and the user should know how the system should behave in response to any particular action. There are many testing types that fall under the black testing strategy such as recovery testing, usability testing, alpha testing, beta testing, and so on.

On the other hand, white box testing, which is classified as a structural analysis for the software, only concern testing the implementation, internal logic, and structure of the code. It should be used in all phases of the development cycle. It is mainly used to find test data that will force sufficient coverage of the structures present in the formal representation (Adrion et al., 1982). The tester has to deal with each code unit, statement, or chuck of code and find out which one is not functioning correctly.

19.7.2 Traditional Manual V&V Testing

Desk checking is going over a program manually by hand. It is the most traditional means for program analysis, and thus, it should be done carefully and thoroughly. This can be done with many techniques such as walk-through, inspections, and reviews. Requirements, specifications, and code always should be hand analyzed by walk-through and/or inspections as it is developed, which requires teamwork directed by a moderator and including the software developer.

19.7.3 Proof of Correctness Testing Strategy

This type of testing is classified as the most complete static analysis. It reduces step-by-step reasoning that was mentioned in the previous method with inspection and walk-through. This method works as a mathematical logic to prove the consistency of the program with its specifications and requirements. Furthermore, for a program to be completely correct, it also should be proved to be terminate. Proof of correctness includes two approaches: a formal approach, which is the mathematical logic and the ability of expressing the notion of computation, and the informal approach, which requires the developer to follow the logical reasoning behind the formal proof techniques, leaving aside the formal proof details.

19.7.4 Simulation Testing Strategy

Simulation is a powerful tool for validation testing that plays a useful role in determining the performance of algorithms, and it is used by all the previously mentioned techniques. It is deployed more in the real-time systems. Simulation as a V&V tool acts as a model of the software behavior that is expected on models of computational and external environments (Adrion, et al., 1982).

Simulation has different representations according to the stage in the development cycle, so it may consist of the formal requirements specification, in the requirement stage, design specification, or the actual code in the design stage, or it may be a separate model of the program behavior. At the deployment stage, the code may be run on a simulation of the target machine under interpretative control because the code sometimes is developed on a host machine different from the target machine.

19.7.5 Software Testing Summary Table

Table 19.3 shows the summary and comparison of several software testing strategies.

19.8 SOFTWARE DESIGN STANDARDS

This section contains several standards that are related to the software design, particularly those that are related to verification and validation process. A software standard

TABLE 19.3 A Summary and Comparison of Several Testing Strategies

Testing strategy	Way of Use	Types	Special Feature(s)
Test data—generation testing	Exercises the software input and provides the expected correct output	Black box, white box, alpha testing, beta testing, usability testing, and recovery testing	Perform functional analysis for software externally, perform structural analysis for software internally
Traditional manual testing	Hand analysis of code, requirements, and specifications	Walk-through, inspection, and review	Done manually, most traditional mean for program analysis
Proof of correctness testing	Mathematical logic to prove the software consistency	Formal, and informal	Most complete static analysis
Simulation testing	Perform model behavior to external environments	Forms of formal specification, design specification, and separate model	Can be used in all testing techniques, and software model performance

TABLE 19.4 Some V&V Standards

Software design standard	Used for
IEEE Std 1012-1986	IEEE Standard for Software Verification and Validation Plans
ISO 9000-3-1992 and IEEE Std 1077-1995 (SDLC Process)	Elaboration of design compromise, and implementation process
IEEE Std. 982. 1-1998	Software reliability
IEEE Std. 982.2-1988	Effective software process
IEEE Std. 1058.1-1987	Project management plans
ISO 9000 series	Software Quality Management
IEEE Std 1016.1-1993	Practice for Software Design Description
IEEE Std 1074-1995 SDLC	Risk Engineering
IEEE Std 1044-1993	Classification for Software Anomalies

prescribes methods, rules, and practices that are used during software development. The standards are presented in table 19.4 which includes the standard, its purpose, and different uses.

Standards originated from many resources such as IEEE (The Institution of Electrical & Electronic Engineers), ISO (International Standards Organization), ANSI (American National Standards Institute) and so on.

The IEEE Std 1012-1986 contains five important parts, which are traceability, design evaluation, interface analysis, test plan generation, and test design generation. The systems development life cycle (SDLC) has three main processes: requirements, design, and implementation. The implementation process has four main tasks, which are the selection of test data based on test plan, the design elaboration or coding, verification and validation, and integration. The key to software reliability improvement

TABLE 19.5 The Differences between the Verification and Validation with the ISO 9001 Standard

ISO 9001 Validation	ISO 9001 Verification
• Design and development validation should be performed in accordance with planned arrangements. • To ensure that the resulting product is capable of meeting the requirements for the specified application or intended use, where known. • Wherever practicable, validation should be completed prior to the delivery or implementation of the product. • Records of the results of validation and any necessary actions should be maintained.	• Verification should be performed in accordance with planned arrangements. • To ensure that the design and development outputs have met the design and development input requirements. • Records of the results of the verification and any necessary actions should be maintained.

is having an accurate history of errors, faults, and defects associated with software failures. The project risk explained here is in terms of an appraisal of risk relative to software defects and enhancements (Paradkar, 2000).

The ISO 9000 series is used for software quality management. We will concentrate on the verification and validation sections of ISO 9001. Table 19.5[14] shows the differences between verification and validation with the ISO 9001 standard.

19.9 CONCLUSION

Many V&V methods and testing strategies were presented in this chapter. The Petri Net method seems to provide practical and effective means for management and maintenance of the domain knowledge system. The hybrid approach can boost the performance and capacity of an SAT/BDD-based symbolic model checking. Moreover, this methodology enables the verification engineer to have much more control over the verification process, facilitating a better debugging environment.

Testing strategies has a general role to uncover software errors and maintain software quality. Testing begins at the module level and works outward toward the integration of the entire system. Different testing techniques are required at different stages of the software life cycle. The most successful technique would be the traditional manual techniques because they are applied to all stages in the life cycle. The cost of finding software errors increases by moving forward in the software development; so for example, when the tester finds an error at an earlier stage, such as the requirements phase, it will be much less costly than finding it at the deployment phase.

Testing strategy has certain problems that might delay or stand against completing the software as assumed. Simulation has a major cost related to customizing it to the verification process, whereas the proof of correctness sometimes has the inability of proving certain practice. Moreover, with the technology advancement, many problems occur for different software environments that software engineers have to know how to deal with to save time and money.

Verification and validation methods verify the software quality and testing the software assures that, whereas V&V standards are available for use to clarify and simplify rules of using any V&V method or any testing strategy.

REFERENCES

Bergeron, J (2003), *Writing Testbenches: Functional Verification of HDL Model*, 2nd ed; Kluwer Academic, Boston, MA.

Civera, P. Conte, G. Del Corso, D. and Maddaleno, F. (1987), "Petri net models for the description and verification of parallel bus protocol," *Computer Hardware Description Languages and their Applications*, M.R. Barbacci and C.J. Koomen (Eds.), Elsevier, Amsterdam, The Netherlands, pp. 309–326.

[14] http://www.platinumregistration.com/kbfiles/VAL_VER-ISO9k_CMMI.pdf.

Clarke, E. Grumberg, O. McMillan, K. and Zhao, X. (1995), "Efficient generation of counterexamples and witnesses in symbolic model checking." Proceedings of 32nd ACM/IEEE Design Automation Conference.

Commoner, F. (1972), Deadlocks in Petri Nets, Report #CA-7206-2311, Applied Data Research, Inc. Wakefield, MA.

Girault, C, and Valk R. (2003), *Petri Nets for System Engineering—A Guide To Modeling, Verification, and Application*, Berlin, Germany, Spring-Verlag.

Hazelhurst, S. (2002), *On Parametric and Characteristic Representations of State Spaces*, Technical Report. 2002-1, School of Computer Science, University of the Witwatersrand Johannesburg, South Africa, ftp://ftp.cs.wits.ac.za/pub/research/reports/TR-Wits-CS-2002-1.ps.gz.

Hazelhurst, S. and Seger, C.-J.H. (1997), Symbolic Trajectory Evaluation. *Formal Hardware Verification:Methods and Systems in Comparison*, Kropf, T. (Ed.), Springer-Verlag, Berlin, Germany, pp. 3–79.

Hazelhurst, Scott Weissberg, Osnat Kamhi, Gila and Fix Limor (2002), "A Hybrid Verification Approach: Getting Deep into the Design." Annual ACM IEEE Design Automation Conference, Proceedings of the 39th Annual Design Automation Conference, New Orleans, LA.

Hekmatmpour, A. and Coulter J. (2003), "Coverage-Directed Management and Optimization of Random Functional Verification," Proceedings of the International Test Conference, pp 148–155.

Kovalyov, A. and McLeod, R. (1998), "New Rank Theorems for Petri Nets and their Application to Workflow Management," IEEE International Conference on Systems, Man, and Cybernetics, San Diego, CA, pp. 226–231.

Kovalyov, A. McLeod, R., and Kovalyov, O. (2000), "Performance Evaluation of Communication Networks by Stochastic Regular Petri Nets," 2000 International Conference on Parallel and Distributed Processing Techniques and Applications (PDPTA' 2000), pp. 1191–1998.

McMillan, K.L. (1993), *Symbolic Model Checking,* Kluwer Academics, Norvell, MA.

Paradkar, A. (2000), "SALT—An Integrated Environment to Automate Generation of Function Tests for APIs," Proceedings of the 11th IEEE International Symposium on Software Reliability Engineering, San Jose, CA, Oct.

Pressman, Roger S. (1997), *Software Engineering—A Practitioner's Approach*, 4th., McGraw-Hill, NewYork.

Adrion, W. Richards Branstad, Martha A. and Cherniavsky, John C. (1982), "Validation, verification, and testing of computer software." *ACM Computing Surveys* Volume *14*, #2 pp 159–192.

Tasiran, S. and Keutzer, K. (2001), "Coverage metrics for functional validation of hardware designs." *IEEE Design and Test of Computers*, Volume *18*, pp 36–45.

Wakefield, Amory (2008), "Early Verification and Validation in Model-Based Design," Proceedings of the Mathworks Automotive Conference.

Daliang, Wang, Dezheng, Zhang, Gao, Li-xin, Jian-ming, Liu, and Zhang Huansheng (2008) "Process Knowledge Verification Method Based on Petri Net," Proceedings of the 1st International Conference on Forensic Applications and Techniques in Telecommunications, Information, and Multimedia and Workshop, Adelaide, Australia.

Yuan, J. Shen, J. Abraham, J. and Aziz A. (1997), "On Combining Formal and Informal Verification." Proceedings of CAV '97, pp. 376–387.

INDEX